AF784637

# Integrated Concepts of Zoology

# Integrated Concepts of Zoology

Edited by **Kenneth Hayes**

SYRAWOOD
PUBLISHING HOUSE
New York

Published by Syrawood Publishing House,
750 Third Avenue, 9th Floor,
New York, NY 10017, USA
www.syrawoodpublishinghouse.com

**Integrated Concepts of Zoology**
Edited by Kenneth Hayes

© 2016 Syrawood Publishing House

International Standard Book Number: 978-1-68286-313-8 (Hardback)

This book contains information obtained from authentic and highly regarded sources. Copyright for all individual chapters remain with the respective authors as indicated. All chapters are published with permission under the Creative Commons Attribution License or equivalent. A wide variety of references are listed. Permission and sources are indicated; for detailed attributions, please refer to the permissions page and list of contributors. Reasonable efforts have been made to publish reliable data and information, but the authors, editors and publisher cannot assume any responsibility for the validity of all materials or the consequences of their use.

The publisher's policy is to use permanent paper from mills that operate a sustainable forestry policy. Furthermore, the publisher ensures that the text paper and cover boards used have met acceptable environmental accreditation standards.

**Trademark Notice:** Registered trademark of products or corporate names are used only for explanation and identification without intent to infringe.

Printed in China.

# Contents

# Preface

Zoology as a discipline specifically caters to the study of animal kingdom. The aim of this book is to present researches that have transformed this discipline and aided its advancement. Detailed studies on the structural, physiological and evolutionary processes of animals are provided in this book. It will serve as a valuable source of reference for graduate and post graduate students of zoology, veterinary sciences and allied disciplines. Those in search of information to further their knowledge in this field will be greatly assisted by this book.

The information contained in this book is the result of intensive hard work done by researchers in this field. All due efforts have been made to make this book serve as a complete guiding source for students and researchers. The topics in this book have been comprehensively explained to help readers understand the growing trends in the field.

I would like to thank the entire group of writers who made sincere efforts in this book and my family who supported me in my efforts of working on this book. I take this opportunity to thank all those who have been a guiding force throughout my life.

**Editor**

1

# Factors affecting the levels of protection transferred from mother to offspring following immune challenge

Christina M Coakley*, Vincent Staszewski, Katherine A Herborn and Emma JA Cunningham

**Abstract**

**Introduction:** The transfer of antibodies from mother to offspring is key to protecting young animals from disease and can have a major impact on responses to infection and offspring fitness. Such maternal effects also allow young that may be exposed to disease in early life to focus resources on growth and development at this critical period of development. Maternally transferred antibodies are therefore an important source of phenotypic variation in host phenotype as well as influencing host susceptibility and tolerance to infection across generations. It has previously been assumed the transfer of antibodies is passive and invariant and reflects the level of circulating antibody in the mother at the time of transfer. However, whether females may vary in the relative amount of protection transferred to offspring has seldom been explored.

**Results:** Here we show that females differ widely in the relative amount of specific blood antibodies they transfer to the embryonic environment (range 9.2%-38.4% of their own circulating levels) in Chinese painted quail (*Coturnix chinensis*). Relative transfer levels were unrelated to the size of a female's own immune response. Furthermore, individual females were consistent in their transfer level, both across different stages of their immune response and when challenged with different vaccine types. The amount of antibody transferred was related to female condition, but baseline antibody responses of mothers were not. However, we found no evidence for any trade-offs between the relative amount of antibody transferred with other measures of reproductive investment.

**Conclusions:** These results suggest that the relative amount of antibodies transferred to offspring can vary significantly and consistently between females. Levels of transfer may therefore be a separate trait open to manipulation or selection with potential consequences for offspring health and fitness in both wild and domesticated populations.

**Keywords:** Maternal antibodies, Maternal effect, Trade-offs, Immunity, Differential allocation, Ecoimmunology

## Introduction

The immune response plays a key role in protecting individuals from disease and this immune defense can be extended to offspring via the transfer of maternal immunity (for review see [1]). Maternal antibodies are produced when mothers are exposed to pathogens in the environment prior to, or during, reproduction but can also be induced by health interventions such as vaccination. The resulting maternal immunity can be both crucial in protecting offspring from disease during early development and in allowing offspring to focus more resources on growth and less on dealing with pathogen exposure during this critical period of development [2]. Maternally-transferred antibodies are therefore an important source of phenotypic variation in a range of offspring traits as well as influencing responses to infection across generations. Such maternal effects, whereby female condition or allocation of resources during pregnancy or egg laying can influence the phenotype of the developing young are now recognized to have a major impact on the relationship between genotype and fitness [3]: Maternal immunity may therefore have both a direct impact on offspring health and fitness and be a powerful selective pressure that can accelerate or limit how host and pathogen traits might change in response to selection [4].

* Correspondence: christina.m.coakley@gmail.com
Centre for Infection, Immunity and Evolution, Institute of Evolutionary Biology, School of Biology, King's Buildings, University of Edinburgh, Edinburgh EH9 3JT, UK

While there has been much work on factors that affect an individual's immune response, far less is known about factors affecting the level of antibody transferred from mothers to offspring between and within individuals over time. The maternal effect of transferring immunity to offspring is widespread throughout different animal groups including mammals, birds, fish and invertebrates [5]. In mammals these substances can be transferred directly from the mother prenatally through the placenta and post-natally through the colostrum (first milk) and continued milk [6]. In birds, most substances are transferred to offspring via the yolk when the egg is in oogenesis [7] prior to other egg components such as the albumen and shell being added. In both these groups, the transfer of protective antibodies can be induced by maternal vaccination and this has been harnessed as strategy to protect young from certain disease causing pathogens e.g. whooping cough in humans [8] and infectious bursal disease (IBD) in chickens [9]. Maternally transferred protection has previously been assumed to be less important in other taxa but ecoimmunology studies have demonstrated similar protection strategies, though these may stem from different types of mechanism [10].

It has previously been assumed that the transfer of maternal immunity is passive and that high antibody responders simply transfer more to their offspring by default [10]. In birds, much of the early literature from poultry science suggests the level of antibody transfer is invariant and that females consistently transfer approximately 20% of their circulating antibody level to each egg produced [11]. This would suggest that simply selecting for high responders would also select for females that transfer a high level of antibodies to their offspring. However, variation in the transfer part of the process has seldom been fully explored at the individual level and whether variation may exist between individuals in the ratio of their circulating antibodies transferred to offspring, or indeed how this relates to the production of their own immune response, remains unclear. Given independent mechanisms are responsible for the production and transfer parts of the process [10] the level of transfer may itself be a trait open to selection [5] or manipulation. Furthermore, recent literature from the field of evolutionary ecology has found that levels of immune transfer may vary in response to a number of factors including mating partner [12], sex of offspring [13] or position in a clutch [14]. This is consistent with life history theory, which would predict that the levels of antibody transferred to different offspring should vary, depending on the likelihood of their success. However, it remains unclear as to whether levels of antibody transferred are actively targeted to different offspring or are simply a consequence of changing levels of antibodies in the mother over time as different eggs in a laying sequence are formed. This latter possibility does not exclude these mechanisms from being adaptive but whether plastic adjustment may also occur, and what factors, if any, may induce such adjustment, remains unclear.

Establishing how patterns of antibody allocation relate to a female's antibody production within and between individuals is key to establishing the implications of these maternal effects and how they might be usefully applied in the protection of young from disease. However, progress has been hampered by a lack of information on how a mother's antibody level varies with levels of allocation to the embryonic environment over the same period as eggs or young are being formed as antibody levels are unlikely to be consistent in a mother over this time. Data from the general immunological literature would suggest that the typical pattern of an antibody response post-challenge is characterised by a slow then increasing rise to peak antibody level, followed by a steady decline as antibodies are catabolised [15]. After acute infection, cells produced by the primary immune response persist as memory cells allowing rapid production of further antibodies to be produced if there is re-infection. Co-infection is common with females potentially producing a range of specific antibodies to a range of different types of pathogen they encounter in the environment [16]. If mounting an immune response is costly, the transfer of large numbers of antibodies to offspring is therefore likely to be an expensive process for the mother, even if the process of transfer is entirely passive. The ability to transfer essential constituents to the developing offspring may therefore be expected to relate to a mothers condition and/or result in trade-offs between other life history traits associated with reproduction or survival [17].

Here we investigate experimentally in Chinese painted Quail (*Coturnix chinensis*) whether there is variation between individual females in the amount of antibodies transferred to their offspring and whether individual females are consistent in the ratio of antibodies passed to their different eggs produced over the course of an antigen challenge and between different types of challenge (a bacterial and viral challenge). We then investigate whether any female traits co-vary with their transfer ability and discuss the potential implications of this variation on evolutionary and ecological processes affecting host parasite responses and how such variation might be usefully harnessed in an applied context.

## Results

### Variation between females in immune responses and levels of transfer

We challenged females (n = 38) at the start of egg laying with one of two inactive vaccines; one viral vaccine against Newcastle Disease virus (Nobilis Paramyxo P201) (n = 19) and one bacterial vaccine against salmonella (Nobilis SalenvacT), n = 19). We then monitored maternal and egg antibody levels over the next 42 days. Of the 38 challenged females, egg sequences that had corresponding maternal sequences were collected for 21 females (n = 7 NDV and n = 14 Salmonella). These females had been reared in a pathogen free environment and had no previous exposure to disease.

Post challenge, females showed a steady increase in antibody level in the days that followed until peak antibody level was reached, followed by a slower decline in antibody level over time (Figure 1). The type of challenge did not influence the timing of peak response; $F_{1,20} = 2.37$, $p = 0.149$ but across both challenge types females varied both in the time they took to reach their peak of circulating antibody level (mean peak day $22.22 \pm 3.425_{SE}$, n = 21) and in the magnitude of their response (mean peak size $1.127 \pm 0.163_{SE}$).

Maternal patterns of antibody production were reflected in the level of antibodies found in eggs produced over the same time period (Figure 1). Both the number of days post challenge and maternal antibody level per day, therefore had a large effect on the level of antibody present in the eggs laid at the corresponding time (Table 1). However there was significant variation around this relationship suggesting other factors may influence the relative level of transfer from mother to egg over and above this response. Therefore, to examine whether females differed in the relative level of antibodies they transferred to their eggs when differences in antibody level due to time were controlled for, a comparison between the models with and without female ID as a random effect was explored. Female ID play a significant role in the model (using log likelihood ratio); $\chi^2_1 = 25.18$, p = <.0001. Within females, the repeatability of the amount transferred over the course of the laying sequence was high, both when we controlled for day of sampling (intraclass correlation coefficient); r = 0.728 (Figure 2) and when day of sampling was not included in the model; r = 0.688. Assuming a female's blood volume equals 10% of her body weight [18] the results suggest that females transfer on average 25.5% of their circulating antibody level to each egg which is, in line with previous findings in poultry [10]. However, the variation between females was high with the lowest responders transferring an estimated 9.21% (range 7-10%) and the highest transferring 38.4% (range 22-58%).

While there was a strong relationship between egg antibody levels and female antibody levels on specific days (Table 1), the mothers with the highest overall response did not transfer more to their eggs overall and peak antibody level of mother had no relationship with egg antibody levels through time (Table 1). Higher antibody responders therefore do not necessarily transfer more antibodies to their offspring due to the high levels of variation between females in their relative transfer rates.

**Effects of maternal traits and maternal investment on maternal Ab level, egg Ab level and relative transfer level**

We then investigated the relationship between female immune responses and the relative amount of antibodies transferred with other female traits. We had selected non-live vaccines for the experiments to ensure any effects we detected were only related to the effect of producing the associated immune response, without any costs associated with the pathology caused by parasitism or infection which, by definition, incurs a direct cost to the host. We found no direct effect of between maternal condition on maternal circulating antibody levels or egg antibody level (Table 1) (or any of the other measures of maternal antibody response (all $p > 0.1$). However, we found a significant interaction between maternal antibody level and maternal condition on the level of antibodies transferred to a female's eggs (Table 1). This was found to be because females in lower body condition transferred a greater proportion of their circulating antibodies to their eggs than females in better condition (Figure 3). No significant relationship was found between the amount of antibody transferred and total egg investment over the period of the experiment, suggesting that females who were transferring a greater ratio of their antibody level were not trading off costs associated with a higher transfer with resources allocated to reproduction in the same time frame. Mean eggs investment also had no relationship on the propensity to transfer antibodies to the egg (Table 1). These findings are confirmed by examining a model testing the effects of these different factors on the proportion of antibody transferred directly (Table 2).

**Maternal, egg and relative transfer levels in response to different and subsequent challenge types**

A subset of females was challenged again at day 42 to look at the repeatability of transfer patterns across different challenge types: individuals that had been in the bacterial challenge group (SalenvacT group) were challenged with the viral NDV vaccine (n = 4) and individuals from the viral challenge group (NDV group) were challenged with the bacterial vaccine (SalenvacT) (n = 3). In these individuals we found a positive relationship between the relative level of antibodies transferred to eggs in response to the first challenge and the relative level transferred in response to the second challenge, irrespective of which treatment was given first: $F_{1,6} = 7.287$, $p = 0.042$ (Figure 4). Again, using the log likelihood ratios (between models with and without female ID as a random effect), there was significantly more variation between individuals ($\chi^2_1 = 6.07$, $p = 0.01$) than within individuals and the estimate of across challenge repeatability (r = 0.577) was again high.

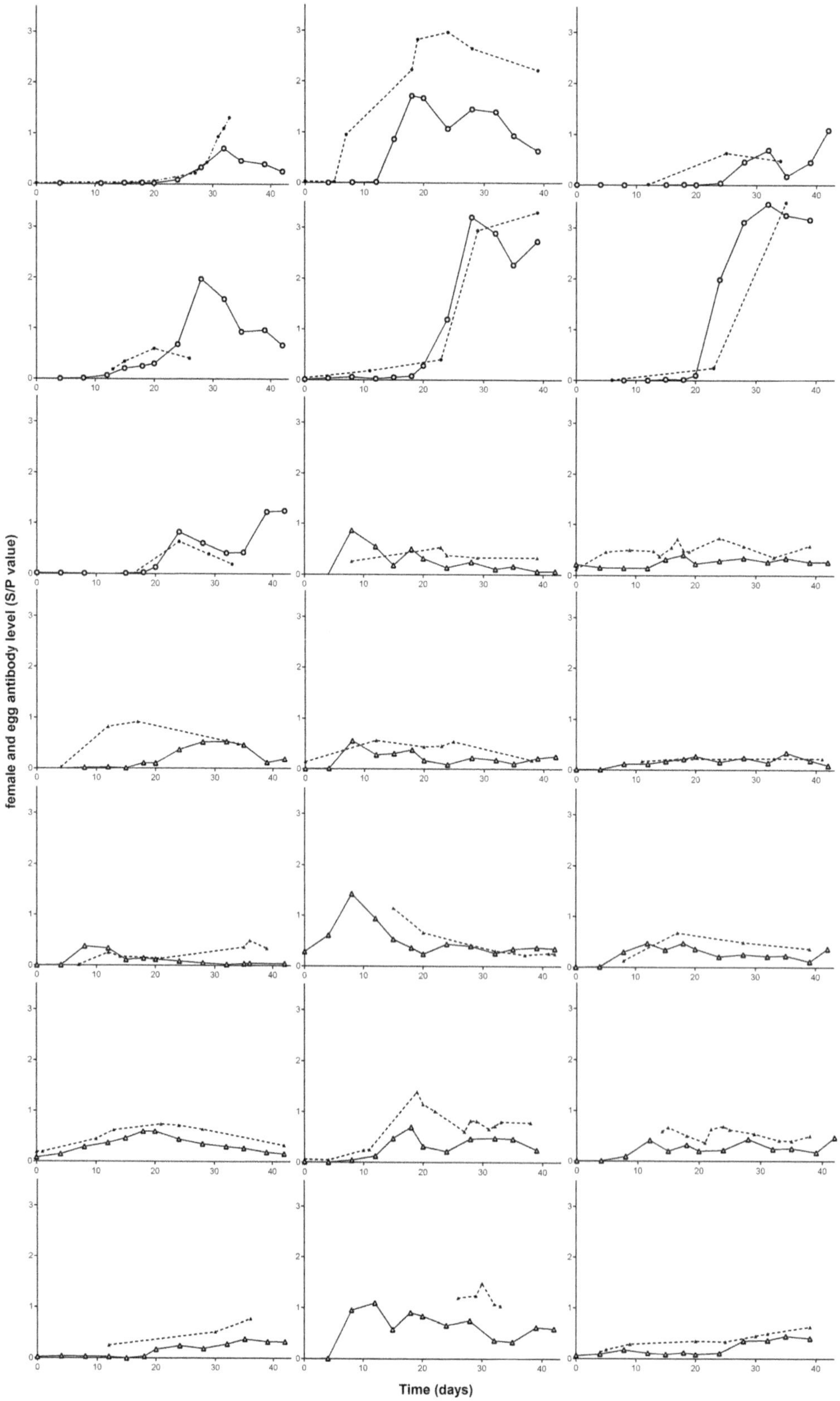


**Figure 1 Individual female and egg antibody levels graphed over time for n = 21 females.** Solid line and open points indicate female antibody concentration (triangles- NDV vaccinated, circles- SalenvacT vaccinated). Dotted line and solid points indicate egg antibody concentration (triangles – NDV vaccinated, circles- SalenvacT vaccinated).

**Table 1 Summary of linear mixed-effects model of egg antibody titre (log)**

| Response | Explanatory | *df* | F | *p* |
|---|---|---|---|---|
| Yolk Ab (log) | **Female Ab (log)** | 1 | 47.85 | **<.0001** |
| | Female mean condition | 1 | 0.94 | 0.345 |
| | Yolk mass | 1 | 2.42 | 0.124 |
| | Egg investment | 1 | 0.88 | 0.360 |
| | Treatment | 1 | 2.65 | 0.121 |
| | Female peak Ab | 1 | 0.05 | 0.814 |
| | Time (day) | 1 | 2.07 | 0.154 |
| | **Time (day)$^2$** | 1 | 6.65 | **0.012** |
| | **Interaction** | | | |
| | **Female Ab (log)*Female mean condition** | 1 | 11.09 | **0.004** |

Results for a linear mixed effects model on egg antibody titre. Significant fixed effects are in bold.

## Discussion

Here we show that while females vary in the timing and strength of their antibody production following a standard antigen challenge, the relative amount of their own antibodies that individual females transfer to eggs over the course of this variable immune response is highly consistent. However, some females consistently transfer more than others. Furthermore, the proportion of antibody transferred by a particular female appeared to be consistent across two different challenge types which utilize different arms of the immune response (a bacterial and viral challenge). The ratio of circulating antibody that was transferred to the egg was independent of the individual's overall strength of the antibody response highlighting that transfer mechanisms can vary independently of production levels. The amount of antibody transferred was negatively related to the female's body condition, but the female's circulating antibody level prior to transfer was not. However, we found no evidence for any trade-off between levels of allocation of immunity to the egg or the female's circulating antibody level with overall levels of reproduction.

The results from this study demonstrate that a female's capacity to transfer antibodies to her eggs is not just related to her ability to produce an antibody response. This has important implications as it suggests that both a female's propensity to produce a particular level of immune response and her ability to transfer it to offspring may be two separate traits open to manipulation or selective processes. In natural populations, the fitness benefits of how protection is allocated between a mother investing in her own immune response and that of her offspring is likely to depend on a number of factors, including the condition a mother finds herself breeding in, the likely pathogen exposure faced and the likelihood a female may breed again in the future as they balance the costs and benefits of allocation strategies over their life history. However, ultimately, any variation in the level of transfer will create a more variable environment for pathogens and therefore impact any ensuing arms race between host and parasite.

A separation of the mechanisms responsible for the generation and transfer of an immune response is also potentially important commercially. Poultry lines, for example, have been established with genetic differences in adult antibody responses to select for genotypes that produce higher levels of antibody [19,20]. If the level of antibody transfer to offspring could separately be selected upon for either high or low transfer levels, this could potentially be advantageous; high transfer levels may be advantageous for some purposes, for example, in the production of antibodies in chicken eggs for therapeutic purposes or in flocks where maternal vaccination to transfer protection to offspring is important in offspring health (whereby mothers are vaccinated to confer protection to offspring against common diseases as a cost effective alternative to vaccinating all

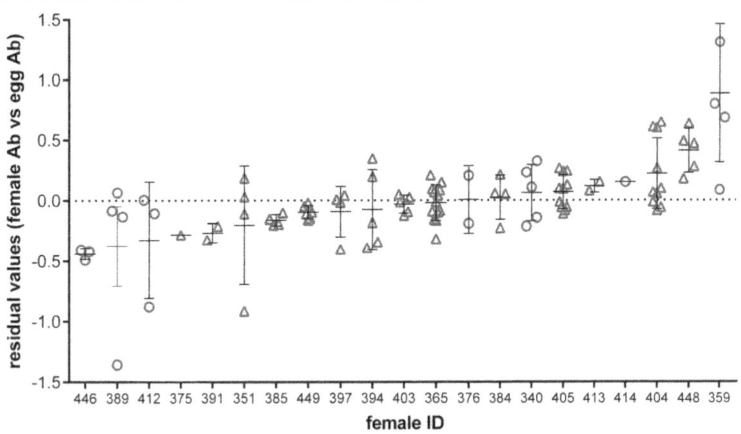


**Figure 2 Inter-individual variability in the capacity to transfer antibodies.** Relationship between residual of the relationship between female and egg antibodies (triangles- NDV vaccinated females, circles- SalenvacT vaccinated females). Each point indicates the residual relationship between female antibody and a single egg at a certain point in time.

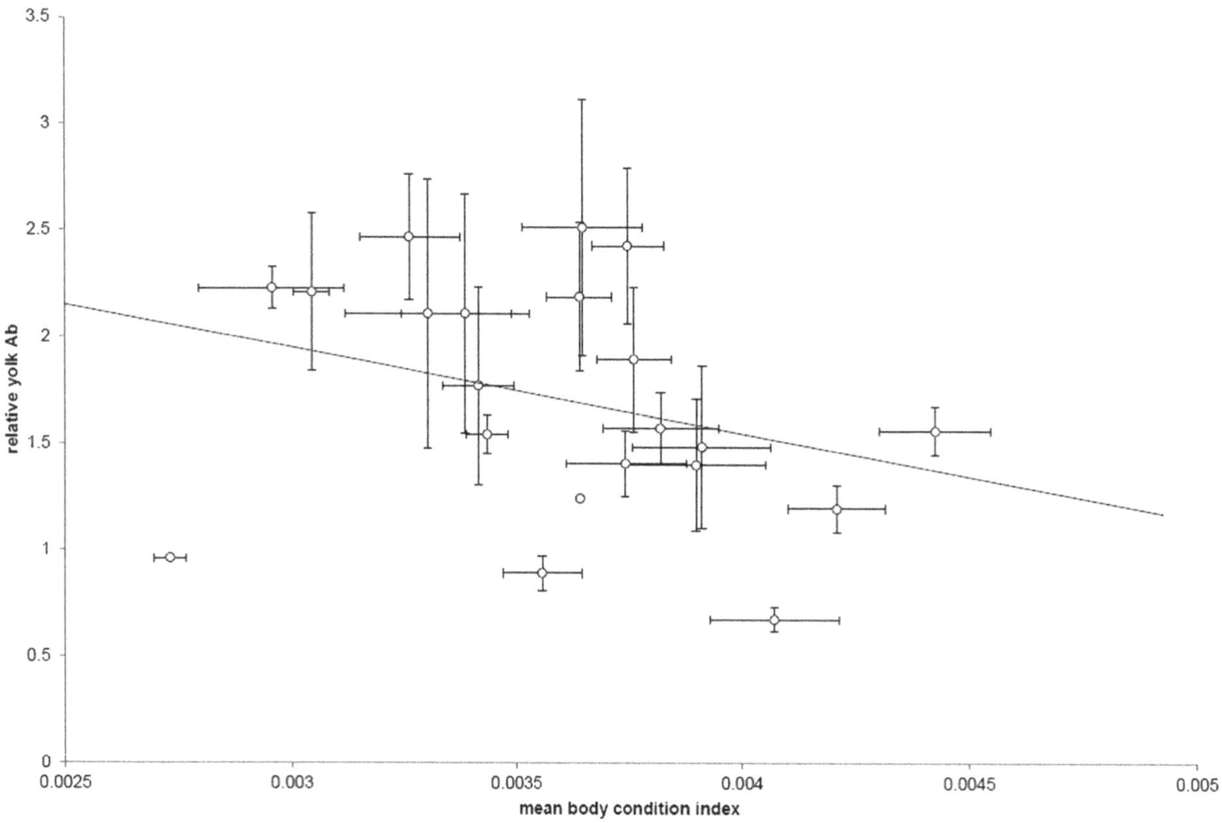


**Figure 3 Egg antibody level and female body condition.** Relative transfer of female antibodies transferred to eggs in relation to female's mean body condition.

offspring in many poultry systems). However, low transfer levels may be advantageous in other situations e.g. *in ovo* vaccination strategies are a common strategy for protection against some diseases and successful uptake by the developing embryo requires maternal antibodies to be low. For example, it is well documented that maternal antibodies impede vaccination procedures for Marek's disease, and can reduce its protection by up to 39% [21].

Our study suggested females are highly repeatable in their transfer levels and that there may be consistency in antibody transfer across two vaccine types suggesting generality of the trait and that it is not pathogen specific.

**Table 2 Linear mixed-effects model of ratio of antibodies transferred to eggs**

| Response | Explanatory | *df* | F | *p* |
|---|---|---|---|---|
| Relative | Yolk mass (g) | 1 | 0.46 | 0.641 |
| | **Female mean condition** | 1 | −2.45 | **0.024** |
| yolk Ab | Female peak Ab | 1 | 0.37 | 0.705 |
| | Treatment | 1 | 1.21 | 0.243 |
| | Time (day) | 1 | 0.197 | 0.844 |

Results for a linear mixed effects model on relative transfer of antibodies. Significant fixed effects are in bold.

Between individual variation estimates and within-individual repeatability estimates for traits can be extremely useful in estimating the potential heritability of traits and their potential to evolve [22]. While some studies have shown genetic differences between domestic lines of chicken in Ab production [23] and there is

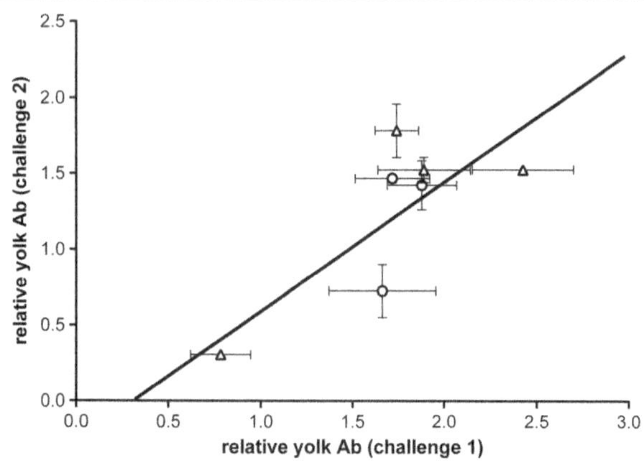


**Figure 4 Positive correlation between mean ratio of antibodies transferred to eggs from females (n = 7) challenged twice.** Circles represent females vaccinated with SalenvacT first, triangles represent females vaccinated with NDV first.

some evidence that genetically different color morphs of pigeon may transfer different amounts of maternal Ab [24], to our knowledge there is very little information on relative between- and within-individual variability in levels of antibody transfer in birds. Similarly high estimates of repeatability have been shown for several key egg constituents that play an important role in avian development such as maternal yolk hormone levels, both across breeding attempts [25,26] and between years [27]. It is unclear whether females that transfer high levels of one maternal component such as antibodies may also transfer high levels of other key egg components in general. This would warrant further investigation to test if some females generally transfer more of all constituents or whether trade-offs occur (for example, more antibodies may be needed to balance hormonal immuno-mediated effects [28]).

The mechanism underlying differences in the ability of females to transfer specific antibodies remain unclear. Transfer of maternal immunity in birds is a two-stage process. Firstly low levels of antibodies are transferred to the yolk as it develops and at oogenesis, then there is a sudden influx of antibodies and other yolk components in the last few days prior to egg formation and laying [10]. This is believed to be receptor mediated and individuals may differ in the number and effectiveness of this transfer process. However, the transfer of antibodies also appear to be directly correlated to the increase in size of yolk as an egg is formed [10] so investment may be tied to allocation of other constituents laid down in the last few days of egg formation (though we foud no evidence that a female's average egg size is related to her general level of transfer). There may also be variation in transfer levels based on variation in antibodies produced; for example, surprisingly, quail have been shown to transfer higher levels of chicken IgY than quail IgY in experimental manipulations [29]. How the molecules of IgY may vary between and within individuals remains unclear and is something seldom considered by immunologists as detection techniques tend to target specific known parts of the molecule making this hard to address (J. Allen pers.com). In birds, eggs are produced on a 24-hour cycle so as one egg is laid, the yolk for the next day's egg has just been formed and packaged and is awaiting fertilization in the reproductive tract and the egg for the day after is still sequestering yolk constituents in the ovary. This sequential pattern of egg production over a series of days explains why eggs vary in their antibody level over the course of an immune response. Understanding these parts of egg development may be important for determining what ultimately controls variation in transfer of maternal antibodies and other yolk components.

In this study the ratio of antibodies transferred to eggs had a significant negative relationship with body condition. We examined whether the effect may arise from low condition females producing smaller eggs but transferring similar amounts of antibody with the result that yolk concentrations of antibodies could be higher. However, this did not appear to explain this result in this case; when yolk size was taken into account in our models, we found no interaction between maternal condition and yolk volume on the amount of antibody transferred. In normal infections a negative relationship between immune response and transfer and maternal condition might not be surprising as females experiencing high levels of infection may be both more likely to produce and transfer a large immune response while simultaneously suffering the costs of associated with infection [30]. However, by purposefully selecting inactive vaccines with no associated pathology we can rule out this possibility. Furthermore by establishing mothers from a clean colony of birds we can also rule out any effects of exposure history. The negative relationship in the level of antibodies transferred and condition might therefore appear counterintuitive as it is often assumed producing an antibody response is costly and that individuals in better condition can "afford" to invest more in transferring egg antibodies to their offspring [31-34]. However, a negative relationship between maternal condition and transfer has also been reported in other systems; in kittiwakes for example, non-food-supplemented females transferred more resources, including antibodies, to offspring compared to food supplemented females [32]. Other studies in birds, however, have found improving maternal condition can increase egg antibodies transfer [32], while others have found no effect of diet manipulation on immune transfer [35] so the relationship may vary depending on the conditions faced. Indeed changing environmental conditions have been shown to reverse the impact of infection for different offspring in some avian species [36]. Alternatively and perhaps more likely given our housing conditions, our findings may simply reflect differences between females that correlate with both their propensity to transfer antibodies and their mass to size ratio. For example, differences in maternal condition are likely to reflect differences in an individual's early conditions and their ability to compete and gain food which could in turn affect either the development of their receptor based system or their ability to utilize this mechanism to maximize transfer. In contrast, female cumulative antibody level alone does not have a relationship with body condition scores, suggesting that female's ability to produce an antibody response is not confined to the same factors as female's propensity to transfer antibodies to eggs. We also found no trade-off between levels of allocation and levels of overall reproductive investment to suggest females might be trading off costs of antibody production or transfer with other costly aspects of reproduction in this system though birds were not under severely food limited conditions (10% lower feed than adlib).

We have demonstrated that the propensity to transfer protection to offspring can be consistent within individuals but vary between individuals. Could there also be adaptive active allocation over and above these effects? There is certainly evidence that levels of antibodies vary with laying sequence [37], brood value [33] and sex of offspring [13]. However, the single most important factor affecting how much antibody was found in the eggs was at what point in the immune response a female was at when a particular egg was laid. For example, if a clutch of eggs was collected from a female in the initial phase of her immune response – perhaps up to 10 days post-challenge – she would likely show a trend of laying eggs with higher levels of antibodies towards the end of the clutch as antibody levels would be increasing on a daily basis and oocytes for successive eggs develop in 24 hour cycles and are laid approximately 24 hours apart. However, if we had collected a clutch of eggs slightly later, we might find females lay eggs with fewer antibodies in the end of the clutch as these are being formed at the time point just after the peak of her immune response, so subsequent eggs would sequester declining levels of antibodies from the female's circulating blood. Females show considerable variation in the magnitude and timing of peak responses, thus making it difficult to predict what an average response in response to challenge (or to natural levels of infection) might be without corresponding measure of female antibody levels over the same time period. In natural systems infection history may not be known and we have demonstrated that females can be very variable in both the magnitude and timing of their immune responses even following a controlled experimental challenge; collecting maternal antibody level throughout a clutch in these types of study would therefore be very informative. Furthermore, the discovery that the distribution of the sexes [38] and extra- pair paternity (EPP) offspring [39] are often non-randomly distributed within a clutch demonstrates the need to ensure these correlated antibody levels with the point in a female's response are considered too. This does not necessarily mean that these patterns are not adaptive – directing particular offspring to a particular part of clutch that may receive more or less antibodies may in itself be an adaptive strategy and a mechanism by which allocation could be adjusted. However, distinguishing between these alternatives is crucial in terms of our understanding of the mechanisms underlying these effects and their implications for offspring fitness as they develop.

## Conclusions

We have shown there is substantial variation between females in the proportion of their own antibodies they transfer to their offspring and that this is unrelated to the overall magnitude of an individual's own immune reponse. This has important implications as it suggests that both a female's propensity to produce a particular level of immune response and her ability to transfer it to offspring may be two separate traits open to manipulation or selective processes. Estimating the transfer component of maternal immunity as a separate trait may therefore allow more informed decisions to be made to increase the production, health and welfare in domesticated animals exposed to disease and increases our understanding of why different offspring may vary in their responses to infection and the consequences maternal traits may have on pathogen host dynamics across generations.

## Materials and methods

### Study system: establishing the maternal generation

Chinese painted quail (*Coturnix chinensis*) are small game birds in the Phasianidae family and have been noted as being an ideal candidate for laboratory experiments due to their small size (40-85g), low aggression and high reproductive success in captivity [40]. *Coturnix* quail are also a commercial species and share many traits with other closely related poultry and are particularly valued for the egg production. The study was conducted using a colony of Chinese painted quail established at the University of Edinburgh to provide parental birds which have no previous exposure to common pathogens. In their natural environment (as wild or domesticated birds) individuals would most likely be in contact with and responding to a range of pathogens. Booster effects of repeated exposure to the same disease and secondary costs of dealing with infection aside from the mounting of an antibody response can therefore impact on measures of both antibody production and measures of host fitness. We therefore controlled for these effects by using inactive as opposed to live vaccines and a clean colony free of the pathogens of interest. By removing these potential confounding effects we were also able detect effects using a smaller sample size to address the specific questions of the study (in line with the 3Rs strategy) and to follow the effects of vaccination on the level of antibody response and transfer over time.

To establish the parental generation, eggs were collected from multiple commercial breeders around the UK and sprayed with Ambicide™ (1% dilution) prior to incubation to prevent any transfer of common environmental pathogens into the colony. Eggs were incubated under standardized conditions (37-38 °C and 40-50% humidity rising to 70% prior to hatching) then brooded for 24 hours prior to transfer to communal cages. Chicks were fed *ad libitum* and kept in large (2,400 × 500 × 375 mm) mixed sex cages (14 birds per cage) until they reached sexual maturity. Cages consisted of wood shavings, multiple feeding stations, covered areas and sand baths to allow birds to follow their full repertoire of natural behavior. A heat lamp was provided in one corner until 10 days post-hatching.

Birds were ringed to allow individual identification at six weeks of age. Standard biosecurity measures to maintain a pathogen free flock were in place throughout the duration of colony establishment and the experiment.

### Experimental protocols

Two weeks prior to the start of the experiment, birds were moved into breeding groups in which they remained for the duration of the experiment. Birds were housed in cages (800 × 500 × 375 mm) in a ratio of three females to one male. Each cage was lined with wood shavings, and contained a nest area for each female and a communal sand bath. Adult birds were maintained on a photoperiod of 16 h:8 h L:D and on a diet of Haith's finch seed, EMP, Prosecto Insectivorous and oystershell in a mix of: 20% protein, 2.5% calcium and 77.5% seed for the duration of the experiment. Total feed per cage was 71.6g based on 17.9g of feed per bird, which is 10% less than how much an adult laying female consumed *ad lib* over a 24hour period (unpublished data).

Once laying commenced, females were observed continuously during the egg-laying window to identify the color morph of each female to enable eggs to be assigned to individual females. Eggs are highly polymorphic in this species in both background color and in the presence/absence and type of markings. No females with similar egg markings were housed in the same cage. Eggs were removed on the same day they were laid to induce females to continue laying over the course of the experiment. Biometrics of females (total body mass (g), tarsus length (mm)) were measured and baseline blood samples were collected on day 0 prior to treatment, along with tarsus and mass measurements.

### Vaccine selection

The response of mothers and subsequent levels of transfer were monitored for two different types of vaccine effective against diseases that are important in both domesticated and wild bird populations. The first was a vaccine (Paramyxo P20, Intervet) against the viral disease NDV (Newcastle Disease Virus); this is a notifiable disease known to infect 200 bird species and has high virulence in poultry [41]. The second was a vaccine (SalenvacT, Intervet) against the bacterial pathogen *Salmonella*; a monitored pathogen of poultry breeding flocks and hatcheries by the European Union (EU) under the Zoonoses Directive 2003/99/EC [41]. Bacterial and viral challenges utilize different arms of the immune system when generating a response in the mother. Both vaccines had previously been demonstrated to produce antibodies that were effective at protecting offspring; NDV [42] and *Salmonella* [43].

### Challenge of treatment groups

Each female in each cage was randomly assigned to one of three treatment groups: a negative control group was injected with 0.1ml of phosphate-buffered saline (PBS) (n = 19), the bacterial challenge group was challenged with 0.1ml inactivated *Salmonella* vaccine (SalenvacT, Intervet) (n = 19) and the viral challenge group was challenged with 0.25ml inactive Newcastle Disease virus (NDV) (Paramyxo P201 Intervet) (n = 18). Each cage therefore contained one bird of each treatment, controlling for cage effects. Blood samples were taken immediately prior to challenge then every 4 days up until 39 days post-vaccination. Samples were collected via puncture of the metatarsal vein using a sterile needle and collection of the blood using 100μl capillary tube. Blood samples were centrifuged on day of collection at 13,400 rpm for 10 minutes and serum collected and stored at -20°C until antibody analyses were conducted. All work was conducted under Home Office License (No PPL 60-4115), with full ethical approval from the University's ethical committee and with veterinary supervision throughout.

### Measurements of maternal traits

The mass and condition of all females were measured on all blood-sampling days. As this was an experimental set up with birds kept under standardized conditions, body condition was calculated by mass/tarsus length$^3$ (see [44] for discussion on the merits of different body condition indices). Eggs were generally laid in the early afternoon and collected on a daily basis throughout the experiment.

### Repeatability of response over different treatments

A subset of individuals was re-vaccinated at 42 days post first treatment (n = 7). Four SalenvacT individuals were vaccinated additionally with 0.25ml NDV at day 42 and three NDV individuals were additionally vaccinated with 0.1 ml SalenvacT. Blood samples and condition scores continued to be collected at four-day intervals and eggs were collected daily from these birds until 76 days post-first vaccination.

### Egg measurements and antibody extraction

Egg length, width and mass were recorded at the time of collection, eggs were then frozen at -20°C for antibody analysis. Yolk mass was recorded after defrosting at the time of antibody extraction.

Antibodies were extracted from egg yolk following the protocol from Mohammed et al. (1986) [45]. The homogenized yolk was diluted 2:1 in Phosphate- buffered saline (PBS) and vortexed for 2 minutes. Chloroform was then added to the egg yolk/PBS mixture at a 1:1 ratio, and vortexed for a further 2 minutes. The mixture was then centrifuged on 13,400 rpm for 10 minutes. After this time the mixture separates into three layers; a top layer containing PBS and supernatant (used for antibody analysis), a deposit of fatty lipids in the middle layer and an organic phase containing chloroform and carotenoids in the bottom layer.

### Antibody analysis: enzyme-linked immunosorbent assay (ELISA)

Specific enzyme-linked immunosorbent assays (ELISAs) were performed for treatment and control groups to detect IgY. For the SalenvacT treated individuals a *Salmonella*-specific ELISA test was performed using FLOCKTYPE® *Salmonella* ELISA kit (Labor Diagnostik Leipzig, Germany). For the Paramyxo P201 treated individuals, a Newcastle disease virus-specific ELISA test was performed using FLOCKTYPE® recNDV ELISA kit (Labor Diagnostik Leipzig, Germany). Both kits were manufactured for chicken serum and plasma but quail antibodies are also detected and a number of studies have used anti chicken antibody to detect antibodies in quails [29,35,46]. Previous pilot work had established the appropriate dilutions to ensure antibodies lay within the bounds of the detectable range of the ELISA kits and dilution curves confirmed these lay within the linear part of the test.

Yolk extractions were diluted 1:62 and blood samples were diluted 1:124 with the buffer provided. The optical density (OD) of the resulting solution was read in a spectrophotometer at 450nm immediately after stopping the reaction, and corrected for using positive and negative controls supplied in the kit. To estimate the repeatability of the method, one sample for each of the treatment groups was tested on all of plates for each ELISA kit. The estimated repeatability within plates (98.35%, $F_{16,17} = 0.898$) and between plates (98.21%, $F_{16,17} = 0.770$) per kit was high. Antibody titre was calculated using the mean values (MV) of the measured optical density (OD) for the negative control (NC) and positive control (PC). The ratio sample to mean PC was then calculated using the following equation.

$$\text{S/P ratio} = \frac{OD_{sample}\text{-}OD(MV)_{NC}}{OD(MV)_{PC}\text{-}OD(MV)_{NC}}$$

Control individuals had readings of $0.0491 \pm 0.0447_{2SD}$ (*Salmonella* ELISA plate) and $0.029 \pm 0.067_{2SD}$ (NDV ELISA plate). Any serum S/P ratio of less than 0.1 was treated as a negative antibody response (within the confidence intervals of the control groups antibody titre values) resulting in 9 birds (1 SalenvacT, 8 NDV) being excluded from the study. Seven birds (4 SalenvacT, 3 NDV) did not produce a full sequence of eggs or ate their eggs before they could be collected and were also excluded from the study. We were therefore able to collect a full sequence of blood and egg antibody levels of the course of a laying sequence for 12 control females, 14 bacterial challenge females and 7 viral challenge females.

### Statistical analyses

Statistical analyses were performed in R (version 2.10.1). We fitted mixed models using the lme4 package. We investigated the effects of a range of factors on the amount of antibody transferred to eggs using linear mixed effects models and linear models and calculated repeatability estimates from the intraclass correlation coefficient. Female antibody titre and egg antibody titre were log transformed to achieve normality. Treating egg antibody level as a dependent variable, we investigated the effects of treatment (vaccination against NDV vs Salmonella), females' circulating antibody level, yolk mass, female condition and number of days since challenge on the amount of antibody transferred to the embryonic environment. Female ID was included in all repeated measures models as a random effect. Blood samples were collected from females every 4 days. To examine how egg antibody levels varied with a female's antibodies over the course of the antibody response we ran a smoothing spline on the female's antibody response to get predicted values for days where blood samples were not collected. Lambda was set at 0.1 to maximize a close fit to the actual data available.

To explore the relative transfer of antibodies within a challenge and across challenges we controlled for the effect of variation between mothers in their own antibody responses to challenge by calculating the ratio of circulating antibody transferred to the egg (egg antibody/female antibody) and using this as our dependent variable. This is a measure of the relative amount of antibodies a female transfer regardless of her own response. This allowed us to investigate how this ratio of antibodies transferred to the second challenge in relation to the first challenge, in this model we also explored which was treatment was given first to determine if there was any order effect.

#### Competing interests

The authors declare that they have no competing interest.

#### Authors' contributions

CMC carried out the experimental work, immunoassay and antibody extraction work. KAH participated with experimental work and egg separation and morphometric data collection. VS participated in the exploration of data and discussion on manuscript. CMC and EJAC were both involved with design of the study and performed the statistical analysis. EJAC conceived of the study, and participated in its design and coordination and helped to draft the manuscript. All authors read and approved the final manuscript.

#### Acknowledgements

This work was funded by The Natural Environment Research Council, UK (studentship to CMC), The Royal Society (to EJAC) and The Wellcome Centre for Infection, Immunity and Evolution (to VS and EJAC). We would also like to thank Gidona Goodman for veterinary advice and research support throughout and Tom Little and Loeske Kruuk for helpful comments on the manuscript.

### References


1. Hasselquist D, Nilsson JA: **Maternal transfer of antibodies in vertebrates: trans-generational effects on offspring immunity.** *Philos Trans Royal Soc B-Biol Sci* 2009, **364:**51–60.
2. Grindstaff JL: **Maternal antibodies reduce costs of an immune response during development.** *J Exp Biol* 2008, **211:**654–660.
3. Mousseau TA, Fox CW: *Maternal effects as adaptations.* New York: Oxford University Press; 1998.
4. Kirkpatrick M, Lande R: **The Evolution of Maternal Characters.** *Evolution* 1989, **43:**485–503.
5. Boulinier T, Staszewski V: **Maternal transfer of antibodies: raising immuno-ecology issues.** *Trends Ecol Evol* 2008, **23:**282–288.
6. Glezen WP: **Effect of maternal antibodies on the infant immune response.** *Vaccine* 2003, **21:**3389–3392.
7. Rose ME, Long PL: **Immunity to Coccidiosis - Protective Effects of Transferred Serum and Cells Investigated in Chick Embryos Infected with Eimeria-Tenella.** *Parasitology* 1971, **63:**299.
8. Mooi FR, de Greeff SC: **The case for maternal vaccination against pertussis.** *Lancet Infect Dis* 2007, **7:**614–624.
9. Muller H, Mundt E, Eterradossi N, Islam MR: **Current status of vaccines against infectious bursal disease.** *Avian Pathol* 2012, **41:**133–139.
10. Kowalczyk K, Daiss J, Halpern J, Roth TF: **Quantitation of Maternal-Fetal Igg Transport in the Chicken.** *Immunology* 1985, **54:**755–762.
11. Brambell FWR: *The transmission of passive immunity from mother to young.* Amsterdam, New York: North-Holland Pub. Co.; American Elsevier; 1970.
12. Saino N, Romano M, Ferrari RP, Martinelli R, Moller AP: **Maternal antibodies but not carotenoids in barn swallow eggs covary with embryo sex.** *J Evol Biol* 2003, **16:**516–522.
13. Martyka R, Rutkowska J, Cichon M: **Sex-specific effects of maternal immunization on yolk antibody transfer and offspring performance in zebra finches.** *Biol Lett* 2011, **7:**50–53.
14. Blount JD, Surai PF, Nager RG, Houston DC, Moller AP, Trewby ML, Kennedy MW: **Carotenoids and egg quality in the lesser black-backed gull Larus fuscus: a supplemental feeding study of maternal effects.** *Proceed Royal Soc London Series B-Biol Sci* 2002, **269:**29–36.
15. Cruse JM, Lewis RE: In *Atlas of immunology.* 2nd edition. Edited by Boca Raton F. London: CRC; 2004.
16. Petney TN, Andrews RH: **Multiparasite communities in animals and humans: frequency, structure and pathogenic significance.** *Int J Parasitol* 1998, **28:**377–393.
17. Sheldon BC, Verhulst S: **Ecological immunology: costly parasite defences and trade-offs in evolutionary ecology.** *Trends Ecol Evol* 1996, **11:**317–321.
18. Fudge AM: *Laboratory medicine : avian and exotic pets. Philadelphia.* London: Saunders; 2000.
19. Cheng S, Lamont SJ: **Genetic-Analysis of Immunocompetence Measures in a White Leghorn Chicken Line.** *Poult Sci* 1988, **67:**989–995.
20. Parmentier HK, Lammers A, Hoekman JJ, De Vries Reilingh G, Zaanen IT, Savelkoul HF: **Different levels of natural antibodies in chickens divergently selected for specific antibody responses.** *Dev Comp Immunol* 2004, **28:**39–49.
21. Witter RL, Burmester BR: **Differential effect of maternal antibodies on efficacy of cellular and cell-free Marek's disease vaccines.** *Avian Pathol* 1979, **8:**145–156.
22. Lynch M, Walsh B: *Genetics and analysis of quantitative traits.* Sunderland, MA: Sinauer; 1998.
23. Abdel-Moneim AS, Abdel-Gawad MMA: **Genetic variations in maternal transfer and immune responsiveness to infectious bursal disease virus.** *Vet Microbiol* 2006, **114:**16–24.
24. Jacquin L, Haussy C, Bertin C, Laroucau K, Gasparini J: **Darker female pigeons transmit more specific antibodies to their eggs than do paler ones.** *Biol J Linn Soc* 2013, **108:**647–657.
25. Gil D, Marzal A, de Lope F, Puerta M, Moller AP: **Female house martins (Delichon urbica) reduce egg androgen deposition in response to a challenge of their immune system.** *Behav Ecol Sociobiol* 2006, **60:**96–100.
26. Tschirren B, Sendecka J, Groothuis TGG, Gustafsson L, Doligez B: **Heritable Variation in Maternal Yolk Hormone Transfer in a Wild Bird Population.** *Am Nat* 2009, **174:**557–564.
27. Eising CM, Pavlova D, Groothuis TGG, Eens M, Pinxten R: **Maternal yolk androgens in European starlings: affected by social environment or individual traits of the mother?** *Behaviour* 2008, **145:**51–72.
28. Postma E, Siitari H, Schwabl H, Richner H, Tschirren B: **The multivariate egg: quantifying within- and among-clutch correlations between maternally derived yolk immunoglobulins and yolk androgens using multivariate mixed models.** *Oecologia* 2014, **174:**631–638.
29. Bae HD, Kitaguchi K, Horio F, Murai A: **Higher incorporation of heterologous chicken immunoglobulin Y compared with homologous quail immunoglobulin Y into egg yolks of Japanese quail (Coturnix japonica).** *Poult Sci* 2009, **88:**1703–1711.
30. Landete-Castillejos T, Garcia A, Gomez JA, Laborda J, Gallego L: **Effects of nutritional stress during lactation on immunity costs and indices of future reproduction in Iberian red deer (Cervus elaphus hispanicus).** *Biol Reprod* 2002, **67:**1613–1620.
31. Hargitai R, Prechl J, Torok J: **Maternal immunoglobulin concentration in Collared Flycatcher (Ficedula albicollis) eggs in relation to parental quality and laying order.** *Funct Ecol* 2006, **20:**829–838.
32. Moreno J, Lobato E, Morales J, Merino S, Martinez De La Puente J, Tomas G: **Pre-laying nutrition mediates maternal effects on offspring immune capacity and growth in the pied flycatcher.** *Oecologia* 2008, **156:**727–735.
33. Pihlaja M, Siitari H, Alatalo RV: **Maternal antibodies in a wild altricial bird: effects on offspring immunity, growth and survival.** *J Anim Ecol* 2006, **75:**1154–1164.
34. Ismail A, Jacquin L, Haussy C, Legoupi J, Perret S, Gasparini J: **Food Availability and Maternal Immunization Affect Transfer and Persistence of Maternal Antibodies in Nestling Pigeons.** *Plos One* 2013, **8.** doi:10.1371/journal.pone.0079942.
35. Grindstaff JL, Demas GE, Ketterson ED: **Diet quality affects egg size and number but does not reduce maternal antibody transmission in Japanese quail Coturnix japonica.** *J Anim Ecol* 2005, **74:**1051–1058.
36. Granroth Wilding H, Burthe S, Lewis S, Reed T, Herborn K, Newell M, Takahashi E, Daunt F, Cunningham E: **Parasitism in early life: environmental conditions shape intra-brood variation in responses to infection.** *Funct Ecol.* In press.
37. O'Brien EL, Dawson RD: **Palatability of passerines to parasites: within-brood variation in nestling responses to experimental parasite removal and carotenoid supplementation.** *Oikos* 2009, **118:**1743–1751.
38. Badyaev AV, Hill GE, Beck ML, Dervan AA, Duckworth RA, McGraw KJ, Nolan PM, Whittingham LA: **Sex-biased hatching order and adaptive population divergence in a passerine bird.** *Science* 2002, **295:**316–318.
39. Magrath MJL, Vedder O, van der Velde M, Komdeur J: **Maternal Effects Contribute to the Superior Performance of Extra-Pair Offspring.** *Curr Biol* 2009, **19:**792–797.
40. Tsudzuki M: **Excalfactoria Quail as a New Laboratory Research Animal.** *Poult Sci* 1994, **73:**763–768.
41. OIE: *Manual of diagnostic tests and vaccines for terrestrial animals.* 7th edition. Paris: Office international des epizooties; 2012.
42. Hamal KR, Burgess SC, Pevzner IY, Erf GF: **Maternal antibody transfer from dams to their egg yolks, egg whites, and chicks in meat lines of chickens.** *Poult Sci* 2006, **85:**1364–1372.
43. Hassan JO, Curtiss R 3rd: **Effect of vaccination of hens with an avirulent strain of Salmonella typhimurium on immunity of progeny challenged with wild-Type Salmonella strains.** *Infect Immun* 1996, **64:**938–944.
44. Galvan I: **Plumage coloration can be perceived as a multiple condition-dependent signal by Great Tits Parus major.** *Ibis* 2010, **152:**359–367.
45. Mohammed HO, Yamamoto R, Carpenter TE, Ortmayer HB: **Comparison of egg yolk and serum for the detection of Mycoplasma gallisepticum and M. synoviae antibodies by enzyme-linked immunosorbent assay.** *Avian Dis* 1986, **30:**398–408.
46. Staszewski V, Siitari H: **Antibody injection in the egg yolk: maternal antibodies affect humoral immune response of the offspring.** *Funct Ecol* 2010, **24:**1333–1341.

2

# The embryoid development of *Strigamia maritima* and its bearing on post-embryonic segmentation of geophilomorph centipedes

Carlo Brena

**Abstract**

**Background:** Many arthropods add body segments post-embryonically, including most of the myriapods. However, geophilomorph and scolopendromorph centipedes are epimorphic, i.e. they form all their segments during embryonic time, although this has never been demonstrated directly. Understanding the similarity between embryonic and post-embryonic segmentation is pivotal to understand the possible evolution from anamorphosis to epimorphosis. We have previously demonstrated that in the geophilomorph centipede *Strigamia maritima* most segments are produced by an oscillatory mechanism operating through waves of expression at double segment periodicity, but that the last-forming (posteriormost) segments are patterned with a different system which might be more similar to post-embryonic segmentation.

**Results:** With a careful analysis of a large number of specimens, I show that the first ("embryoid") phase of post-embryonic development is clearly distinct from the following ones. It is characterized by more moults than previously reported, allowing me to define and name new stages. I describe these embryoid stages and the first free-leaving stage in detail, providing data on their duration and useful identification characters. At hatching, the prospective last leg-bearing segment is limbless and the genital segments are added in the following stages, indicating a residual anamorphosis in *Strigamia* segmentation. I demonstrate directly for the first time that at least the leg-bearing segments are in general produced during embryonic life, although in some individuals the external delineation of the last leg-bearing segment may be delayed to post-embryonic time, a possible further residual of anamorphic development. Additionally, I show that the development of the poison claws during this post-embryonic phase may have some element of recapitulation.

**Conclusions:** The data presented in this paper show that the embryoid phase of post-embryonic development of geophilomorph centipedes may represent an extension of embryonic development, possibly in correlation with the evolution of epimorphic development from an anamorphic ancestor, accomplished without completely losing post-embryonic segmentation activity. This continuity in the segmentation process across the embryonic/postembryonic divide may concur to the evolvability of this developmental process.

**Keywords:** Myriapods, Chilopod, Staging, Post-embryonic, Epimorphosis, Anamorphosis, Arthropod evolution, Forcipule, Recapitulation

Correspondence: carlobrena@hotmail.com
Laboratory for Development and Evolution, Department of Zoology, University of Cambridge, Downing Street, Cambridge CB2 3EJ, UK

## Introduction

Metamerism is a key morphological character of arthropods. To understand its evolution, it is fundamental to understand how this reiteration of modules along the anteroposterior (AP) axis is laid down during development. In the great majority of arthropods with a high number of segments, these are mostly formed post-embryonically, differentiating from a mass of posterior undifferentiated tissue, with the process not being visible externally from the morphology of the exoskeleton. On the contrary, the first segments of those arthropods - and all segments in arthropods like the insects - are laid down embryonically through a molecular patterning process that can precede any morphological differentiation. This process extends to several segments (sequentially segmenting arthropods) or even to all segments (long germ arthropods like *Drosophila*). It is not clear whether this embryonic process is in continuity with the post-embryonic phase of segmentation, and if so to what extent, although understanding this would be pivotal to understanding the evolutionary transition from anamorphosis to epimorphosis.

Geophilomorph centipedes are the only arthropods with a high and, in most cases, intraspecifically variable number of segments (up to 191 leg-bearing segments, LBS) known to form all of their segments embryonically [1] (but see below). Geophilomorphs are epimorphic as are their sister group, the scolopendromorphs, although the latter have an intraspecifically-invariant and smaller number of segments (21–23 LBS) (with two exceptions [2,3]). This is in contrast with the other centipedes, or Chilopoda (Craterostigmomorpha, Lithobiomorpha and Scutigeromorpha) and all other myriapods (Symphyla, Pauropoda and Diplopoda, or millipedes) all of which have anamorphic development, with some, or, in the case of the majority of millipedes, most of the segments formed post-embryonically. The epimorphic nature of development has been always considered a well defined trait of Geophilomorpha and Scolopendromorpha and has been recognised as a major apomorphy, to the extent that their clade is called Epimorpha. Understanding how this character has evolved and fixed represents a major step towards understanding the evolution of the clade.

I have been studying *Strigamia maritima* as a representative of the geophilomorph centipedes. In these centipedes the many segments laid down along the whole germ band are fully available for molecular and developmental analysis as a whole. Recent studies [4] have confirmed that the great majority of these segments are originally produced as morphogenetic units by a sort of segmentation clock, through a cycling activity of segmentation genes operating by waves of expression acting at a double segment periodicity. In contrast, the last part of the segmentation process, although in continuity with the main one and involving most of the same genes, is associated with a change in the molecular process. Here single segmental molecular patterning appears periodically at the anterior margin of the still active growth zone, where conditions are definitely more static and patterning of the localized undifferentiated tissue appears to be associated more with cell proliferation [4]. These conditions might be more similar to the ones which characterise postembryonic segmentation.

In fact, the embryonic/post-embryonic divide is a problematic issue per se. Indeed, it is clear that at hatching, a geophilomorph centipede is little more than an embryo without an egg shell, still brooded by the mother for an additional long period. In this animal, the rupture of the egg chorion does not represent a clear-cut change from an embryonic to a post-embryonic state [5,6]. These non-feeding post-embryonic stages reside with the mother, who stays coiled around them in the brood chamber, a small cavity in the soil matrix. All these brood stages have been termed 'larval stages' by several authors [7-9] (in particular Lewis in his review book [10]) to distinguish them from the later juvenile instars ('post-larval stages,' in Lewis' terms). (In this paper I will use the term 'juvenile' as a generic term to indicate all post-embryonic immature stages). The use of the term 'larva' to describe these brooded stages is highly debatable, and I agree with Minelli and Fusco [11] in suggesting that we should be cautious using this word. I refer to them for a comparative discussion on this topic throughout the arthropods. Consequently, I will call these early post-embryonic stages (proembryoid I to III, peripatoid, foetus; see below) 'embryoid', a term already used for similar instars in several arthropods (see e.g. [11]), because of their similarity with actual embryos. Post-embryoid stages start with the first free-living and feeding stage, the adolescens I, which leaves the mother some time after the moult and has most of the adult characters.

Besides earlier accounts [12-15], the first proper description of these embryoid stages was provided by Verhoeff [16], describing two stages: a peripatoid and a foetus, and those terms have been used in subsequent studies on different species (e.g. [7-9,17,18]). As far as development is concerned, *Strigamia maritima* is by far the best described geophilomorph species. Yet, while we have a reasonably good description of the postembryoid stages and their timing [18-20], previous attempts to describe in part [20] or specifically [21] these early post-embryonic stages have somehow failed to recognise the number and timing of moults precisely, leading to some confusion and inconsistency in recognising, describing, or naming the stages, an issue which is common to the whole of Geophilomorpha.

To try to better understand and define it, I carried out a detailed analysis of this phase of post-embryonic development, based on the improved techniques of culturing eggs and juveniles in mineral oil [22]. This analysis follows up published work on the embryonic development of *Strigamia maritima* where we have defined, described and timed the several stages [6]. Following a rather traditional approach, in this paper I will use a moult, or more precisely an ecdysis, i.e. the actual shedding of the exuvia, as a well defined, recognisable and, most importantly, discontinuous point in time to define the passage from one stage to another. Nevertheless, I do not consider it necessarily a major overall change in the developmental state of the whole animal, as we have previously pointed out [5] (see [11] for a more detailed account and general discussion on the use and definition of the word 'stage').

My analysis has allowed me to identify three previously undescribed moults and their associated embryoid stages, thus I present here a better description of known stages and their timing of development covering the previous gap of information between embryonic and later juvenile stages. This should help understand the transition between these two phases both in development and in evolution.

While analysing specimens for descriptive reasons during their larval development, I found an individual that was apparently still adding leg-bearing segments. Based on weak evidence, some authors have claimed that a few segments might be added after hatching [23,24], although other authors have dismissed that claim [8,25,26]. In fact, all these studies are based on indirect evidence, mostly analyzing the distribution of segment number between different age classes or based on the lack of germinal tissue in late embryoid stages. Indeed, the epimorphic nature itself of segment formation in geophilomorphs, although always assumed to be a fact, has never been based on direct observations on the very first post-embryonic stages. Published data cannot rule out the possibility of an occasional delaying of segment formation into the post-embryonic phase. This possibility may be rather important from an evolutionary point of view, as it would show the potentiality for a shift from one segmentation mode to another. To try to identify this possibility, I have followed a workably high number of specimens individually through the embryoid stages to check whether there was any addition of segments.

Considering that this study is based on the analysis of live specimens, I should stress that, in general, 'segment' in this paper means an identifiable subdivision of the cuticle, externally recognisable as a full serial unity along the antero-posterior (AP) axis of the animal. This in fact is the standard approach when defining the number of segments, for example, of the different instars of anamorphic myriapods, independent of the more advanced state of development of the internal metameric tissues or organs.

I have found that almost no individual added leg-bearing segments after hatching, although a very small fraction have delayed the complete production of the last leg-bearing segment. In any case, all specimens added the two genital segments during embryoid (post-embryonic) development. On one hand, this has confirmed, directly for the first time, the general epimorphic nature of leg-bearing segment addition in *Strigamia*. On the other hand, though, it has shown that indeed a small remnant of anamorphic segmentation is still present.

## Results

### Embryoid stages in *Strigamia*

The following description of stages is based on the best conditions under which live embryoid specimens can be observed, i.e. immersed, mostly in mineral oil, with their exuviae artificially removed. Those exuviae will stay attached to the rear end of the animal in any culturing conditions, although it is not clear when and how exactly the young *Strigamia* dispose of them in nature and if and how the mother helps them. Although described by previous authors [18,21], I don't think movement of any kind can be a real discriminatory trait between stages. Indeed, there is a continuum in the acquisition of motility of *Strigamia* embryoids, from the strong peristaltic movement already present in late embryonic stages [6] to writhing movements in late foetus stages. Moreover, presumably there is an increase of this activity within a single stage, as the animal gets older.

It is impossible to define a general state of development of a generalized leg (i.e. a leg representative of trunk segments), because from the appearance of the first segment during embryonic development to the last embryoid stage there is a smooth gradient in the level of development along the animal's AP axis. This implies that it is equally impossible to define precisely how many final legs are in a given state of development, e.g. externally segmented as in proembryoid III (see below).

Given my introduction of new early stages and the inconsistency of previous descriptions, I have decided to use the known term 'peripatoid' for the longer stage preceding the foetus stage, in accord with [27].

See Table 1 for the defining characters of each embryoid and adolescens I stage.

### Hatched stage 8 embryo

***Approx duration: 2 days***

*Strigamia* eggs hatch after approximately 48 days at 13°C, a temperature close to the average June temperature at the field site for this population [6]. The rupture of the chorion is probably promoted by internal pressure and by movements of the embryo which are characteristic of the last embryonic stage (8) [6]. The opening is presumably triggered by the action of the so called 'egg tooth': this is a sharp sclerotised tip on the embryonic cuticle of the second maxillae. The cuticle is already free, as the apolysis

**Table 1 Summary of the defining characters of early post-embryonic stages in *S. maritima*: embryoid stages (Proembryoid I to Foetus) and first free living juvenile stage (adolescens I) (ec: embryonic cuticle)**

| Strigmia marittima embryoids and adolescens I STAGING TABLE | | | | | |
|---|---|---|---|---|---|
| **Stages** | **Shape of the body/trunk** | **Last leg-bearing segment** | **Genital segment(s)** | **Antennae** | **Additional characters** |
| Ecdysis ec | | | | | |
| Proembryoid I | Sharp angle between anterior and posterior trunk anterior half much larger, with ventral side concave | Limbless | terminal trunk undifferentiated, barely developed | Ventral (slightly anterior), 14 unclear antennomeres, depressed | |
| Ecdysis e1 and e2 | | | | | |
| Proembryoid II and III | Angle between anterior and posterior trunk anterior half slightly larger, with ventral side flat | Limb buds just appearing | terminal trunk undifferentiated, extended | Ventral orthogonal, 14 antennomeres slightly elongated | |
| Ecdysis e3 | | | | | |
| Peripatoid | Uniform circular section of trunk along AP axis | Limb buds extended but barely, if not, segmented | single genital segment defined (single segmental furrow present) | Ventral forward more than 90 degrees from the main head/trunk axis, 14 well defined antennomeres | |
| Ecdysis e4 | | | | | |
| Foetus | Trunk slightly dorsoventrally depressed | Limbs segmented, although less developed than the rest of the legs | subdivision in segments, but number and shape unclear because not clearly defined | Elongated and divaricated further, extending forward, almost on the trunk plane | Appearance of setae, maxillipeds/ forcipules tanned and multisegmented |
| Adolescens I | Trunk dorsoventrally depressed | Legs fully differentiated but distinguished from the rest of the legs | differentiated with several sclerites in two genital segments and terminal anal valves | Tanned in stripes, rich in setae and sensilla | Well developed and abundant setae, tanning in particular of head and forcipular segment |

has started already at the transition between stage 7 and 8 of embryonic development [6] (Figure 1A,B,B1). Under culture conditions, the chorion of some eggs does not open; in this case those embryos, although constrained and not able to extend, will somehow continue development for a while within the intact chorion, going through several moults (data not shown).

In general, the chorion is not fully shed, and at least half of it may stay attached to the posterior end of the animal for several stages (see below). The hatched specimen is basically the stage 8 embryo, with the embryonic cuticle (ec in Figure 1) still covering the whole body but largely detached (see on the tip of antennae in Figure 1B1). The animal is still fully folded in two by the dorso-ventral flexure, head facing the proctodeal opening, with the juxtaposed anterior and posterior halves of the ventral side still apparently kept in position by lateral cuticle extensions (see [6]).

After a couple of days, the embryo starts undergoing ecdysis, cracking the embryonic exuvia just posterior to the head. Once detached, this exuvia shows the entire ectoderm pattern associated with the stage 7 embryo, including the undifferentiated ectoderm covering the surface dorsal to the hemitergites (htb in Figure 1C1), which have not yet extended to complete dorsal closure. Once joined, the left and right hemitergites will give rise to the full tergite, as seen in the proembryoid I (t in Figure 1D) (see [27] for full description).

Throughout this stage, and indeed from before hatching, the gut of the animal shows extensive peristaltic contractions which redistribute the mass of pink pigmented yolk contained within it (see Additional file 1: movie 1). This

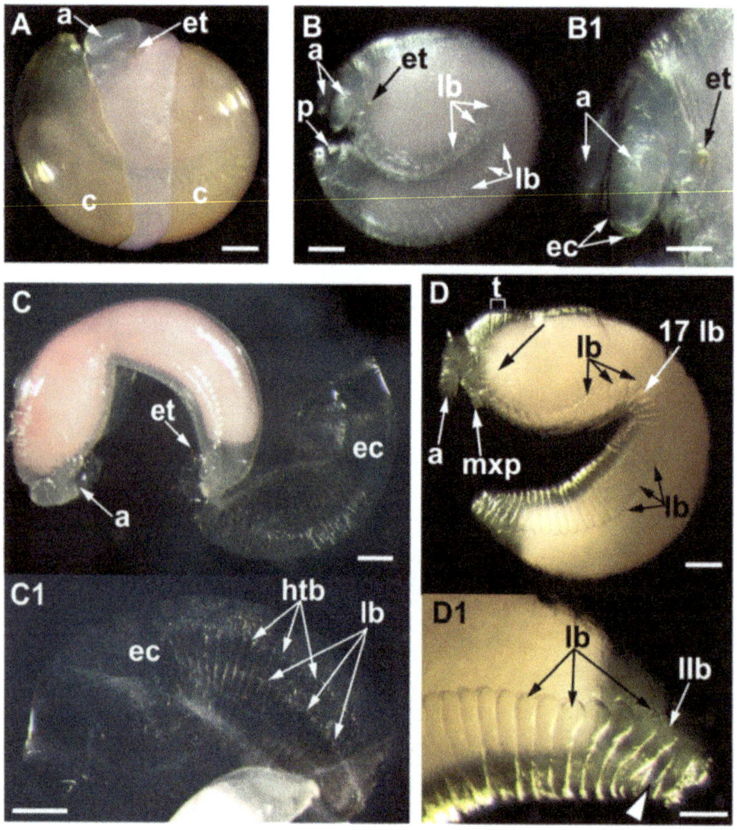


**Figure 1 Hatched egg and proembryoid I.** Lateral view; live specimens immersed in LES **(A-C)** or in mineral oil **(D)**. In this figure anterior is to the left and the dorsal side is to the top, at least for the anterior half of the animal (see Figure 2 in [27] to better understand orientation). Panels with the same letter indicate different views of the same specimen. (The same conventions apply to all other figures, unless stated otherwise). **A**: Hatched egg showing what is essentially a final (late stage 8) embryo lying within the cracked chorion (c). **B**: Hatchling of the same stage with the chorion manually removed. B1 is a high magnification of **B**, showing in more detail the egg tooth (et) on the second maxillae, which is used to crack the chorion. Note that the embryonic cuticle (ec) on the antennae (a) has already detached from the underlying epithelium (i.e. apolysis has occurred). **C-D**: proembryoid I, just after moulting in **C**, with the embryonic cuticle (ec) still attached to the rear of the animal; this exuvia has been manually removed in **D**. C1: the detached exuvia showing all cuticular traits as laid down in stage 7 embryo. The proembryoid I is still sharply folded at the level of the 17$^{th}$ LBS. The limb buds (lb) appear as lateral protrusions of the ventral segmental units, and are differentially developed along the anteroposterior axis, from the more differentiated antennae (a) to the last formed segment (eventually the last leg-bearing segment), which lacks the limb buds altogether. D1 is a higher magnification of the specimen in D, turned upside down. htb: semicircular dorsal border of the hemitergites; llb: last limb bud; mxp: maxilliped; p: proctodeum; t: tergite. Scale bars: **A**, **B**, **C**, C1, **D**: 200 μm; B1, D1: 100 μm.

movement and the associated contraction of the whole body may promote the following ecdysis (see Additional file 2: movie 2).

### Proembryoid stages

I introduce the term 'proembryoid' to describe the very first post-embryonic stages, unrecognised by previous authors. This phase is characterized by two ecdyses, with the third ecdysis giving rise to the following peripatoid stage. Those two ecdyses happen in a short interval, to the extent that the second one starts at the anterior part before the first is completed at the posterior part (see Figure 2). The near simultaneity of these two ecdyses did not allow for the characterization of the stage between them, as I could directly observe only stages proembryoid I and proembryoid III. For this reason, proembryoid stages II and III will be discussed together. Finer future studies may provide discriminating traits between proembryoid II and III.

### Proembryoid I

*Approx duration: 2 days*

This stage basically corresponds to the last embryonic stage 8, without the embryonic cuticle (the ecdysis of which starts at the beginning of embryonic stage 8 [6]) and with the yolk in the gut still occupying most of the internal volume. The anterior half is still markedly larger than the posterior one. The anterior and posterior halves of the ventral surface are now well separated and form a rather sharp angle. This position seems to be rather fixed and in general not correlated with the variable final number of segments. Indeed, in 85% of cases (n = 66), the limb bud just before or in the middle of the bending point belongs to the 17th LBS (the 16th in 6% and the 18th in 9% of cases). In particular, there does not seem to be a correlation between this position and the sex of the animal, as among those with the bending at the 17th LBS, 50% have 47 LBS and 50% 49 LBS, mirroring the expected 50:50 sex ratio of the Brora population where 86% of the animals are either males with 47 LBS or females with 49 [28]. Having said that, some very minor, but consistent, correlations may be present at a finer level, as the 17th limb bud precedes the bending point in 46% of the specimens with total 47 LBS and only 8% of those with 49 LBS total. Additionally, extreme lower values at the bending point never correspond to a high value of total LBS number (there is no 16th LBS at the bending point in specimens with a total of 49 or 51 LBS and no 18th at the bending point in specimens with total 45 LBS).

The developmental gradient along the AP axis is particularly evident, in particular in the development of the limb buds. In this sense, most of the anterior limb buds including the antennae, maxillae and maxillipeds (forcipules) show a good degree of segmentation (Figure 1D), while the most posterior limbs are just unsegmented buds. In live material and in my preparations it is difficult to ascertain the specific number of segments in the limbs, but at least the antennae appear to be articulated into their final

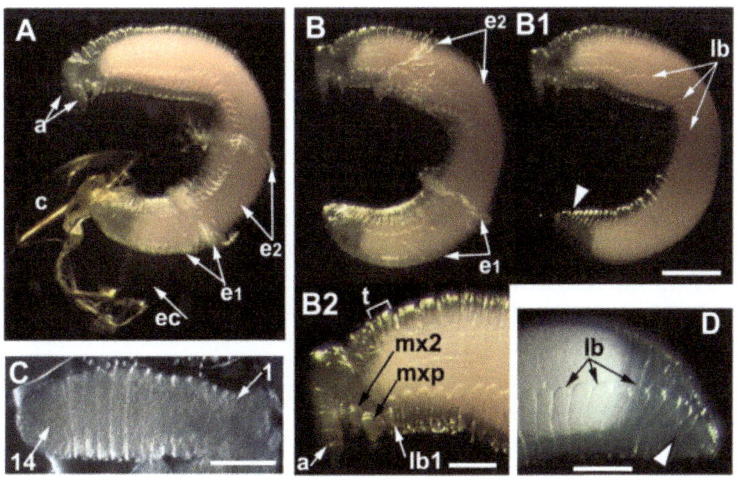


**Figure 2 Proembryoid II and III.** **A**, **B**, **D**: Lateral views; live specimens immersed in mineral oil; **C**: unfixed preparation mounted on a slide and viewed at the compound microscope with Nomarski optics. **A**: the proembryoid III has partially shed the II exuvia (e2) while the embryonic cuticle (ec) and the chorion (c) are still attached at the rear end of the animal and the proembryoid I exuvia (e1) still covers part of the trunk, overlying the e2 exuvia. In **B** the chorion and embryonic cuticle, and in B1 also the proembryoid I and II exuviae (e1, e2) have been removed to show all limb buds (lb) of the proembryoid III stage. The limb buds of the last LBS is now visible (arrowhead); note that the specimen in **B** is slightly younger than the one in **A**, with the e2 exuvia still covering most of the trunk. B2: high magnification of **B**, showing the extent of differentiation of the most anterior limbs including the antenna (a), second maxilla (mx2), maxilliped (mxp), first leg (lb1) and the tergite (t). **C**: high magnification of an antenna, showing the differentiation of all 14 antennomeres. **D**: high magnification of terminal part of the trunk, showing the initial development of the last limb bud (arrowhead). Scale bars: **A**, **B**, B1: 500 μm; B2, **D**: 200 μm; **C**: 100 μm.

number of 14 antennomeres (Figure 1D). The most noteworthy fact is that the last formed segment, which will be eventually the last LBS, is limbless at this stage (arrowhead in Figure 1D1). Posterior to that segment, genital segment (s) are not developed yet, at least at the level of the cuticle/ external morphology and general extension of the region.

### Proembryoid II and III

*Approx duration: 4 days*

By this stage the animal has dramatically extended, mostly thanks to the elongation of the posterior part with all segments along the body now with similar extension. The anterior part of the body is only slightly larger than the posterior one. The midgut with the yolk in its lumen starts to retract from the posterior end of the trunk. If not manually removed, the first three exuviae, i.e. the embryonic cuticle and proembryoid I and II exuviae (ec and e1-2 in Figure 2) are still covering a decreasing part of the animal (Figure 2A). The anterior part of the trunk is now almost straight ventrally and the sharp angle between it and the posterior part is reduced. The differentiation of the tergites into pretergites and metatergites is now well marked.

Segmentation of the anterior limbs is further developed and by now the final number of 14 antennomeres can be precisely identified (Figure 2B2 and C), although the antennae as a whole are only slightly elongated in comparison with the previous stage. The antennae form now a 90° angle with the major head axis. The few last limbs are still un-segmented and the very last leg buds has now just started to develop as a lateral protrusion of the ventral portion of the last LBS (arrowhead in Figure 2B1 and D). The genito-anal area of the trunk is now more extended but no subdivision of the cuticle is recognizable.

### Peripatoid

*Approx duration: 4 days*

The peripatoid appears after the 4th ecdysis, and all 4 exuviae tend to remain attached to the rear of the animal, each included in the previous one, hence requiring manual removal to get the entire body clean (Figure 3A).

The trunk is now elongated, with a roughly circular cross-section, more or less uniform along the AP axis (Figure 3B). The antennae extend ventrally forward more than 90 degrees from the main head/trunk axis. The antennomeres are now very distinct. The second maxillae start to differentiate the telopodite, while the maxillipeds, faintly multi-segmented, present a sharp tip (Figure 3C). Legs up to mid-trunk show a good level of differentiation, although the identity of the single podomeres is unclear (Figure 3D). The last leg buds are still almost unsegmented (Figure 3E). For the first time a clear cuticle furrow defines a genital segment (Figure 3E).

### Foetus

*Approx duration: 28 days*

At this stage the embryoid has further elongated, shows more defined/separated trunk segments and starts to be slightly compressed dorso-ventrally (Figure 4A,C). Foregut and hindgut, which lack the pink yolk confined to the midgut, are more extended, but their extension depends on the age of the animal and probably on the state of body contraction of the specimen (compare Figure 4A with Figure 4C).

The antennae, further elongated and divaricated, now extend forward, almost in the same plane as the trunk (Figure 4A-C). The maxillipeds, extended forward at a 45° angle, are now properly developed, with extended, sclerotized and partly tanned tips; they appear to be constituted by up to 7 segments, although the precise number and identity of forming podomeres is far from clear in my preparations (Figure 4B and Additional file 3: Figure S1). Segmentation of most of the legs is now well defined at the cuticular level and, although of similar length, the different podomeres have reached their final number and can be tentatively distinguished as coxa, trochanter, prefemur, femur, tibia, tarsus and a final claw (or pre-tarsus) (Figure 4D, the coxa is not distinguishable here). The first small setae appear, although they are clearly visible only with careful focusing at the microscope, in particular on antennae and limbs (they are barely visible around the tip of the prospective pre-tarsus in Figure 4D). The terminal legs, although less developed, are now clearly segmented. The genital segment(s) is further developed, although a subdivision of the cuticle into more segments or segmental structures is not clearly detectable (Figure 4E). In cultured specimens, all 5 exuviae and the chorion can still be attached to the rear of the animal (Figure 4F see also Figure 1C in [5]) but not when the embryoids are in the brood chamber with the mother (Figure 4G-H), which may indicate her active role in disposing of those materials. Foetuses tend to show pronounced writhing movements, both in immersion in mineral oil (Figures 4A-C), and in their brood chamber (Figure 4H).

### Adolescens I

*Beginning around 40 days after hatching*

With the 6th moult, the animal has now all the characters that in a more developed state will typify the adult. The trunk is dorsoventrally depressed with well developed legs extending laterally. The juvenile is very active, contracting in twisting movements in immersion (Figure 5A); in these conditions, the animal survives for only few days. The cuticle, although mostly transparent, is relatively well sclerotized and starts to be tanned. Particularly tanned is the cuticle of the head and of the first trunk segment, i.e. the forcipular segment (Figure 5A,C-F). Gnathal limbs are

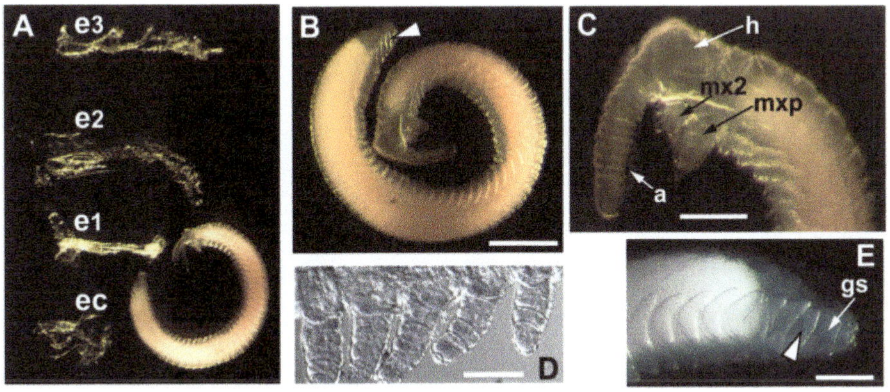


**Figure 3 Peripatoid.** **A**, **B**, **C**, **E**: Lateral view; live specimens immersed in mineral oil. **D**: unfixed preparation mounted on a slide and viewed at the compound microscope with Nomarski optics. **A**: specimen with its 4 manually removed exuviae (the embryonic cuticle, ec, and the three proembryoid exuviae, e1-3) originally attached to the rear end of the animal; **B**: the peripatoid tends to keep a coiled position with the head in proximity of the posterior end of the animal; all limb buds are more or less developed up to the last one (arrowhead); **C**: high magnification of head (h) and anterior trunk showing elongation of the antennae (a) and development of second maxillae (mx2) and maxillipeds (mxp); note that in this specimen the apolysis of the next moult has already taken place in some anterior regions, as shown by the detachment of the cuticle from the tip of the maxillipeds. **D**: high magnification of mid-trunk limbs showing clear signs of segmentation of the cuticle, although the identification of single podomeres is not possible; E: high magnification of the terminal part of the trunk, showing the initial development of the last limb buds (arrowhead) and a clear delineation of a genital segment (gs). Scale bars: **A**, **C**: 500 µm; **B**, **E**: 200 µm; **D**: 100 µm.

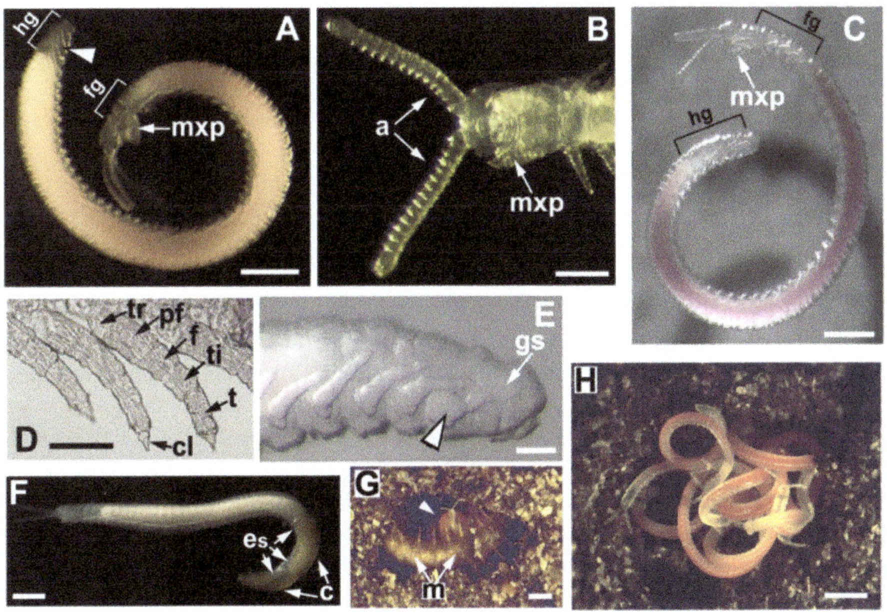


**Figure 4 Foetus.** Specimens immersed in mineral oil (**A**, **B**, **C**, **E**: alive; **F**: dead); **A**, **C**, **E**, **F**: lateral view; **B**: ventral view; **D**: unfixed preparation mounted on a slide and viewed at the compound microscope with Nomarski optics. **A**: The foetus tends to contract, putting the head close to the posterior end of the animal; all legs are developed, including the last one (arrowhead); maxillipeds (mxp) stretch forward ventrally; brackets show extension of foregut (fg) and hindgut (hg); **B**: high magnification of the head (a: antennae; mxp: maxillipeds); **C**: specimen with light-coloured background, to get a more natural appearance, including the brownish tip of the maxillipeds (mxp); **D**: high magnification of mid-trunk legs with tentative identification of podomeres (tr: trochanter, pf: prefemur, f: femur, ti: tibia, t: tarsus, cl: final claw); **E**: high magnification of the terminal part of the trunk, showing the last leg (arrowhead), by now partially developed and segmented, and the genital segment (gs) (the wrinkled aspect of the last legs is probably a culturing condition artefact); **F**: specimen in which chorion (c) and exuviae (es) have not been removed but still cover the terminal part of the trunk; **G**: window on the beach matrix, partially showing a brood chamber: the mother (m) is still coiled around a brood of foetus juveniles (arrowhead) with her ventral surface facing outside; **H**: brood of foetus juveniles of panel G once the mother has escaped following the opening of the brood chamber. Scale bars: **A**, **C**, **F**: 500 µm; **B**: 300 µm; **D**, E: 100 µm; **G**, **H**: 1 mm.

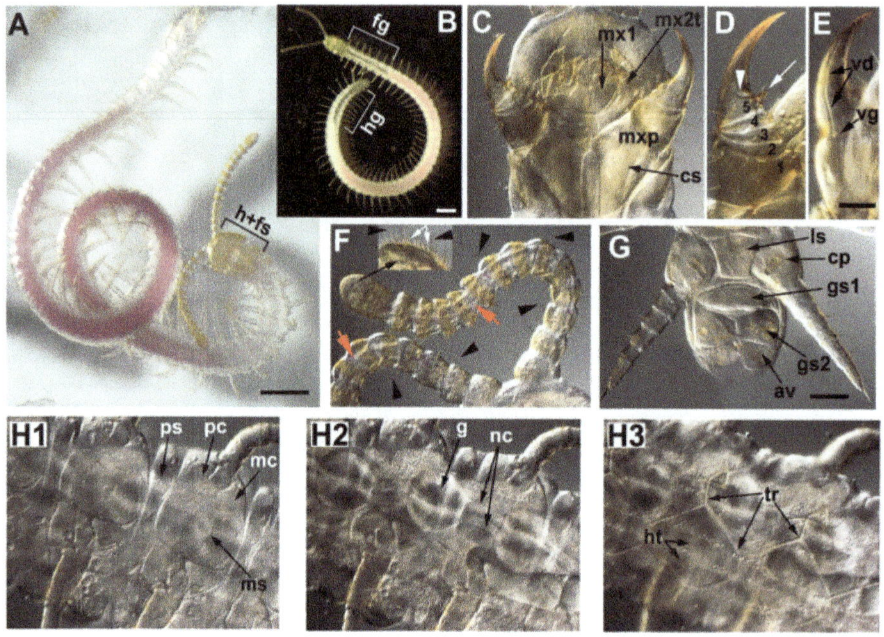


**Figure 5 Adolescens I. A**, **B**: live specimens in mineral oil, dorsal view; **C-H**: unfixed specimens viewed at the compound microscope with Nomarski optics, ventral view. **A**: adolescens I, showing typical torsion movements, viewed on white background to highlight the tanning of the head cuticle (h) and maxilliped/forcipular segment (fs) and intensity of yolk pink pigment associated with the mid-gut. **B**: specimen showing relative extension of foregut (fg) and hindgut (hg). **C**: high magnification of the head, showing complete differentiation of the gnathal limbs: first maxillae (mx1), second maxillae (mx2t: telopodite of the second maxillae) and forcipules/maxillipeds (mxp). **D**: in addition to the coxosternite (cs in **C**) the forcipule appears to be composed of additional 5 articles (1–5) (see text for details); **E**: middle focal plane of a forcipule showing venom duct (vd) and venom gland (vg); **F**: high magnification of the antennae covered by setae (arrowheads) and, on the last antennomere, also by basiconic sensilla (white arrows in the high magnification insert); muscle fibres are visible along the main axis (red arrows); **G**: high magnification of the terminal segments of a female, showing the sternites of the last LBS (ls) and of the first (gs1) and second genital segment (gs2), followed by the anal valves (av); appearance on the surface of the coxal pore (cp) canal is an artefact of focal planes stacking; **H**: high magnification of mid trunk, viewed at different focal planes, progressively away from the viewer (H1, H2, H3); H1: surface focal plane, with metasternite (ms), presternite (ps), metacoxa (mc) and procoxa (pc); H2: focal plane just under the surface with double nerve cord (nc) and ganglia (g); H3: focal plane just below the dorsal surface, with heart (ht) and tracheae (tr). Scale bars: **A**, **B**: 500 μm; **C**, **F-H**: 100 μm; **D-E**: 50 μm.

developed with their specific components, e.g. the telopodites of the second maxillae, and hence should be by now fully functional (Figure 5C). In particular, the maxillipeds/forcipules are well developed, and each appears to be constituted by a basal large coxosternite (cs in Figure 5C), and 5 other segments (1–5 in Figure 5D), although I don't know whether there is articulation between all of them. This subdivision has been more or less clearly observed in all good quality flat-mounted preparation of adolescens I (n = 20). The identification of the individual articles is uncertain because in the adult they are only 4 (see Discussion). I tentatively interpret them as being 1: trochanter, 2: prefemur, 3: femur, 4: tibia, and 5: tarsungulum, the latter with its well formed denticle (arrow in Figure 5D) (a character found in many geophilomorph lineages) and transverse sulcus (arrowhead in Figure 5D). The forcipules are equipped with a venom gland with its well-formed duct, opening close to the tip of the claw (Figure 5E). Judging from the intensity of the pink colour, the midgut is still rather full of yolk, but the transparent foregut and hindgut are now much more extended than in previous stages (Figure 5A-B).

The antennae, provided with muscle fibres all along their length, are covered with setae, the last antennomere with small basiconic sensilla as well. The antennae show a pattern of alternating bands due to the partial tanning of the cuticle within each antennomere (Figure 5F).

The last legs are now fully developed as the other ones, but with some peculiarities, particularly the coxae, enlarged and provided with a single coxal pore at its centre (Figure 5G); those legs appear already at this stage not be used for walking, although they are not yet sexually differentiated as they are in the adult (see [29]). The terminal area is now for the first time completely differentiated into two genital segments and terminal anal valves; in males the sternite of the first genital segment starts to differentiate two medio-lateral protusions (see also [20,22,29]), which will become the two gonopods in the adult. Not only the exoskeleton is fully differentiated into its different components, like the ventral meta- and pre-sternite and meta- and pro-coxa (Figure 5H1), but also internal major organs/apparatuses are fully developed, like the double nerve cord with segmental ganglia (Figure 5H2) (one single suboesophageal ganglion for the gnathal segments, and one for each trunk segment, forcipular segment included). Two smaller ganglia are associated with the two genital segments (data not shown). Circulatory and respiratory systems appear fully functional, at least according to the development of heart and tracheae (Figure 5H3).

### Timing of embryoid development

Figure 6 shows a schematic representation of the stages described above, on a timeline for *Strigamia* development, with the duration of stages expressed in days, based on my analysis at 13°C on single specimens (n = 23) and whole clutches as described in M&M. The timing of development is probably in part affected by the culturing conditions, in particular if one considers that my specimens have been constantly immersed in mineral oil for three months to reach the adolescens I stage. Additionally, precise timing has been affected by limited observations (frequent observations would have further jeopardized the health of the juveniles, as happens with eggs [6]) and by inter-individual variability, increasingly higher as post-embryonic time progresses. Consequently, the reported timing should be mostly useful to perceive the relative duration of the different stages. The only 'natural' case I observed, i.e. the clutch of eggs with their mother of Figure 4G-H (constantly at 13C, undisturbed), fits well with the reported time frame. Indeed, those foetuses, well developed, were observed 90 days after collection (early in the season, very close to deposition time), i.e. at 40 days of post-embryonic development, considering an embryonic development of around 50 days [6].

The overall duration of embryoid development is around 40 days of which the great majority (70%) is represented by foetus development, with embryoid and peripatoid stages representing, respectively, only 15% and 10% of the total duration.

### Post-embryonic segmentation

To check the epimorphic nature of development in *Strigamia*, in particular the possible post-embryonic addition of leg-bearing segments, I followed 148 specimens from early post-embryonic stages to adolescens I stage (see Materials and Methods for details). 33% of them could not be used because they either died before reaching adolescens I stage (10% of the total), or developed abnormalities (like helicomerism) (13%) or simply the condition of the terminal trunk in the picture would not allow identification of segments (10%). Almost none (see below) of the juveniles that reached the adolescens I stage (n = 98) increased the number of LBS during embryoid development, although the limbs of the last LBS and the following genital segments are indeed added post-embryonically, as described above. The distribution of the final number of LBS was: 4% with 45 LBS, 45% with 47 LBS, 47% with 49 LBS and 4% with 51 LBS, reasonably close to the distribution of the number of LBS in the wild Brora adult population (see [28]). In 2 cases of proembryoid I, like the specimen shown in Figure 7A-A1, it was impossible to identify traces of segmentation on the cuticle posterior to the 48th LBS, although the latter appears to be normally developed, with the associated limb buds, as expected for that segment at this stage. Once they reached the adolescens I stage, those two individuals had 49 normally developed LBS.

## Discussion

### Embryoid stages in *Strigamia maritima*

In this work I have analysed, through direct observation on cultured material, the early post-embryonic development of *Strigamia maritima*, at the time the juveniles are brooded by the mother. Anticipated by the publication of details of epithelium structure [27], I have identified for the first time several previously overlooked moults and clarified the post-embryonic addition of segments (see below). This analysis has allowed me to 1) identify in detail a new developmental phase, the proembryoid, further subdivided into three stages, 2) better identify the previously inconsistently described stages of peripatoid and foetus,

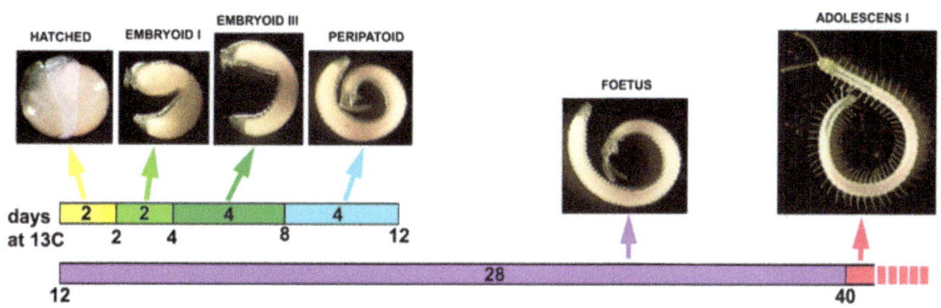


**Figure 6 Schematic representation of *Strigamia* embryoid development.** *S. maritima* development is represented by a series of photographs exemplifying each embryoid stage as indicated. Bars of different colours, are scaled to represent the duration of each developmental stage in days at 13°C. Under the bars, the approximate cumulative duration in days of the overall developmental period at the end of each stage is indicated.

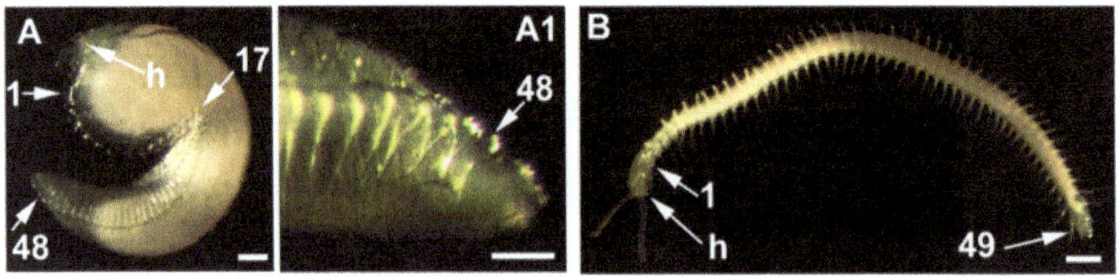


**Figure 7 Development of last leg-bearing segment in a single animal.** Specimen immersed in mineral oil, still alive in **A**, A1, dead in **B**; **A**: Lateral view, **B**: ventral view. A *Strigamia* proembryoid I with 48 well-formed limb buds on as many segments; A1: high magnification of the terminal part of the trunk, showing no apparent external segmentation posterior to the 48th LBS. **B**: same individual as the one in **A**, at the adolescens I stage, with 49 well formed legs and associated segments, followed by genital segments. Numbers indicate the number of the LBS and associated limb buds; h: head. Scale bars: **A**: 200 μm; A1: 100 μm; **B**: 500 μm.

and 3) highlight some characters of the following first free-living stage (adolescens I).

I have clearly shown that this first phase of post-embryonic development is markedly distinguished from the following phase. In particular, specific embryoid characters can be summarised as follows:

1. There is a gradient in the state of development along the AP axis, where the last segments are still in an 'embryonic' state up to the end of this phase.
2. Not all segments are externally visible (see below for discussion).
3. The way of life is different from the post-embryoid juveniles and adults, with the embryoid juveniles brooded by the mother and relying solely on yolk.
4. Adult structures/organs are not (fully) developed.

It is interesting to note that the main reason why the anamorphic phase of development in more basal centipedes (scutigeromorphs: 6 stages; lithobiomorphs:in general, 5 stages) is traditionally called a 'larva' relies basically only on character 2) above. This 'larva' is already a free-living juvenile and, somehow, just a small version of the adult, as are post-embryoid stages in Epimorpha centipedes (and indeed this use is disputed by several authors, see [11]). I think the 'embryoid' terminology should be applied also to the scolopendromorphs, as they are characterized by three post-hatching stages brooded by the mother, separated by moults. In *Scolopendra*, the hatching instar, clearly corresponding to the last embryonic stage as in *Strigamia*, is called first embryonic stage by Heymons [30] and postembryonic stage I by Dugon et al. [31]. The following two stages, i.e. the proper post-embryonic stages, are called, respectively, second embryonic stage and foetus by Heymons [30], postembryonic stage II and fetus by Dugon et al. [31], and, rather confusingly, third embryonic stage and adolescens I by Lawrence [32] and Brunhuber [33]. It is of course possible that even in this case some moults might have not been recognised, and that the situation might be more complex than it appears. Consequently, the homology of these stages with those of *Strigamia* might not be straightforward, although according to their general level of development, it would seem quite plausible that Dugon's [31] postembryonic stage II in *Scolopendra* could match with *Strigamia*'s peripatoid, and that the foetus in the two species would correspond. Given the extremely limited, and in general dated, literature on centipede development and considering the major differences between the epimorphic model of embryoid and the anamorphic model of 'larva', I consider it unwise at the moment to try to homologise further early post embryonic stages between Epimorpha and the more basal centipedes. Indeed, although Verhoeff [16] called the hatchling of lithobiomorphs, i.e. the first free living instar, 'foetus', there are no reasons to consider that stage specifically homologous to the Epimorpha foetus, and indeed it has been called by more recent authors simply larva I [34,35] or not named at all [36].

The whole embryoid development lasts around 40 days at 13°C (probably with a rather high level of variability) representing 80% of the time it takes for embryonic development at the same temperature [6]. I don't have the exact timing for the natural population in Brora (Scotland), but in natural populations in southern England adolescens I juveniles appear around early August, two months after egg-laying [18]. The shorter developmental time is probably caused by the higher average temperature compared with my culturing conditions. A similar total brooding time of 8 weeks has been reported for the west coast of Norway, near Bergen [9].

The transformation from embryo to free-living adolescens I with the associated drastic elongation is accomplished through 6 moults, although not all of these are accompanied by major morphological changes. The differences between the first stages (hatched stage 8 embryo to peripatoid) are limited, which is expected given the short duration (2–4 days). Nevertheless, there are defined character changes that justify the introduction of my new staging subdivisions, important for their developmental value (the addition and/or maturation of some segments -

see below) and practically useful for easy identification, a necessary step for a better timing (see Table 1).

The most evident character distinguishing the proembryoid in general from the peripatoid is the <90° ventral angle between the anterior and the posterior part of the trunk. This is the remnant of the point of bending of the embryo after dorso-ventral flexure (embryonic stage 6) [6]. We have previously shown that the number of LBS in *Strigamia* is determined early in embryonic development [29], well before the end of segmentation and at least before dorso-ventral flexure, but possibly much earlier. If the flexure position depends on the positioning along the AP axis, one would have expected LBS values for the bending point reflecting the intraspecific variability and in particular the modal distribution found in the presinking germ bands and in the adults. In fact, the bending point seems to be rather stable, in most cases at 17 LBS, independently from sex and number of leg pairs. From a practical point of view, this means that one cannot use the LBS number at the bending point to estimate the sex of the embryo/proembryoid.

Whatever the initial internal trigger driving the bending in a specific position, the result is that there is for quite some time during development, from embryonic stage 6 to proembryoid III included, a change in the shape of trunk segments positioned at approximately 35% of the LBS number (considering modal numbers in the adult of either 47 or 49 LBS). This is strikingly similar to the position of the so called 'midbody anomaly', typical of centipede trunk, which in geophilomorphs is rather subtle and is represented by a change in the structure of the otherwise homonomous trunk, at about the two fifths of its length (i.e. 40%) [37]. I wonder if this phenomenon might be somehow just a by-product of the dorso-ventral flexure in geophilomorphs, although this would be impossible to test directly in *Strigamia* because the 'anomaly' in this species is altogether absent [37].

Although rather similar to the proembryoid, the peripatoid has a uniform cylindrical trunk. It has still a limited differentiation and sclerotisation of the cuticle and a very limited degree of limb segmentation. Its shape recalls that of an onychophoran, like a *Peripatus*, from which, indeed, it derives its name.

Of course the cuticle is just a good footprint of the underneath epithelium at the time of hardening, but can be nevertheless used as a convenient developmental snapshot [38]. Indeed, the first shed exuvia, whose apolysis marks the end of embryonic stage 7, still shows separated bilateral hemitergites (see Figure 1), and consequently is a good marker of the timing of the dorsal closure which must happen during embryonic stage 8. Ortega-Hernández and Brena [27] have shown that the actual epidermal anatomy, as shown through nuclear staining, is more distinct than the overlaying cuticle for a given stage and subdivision of limbs or appearance of other structures show up much earlier. For example, the complete array of 14 antennomeres is apparently formed already at embryonic stage 8 (see Figure 4C in [27]), although it will be clearly defined on the cuticle only starting with the proembryoid II and III. By the foetus stage, as shown in detail in [27], all the exoskeletal elements have acquired the characteristic subdivisions, as in the case of the 7 podomeres of the walking legs. The development from embryo to foetus is a rather smooth process, with limited changes in general morphology, although slightly more marked from peripatoid to foetus: this is, of course, correlated with the short duration of the stage. Indeed all the first 4 stages (from hatched embryo 8 to foetus) account for only 30% of the total embryoid developmental time. In contrast, the foetus itself accounts alone for 70% of the total duration, and this is correlated with the fact that it is during this stage that the transformation from an embryo-like animal, relying only on yolk, to a free living individual takes place.

In contrast with all previous stages, the adolescens appears to have all the different structures and apparatuses required for independent life, although the large quantity of yolk apparently still available (at least if one considers the pink pigment a good estimate of the yolk content) would indicate that this juvenile is actually able to rely on yolk for quite some time. As in the adult, its relation with the external world will rely on well formed sensilla. In particular, the well developed basiconic sensilla - which may be thermohygroreceptors (see [39]) - would have a pivotal function in an animal so sensitive to water loss. Tracheal pits, appearing for the first time during early embryonic stage 7 [6], and further developed during the embryoid stages [27], are now open to the exterior, and the respiratory system should be by now fully functional. The feeding apparatus is likewise completely formed, with the forcipules/maxillipeds fully functional, with the associated venom gland developed (see further Discussion below). Free walking is allowed for the first time by the well sclerotised and partly tanned exoskeleton, specifically articulated into a mature array of sclerites, the result of a specific sequential developmental process during the previous stages, as shown in detail elsewhere [27]. Male gonopods are just appearing on the first genital segment, allowing the assessment of sex in good preparations (see also [20,22,29]). These appendages will require the following juvenile stages to reach full development. Indeed, in particular in these last segments, there are still structures or organs developing further. For example, the coxal pore number of the last pair of legs increases from 1 in the adolescens I to ten in the maturus senior stage [20]. The heart also develops further, not just in size but also in internal organization [9].

### Forcipule differentiation and evolution

Probably the main apomorphy of centipedes is the forcipules, a kind of maxillipeds used to grab the prey and inject venom into it. Traditionally, forcipules have been always viewed as resulting from a modification of the first pair of the trunk's walking legs which may be associated early in development with a specific combination or exclusive expression of *Hox* [40] and *wnt* [41] genes. In comparison with the legs, adult forcipules are formed by a lower number of podomeres, which are named using a modified leg nomenclature, with a clear homology intent (see Figure 3D in [42] for a SEM view of adult *Strigamia maritima* forcipules). Besides the large basal fused coxosternite, there are 4 articles: a large trochantero-prefemur, a much reduced femur, a tibia and a terminal tarsungulum. Possible evolutionary scenarios have been discussed elsewhere from both a morphological [42,43] and a palaeontological point of view [44]. A detailed analysis of development from embryo to adult has only been done, at least in part, for *Scolopendra*, where the foetus appears to have the same forcipular articles as the adult [31].

In *Strigamia*, although the forcipules are already multi-segmented by embryonic stage 8 (see also Figure 4C in [27]), is has been impossible in my material to identify with accuracy the segments forming the forcipule throughout the following stages of development, before it appears well sclerotized in the adolescens I. Surprisingly, in this stage the forcipule seems to have an additional intermediate segment, comparable in size with femur and tibia, for a total of 5 instead of 4 articles (coxosternum excluded). I don't know whether the additional article is really articulated or just a distal modification of the more basal article, but it is interesting to note that, although my material is far from clear, in the previous stage, the foetus, there could be up to 6 segments (Additional file 3: Figure S1). If this subdivision is real, the last two articles would represent the tarsus and the ungulum/claw, later fused to form the tarsungulum, as the evident transverse sulcus in the adolescens I and in *Scolopendra* [31] would confirm (see [43] for additional evidence and a detailed discussion). With the same argument, the additional intermediate article could be interpreted as a developmental remnant of the prefemur to be fused with the tronchanter, later in development. If that is really the case, this would be a quite straightforward case of recapitulation (as shown by the venom duct development in Scolopendra [31]), although a more detailed analysis, particularly with molecular markers, would be clearly needed to clarify this point.

### Embryoid stages in evolutionary scenarios

According to the most recent phylogenies (e.g. [45,46]), longer body with a higher number of segments is a derived character within Myriapoda, expressed by both the Geophilomorpha within Chilopoda and the Helminthomorpha (which includes several kind of 'long' millepedes, like Platydesmidae and Julidae) within Diplopoda. As myriapods are sequentially segmenting arthropods, an increase in the number of segments implies an extension of the segmentation time. This is achieved in Helminthomorpha by extending, in some species indefinitely, the post-embryonic addition of segments that was probably typical of the myriapod ancestor, and by the Geophilomorpha by protracting, and possibly accelerating, the embryonic phase of segment addition. The delaying of complex organogenesis and further development of the body to early post-embryonic stages (embryoid stages) might have been a necessary consequence of this exploitation of the segmentation patterning machinery during embryonic time. In this sense, I think that *Strigamia* embryoids, brooded by the mother for a considerable time, could be considered nothing more than an 'extended embryonic phase'. Although probably the results of different evolutionary pressures, a newborn embryoid-like, i.e. not free-living and not fully developed, as are *Strigamia* embryoids, is actually present in many arthropod clades, and is indeed common to many myriapods, where one or two such stages are found and are called pupoid in Diplopoda and Pauropoda, and prelarva in Symphyla (see [5,11] for a detailed general comparative discussion).

This is arguably a case of heterochrony. Indeed, in *Strigamia*, as the embryonic development seems to have extended across the post-embryonic divide, on the contrary it would seem that some moulting cycles possibly link to more ancestral post-embryonic development than might have been anticipated, and then condensed. This could provide a possible explanation for the frequent moults of *Strigamia* embryoid instars (but see also Discussion below), so closely packed that the second and third moults happen almost simultaneously. This is not a unique case, and indeed the simultaneous shedding of multiple cuticles (associated with 'virtual stages') has been observed, for example, in some leiodid beetles and some phorid flies, insects with extremely abbreviated development (see [11]), a clear situation of 'condensation' of more ancestral stages.

### The residual anamorphic nature of *Strigamia* post-embryonic development

Lacking a direct analysis of early post-embryonic development, geophilomorphs and scolopendromorphs have been always considered epimorphic, commonly reported as having all body segments and all legs present at hatching. In this work I have shown that this is not strictly the case, as in the first post-embryonic stage(s) not all segments are present, because the genital segments are not formed yet (see below), and not all legs are present, because the last LBS is limbless. Additionally, my analysis has shown that a small fraction (2 out of 98 specimens) of specimens fail to show on the cuticle traces of the last LBS, although this condition may be the result of the low quality of the live material or an artefact of the culturing conditions (although the resulting adolescens I were actually normally developed).

The last LBS is in general molecularly determined already at embryonic stage 7 at the level of expression of the gene *engrailed* (*en*) [27]. This leaves open the possibility that the lack of external appearance of the last LBS is the result of a delay into postembryonic time either of the actual initial molecular patterning, or, at the very least, of the gross epidermal morphology of the segment at the moment of the formation of the cuticle.

Certainly, posterior to the last LBS, the 2 last segments, i.e. the genital segments, are added post-embryonically, apparently sequentially, and appear well defined externally only with the last embryoid stage, the foetus.

The interpretation of how many genital segments are present in Chilopoda is a matter of discussion. I here consider them two, as most authors maintain, relying on the presence of 2 genital nervous ganglia (in the adult, smaller than the LBS ones; Brena, unpublished data). Indeed, morphologically, the boundaries of the genital segments are unclear in my foetus material, but are already too complex in the adolescens I. There are no additional *en* stripes after the last LBS in embryo 8 of *Strigamia*, and attempts to mark with the gene *en* post-embryonic stages and, hence, the appearance of the genital segments, have failed [27].

In conclusion, the last LBS can appear, at least externally, post-embryonically and the genital segments are both determined molecularly and appear morphologically postembryonically. For this reason *Strigamia* (and possibly also all Epimorpha) is not fully epimorphic.

#### The variable nature of segmentation across development and evolution

We have previously shown that most segments in geophilomorphs are initially produced by an early double segment patterning system, operating through oscillation of expression of genes like *even-skipped*, activated at the posterior of the germ band [4]. This initial pair-rule system is not present in other studied myriapods such as the centipede *Lithobius* [36] and the millipede *Glomeris* [47]; consequently, we hypothesised that this pair rule system could have facilitated the formation of the much higher number of segments typical of geophilomorphs. Nevertheless, the last 8 leg-bearing segments or so are patterned differently, at single segment periodicity with no indication that the whole tissue is undergoing dynamic cycles of gene expression [4]. It is possible that this final phase of segment addition depends on a progressive wave of spatial patterning that is triggered by signals emanating from the maturing segments that it leaves behind. Patterning by such a progressive front would be a fundamentally different mechanism from the clock-like process of posterior oscillation, but both share the characteristic feature that, at the level of genetic control, they are intrinsically repetitive processes, requiring the same set of gene interactions to make each segment. Either would allow indefinite sequential repetition, to make a trunk of indeterminate length. It is possible, though, that local signalling might be more easily initiated and repressed in co-ordination with the moult cycle than would early global oscillation, where cell-cell coordination may be based on Notch-signalling [4]. Such late co-ordination with the moulting cycle would almost certainly have characterised segment addition during post-embryonic trunk elongation in ancestral arthropods and may still do so in most myriapods during their anamorphic segmentation phase, including, perhaps, the residual 'anamorphic' phase of genital segments formation in *Strigamia*.

In this context, the early embryoid of *Strigamia* moults frequently, perhaps more as a byproduct of the periodic expression of ecdysone, involved originally in the hormonal regulation of the last phases of segmentation. Note that, at least in insects, not only morphogenetic processes and cuticle deposition but also elongation and segmentation of the germ band may be regulated by ecdysteroids present in the yolk [48-50]: in this sense, *Strigamia*'s persistent and abundant yolk may have not only a nutritional function, but also a timing regulatory function across both embryonic and embryoid time.

If the formation of the genital segments may fully represent a vestigial anamorphic phase of segmentation in *Strigamia*, the formation of the last LBS appears to represent the transition from the last phase of embryonic segmentation (the last 8 segments or so) [4] to post-embryonic segmentation. Indeed, the last LBS are molecularly patterned before the first embryonic moult [27], but morphologically appear only after it [6], or even, in some rare cases, post-embryonically altogether [this article].

In other words, there appears to be somehow a continuum in the segmentation process across the different developmental stages of *Strigamia*, mid-embryonic, late embryonic, and post-embryonic included. This, in turn, illustrates the variability of the segmentation process which may significantly concur to its evolvability, as exemplified by the shift from anamorphic to epimorphic development that has characterized the evolution of these myriapods.

## Materials and methods

*Strigamia maritima* eggs were collected from a previously studied population near Brora (NE Scotland). Individual clutches of eggs were taken to the laboratory, and cultured at 13°C, in most cases under a shallow layer of mineral oil (Sigma M8410, embryo tested), but, occasionally, in LES (Locust Embryo Saline) buffer, as described in [6].

Under these conditions, *Strigamia* eggs can develop, hatch and develop further to the adolescens I stage, more than 2 months after collection [22].

For the description of development I used 23 specimens from 7 clutches collected in June 2009, which I followed individually as they developed, photographing each embryoid every two days initially, and then every time I identified a change in development (e.g. an ecdysis). As we did for the embryonic stages [6], I compared the timing of these specimens to estimate the average duration of development for each embryoid stage. I compared and confirmed this estimate by tracking the embryoid development of 5 clutches collected in June 2010, followed as a whole. For the experiment aimed at checking the possible post-embryonic addition of segments I used 148 hatched and very early embryoid juveniles collected as eggs in June 2010. I photographed them at the beginning, at mid embryoid stages (proembryoid III or peripatoid stage) and then at the adolescens I stage, limiting interference as much as possible, to reduce any effect that this might have on development rate and survival (as shown, at least for embryonic development, by [6]). In one single case, I was able to collect in the field a clod of substrate which included a brood chamber containing the mother with her clutch of eggs and kept it in a wet chamber in culture at 13°C up to the foetus stage (see Figure 4G-H).

After hatching, the chorion and the following exuviae tend to stay attached to the posterior end of the animal up to the foetus stages, at least in cultured specimens (see [5,7,21]. To better identify the morphology of the different stages, chorion and exuviae were manually removed with fine tweezers from some of the live specimens (see Figures 2AB and 3A).

In some of my first experiments, eggs were cultured on filter paper moistened with LES. These specimens were photographed with the live animal immersed in LES (see e.g. Figure 1A,B,C). Cultured like this, the chorion and the exuviae (at least the first one) tend to detach from the animal (see Figure 1C). However, under these conditions I was not able to keep specimens alive for more than a few days after hatching.

Pictures of whole juvenile specimens and whole clutches (see Figure 4G-H) were taken using a Leica MZFLIII stereomicroscope with a Leica DFC 500 Camera (and Leica Firecam software), using lateral light and, generally, a black background. Flat mounted preparations of unfixed specimens were prepared as described in [22] and photographed with a Zeiss Axiophot compound microscope (Zeiss, Jena, Germany) with a Leica FC3 FX camera. Contrast and colour of the photos were adjusted using Adobe Photoshop. Stacks of images at multiple focal depths were combined with Helicon Focus software (Helicon Soft Ltd., Kharkov, Ukraine).

## Additional files

**Additional file 1: Gut peristaltic movements of hatchling.** Hatched *Strigamia*, alive, immersed in LES, lateral view, showing peristaltic contractions of the gut still full and swollen with yolk. This specimen, still partially wrapped in the chorion, has just started the first ecdysis.

**Additional file 2: Gut and body contractions of a moulting proembryoid I. *Strigamia* proembryoid I, immersed in LES, lateral view, showing peristaltic movement of the gut and contraction of the whole body.** This specimen is undergoing the first ecdysis, with the embryonic cuticle partially floating in the medium and still covering the posterior half of the trunk; the hind trunk is still included in the chorion.

**Additional file 3: Figure S1.** High magnification of maxillipedes/forcipules of foetus. Preparation on slide of a dead unfixed specimen in mineral oil, viewed at the compound microscope with Nomarski optics, ventral view. Numbers mark potential maximum number of segments constituting the limb, although their nature as separate podomeres is far from clear. Numbering highlight possible correspondence with numbering in Figure 5D; with 0 tentatively corresponding to the coxosternum and 6 corresponding to the apical claw, fused with 5 in the adolescens I to form the tarsungulum (see main text for details). Scale bar: 50 μm.

### Abbreviation
LBS: Leg bearing segment.

### Competing interests
The author declares that he has no competing interests.

### Acknowledgements
The author wishes to thank Javier Ortega-Hernández for kindly providing 3D and 4D figures, Michael Akam for providing laboratory space, facilities and freedom of research, and valuable comments on a early version of the draft, and Wallace Arthur, Giuseppe Fusco, Alessandro Minelli and an anonymous referee for very helpful comments on the draft; the author is particularly grateful to Giuseppe Fusco for providing a solution for the nomenclature of the early post-embryonic stages. This work was funded by BBSRC grant No. BBS/B/07519 to Michael Akam, and by a University of Cambridge Value in People award to CB.

### References
1. Edgecombe GD, Giribet G: **Evolutionary biology of centipedes (Myriapoda: Chilopoda).** *Annu Rev Entomol* 2007, **52:**151–170.
2. Chagas-Júnior A, Edgecombe GD, Minelli A: **Variability in trunk segmentation in the centipede order Scolopendromorpha: a remarkable new species of *Scolopendropsis* Brandt (Chilopoda: Scolopendridae) from Brazil.** *Zootaxa* 2008, **1888:**36–46.
3. Schileyko AA: **Redescription of *Scolopendropsis bahiensis* (Brandt, 1841), the relations between *Scolopendropsis* and *Rhoda*, and notes on some characters used in scolopendromorph taxonomy (Chilopoda: Scolopendromorpha).** *Arthropoda Sel* 2006, **15:**9–17.
4. Brena C, Akam M: **An analysis of segmentation dynamics throughout embryogenesis in the centipede *Strigamia maritima*.** *BMC Biol* 2013, **11:**112.
5. Minelli A, Brena C, Deflorian G, Maruzzo D, Fusco G: **From embryo to adult-beyond the conventional periodization of arthropod development.** *Dev Genes Evol* 2006, **216:**373–383.
6. Brena C, Akam M: **The embryonic development of the centipede *Strigamia maritima*.** *Dev Biol* 2012, **363:**290–307.
7. Weil E: **Zur Biologie der einheimischen Geophiliden.** *Z Angew Ent* 1958, **42:**173–209.
8. Prunescu C, Capuse I: **Nouvelles données sur le début du développment postembryonnaire chez les myriopodes de l'ordre Geophilomorpha.** *Trav Mus Hist Nat Gr Antipa* 1971, **11:**111–119.
9. Økland S: **Changes in heart ultrastructure during development of *Strigamia maritima* Leach (Myriapoda, Chilopoda, Geophilidae).** *Int J Insect Morphol Embryol* 1984, **13:**233–245.

10. Lewis JGE: *The Biology of Centipedes.* Cambridge: Cambridge University Press; 1981.
11. Minelli A, Fusco G: **Arthropod post-embryonic development.** In *Arthropod Biology and Evolution.* Edited by Minelli Boxshall G, Fusco G. Berlin: Springer; 2013:91–122.
12. Newport G: **Monograph of the Class Myriapoda, Order Chilopoda; with observations on the general arrangement of the Articulata.** *Trans Linn Soc Lond* 1844, **19:**265–302.
13. Latzel R: *Die Myriopoden Der Österreichisch-Ungarischen Monarchie: Die Chilopoden*, Volume 1. Wien A. Hölder; 1880.
14. Sograff N: **Materialen zur Kenntnis der Embryonalentwicklung von *Geophilus ferrugineus* L. K. und *Geophilus proximus* L. K.** *Nachr Ges Freunde Naturkunde Anthr Ethnol Moskau* 1883, **43:**1–77.
15. Metschnikoff E: **Embryologisches über *Geophilus*.** *Z Wiss Zool* 1875, **25:**312–322.
16. Verhoeff KW: **Myriapoda.** In *Klassen Ordnungen Tierreiches. Volume 5.* Edited by Bronn HG. Leipzig: Akademische Verlagsgesellschaft; 1902 [vol. 2].
17. Lewis JGE: *The Life History and Ecology of the Littoral Centipede Strigamia Maritima (Leach). PhD Thesis.* London: University of London; 1960.
18. Lewis JGE: **The life history and ecology of the littoral centipede *Strigamia* (=*Scolioplanes*) *maritima* (Leach).** *Proc Zool Soc Lond* 1961, **137:**221–248.
19. Johnstone J, Arthur W: **Observations on the eggs and early postembryonic stages of *Strigamia maritima*.** *Bull Brit Myr Iso Group* 2002, **18:**33–40.
20. Horneland EO, Meidell BA: **Postembryonic development of *Strigamia maritima* (Leach, 1817) (Chilopoda: Geophilomorpha: Linotaeniidae) with emphasis on how to separate the different stadia.** *Soil Org* 2009, **81:**373–386.
21. Vedel V, Arthur W: **Character changes during the early post-embryonic development of the centipede *Strigamia maritima* (Leach, 1817) (Chilopoda: Geophilomorpha).** *Int J Myriapodology* 2009, **2:**53–61.
22. Vedel V, Apostolou Z, Arthur W, Akam M, Brena C: **An early temperature-sensitive period for the plasticity of segment number in the centipede *Strigamia maritima*.** *Evol Dev* 2010, **12:**347–352.
23. Archey G: **Revision of the Chilopoda of New Zealand.** *Rec Auckl Mus* 1936, **2:**43–70.
24. Misioch M: **Variation of characters in some geophilid chilopods.** *Abh Verhandlungen Naturwissenschaftlichen Ver Hambg* 1978, **21/22:**55–62.
25. Minelli A: **Post-embryonic development and the phylogeny of geophilomorph centipedes (Chilopoda).** *Bijdr Dierkd* 1985, **55:**143–148.
26. Horneland EO, Meidell BA: **The epimorphosis of *Strigamia maritima* (Leach, 1817) (Chilopoda: Geophilidae).** *Insect Syst Evol* 1986, **17:**127–129.
27. Ortega-Hernández J, Brena C: **Ancestral patterning of tergite formation in a centipede suggests derived mode of trunk segmentation in trilobites.** *PLoS One* 2012, **7:**e52623.
28. Kettle C, Arthur W: **Latitudinal cline in segment number in an arthropod species, *Strigamia maritima*.** *Proc R Soc Lond Ser B-Biol Sci* 2000, **267:**1393–1397.
29. Brena C, Green J, Akam M: **Early embryonic determination of the sexual dimorphism in segment number in geophilomorph centipedes.** *EvoDevo* 2013, **4:**1–9.
30. Heymons R: **Die Entwicklungsgeschichte der Scolopender.** *Zool Stuttg* 1901, **13:**1–244.
31. Dugon MM, Hayden L, Black A, Arthur W: **Development of the venom ducts in the centipede *Scolopendra*: an example of recapitulation.** *Evol Dev* 2012, **14:**515–521.
32. Lawrence RF: **Some observations on the post-embryonic development of the Natal Forest Centipede, *Cormocephalus multispinus* (Kraep.).** *Ann Natal Mus* 1947, **11:**139–156.
33. Brunhuber BS: **Egg laying, maternal care and development of young in the scolopendromorph centipede, *Cormocephalus anceps anceps* Porat.** *Zool J Linn Soc* 1970, **49:**225–234.
34. Scheffel H: **Untersuchungen über die hormonale Regulation von Häutung und Anamorphose von *Lithobius forficatus* (L.) (Myriapoda, Chilopoda).** *Zool Jahrb Abt Allg Zool Physiol Tiere* 1969, **74:**436–505.
35. Voigtländer K: **The life cycle of *Lithobius mutabilis* L. Koch, 1862 (Myriapoda: Chilopoda).** *Bonn Zool Beitr* 2006, **55:**9–25.
36. Hughes CL, Kaufman TC: **Exploring myriapod segmentation: the expression patterns of *even-skipped*, *engrailed*, and *wingless* in a centipede.** *Dev Biol* 2002, **247:**47–61.
37. Eason EH: *Centipedes of the British Isles.* London: Warne; 1964.
38. Fusco G, Brena C, Minelli A: **Cellular processes in the growth of lithobiomorph centipedes (Chilopoda: Lithobiomorpha). A cuticular view.** *Zool Anz* 2000, **239:**91–102.
39. Minelli A: **Chilopoda.** In *Microsc Anat Invertebr Vol 12 - Onychophora Chilopoda Lesser Protostomata.* Edited by Harrison FW. New York: Wiley-Liss; 1993:57–114.
40. Brena C, Chipman AD, Minelli A, Akam M: **Expression of trunk *Hox* genes in the centipede *Strigamia maritima*: sense and anti-sense transcripts.** *Evol Dev* 2006, **8:**252–265.
41. Hayden L, Arthur W: **Expression patterns of Wnt genes in the venom claws of centipedes.** *Evol Dev* 2013, **15:**365–372.
42. Dugon MM, Black A, Arthur W: **Variation and specialisation of the forcipular apparatus of centipedes (Arthropoda: Chilopoda): A comparative morphometric and microscopic investigation of an evolutionary novelty.** *Arthropod Struct Dev* 2012, **41:**231–243.
43. Maruzzo D, Bonato L: **Morphology and diversity of the forcipules in *Strigamia* centipedes (Chilopoda, Geophilomorpha).** *Arthropod Struct Dev* 2014, **43:**17–25.
44. Haug JT, Haug C, Schweigert G, Sombke A: **The evolution of centipede venom claws - Open questions and possible answers.** *Arthropod Struct Dev* 2014, **43:**5–16.
45. Shear WA, Edgecombe GD: **The geological record and phylogeny of the Myriapoda.** *Arthropod Struct Dev* 2010, **39:**174–190.
46. Miyazawa H, Ueda C, Yahata K, Su Z-H: **Molecular phylogeny of Myriapoda provides insights into evolutionary patterns of the mode in post-embryonic development.** *Sci Rep* 2014, **4:**4127.
47. Janssen R, Budd GE, Prpic NM, Damen WG: **Expression of myriapod pair rule gene orthologs.** *EvoDevo* 2011, **2:**5.
48. Lanot R, Dorn A, Gunster B, Thiebold J, Lagueux M, Hoffmann JA: **Functions of ecdysteroids in oocyte maturation and embryonic development of insects.** In *Ecdysone from Chemistry to Mode of Action.* Edited by Stuttgart KJ. Georg Thieme; 1989:262–270.
49. Chávez VM, Marqués G, Delbecque JP, Kobayashi K, Hollingsworth M, Burr J, Natzle JE, O'Connor MB: **The *Drosophila disembodied* gene controls late embryonic morphogenesis and codes for a cytochrome P450 enzyme that regulates embryonic ecdysone levels.** *Development* 2000, **127:**4115–4126.
50. Sonobe H, Yamada R: **Ecdysteroids during early embryonic development in silkworm *Bombyx mori*: metabolism and functions.** *Zoolog Sci* 2004, **21:**503–516.

3

# The metameric pattern of *Hypsibius dujardini* (Eutardigrada) and its relationship to that of other panarthropods

Frank W Smith[1,2*] and Elizabeth L Jockusch[1]

**Abstract**

**Introduction:** Tardigrades are an ancient lineage of microinvertebrates with a unique metameric pattern consisting of a head and four lobopodal leg-bearing segments. While their close relationship to Onychophora and Arthropoda is well established, many questions remain about the structure and origin of the tardigrade metameric pattern. For example, the relationship of the tardigrade head to that of Arthropoda and Onychophora remains a contentious issue. One source of contention stems from disagreement about the structure of the tardigrade brain. The availability of developmental tools for the tardigrade *Hypsibius dujardini* give this species the potential to clarify questions regarding the relationship of tardigrade segmental patterns to those in Arthropoda and Onychophora. Here we investigate the nervous system, muscle system, and cuticle anatomy of *H. dujardini* using high-resolution microscopy methods.

**Results:** We characterized nervous system anatomy of *H. dujardini* using a combination of anti-β -tubulin staining and DAPI staining and muscle system anatomy using phalloidin staining. We identified several brain lobes: paired outer lobes, paired inner lobes, and a single horseshoe-shaped ventral lobe. We also characterized similarities and differences in the nervous system and muscle system anatomy of the four body segments. Based on these, we detect distinct morphological identities for each segment in this species.

**Conclusions:** Based on comparisons of our results to previous reports, we find support for an ancestral tardigrade brain exhibiting architecture similar to that of *H. dujardini*. Comparisons to other tardigrade species suggest that each segment in the ancestral tardigrade possessed a unique morphological identity, rather than exhibiting strictly homonomous segmentation, and thus that differentiation of anterior segment identities arose prior to the diversification of panarthropodan lineages. This hypothesis can be further tested by examination of the expression boundaries of anterior *Hox* genes, which differentiate anterior segments through a mechanism conserved between onychophorans and arthropods. Our investigation of *H. dujardini* segmental morphologies will facilitate developmental genetic studies in this species that promise to illuminate the relationship of the tardigrade metameric pattern to that of other panarthropods.

**Keywords:** Evolution, Homology, Tardigrada, Panarthropoda, Tagmosis

## Introduction

The taxon Tardigrada is composed of microscopic invertebrates known for their remarkable ability to survive hostile environments, including extremes in temperature, pressure, salinity, radiation, and even the vacuum of outer space [1-8]. Tardigrada is an ancient lineage, with a fossil record extending back to the Cambrian period [9]. Extant tardigrade species inhabit marine, freshwater, and terrestrial habitats [10,11]. The metameric pattern of tardigrades consists of a head, housing a multilobed brain [12-18], and four segments bearing lobopodal legs [10,11,19,20]. Based on morphological similarities, a close relationship of Tardigrada with two other ancient lineages, Onychophora and Arthropoda, has long been acknowledged; the taxon Panarthropoda was erected in recognition of this relationship [10,12,20-23]. A recent taxonomically well-sampled phylogenomic analysis recovered a highly supported

* Correspondence: frank.w.smithiii@gmail.com
[1]Department of Ecology & Evolutionary Biology, University of Connecticut, 75 N. Eagleville Road, U-3043, Storrs, CT 06269-3043, USA
[2]Current address: Biology Department, University of North Carolina, 616 Fordham Hall, Chapel Hill, NC 27599-3280, USA

monophyletic Panarthropoda and suggested that the earliest divergence within this lineage was between Tardigrada and the ancestor of Arthropoda + Onychophora [24]. A recent analysis that included morphological data from extant and extinct species recovered the same topology [25]. However, other studies have found support for Arthropoda + Tardigrada [26] or Tardigrada + Onychophora [27].

Both arthropods and onychophorans exhibit clear head tagmata [10,28,29], which house the brains of these animals. In one model of panarthropod nervous system evolution, ancestrally each segment housed a ganglion [30]. In this model, the three parts of the arthropod brain, the protocerebrum, deutocerebrum, and tritocerebrum, are derived from ancestral ganglia of the anterior three segments. Onychophorans possess a head tagma composed of only three morphologically and functionally distinct segments, and a trunk composed of many morphologically similar segments [10]. The onychophoran brain is derived from nervous tissue of the two anterior-most segments [31]. Fossil evidence that the common ancestor of arthropods and onychophorans lacked tagmosis [32,33], along with differences in tagmosis patterns in arthropods and onychophorans, suggest that tagmosis evolved independently in these two lineages [27].

The relationship of the tardigrade head to the heads of arthropods and onychophorans has important implications for the evolution of tagmosis in Panarthropoda. However, how the segments of tardigrades align to those of arthropods and onychophorans is a contentious issue. This debate stems in part from disagreements concerning the relationship of the tardigrade brain to the arthropod brain. In the tripartite hypothesis of tardigrade brain composition, paired outer, inner, and ventral brain lobes are considered to be serially homologous to segmental ganglia [17-20,34]. This is based on similarities in structure between lobes and ganglia, such as the presence of commissures. Under this hypothesis, there is direct homology between tardigrade brain lobes and the three segments of the arthropod brain [17-20,34]. Generally, the inner lobe of the tardigrade brain is aligned with the protocerebrum, the outer lobe with the deuterocerebrum, and the ventral lobe with the tritocerebrum [20,34]. There are several important evolutionary implications of this hypothesis. If the tardigrade brain is composed of three ancestral segmental ganglia, then the tardigrade head originated by fusion of three ancestral segments. Tardigrades would possess a head tagma under this hypothesis. Under the hypothesis that Onychophora is the sister group of Arthropoda, it suggests that the bipartite brain of Onychophora is derived from an ancestral tripartite brain. Alternatively, if tardigrades are the sister group of arthropods, then the tripartite brain may be a synapomorphy of this clade. By contrast, the protocerebral hypothesis of brain composition asserts that the entire tardigrade brain is directly homologous to the protocerebrum of arthropods [15,35], and therefore that the tardigrade head is composed of a single segment and the tardigrade brain is derived from the ganglion of that segment. Under this hypothesis, the tripartite brain is a synapomorphy of Arthropoda [31].

Proponents of the protocerebral hypothesis have presented high-resolution confocal microscopy investigations of brain anatomy of two species of *Macrobiotus* (Eutardigrada) [15,35], *Echiniscus testudo* (Heterotardigrada) [36], *Batillipes pennaki* (Heterotardigrada) [37], and *Actinarctus doryphorus ocellatus* (Heterotardigrada) [37]. Proponents of the tripartite hypothesis have presented confocal microscopy investigations of *Halobiotus crispae* (Eutardigrada) [17] and *Actinarctus doryphorus* (Heterotardigrada) [18]. The above studies are generally in agreement that tardigrades possess a multilobed brain, but how many lobes are present in the brain and the relationship of brain lobes across Tardigrada are not well resolved. Moreover, investigations of brain lobes and commissures alone have been unable to resolve the relationship of the tardigrade brain to that in other panarthropods.

Several additional aspects of anterior nervous system anatomy have been evaluated in attempting to homologize tardigrade brains to those of other panarthropods. According to the tripartite hypothesis of tardigrade brain architecture, an arthropod-like subesophogeal ganglion is also present in the tardigrade head. By contrast, under the protocerebral hypothesis, there is no subesophogeal ganglion, because its existence would require the brain to be composed of at least two ganglia. A subesophageal ganglion was included in the brain anatomy interpretations of several studies [13,17,18,20,34]. However, most recent studies of tardigrade brain anatomy have not identified a subesophogeal ganglion [15,35-37]. In arthropods, the stomatogastric ganglion is innervated by the tritocerebrum [38]. Mayer et al. [15] interpreted a structure above the ventral ganglion of the second leg-bearing segment of *Macrobiotus harmsworthi* as a stomatogastric ganglion, suggesting that the third segment of *M. harmsworthi* aligns with the tritocerebral segment of arthropods, and by extension, that the brain-housing segment of *M. harmsworthi* aligns with the protocerebrum of arthropods [15]. However, a candidate stomatogastric ganglion has not been identified in other tardigrade species.

One explanation for these conflicting conclusions is that there is variation in brain anatomy across Tardigrada. Alternatively, the putative differences may instead reflect different evolutionary interpretations of similar brain morphologies by workers in the field. Although it seems possible that nervous system anatomies have diverged among tardigrade lineages, segmental patterns of muscle anatomy are highly conserved across tardigrades [39], suggesting that, at the very least, all tardigrades share the same complement of directly homologous segments.

The value of developmental studies for discriminating between the protocerebral and tripartite hypotheses has been acknowledged by proponents of both interpretations of tardigrade brain morphology [15-17,35,37]. Such data have put to rest long-standing controversy regarding the composition of the chelicerate head [40-42]. The tardigrade *Hybsibius dujardini* (Parachela, Eutardigrada) is an ideal system for testing these and other hypotheses related to the evolution of tardigrades with developmental data. First, *H. dujardini* is easily cultured, providing unlimited access to embryos for experimentation [43]. Second, the feasibility of embryonic studies of gene expression [44,45] and gene function through RNAi [46] has been demonstrated for this species. Third, an *H. dujardini* genome sequencing project is in progress (Goldstein, pers. comm.). Mayer et al. [16] previously investigated the organization of the nervous system of *H. dujardini*. In order to investigate segmental patterns in *H. dujardini* in further detail, we stained the nervous system with a β-tubulin antibody and the muscle system with phalloidin, and then analyzed the segmental structure of these systems with confocal microscopy. We also examined the cuticle of *H. dujardini* using scanning electron micrographs. Our results reveal that *H. dujardini* retains highly conserved segmental muscle and nervous system anatomies, supporting the suitability of this species as a developmental model for Tardigrada. Our results support an ancestral tardigrade brain composed of paired outer lobes, paired inner lobes, and a ventral lobe. We propose that the ventral lobe of *H. dujardini* is directly homologous to both the ventral lobe and subesophogeal ganglion identified in previous studies [17,18]. Furthermore, these results suggest that internal morphological aspects of the anterior segments were already differentiated in the panarthropod ancestor. Thus, although the panarthropod common ancestor is characterized as being homonomously segmented [19,22,23,32,47-50], these comparative data suggest that segmental diversity may have evolved prior to the divergence of the three panarthropod lineages. The developmental mechanism responsible for differentiating anterior segments in Onychophora and Arthropoda is conserved [51,52]. The discovery of segmental diversity in tardigrades raises the question of whether this mechanism is even evolutionarily more ancient.

## Results

### General morphology

Like other tardigrades, *H. dujardini* possesses a head and four trunk segments. A pair of ventral legs is found in each trunk segment (Figure 1A). A series of claws attach to the distal tip of each leg (c1–c4, Figure 1A, D). The head lacks legs, but has an internal pharyngeal apparatus, which has been suggested to be derived from an ancestral pair of head appendages [20,53]. The pharyngeal apparatus includes a muscular pharyngeal bulb (ph) and stylets (st), which are used in feeding (Figure 1C). The mouth is found at the anterior end of the head (mo, Figure 1A). Below we describe the nervous and muscular systems of *H. dujardini*. We identify nervous system anatomy and muscle system anatomy with terminology based on previous confocal microscopy analyses [15,17,35,53,54].

### Nervous system anatomy

We used a combination of DAPI-staining and anti-β-tubulin immunohistochemistry to characterize the nervous system of *H. dujardini*. The *H. dujardini* nervous system is composed of an anterior brain, a pair of ventral nerve cords, and a peripheral nervous system composed of many smaller nerves. In DAPI-stained specimens, we identified several clusters of somata in the head, which we refer to as brain lobes. Two bilaterally symmetrical pairs of brain lobes, outer lobes (ol) and inner lobes (il), were detected, along with a ventral lobe (vl) that spanned the midline (Figure 2B–E). In anti-β-tubulin stained specimens, two nerves, the outer connective (oc) and the inner connective (ic), could be seen extending from both the right and left anterior region of the first trunk ganglion into the brain (Figure 2A). The outer connective extends dorsally to the dorsal cluster (dc) near the eye. The inner connective extends through the ventral lobe (vl) into a region of dense nervous tissue (ne) innervating the mouth (Figures 2A and 3A). Anterior clusters (ac) of nerves innervate the papillae cephalica (Figures 2A and 3A, D). A dorsal commissure (dco) connects the outer brain lobes (Figure 2A). An inner commissure (ico) connects the inner brain lobes (Figure 2A). A commissure (co*) also extends between the inner connectives beneath the ventral lobe (Figure 3A, C, E). Several nerves extend posteriorly from the brain along the sides of the pharyngeal bulb (labeled 'pn' in Figures 2A and 3D).

Four ganglia are found along the ventral nerve cords, one for each trunk segment (ga1–ga4, Figure 3A). Each ganglion includes a single dense cluster of cell bodies that does not show anti-β-tubulin expression. The ventral nerve cords extend ventrolaterally across the cell clusters. Within each ganglion, the ventral nerve cords swell to roughly three times the width of the connectives between the ganglia. The ventral nerve cords are connected by fibrous commissures that extend across the cell cluster of each ganglion (co, Figure 3A, C–E, G). Nerves extend from each ganglion into the legs of the segment in which the ganglion is found. The first three ganglia are wider than the fourth ganglion (Figure 3A, G). For example, in the specimen shown in Figure 3A, from ganglion 1 to 4, the maximum widths of the ganglia

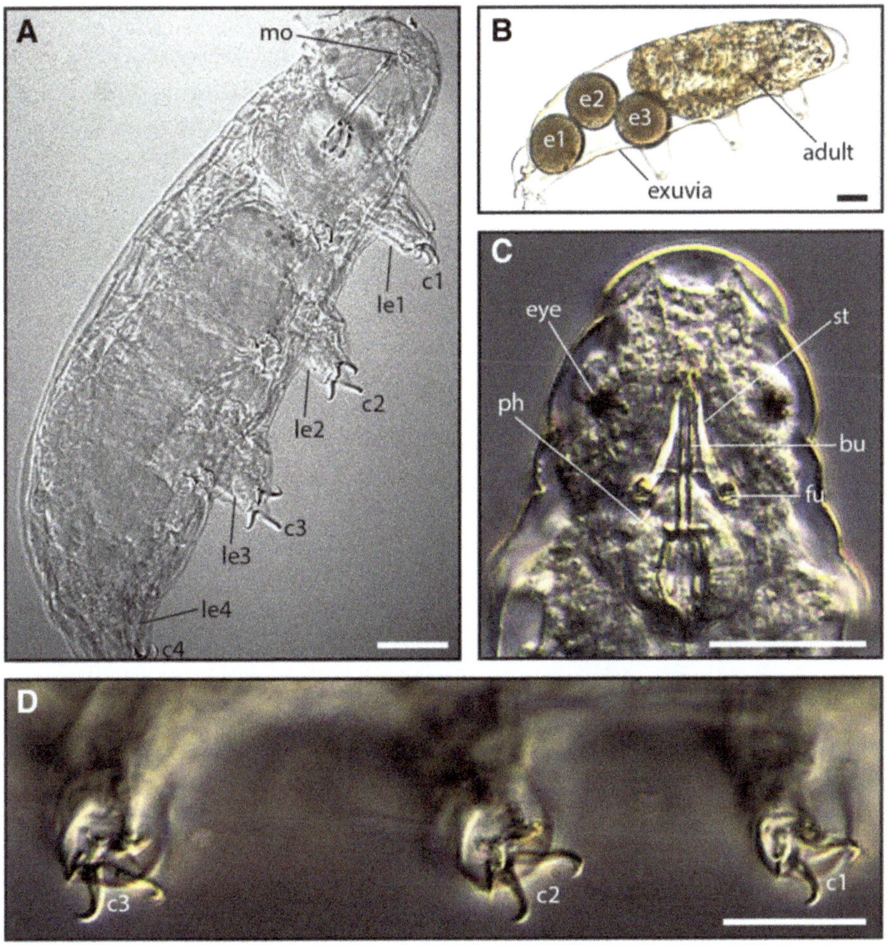


**Figure 1 General anatomy of *H. dujardini*.** Scale bars equal 25 μm **(A)** Differential interference micrograph showing ventrolateral view of the left side of an adult *H. dujardini* specimen. The anteroposterior axis includes a head and four trunk segments, each with a pair of legs (left legs are in focus). **(B)** Light micrograph of an adult laying eggs in its exuvia during molting. **(C)** DIC micrograph of the ventral head of an adult *H. dujardini* specimen. **(D)** DIC micrograph of the leg claws of the anterior three right legs. Abbreviations: bu, buccal tube; c, claw; e, eggs; fu, furca; le, leg; mo, mouth; ph, pharyngeal bulb; st, stylet.

measured 12.4 μm, 12.8 μm, 13.4 μm, and 8.0 μm, respectively. Anti- β-tubulin expression in the stomach in Figure 3A and G is autofluorescence due to algal gut content (see Additional file 1: Figure S1).

### Ventral and lateral musculature

Muscles were visualized using phalloidin staining. A pair of bilaterally symmetric muscle strands extends longitudinally throughout much of the ventral axis (ve, Figure 4A). Each of these muscle strands is interrupted by 7 intermediate attachment sites (a–g, Figures 4A and 5A). At its corresponding attachment site c, each muscle strand bifurcates and extends in an anterior direction past the pharyngeal bulb. For both ventral longitudinal muscles, one strand of each bifurcation extends to intermediate attachment site a, and the other to intermediate attachment site b (Figure 5A).

Seven ventral median muscle attachment sites are found in *H. dujardini* and other tardigrade species (1–7, Figure 4A). These attachment sites are visible as depressions in the ventral cuticle of *H. dujardini* (Figure 4B). Muscles attaching at the ventral median attachment sites extend to the base of the legs or into the legs, run along the ventral muscle strands, extend dorsoanteriorly, extend anteriorly along the ventral muscle bands, or run dorsally to lateral muscle attachment sites. Lateral muscle attachment sites are also visible as depressions in the exterior cuticle (Figure 5B). We label lateral muscle attachment sites following the scheme of Halberg et al. [53]: from the anterior T0, τ1, T1, τ2, T2, etc. We analyzed the structures of the ventral attachment sites in seven phalloidin-stained *H. dujardini* specimens; morphology of ventral median attachment sites was consistent across specimens. All ventral median attachment sites are

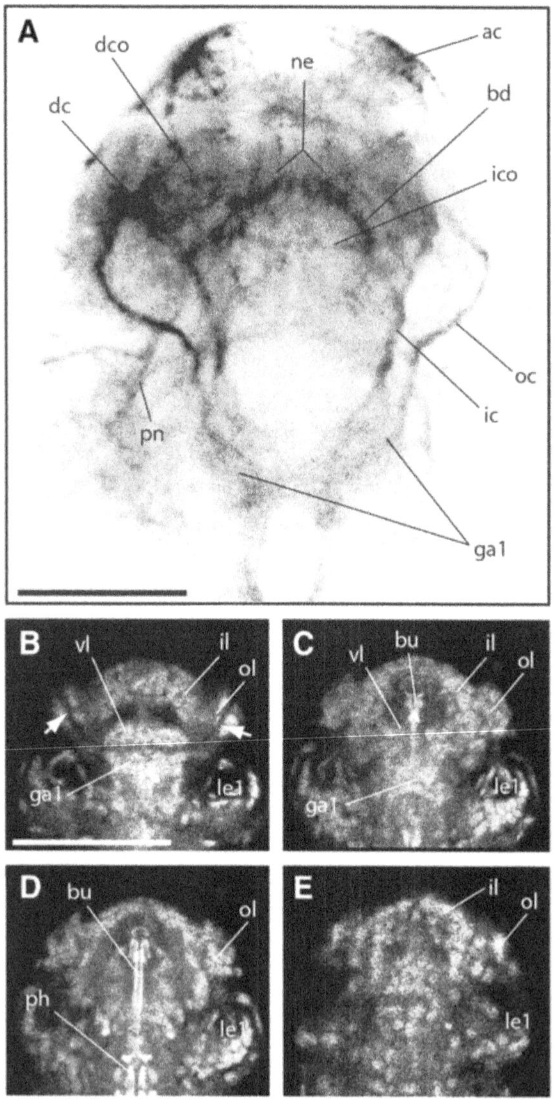


**Figure 2 Structure of the *H. dujardini* head.** All images are of ventral-mounted specimens. Scale bar equals 25 μm. **(A)** Maximum projection of confocal Z-series. **(B–D)**. Increasingly deeper slices of a DAPI-stained specimen. **(B)** The ventral lobe appears continuous across the midline. Note the groove that runs through the outer (arrows) brain lobes. **(C)** The ventral lobe extends on both sides of the buccal tube, suggesting that it is horseshoe shaped. **(D)** Few cell bodies surround the buccal tube in the interior of the head. **(E)** Dorsal brain. Abbreviations: ac, anterior cluster; bd, neurite bundles; bu, buccal tube; dc, dorsal cluster; dco, dorsal commissure; ga, ganglion; ic, inner connective ico, inner commissure; il, inner brain lobe; le, leg; ne, neurites of the mouth; oc, outer connective; ol, outer brain lobe; ph, pharyngeal bulb; pn, posterior nerves; vl, ventral brain lobe.

bilaterally symmetrical, and unless indicated below, muscles they support do not cross the midline. Here we provide anatomical details for one side of each ventral attachment site.

The first ventral median attachment site includes 7 muscle pairs that do not make contact medially (Figure 4A inset). Muscle strand 1a at this site extends anteriorly to an intermediate attachment site (a) of the ventral longitudinal muscle strand. Muscle strand 1i extends into the first leg. A thin muscle strand (1i*) extends across the midline into the first leg of the other side. This is the only instance of a ventral muscle strand crossing the midline. Two thin muscle strands (1ii) extend dorsally to lateral longitudinal attachment site T0 (Figure 5A). Next, two thick muscle strands (1iii) extend to the base of the second leg (Figure 4A, see leg muscles section below).

The second ventral median attachment site includes three pairs of muscles that make contact medially (Figure 4A). The anterior-most muscle strand (2i) extends into the first leg, where it forms a sheet muscle along the poster wall of the leg (see leg muscles section). Next a bifurcated muscle strand (2ii) extends dorsally along the body wall to lateral longitudinal attachment site τ1 (Figures 4A and 5A). The posterior-most muscle extends to the base of the second leg (2iii, Figure 4A). The third ventral median attachment site includes five muscle pairs that do not make contact medially (Figure 4A). The anterior-most muscle strand (3a, Figure 4A) extends anteriorly to an intermediate attachment site (c) of the ventral longitudinal muscle strand. This muscle is followed by a muscle strand that extends into the second leg (3i). Two thin muscle strands (3ii) extend dorsally along the body wall to longitudinal attachment site T1 (Figures 4A and 5A). These are followed by a muscle strand that extends to the base of the third leg (3iii, Figure 4A).

The architecture of ventral median attachment site 4 is identical to that of attachment site 2 (Figure 4A). In this case, the anterior-most muscle strand (4i) runs into the second leg, muscle strand 4ii extends to longitudinal attachment site τ2, and the posterior-most muscle strand (4iii) runs to the base of third leg (Figures 4A and 5A). Ventral median attachment site 5 is similar to attachment site 3. In this case, muscle strand (5i) extends from attachment site 5 laterally into the third leg and two thin muscle strands (5ii) extend to longitudinal attachment site T2 (Figures 4A and 5A). Attachment site 5 differs from attachment site 3 in that the anterior-most muscle strand (5a) runs dorsoanteriorly to a lateral position (Figure 5A) and continues in an anterior direction and no posteriorly-directed muscle strand is present.

Two muscle pairs are attached to ventral median attachment site 6 (Figure 4A); these come into contact medially. The anterior muscle strand (6i) extends laterally into the third leg and forms a sheet muscle along the posterior of the leg (see leg muscles section). The posterior muscle strand (6ii) splits into two muscle bands, which run dorsally along the body wall to longitudinal attachment site τ3 (Figure 5A). Four muscle pairs that do not make contact medially extend from ventral

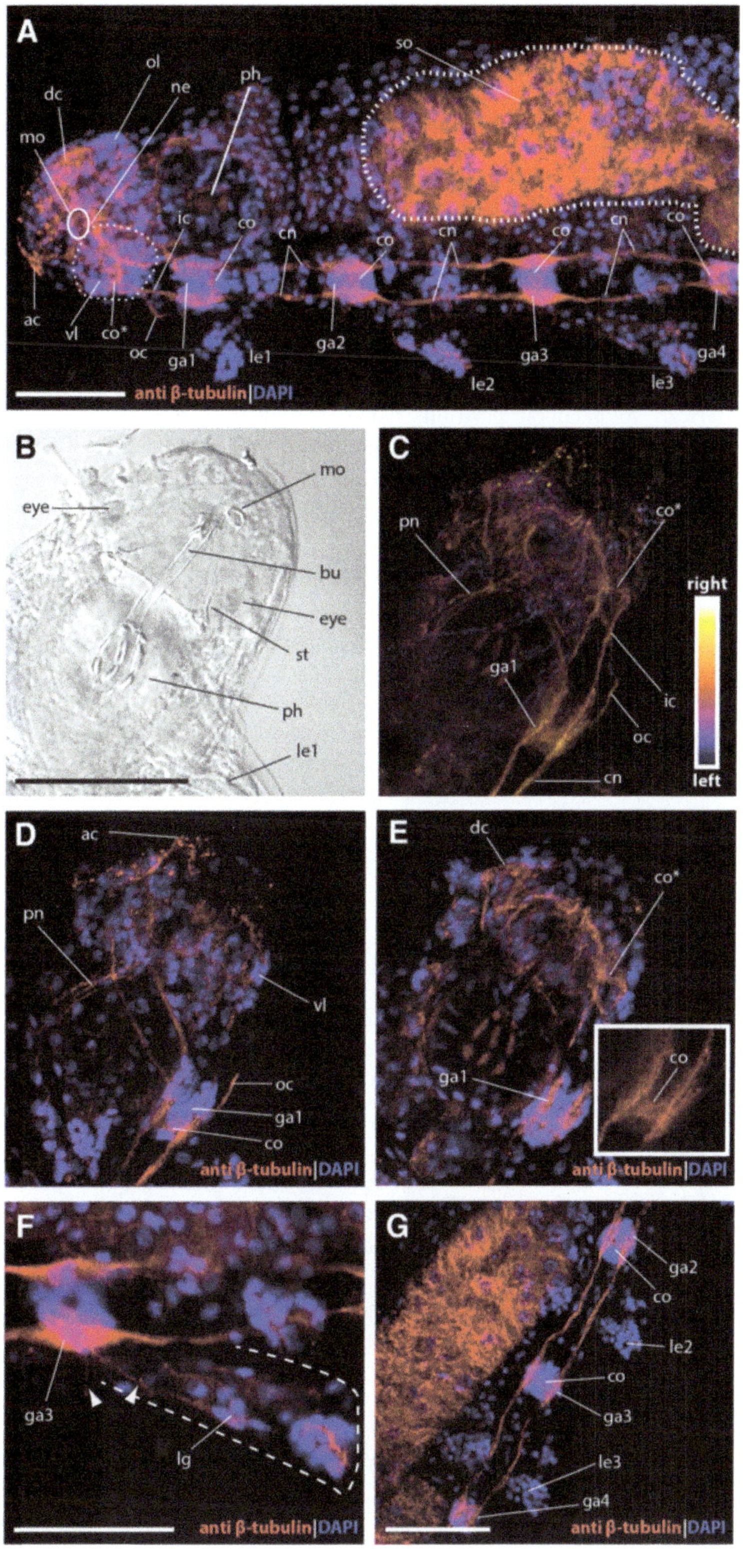


**Figure 3** (See legend on next page.)

(See figure on previous page.)
**Figure 3 "Rope ladder-like" ventral nervous system of *H. dujardini*.** Anti-β-tubulin immunolabeling is shown in **A, C–G**. DAPI counterstaining is shown in **A, D–G**. Scale bars equal 25 μm. **(A)** Maximum projection of brain and ventral nervous system. Each body segment includes a ventral ganglion. The inner connective extends from the anterior-most ventral ganglion to the ventral brain lobe, where a commissure (co*) transverses the midline. The position of the mouth is indicated. **(B)** DIC micrograph of the head. **(C)** Depth-coded maximum-projection of confocal Z-series of specimen shown in B. The brain and first ganglion can be seen. **(D–E)** Individual slices of the Z-series used to produce **C**. Inset in E shows a maximum projection magnification of the ganglion; fibrous commissures extend across the ganglion. **(F)** Maximum projection of the ganglion and a leg of the third trunk segment (same specimen as shown in **A**). The right leg is outlined. Arrowheads point to nerves extending from the third ganglion. **(G)** Maximum projection of the second, third, and anterior fourth segments (same specimen as shown in **B–E**). Ganglion 4 is smaller than ganglion 2 or 3. Abbreviations: ac, anterior cluster; bu, buccal tube; cn, connective; co, commissure; dc, dorsal cluster; ga, ganglion; ic, inner connective; le, leg; lg, leg ganglion; mo, mouth; ne, neurites of the mouth; oc, outer connective; ol, outer lobe; ph, pharyngeal bulb; pn, posterior nerves; so, stomach; st, stylet; vl, ventral lobe.

median attachment site 7 (Figure 4A). The anterior muscle strand (7a) extends dorsoanteriorly to a lateral position and continues in an anterior direction (Figure 5A). Two thin muscle strands (7ii) extend dorsally along the body wall to longitudinal attachment site T3 (Figure 5A). Finally a muscle strand (7iii) extends to the base of the posterior-most leg (Figure 4A).

### Dorsal musculature

Two pairs of longitudinal dorsal muscle strands extend from in front of the pharyngeal bulb (ph) along the dorsum to the posterior end (Figure 6A). Several muscle attachment sites are found along the dorsal muscle strands, which by convention are given capital letter designations. The outer strand of each pair extends from attachment site A to attachment site W. The inner strand extends from attachment site A* to dorsal attachment site V. Longitudinal muscle stands of each pair appear to fuse and split at several positions along their lengths. Several muscle attachment sites are located along these muscle strands. Muscles extending from near the pharyngeal bulb attach to the outer longitudinal muscle strand at attachment sites B and C. At the next several attachment sites, the longitudinal muscle strands connect to muscles extending

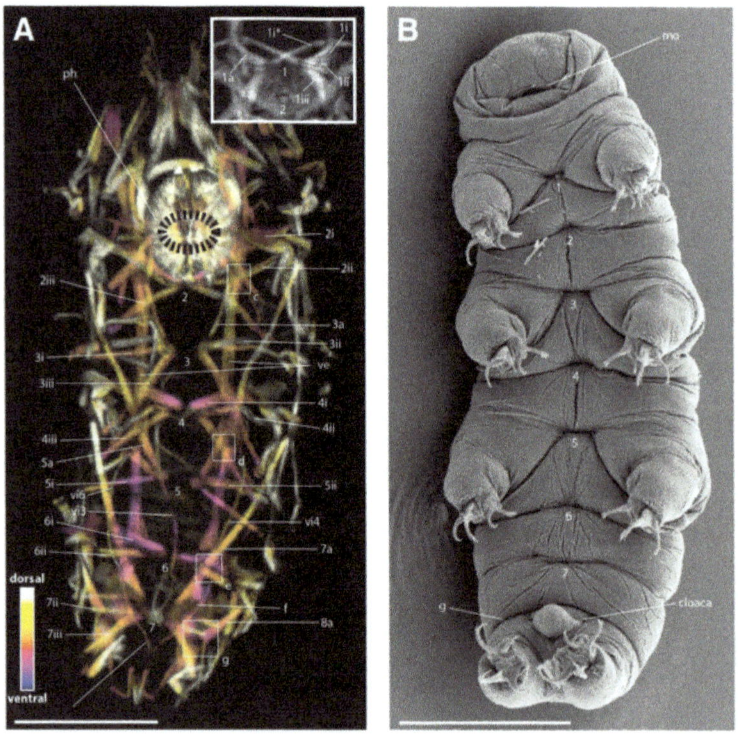


**Figure 4 Ventral muscle anatomy in *H. dujardini* as revealed by phalloidin staining.** Specimens are oriented ventral side up. Anterior is to the top. Scale bars equal 25 μm. **(A)** Depth-coded maximum projection of ventral slices of confocal Z-series. Ventral median attachment sites are numbered 1–7. Muscle strands are labeled following the system of Schmidt-Rhaesa and Kulessa [54]. The first attachment site is out of focus below the pharyngeal bulb. We have outlined its position. Inset shows a maximum projection of the first attachment site. Intermediate attachment sites along one ventral longitudinal muscle strand are boxed. **(B)** Scanning electron micrograph showing cuticular depressions associated with ventral attachment sites. Abbreviations: mo, mouth; ph, pharyngeal bulb; ve, ventral longitudinal muscle; vi, visceral muscle.

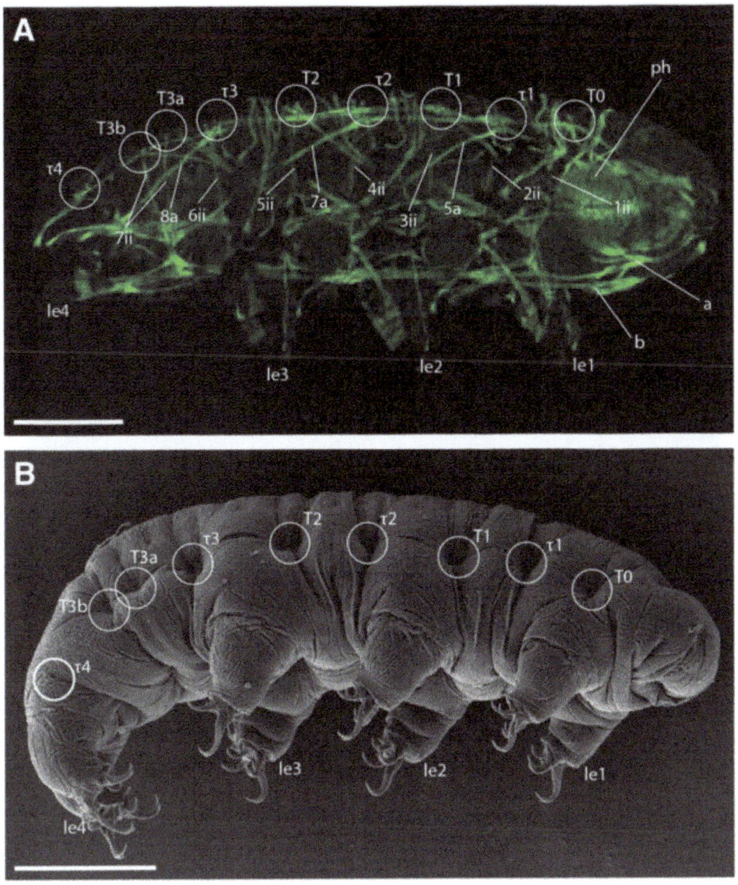


**Figure 5 Lateral muscle anatomy of *H. dujardini* as revealed by phalloidin staining.** Lateral muscle attachment sites are labeled following the system of Halberg et al. [54]. Anterior is to the right. Scale bars equal 25 μm. **(A)** View of longitudinal musculature. Lateral attachment sites are circled. Muscles labeled 'ii' extend from ventral attachment sites to longitudinal attachment sites. Intermediate attachment sites a and b are visible in the ventral anterior and muscles labeled 'a' attach to these. **(B)** Scanning electron micrograph showing cuticular depressions associated with longitudinal attachment sites. Abbreviation: le, leg; ph, pharyngeal bulb.

from the lateral region of the legs (see leg muscles section). Attachment sites D, E, H, I, N, O, S, U and W are on the outer dorsal muscle strand. Attachment sites F, G, K, L, P and Q are connected to leg muscles that cross over the outer longitudinal muscle strand and connect to the inner stand. F*, L* and Q* represent dorsal attachment sites that are off of the longitudinal muscle stands, but connect to leg muscles like most other attachment sites. Attachment sites M and R are points where longitudinal muscle strands bifurcate. Dorsal attachment points are visible as depressions in the dorsal cuticle of *H. dujardini* and cause folding of the cuticle (Figure 6B).

## Visceral musculature

The visceral musculature of tardigrades functions in food acquisition and digestion. The muscular pharyngeal bulb (ph, Figures 4A, 5A and 6A) of *H. dujardini* sits above the ganglion of the first leg-bearing segment (Figure 3A). The stylets (st) sit in front of the pharyngeal bulb (Figure 1C). A total of 10 muscles are attached to the stylets of the pharyngeal apparatus near the dorsoanterior region of the phanyngeal bulb (Figure 6A). 2 muscle pairs (m1, m6) that extend from the furca (fu) to the anterior end of the buccal tube, 2 muscle pairs (m2, m3) that extend from the pharyngeal bulb to the furca, and 2 muscles (m4, m5) that extend from the furca of the left stylet to the furca of the right stylet. We detected an additional six visceral muscles in the gut cavity (vi1–6, Figures 4A and 6A), which likely function in digestion.

## Leg muscles

In terms of leg musculature, the first three pairs of legs are much more similar to each other than they are to the fourth pair of legs. The medial muscles of the first three leg pairs are connected to ventral median attachment sites and are labeled as such (Figure 7A). Each of these legs includes a posterior sheet muscle (sh) that extends from the ventral median attachment site just posterior to it (2i, 4i, 6i). Each of these legs also includes a muscle that extends to its distal tip from an adjacent

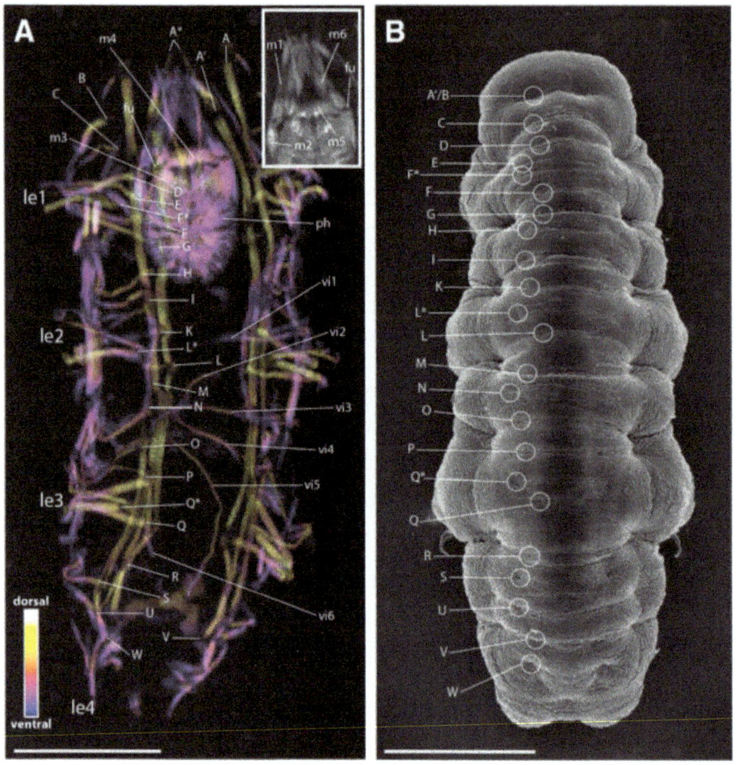


**Figure 6 Dorsal muscle anatomy and visceral musculature of *H. dujardini* as revealed by phalloidin staining.** Anterior is to the top. Scale bars equal 25 μm. **(A)** Depth-coded maximum projection of dorsal slices of confocal Z-series. Same specimen as in Figure 4A. Dorsal attachment sites are labeled with capital letters. Inset shows a maximum projection of stylet muscles not visible in the depth-coded image. **(B)** Scanning electron micrograph showing cuticular depressions and wrinkles associated with dorsal attachment sites. Abbreviations: fu, furca; le, leg; ph, pharyngeal bulb; vi, visceral muscle.

ventral median attachment site (1i, 3i, 5i). Muscles of the iii variety attach to leg pairs two (1iii, 2iii) and three (3iii, 4iii) from the corresponding ventral attachment sites anterior to these legs (Figure 7A). No iii muscles extend to the first leg pair. Unlike the other leg pairs, a muscle extends across ventral attachment site (1) between the first leg pair into the contralateral leg (1i). For the lateral leg, muscles are named after the attachment sites they extend to. As with the medial muscle patterns, there are similar lateral muscle patterns between the legs of the first three leg pairs (Figure 7B, C). Each leg of the first three leg pairs includes a muscle (t1, t2, t3b, respectively) that extends in a dorsoposterior direction to a lateral attachment site. Each leg of the first three leg pairs also includes a muscle that extends dorsally to a lateral attachment site (τ1, τ2, τ3, respectively). Finally, each of these legs also includes several muscles that extend to dorsal longitudinal attachment sites, including two muscles, each (f*1 and 2, l*1 and 2, q*1 and 2, respectively), that extend to dorsal attachment sites that are off of the dorsal longitudinal muscle strands. In contrast to the three anterior leg pairs, very few muscles extend into the posterior-most legs (Figure 7C). A ventral longitudinal muscle strand extends into each leg of this pair, where it forms a sheet muscle (sh). A muscle (7iii) extends from each posterior leg to ventral attachment site 7. Finally, a muscle extends from each posterior leg (w1) to dorsal attachment site W.

## Discussion

### Segmental patterns in *H. dujardini*

Based on muscle and nervous system anatomy, we detected five segments in *H. dujardini*, each exhibiting a distinct morphological identity (summarized in Figure 8A). Characterization of distinct identities is valuable for developmental analyses and also informs our understanding of panarthropod and tardigrade evolution. The head is distinct, with the presence of the brain, stylets, and stylet musculature. In our model of segmentation, a pair of legs and the associated ganglion define each segment posterior to the head; ventral attachment sites are considered to be part of the same segment as the adjacent ganglion. Attachment site 1 is part of the first leg-bearing segment, along with the first ventral ganglion. Attachment site 1 exhibits a unique morphology compared to other ventral attachment sites, and the first leg-bearing segment is unique in housing just a single attachment site. The arrangement of leg muscles in the first leg segment was also unique.

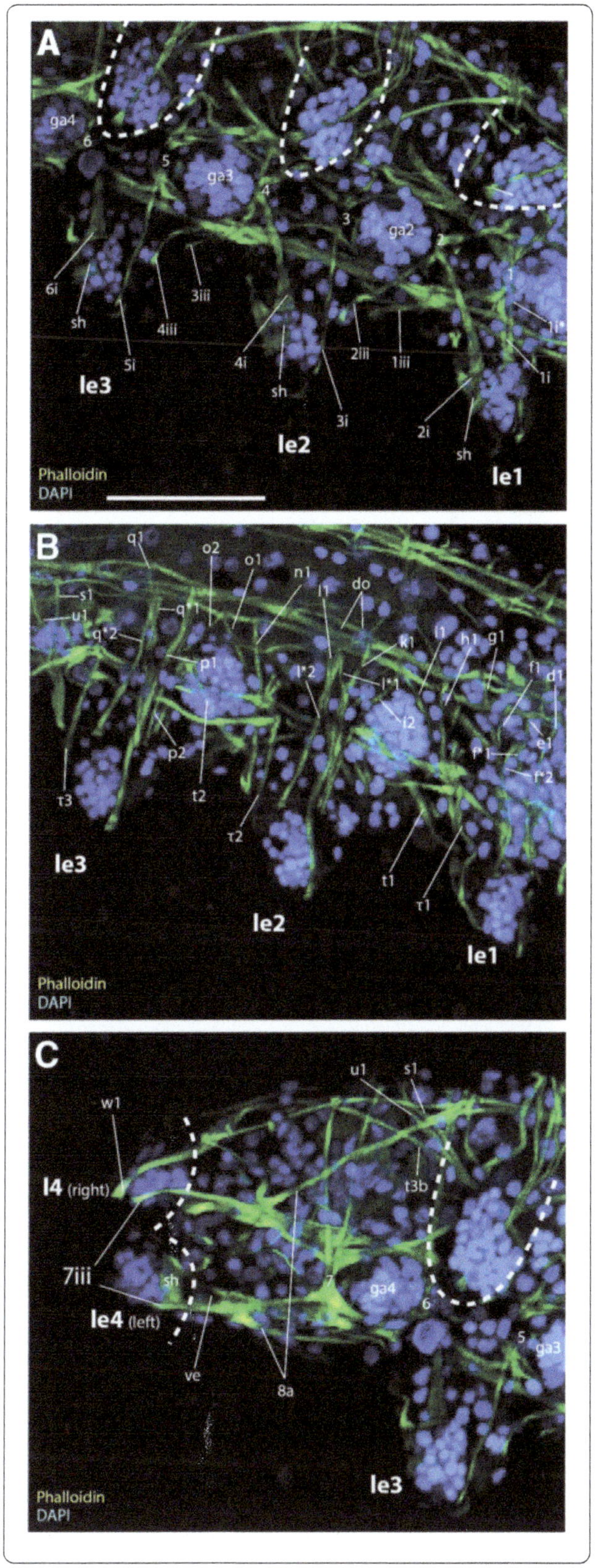


**Figure 7 Side-mounted *H. dujardini* specimen showing leg muscle anatomy as revealed by phalloidin staining.** Anterior is to the right. Scale bar equals 25 μm. **(A)** Maximum projection of the medial muscles of the left three anterior-most legs. All medial leg muscles extend to ventral median attachment sites. Ventral median attachment sites are numbered 1–7. Right legs are outlined. **(B)** Maximum projection of the lateral muscles of the left three anterior-most legs. Lateral leg muscles extend to dorsal attachment sites (lower case letters correspond to letter designation of dorsal attachment site) or to lateral attachment sites (muscles labeled t, τ). **(C)** Maximum projection of the lateral (right leg) and medial (left leg) muscles of the posterior-most legs. Dashed lines in the posterior denotes boundary between legs and body wall. A dashed line also outlines the right third leg. The posterolateral muscle strands of the third leg are labeled (u1, s1). Dorsolaterally directed muscle 8a is labeled. Abbreviations: ga, ganglion; le, leg; sh, sheet muscle; ve, ventral longitudinal muscle strand.

Attachment sites 2 and 3 belong to the second leg-bearing segment. These attachment sites circumscribe the ventral ganglion of this segment (ganglion 2). Attachment sites 4 and 5 are parts of the third trunk segment, and circumscribe ganglion 3. Attachment site 4 shares an identical muscle arrangement with attachment site 2, while the anterior of attachment site 5 is similar to the anterior of ventral attachment site 3. Segments 2 and 3 share leg muscle architecture as well as attachment site architecture. However, they are distinguished by the muscle arrangement in the posterior of attachment sites 3 and 5. Attachment sites 6 and 7 are parts of the fourth trunk segment; the morphologies of both are unique compared to those of other attachment sites. These sites circumscribe ganglion 4. The ganglion of this segment is the smallest of the segmental ganglia. Based on muscle anatomy in the legs, we detected three distinct leg identities. Both appendage pairs one and four have distinct morphologies, while appendage pairs two and three are nearly identical. Therefore, while we could detect distinct segmental identities among the leg-bearing segments, based on morphology, it is clear that leg-bearing segments 2 and 3 are the most similar. Below, we compare our results to those from other tardigrade species to reconstruct the antiquity in Tardigrada of the segmental patterns we detected in *H. dujardini.*

#### Nervous system anatomy of *H. dujardini* and its relationship to that of other tardigrade species

The tardigrade brain is composed of several lobes, which are at least party composed of neuronal somata [15,16,18,35-37]. The brain of *H. dujardini* is composed of two paired lobes (outer and inner) and one unpaired lobe (ventral). Outer and inner lobes have been identified in brains across Tardigrada [15,17,18,35,36]. The similar location of these brain structures across Tardigrada supports their direct homology in this lineage. Ventral lobes and a subesophogeal ganglion

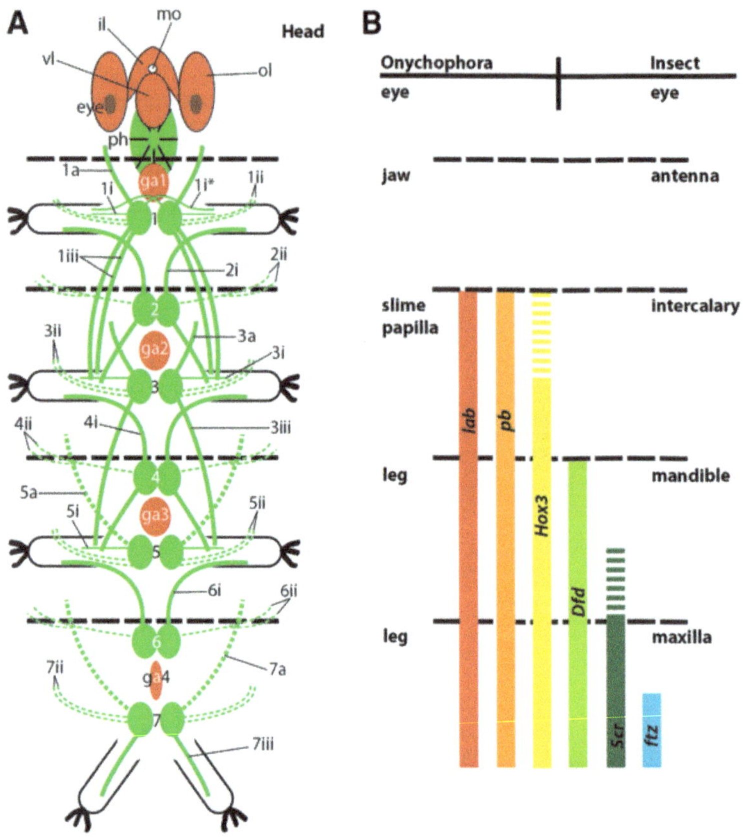


**Figure 8 The metameric pattern of *H. dujardini* includes five distinct segment identities. (A)** A model *H. dujardini* segmental anatomy based on results of the investigations of ventral nervous and muscle systems presented here. Dashed lines demarcate the hypothesized boundaries between segments. Nervous system structures are depicted in red. Muscle system structures are depicted in green. Ventral attachment sites are numbered. Dashed lines represent muscles that extend dorsally. **(B)** Hypothesized anterior *Hox* gene expression domains of the onychophoran-arthropod ancestor based on [63]. The identities of onychophoran segments are provided on the left and the identities of insect segments are provided on the right. Dashed expression domains represent uncertainty in reconstruction of ancestral anterior expression boundaries. The anterior-most expression boundary of the *Hox* genes is the anterior boundary of the third segment. This pattern in *H. dujardini* would support direct alignment of tardigrade segments with arthropod and onychophoran segments and would suggest that the tardigrade head is composed of a single segment. Abbreviations: *Dfd, Deformed*; *ftz, fushi tarazu*; ga, ganglion; *lab, labial*; il, inner lobe; mo, mouth; ol, outer lobe; ph, pharyngeal bulb; *pb, proboscipedia*, *Scr, Sex combs reduced*; vl, ventral lobe.

have been identified in the eutardigrade species *Halobiotus crispae* [17]. Corresponding structures were not identified in two species of *Macrobiotus* [15,35], which are also eutardigrades. Based on these studies, it was suggested that a subesophogeal ganglion is not part of the tardigrade brain [15,35]. While we did not detect a discrete subesophogeal ganglion in *H. dujardini*, we suggest that the ventral brain lobe of this species can be homologized with the ventral brain structures of *H. crispae* [17]. The structure that we identified in *H. dujardini* as the ventral brain lobe (Figures 2B, 3A, 3D) is horseshoe-shaped. The ventral body of this structure is found in a position that corresponds to the position of the subesophogeal ganglion in *H. crispae*, while the branches of the ventral lobe extend around either side of the buccal tube (Figure 2C) to positions that correspond to the position of the paired ventral lobes in *H. crispae*. Based on similarities in size and position of the subesophogeal ganglion and ventral lobes of *H. crispae* to the ventral body and branches of the ventral lobe in *H. dujardini*, we propose that these structures are directly homologous. Based on our homology assignments and the fact that *H. dujardini* is more closely related to *Macrobiotus* than to *H. crispae* [55], we conclude that the ancestor of Eutardigrada included brain regions homologous to the outer, inner, and ventral brain lobes of *H. dujardini*. The interpretation of the brain anatomy of the heterotardigrade *Actinarctus doryphorus* by Persson et al. [18] is consistent with their interpretation of *H. crispae*. In this case, we interpret the ventral lobe of *H. dujardini* as homologous to the subesophogeal ganglion and third brain lobe in *A. doryphorus*. The fact that *H. dujardini* and *A. doryphorus* span the basal split of Tardigrada suggests that the tardigrade ancestor also had a brain exhibiting this architecture.

Additional studies are needed to resolve the debate concerning the morphology of the ventral brain component of tardigrades.

While an understanding is emerging of how the brains of different tardigrade species relate to each other, the question remains of how tardigrade brains relate to the brains of other panarthropods. The tripartite hypothesis is based on the number of brain lobes and commissures found in the tardigrade brain, which are roughly consistent with a brain composed of three ganglia, if each pair of lobes connected by a commissure defines a ganglion. In further support of the tripartite hypothesis, Persson et al. [18] pointed out that differences in the number of sensory appendages found on the heads of arthrotardigrades compared to the leg-bearing segments is consistent with a head composed of multiple segments. However, Zantke et al. [35] noted that the protocerebrum of arthropods is itself composed of multiple lobes, suggesting that brain lobes do not indicate segmental status. Likewise, Mayer et al. [15] noted that a single ganglion can have more than one commissure. Approaches using data other than morphology may be necessary to reveal the relationship of the tardigrade brain to that of other panarthropods (see below).

As with brain anatomy, there are disagreements among various studies regarding nervous system architecture in the leg-bearing segments. However, phylogenetic reconstructions allow us to unambiguously interpret plesiomorphic conditions for Tardigrada. For instance, Zantke et al. [35] did not identify ventral commissures in the ganglia of leg bearing segments of *M. hufelandi*. However, ventral commissures were identified in the ganglia of both heterotardigrade [18,36,37] and eutardigrade species [16,17], including *M. harmsworthi* [15], suggesting that ganglionic ventral commissures are ancestral for tardigrades. In this respect, *H. dujardini* appears to retain the ancestral condition. Interestingly, Mayer et al. [16] detected extra-ganglionic interpedal commissures in *H. dujardini, M. harmsworthi*, and *Paramacrobiotus richtersi* using combined anti-tyrosinated and anti-acetylated α-tubulin immunolabeling. These structures are also present in *E. testudo* [36]. Presence of these structures in taxa that span the basal split of Tardigrada argues that they are ancestral components of tardigrade nervous systems [16].

We did not detect obvious morphological differences between the ganglia of the first three leg-bearing segments. However, these three ganglia are noticeably wider than the ganglion of the posterior-most segment (Figure 3A). Segmental differences in nervous system anatomy have also been reported for other tardigrade species. As in *H. dujardini*, the fourth ganglion of *Halobiotus crispae* is the most morphologically distinct. In this species, the first three trunk ganglia each have two transverse commissures, while the fourth trunk ganglion has only one [17]. Detailed investigations of ganglion morphology are also available for two species of *Macrobiotus*. Zantke et al. [35] reported differences in the number of cells composing each ganglion. In this species, the anterior-most ganglion was consistently composed of the most cells and the posterior-most ganglion was consistently composed of the fewest cells, while the 2$^{nd}$ and 3$^{rd}$ ganglia exhibited similar intermediate cell numbers. As in *H. dujardini*, in *M. harmsworthi*, the first three ganglia are similar in size, while the posterior-most ganglion is much smaller [16]. Therefore, in terms of size, is seems likely that a differentiated fourth ganglion is the ancestral condition for Eutardigrada. Ganglion morphology has been investigated in more detail in *M. harmsworthi* using RFamide immunostaining, which labels a class of neuropeptides. This approach revealed distinct and consistent morphological differences between all segmental ganglia [15]. Schulze and Schmidt-Rhaesa [36] showed that the first and fourth ganglia are narrower than the second and third in *E. testudo*, a heterotardigrade species. This suggests that a differentiated fourth ganglion is ancestral for Tardigrada, and raises the possibility that the first ganglion was also differentiated. The functional significance, if any, for different sizes among ganglia remains unclear.

### Muscle system anatomy of *H. dujardini* and its relationship to that of other tardigrade species

Muscle anatomy has been investigated in two heterotardigrade species, *Batillipes bullacaudatus* and *Echiniscus testudo* [39], and six eutardigrade species, an unidentified *Hypsibius* species [54], *H. crispae* [53], *Milnesium tardigradum* [30,54], *Bertolanius volubilis* [39], *Dactylobiotus parthenogeneticus* [39], and *Richtersius coronifer* [56]. These studies illuminate highly conserved aspects of segmental muscle patterns across tardigrade species (see [33] for reconstructions of ancestral muscle anatomies). For example, the head of all tardigrades, including *H. dujardini* (Figure 6A), includes muscles associated with the stylets. Interestingly, the stylets and stylet glands of the pharyngeal apparatus have been proposed to be serially homologous to the claws and claw glands of tardigrade legs, suggesting that the stylets and associated structures may be derived from ancestral legs of the head [20]. This is supported by the fact that only leg muscles and stylet muscles exhibit cross-striation patterns in the tardigrade species *H. crispae* [53]. We did not investigate whether these two muscle types are striated in *H. dujardini*.

Of particular interest is the arrangement of ventral attachment sites in the leg-bearing segments among tardigrade species. Seven attachment sites is a conserved feature of tardigrades. In both *Hypsibius* (Figure 4A) and *M. tardigradum* [54] the left and right halves of ventral attachment sites make contact at the midline in even-numbered attachment sites, but not in odd-numbered attachment sites. These two taxa span the basal eutardigrade node in a recent molecular phylogeny [55], suggesting that they likely inherited this pattern from the eutardigrade common ancestor. Two exceptions to this rule are found within Eutardigrada. First, in *H. crispae*, the left and right halves make contact at every ventral median attachment site except for site 5 [53]. Second, in *R. coronifer*, the left and right halves of attachment site 1 make contact at the midline [56]. Whether or not the pattern revealed in *H. dujardini* and *M. tardigradum* exists among heterotardigrade species has not been reported, leaving open the question of whether this is the ancestral state for Tardigrada.

The first leg-bearing segment differs from more posterior ones in having only a single ventral median attachment site in both *H. dujardini* (Figures 4A and 8A) and other tardigrade species [39,53,54]. A thin muscle crosses the midline of attachment site 1 and extends into the contralateral leg in both eutardigrades, including *H. dujardini* (Figures 4A and 8A), and heterotardigrades [39]; as a rule, muscles do not cross the midline at more posterior ventral attachment sites in tardigrade species investigated. The only exception to this rule is found in *H. crispae*; in this species, the posteriorly directed muscles extending from attachment site 5 cross the midline [53]. More posterior segments usually possess two attachment sites each. The exception to this rule is the heterotardigrade species *B. bullacaudatus*, in which ventral attachment site 3 is divided into two separate nodes, resulting in three attachment sites in the second segment [39]. However, because the heterotardigrade species *E. testudo* has a single attachment site 3 [39], like eutardigrades, including *H. dujardini* (Figures 4A and 8A), it is inferred that ancestrally, in Tardigrada, leg-bearing segments 2–4 each contained two attachment sites. Ventral attachment site 2 is identical to attachment site 4 in eutardigrade species investigated [39,53,54], including *H. dujardini* (Figures 4A and 8A). This pattern is also found in the heterotardigrade *E. testudo* [39], suggesting that it is the ancestral condition for tardigrades. Interestingly, in the heterotadigrade species *B. bullacaudatus*, ventral attachment site 2 is morphologically distinct from attachment site 4 [39]. As with their paired 3$^{rd}$ ventral attachment sites, this is likely a derived condition. Additional investigations of heterotardigrade species should be performed to test the robustness of these reconstructions of ancestral muscle anatomy.

With the exception of *B. bullacaudatus*, in all tardigrades investigated, including *H. dujardini* (Figures 4A and 8A), ventral attachment site 3 resembles attachment site 5 [39,53,54]. However, in eutardigrade species, including *H. dujardini*, attachment site 3 includes a muscle band that extends to the base of the third leg; a corresponding muscle is lacking in attachment site 5 [39,53,54]. In all species investigated, ventral median attachment sites 6 and 7 show distinct architectures compared to more anterior ventral attachment sites; these sites also exhibit variation between species, making it difficult to infer their ancestral morphologies. Finally, as with *H. dujardini* leg muscle anatomy (Figure 7), muscle anatomy of the fourth leg pair is always the most differentiated in other tardigrade species; these legs generally possess fewer muscles than other legs [39,53,54]. Of the more anterior leg pairs, the anatomy of the first leg pair is the most differentiated (Figure 7A, B) [39,53,54]. While the lateral musculature is very similar among the anterior three leg pairs, the first leg pair lack iii muscles, which attach to the anterior bases of the 2$^{nd}$ and 3$^{rd}$ leg pairs (Figures 7A and 8A).

**The segmental regionalization pattern of tardigrades and its relationship to that of other panarthropods**

Although the relationship among the segments of tardigrades is not a contentious issue, the relationship of their segments to those of other panarthropods remains problematic. It seems unlikely that investigations of the neural anatomies of additional species will allow distinguishing between the protocerebral and tripartite hypotheses of brain anatomy. These hypotheses are based on different evolutionary interpretations of very similar morphologies; i.e., additional data are unlikely to sway opinions in either direction. This debate may seem familiar; it is quite similar to the historic debate among arthropod biologists concerning the homology of head segments of Chelicerata to head segments of mandibulate arthropods (see [30]). In this case, expression patterns of *Hox* genes revealed a conserved pattern of head segmentation across arthropods [40-42,57-62]. Moreover, expression patterns of *Hox* genes in onychophorans have revealed the relationship of their head segments to those of arthropods [51,63]. One conclusion of these studies is that the anterior expression boundaries of anterior *Hox* genes evolved in an ancestor of arthropods + onychophorans and have been highly conserved.

These conserved expression patterns suggest that the developmental mechanisms that differentiate the anterior segments of arthropods already existed in the shared ancestor of Onychophora and Arthropoda, even though the ancestor is thought to have exhibited homonomous segments [32,33]. Why would the ancestor of these two lineages, which is hypothesized to have lacked obvious morphological distinctions between segments, already possess the developmental machinery to produce morphologically distinct anterior segments? Our model of tardigrade segmentation presents the possibility that the muscle system and nervous system had already evolved distinct segmental morphologies in the panarthropod ancestor, long before the evolution of conspicuously different segment morphologies and tagmatization patterns found in extant arthropods and onychophorans. Therefore, the developmental mechanisms differentiating the anterior segments of both onychophorans and arthropods may have been responsible for differentiation of the nervous system and muscular system in the anterior segments of the panarthropod ancestor, explaining the similarity of these mechanisms between Onychophora and Arthropoda. If so, arthropod-like segmental expression patterns of *Hox* genes might explain the subtle segmental differentiation of tardigrades. Functional and expression studies of the *Hox* genes in *H. dujardini* would illuminate ancestral anterior segment patterning mechanisms and whether these genes underlie the production of segmental differentiation in tardigrades. Especially relevant to arguments about the segmental composition of the tardigrade head, the discovery of conserved arthropod-like segmental *Hox* expression patterns in *H. dujardini* would suggest that the tardigrade head is composed of a single segment that is directly homologous to the arthropod and onychophoran protocerebral segment (Figure 8B).

## Conclusions

Tardigrades are an enigmatic group closely related to arthropods with a unique metameric pattern consisting of a head and four leg-bearing segments. We provide a detailed description of segmental differentiation in the emerging developmental genetic model tardigrade *H. dujardini* (Figure 8A), based on examination of muscle and nervous system anatomy. Although morphological differentiation between some segments is subtle, these subtle differences are shared with distantly related tardigrades. This provides evidence that the ancestor of tardigrades already possessed segmental differentiation. Furthermore, unique segmental identities are present in the anterior segments of onychophorans and arthropods, and segmental differentiation in these taxa is the product of shared developmental mechanisms (Figure 8B). Together, these data suggest that the ancestor of Panarthropoda already had a series of differentiated anterior segments, and lead to the hypothesis that the developmental mechanisms producing segmental differentiation predate the divergence of the three panarthropod lineages.

## Materials and methods

### Maintaining tardigrade cultures

Our line of *H. dujardini* was originally collected from a benthic pond sample in Darcy Lever, Bolton, Lancashire, England [43], and has been maintained in culture since 1987. For this study, tardigrade cultures were maintained in 60 mm polystyrene petri dishes in 12.5 ml of Poland Spring brand water. Tardigrades were fed a diet of *Chlorococcum sp.* algae. Cultures were inspected weekly, and fresh water was added to offset evaporation. Subcultures were prepared at least once a month and parent cultures were allowed to expire. We have maintained our *H. dujardini* cultures for over 3 years with this method.

### Specimen relaxation and fixation

Juvenile and reproductively mature *H. dujardini* specimens were collected for this study. Tardigrades tend to contract their muscles during the fixation process, making it difficult to interpret anatomy. To prevent contraction, specimens were relaxed in carbonated water for 1 hour before fixation. At the end of this hour, the bodies of specimens appeared turgid and their appendages were completely extended. Specimens were fixed in 4% EM Grade Paraformaldehyde (Electron Microscopy Sciences) in PB-Triton (1X phosphate-buffered saline, 0.1% Triton X-100, pH 7.4) for 15 minutes at room temperature. Specimens were used immediately for staining of the muscle or nervous systems, as described below.

### Anti-β-tubulin immunostaining

We stained the nervous system of *H. dujardini* using a monoclonal anti-β-tubulin antibody (E7, Developmental Studies Hybridoma Bank) [64], following a slightly modified version of a protocol previously used to stain α-tubulin in a different tardigrade species [29]. Fixation was followed by three 5 minute and four 30 minute rinses in PB-Triton. At this point, the specimens were in a 1.5 ml microcentrifuge tube filled with 1 ml of PB-Triton. Specimens were then permeabilized by sonicating them for 25 seconds with an FS30 sonicater (Fisher Scientific) filled with 850 ml of water. This was followed by three 10 minute and four 30 minute washes in PB-Triton with 0.2% Bovine Serum Albumin and two 30 minute washes in PB-Triton with 5% normal goat serum (NGS). Specimens were then incubated at 4°C overnight in a 1:100 dilution of β-tubulin antibody in PB-Triton + NGS. This incubation period was followed by three 5 minute and four 30 minute washes in PB-Triton. Specimens were then washed twice for 30 minutes in PB-Triton + NGS. This was followed by an overnight incubation at 4°C in a 1:200 dilution of a goat anti-mouse Cy3-conjugated secondary antibody (Jackson ImmunoResearch). This incubation period was followed by three 5 minute and six 30 minute washes in PB-Triton.

### Phalloidin staining

We stained muscle cells with Oregon Green 488 phalloidin (Molecular Probes), which binds actin filaments, following a slightly modified version of the protocol used by Halberg et al. [53]. Fixation was followed by four 15 minute and one overnight wash in PB-Triton with 0.1% $NaN_3$. Specimens were then incubated at room temperature for 27 hrs in a 1:40 dilution of phalloidin in PB-Triton with 0.1% $NaN_3$. Following phalloidin incubation, specimens were rinsed three times for 5 minutes in PB-Triton.

### Specimen preparation, imaging, and analysis

The scanning electron micrograph protocol was previously published [44]. For other imaging techniques, specimens were taken through a glycerol series, which included 1 hour washes in 5%, 10%, 20%, and 40% glycerol in PB-Triton. Specimens were then mounted on slides in Fluoromount-G (SouthernBiotech) with 1.5 μg/ml 4′, 6-diamidino-2-phenylindole (DAPI) added to counterstain nuclei. Differential interference contrast (DIC) micrographs were taken on a Zeiss Axioskop or a Nikon A1R Spectral Confocal Microscope. Z-series of phalloidin (N = 8) and anti-β-tubulin (N = 3) stained specimens were taken on a Nikon A1R Spectral Confocal Microscope. Z-series for 2 additional anti-β-tubulin stained specimens were acquired with a Nikon Eclipse Ti spinning disk confocal microscope. Figi [65-67] was used for image analysis and processing. The Merge function in Figi was used to produce images showing data from multiple confocal channels. Depth-coded images were produced using the Temporal-Color Code function in Figi. Maximum projection images were produced using the maximum intensity option of the Z Project function in Figi. Image levels were adjusted in Adobe Photoshop CS4.

## Additional file

**Additional file 1: Figure S1.** Autofluorescence of algal gut content. The gut is outlined. (A) Unstained specimen with algal gut content. The algal gut content exhibits a high degree of autofluorescence. (B) DIC image of the same specimen. Algae can be seen in the gut.

#### Competing interests

The authors declare that they have no competing interests.

#### Authors' contributions

FWS and ELJ conceived of the study, designed the study, and edited the manuscript. FWS performed all wet lab work, collected data, analyzed data and wrote the first draft of the manuscript. Both authors have read and approved the manuscript.

#### Acknowledgments

We thank Carl D. Schlichting, David L. Wagner, Steven Q. Irvine, and two anonymous reviewers for valuable comments on the manuscript. We are incredibly grateful to Bob Goldstein for providing a *H. dujardini* culture and SEM images for this study. Vicky Nelson assisted in acquisition of SEMs. Karolina Fučíková provided advice for culturing algae. Louise Lewis provided culturing media for algae. Carol Norris of the University of Connecticut Confocal Microscopy Facility provided training and assistance on the confocal scopes used during this study. This research was partially funded by the following awards to FWS: 1.) Betty Decoursey Endowment and the James A. Slater Endowment Fund of the University of Connecticut Ecology and Evolutionary Biology Department and the Connecticut State Museum of Natural History; 2.) a Society for the Study of Evolution (SSE) Rosemary Grant Graduate Student Research Award; 3.) an Extraordinary Expense Award from the University of Connecticut Graduate School.

#### References

1. Horikawa DD, Sakashita T, Katagiri C, Watanabe M, Kikawada T, Nakahara Y, Hamada N, Wada S, Funayama T, Higashi S, Kobayashi Y, Okuda T, Kuwabara M: **Radiation tolerance in the tardigrade *Milnesium tardigradum*.** *Int J Radiat Biol* 2006, **82**:843–848.
2. Jönsson KI, Rabbow E, Schill RO, Harms-Ringdahl M, Rettberg P: **Tardigrades survive exposure to space in low earth orbit.** *Curr Biol* 2008, **18**:R729–R731.
3. Hengherr S, Worland MR, Reuner A, Brümmer F, Schill RO: **High temperature tolerance in anhydrobiotic tardigrades is limited by glass transition.** *Physiol Biochem Zool* 2009, **82**:749–755.
4. Seki K, Toyoshima M: **Preserving tardigrades under pressure.** *Nature* 1998, **395**:853–854.
5. Rebecchi L, Altiero T, Guidetti R, Cesari M, Bertolani R, Negroni M, Rizzo MA: **Tardigrade resistance to space effects: first results of the experiment of LIFE – TARSE mission on FOTON-M3 (September 2007).** *Astrobiology* 2009, **9**:581–591.
6. Persson D, Halberg KA, Jørgensen A, Ricci C, Møbjerg N, Kristensen RM: **Extreme stress tolerance in tardigrades: Surviving space conditions in low earth orbit.** *J Zool Syst Evol Res* 2011, **49**:90–97.
7. Halberg KA, Persson D, Ramløv H, Westh P, Kristensen RM, Møbjerg N: **Cyclomorphosis in Tardigrada: adaptation to environmental constraints.** *J Exp Biol* 2009, **212**:2803–2811.
8. Jönsson KI, Harms-Ringdahl M, Torudd J: **Radiation tolerance in the eutardigrade *Richtersius coronifer*.** *Int J Radiat Biol* 2005, **81**:649–656.
9. Maas A, Waloszek D: **Cambrian derivatives of the early arthropod stem lineage, pentastomids, tardigrades and lobopodians—an 'Orsten' perspective.** *Zool Anz* 2001, **240**:451–459.
10. Brusca RC, Brusca GJ: **From the emergence of the arthropods: onychophorans, tardigrades, trilobites, and the arthropod bauplan.** In *Invertebrates.* 2nd edition. Sunderland MA: Sinaur Associates; 2003:461–510.
11. Kinchin IA: *The Biology of Tardigrades.* London: Portland Press Limited; 1994.
12. Dewel RA, Dewel WC: **The brain of *Echiniscus viridissimus* Peterfi, 1956 (Heterotardigrada): a key to understanding the phylogenetic position of tardigrades and the evolution of the arthropod head.** *Zool J Linn Soc* 1996, **116**:35–49.
13. Kristensen RM, Higgins RP: **A new family of Arthrotardigrada (Tardigrada: Heterotardigrada) from the Atlantic coast of Florida, USA.** *Trans Am Microsc Soc* 1984, **103**:295–311.
14. Kristensen RM, Higgins RP: **Revision of *Styraconyx* (Tardigrada: Halechiniscidae) with descriptions of two new species from Disko Bay, West Greenland.** *Smithson Contrib Zool* 1984, **391**:1–40.
15. Mayer G, Kauschke S, Rüdiger J, Stevenson PA: **Neural markers reveal a one-segmented head in tardigrades (water bears).** *PLoS ONE* 2013, **8**:e59090.
16. Mayer G, Martin C, Rudiger J, Kauschke S, Stevenson P, Poprawa I, Hohberg K, Schill RO, Pflüger H-J, Schlegel M: **Selective neuronal staining in tardigrades and onychophorans provides insights into the evolution of segmental ganglia in panarthropods.** *BMC Evol Biol* 2013, **13**:230.
17. Persson DK, Halberg KA, Jorgensen A, Mobjerg N, Kristensen RM: **Neuroanatomy of *Halobiotus crispae* (Eutardigrada: Hypsibiidae): tardigrade brain structure supports the clade Panarthropoda.** *J Morphol* 2012, **273**:1227–1245.
18. Persson DK, Halberg KA, Jørgensen A, Møbjerg N, Kristensen RM: **Brain anatomy of the marine tardigrade *Actinarctus doryphorus* (Arthrotardigrada).** *J Morphol* 2014, **275**:173–190.
19. Manton SM: *The Arthropoda: Habits, Functional Morphology and Evolution.* Oxford: Clarendon Press; 1977.
20. Nielsen C: *Animal Evolution: Interrelationships of the Animal Phyla.* New York: Oxford University Press; 2001.
21. Budd GE: **The morphology of *Opabinia regalis* and the reconstruction of the arthropod stem-group.** *Lethaia* 1996, **29**:1–14.
22. Budd GE: **Tardigrades as 'stem-group arthropods': the evidence from the Cambrian fauna.** *Zool Anz* 2001, **240**:265–279.
23. Budd GE: **Why are arthropods segmented?** *Evol Dev* 2001, **3**:332–342.
24. Campbell LI, Rota-Stabelli O, Edgecombe GD, Marchioro T, Longhorn SJ, Telford MJ, Philippe H, Rebecchi L, Peterson KJ, Pisani D: **MicroRNAs and phylogenomics resolve the relationships of Tardigrada and suggest that velvet worms are the sister group of Arthropoda.** *Proc Natl Acad Sci USA* 2011, **108**:15920–15924.
25. Legg DA, Sutton MD, Gregory DE: **Arthropod fossil data increase congruence of morphological and molecular phylogenies.** *Nat Commun* 2013, **4**:2485.
26. Dewel RA, Dewel WC: **The place of tardigrades in arthropod evolution.** In *Arthropod Relationships.* Edited by Fortey RA, Thomas RH. London: Chapman and Hall; 1997:109–123.

27. Ma X, Edgecombe GD, Legg DA, Hou X: **The morphology and phylogenetic position of the Cambrian lobopodian *Diania cactiformis*.** *J Syst Palaeontol* 2014, **12**:445–457.
28. Angelini DR, Kaufman TC: **Comparative developmental genetics and the evolution of arthropod body plans.** *Annu Rev Genet* 2005, **39**:95–119.
29. Fusco G, Minelli A: **Arthropod segmentation and tagmosis.** In *Arthropod Biology and Evolution.* Edited by Fusco G, Minelli A. Berlin: Springer-Verlag; 2013:197–221.
30. Scholtz G, Edgecombe GD: **The evolution of arthropod heads: reconciling morphological, developmental and palaeontological evidence.** *Dev Genes Evol* 2006, **216**:395–415.
31. Mayer G, Whitington PM, Sunnucks P, Pfluger HJ: **A revision of brain composition in Onychophora (velvet worms) suggests that the tritocerebrum evolved in arthropods.** *BMC Evol Biol* 2010, **10**:255.
32. Ma X, Hou X, Bergstrom J: **Morphology of *Luolishania longicruris* (Lower Cambrian, Chengjiang Lagerstatte, SW China) and the phylogenetic relationships within lobopodians.** *Arthropod Struct Dev* 2009, **38**:271–291.
33. Ou Q, Degan S, Mayer G: **Cambrian lobopodians and extant onychophorans provide new insights into early cephalization in Panarthropoda.** *Nat Commun* 2012, **3**:1261.
34. Kristensen RM: **The first record of cyclomorphosis in Tardigrada based on a new genus and species from Arctic meiobenthos.** *J Zool Syst Evolut-Forsch* 1982, **20**:249–270.
35. Zantke J, Wolff C, Scholtz G: **Three-dimensional reconstruction of the central nervous system of *Macrobiotus hufelandi* (Eutardigrada, Parachela): Implications for the phylogenetic position of Tardigrada.** *Zoomorphology* 2008, **127**:21–36.
36. Schulze C, Schmidt-Rhaesa A: **The architecture of the nervous system of *Echiniscus testudo* (Echiniscoidea, Heterotardigrada).** *J Limnol* 2013, **72**:44–53.
37. Schulze C, Neves RC, Schmidt-Rhaesa A: **Comparative immunohistochemical investigation on the nervous system of two species of Arthrotardigrada (Heterotardigrada, Tardigrada).** *Zool Anz* 2014, **253**:225–235.
38. Bitsch J, Bitsch C: **The tritocerebrum and the clypeolabrum in mandibulate arthropods: segmental interpretations.** *Acta Zool-Stockholm* 2010, **91**:249–266.
39. Marchioro T, Rebecchi L, Cesari M, Hansen JG, Viotti G, Guidetti R: **Somatic musculature of Tardigrada: phylogenetic signal and metameric patterns.** *Zool J Linn Soc Lond* 2013, **169**:580–603.
40. Damen WGM, Hausdorf M, Seyfarth E-A, Tautz D: **A conserved mode of head segmentation in arthropods revealed by the expression pattern of *Hox* genes in a spider.** *Proc Natl Acad Sci USA* 1998, **95**:10665–10670.
41. Jager M, Murienne J, Clabaut C, Deutsch J, Le Guyader H, Manuel M: **Homology of arthropod anterior appendages revealed by Hox gene expression in a sea spider.** *Nature* 2006, **411**:506–508.
42. Telford MJ, Thomas RH: **Expression of homeobox genes shows chelicerate arthropods retain their deutocerebral segment.** *Proc Natl Acad Sci USA* 1998, **95**:10671–10675.
43. Gabriel WN, McNuff R, Patel SK, Gregory TR, Jeck WR, Jones CD, Goldstein B: **The tardigrade *Hypsibius dujardini*, a new model for studying the evolution of development.** *Dev Biol* 2007, **312**:545–559.
44. Gabriel WN, Goldstein B: **Segmental expression of Pax3/7 and Engrailed homologs in tardigrade development.** *Dev Genes Evol* 2007, **217**:421–433.
45. Smith FW: *Conservation and Diversification of Appendage Identity Specification Mechanisms along the Anteroposterior and Proximodistal Axes in Panarthropoda*, PhD dissertation. Connecticut: University of Connecticut, Department of Ecology and Evolutionary Biology; 2013.
46. Tenlen JR, McCaskill S, Goldstein B: **RNA interference can be used to disrupt gene function in tardigrades.** *Dev Genes Evol* 2013, **223**:1–11.
47. Budd GE, Telford MJ: **The origin and evolution of arthropods.** *Nature* 2009, **457**:812–817.
48. Grenier JK, Garber TL, Warren R, Whitington PM, Carroll S: **Evolution of the entire arthropod *Hox* gene set predated the origin and radiation of the onychophoran/arthropod clade.** *Curr Biol* 1997, **7**:547–553.
49. Shubin N, Tabin C, Carroll S: **Fossils, genes and the evolution of animal limbs.** *Nature* 1997, **388**:639–648.
50. Snodgrass RE: *Principles of Insect Morphology.* Ithaca: Cornell University Press; 1935.
51. Eriksson BJ, Tait NN, Budd GE, Janssen R, Akam M: **Head patterning and *Hox* gene expression in an onychophoran and its implications for the arthropod head problem.** *Dev Genes Evol* 2010, **220**:117–122.
52. Eriksson BJ, Samadi L, Schmid A: **The expression pattern of the genes *engrailed, pax6, otd* and *six3* with special respect to head and eye development in *Euperipatoides kanangrensis* Reid 1996 (Onychophora: Peripatopsidae).** *Dev Genes Evol* 2013, **223**:237–246.
53. Halberg KA, Persson D, Mobjerg N, Wanninger A, Kristensen RM: **Myoanatomy of the marine tardigrade *Halobiotus crispae* (Eutardigrada: Hypsibiidae).** *J Morphol* 2009, **270**:996–1013.
54. Schmidt-Rhaesa A, Kulessa J: **Muscular architecture of *Milnesium tardigradum* and *Hypsibius* sp. (Eutardigrada, Tardigrada) with some data on *Ramazottius oberhaeuseri*.** *Zoomorphology* 2007, **126**:265–281.
55. Guil N, Giribet G: **A comprehensive molecular phylogeny of tardigrades—adding genes and taxa to a poorly resolved phylum-level phylogeny.** *Cladistics* 2012, **28**:21–49.
56. Halberg KA, Jorgensen A, Mobjerg N: **Desiccation tolerance in the tardigrade *Richtersius coronifer* relies on muscle mediated structural reorganization.** *PLoS One* 2013, **8**(12):e85091.
57. Abzhanov A, Kaufman TC: **Homeotic genes and the arthropod head: expression patterns of the *labial, proboscipedia*, and *Deformed* genes in crustaceans and insects.** *Proc Natl Acad Sci USA* 1999, **96**:10224–10229.
58. Damen WGM, Tautz D: **A Hox class 3 orthologue from the spider *Cupiennius salei* is expressed in a *Hox*-gene-like fashion.** *Dev Genes Evol* 1998, **208**:586–590.
59. Hughes CL, Kaufman TC: **Hox genes and the evolution of the arthropod body plan.** *Evol Dev* 2002, **4**:459–499.
60. Janssen R, Damen WG: **The ten Hox genes of the millipede *Glomeris marginata*.** *Dev Genes Evol* 2006, **216**:451–465.
61. Manuel M, Jager M, Murienne J, Clabaut C, Le Guyader H: **Hox genes in sea spiders (Pycnogonida) and the homology of arthropod head segments.** *Dev Genes Evol* 2006, **216**:481–491.
62. Sharma PP, Schwager EE, Extavour CG, Giribet G: **Hox gene expression in the harvestman *Phalangium opilio* reveals divergent patterning of the chelicerate opisthosoma.** *Evol Dev* 2012, **14**:450–463.
63. Janssen R, Eriksson BJ, Tait NN, Budd GE: **Onychophoran Hox genes and the evolution of arthropod Hox gene expression.** *Front Zool* 2014, **11**:22.
64. Chu DT, Klymkowsky MW: **The appearance of acetylated alpha-tubulin during early development and cellular differentiation in *Xenopus*.** *Dev Biol* 1989, **1989**(136):104–117.
65. Abràmoff MD, Magalhães PJ, Ram SJ: **Image processing with ImageJ.** *Biophoton Int* 2004, **11**:36–42.
66. Rasband WS: *1997–2012. Image J.* Bethesda, Maryland, USA: U.S. National Institutes of Health; imagej.nih.gov/ij/.
67. Schneider CA, Rasband WS, Eliceiri KW: **NIH image to ImageJ: 25 years of image analysis.** *Nat Methods* 2012, **9**:671–675.

# 4

# Reproductive cessation and post-reproductive lifespan in Asian elephants and pre-industrial humans

Mirkka Lahdenperä[1*], Khyne U Mar[2] and Virpi Lummaa[2]

**Abstract**

**Introduction:** Short post-reproductive lifespan is widespread across species, but prolonged post-reproductive life-stages of potential adaptive significance have been reported only in few mammals with extreme longevity. Long post-reproductive lifespan contradicts classical evolutionary predictions of simultaneous senescence in survival and reproduction, and raises the question of whether extreme longevity in mammals promotes such a life-history. Among terrestrial mammals, elephants share the features with great apes and humans, of having long lifespan and offspring with long dependency. However, little data exists on the frequency of post-reproductive lifespan in elephants. Here we use extensive demographic records on semi-captive Asian elephants ($n = 1040$) and genealogical data on pre-industrial women ($n = 5336$) to provide the first comparisons of age-specific reproduction, survival and post-reproductive lifespan in both of these long-lived species.

**Results:** We found that fertility decreased after age 50 in elephants, but the pattern differed from a total loss of fertility in menopausal women with many elephants continuing to reproduce at least until the age of 65 years. The probability of entering a non-reproductive state increased steadily in elephants from the earliest age of reproduction until age 65, with the longer living elephants continuing to reproduce until older ages, in contrast to humans whose termination probability increased rapidly after age 35 and reached 1 at 56 years, but did not depend on longevity. Post-reproductive lifespan reached 11–17 years in elephants and 26–27 years in humans living until old age (depending on method), but whereas half of human adult lifespan (of those reproductive females surviving to the age of 5% fecundity) was spent as post-reproductive, only one eighth was in elephants. Consequently, although some elephants have long post-reproductive lifespans, relatively few individuals reach such a phase and the decline in fertility generally parallels declines in survivorship in contrast to humans with a decoupling of senescence in somatic and reproductive functions.

**Conclusions:** Our results show that the reproductive and survival patterns of Asian elephants differ from other long-lived animals exhibiting menopause, such as humans, and extreme longevity alone does not promote the evolution of menopause or post-reproductive lifespan, adding weight to the unusual kin-selected benefits suggested to favour such traits in humans and killer whales.

**Keywords:** Ageing, Age-specific fertility, Reproduction, Senescence, Survival

* Correspondence: mirkka.lahdenpera@utu.fi
[1]Section of Ecology, Department of Biology, University of Turku, Turku FIN-20014, Finland
Full list of author information is available at the end of the article

## Introduction

Senescence is commonly observed as age declines in both fertility and survival [1-3]. The classical evolutionary theory predicts selection against decoupling of senescence patterns between somatic and reproductive maintenance, leading to similar ageing declines in both fitness traits [2,4]; but see [5] for contradictory empirical evidence. The existence of menopause and post-reproductive lifespan across species has attracted considerable recent interest, first, due to medical and evolutionary attention on mid-life menopause in women and, second, due to accumulating evidence of a detectable, albeit limited, post-reproductive lifespan also in shorter-lived species such as small mammals [6], birds [7,8], fish [9], insects [10] and nematodes [8,11]. In contrast, studies on post-reproductive lifespan in long-lived species other than humans are rare because the difficulty in obtaining longitudinal data on these species limits the sample size available on old, potentially post-reproductive individuals. Consequently, there has been long debate over whether other long-lived species show similar kind of menopause and post-reproductive lifespan as women e.g. [6,12].

So far, the most convincing evidence of post-reproductive lifespan in non-human long-lived species comes from marine mammals such as short-finned pilot whales (*Globicephala macrorhyncus*) and killer whales (*Orcinus orca*). These species exhibit similar life-history to humans with reproductive capacity ending around 36 and 50 years and maximum lifespan reaching over 60 and 80 years, respectively, resulting in decades-long post-reproductive phases in both species [13,14]. Recent evidence suggests that similarly to humans [15], at least in killer whales, such a life-history may result from adaptive benefits that old post-reproductive females bring to their adult offspring [16]. The evidence for long post-reproductive lifespan (or lack of it) in terrestrial mammals is far less compelling. For example, female primates appear to retain fertility close to death [17-19], resulting in maximum post-reproductive lifespans of a few years [6].

Elephants are particularly interesting because similarly to whales, great apes and humans, they have well-defined social networks, large brains, offspring with long dependency (weaning at 3–5 years), and are extremely long-lived [20,21]. Their lifespan represents the upper end recorded for terrestrial mammals along with humans, with survival commonly into 60s in the wild and maximum known age >80 yrs [22]. Some have suggested that elephants experience menopause [23] and have similar post-reproductive lifespan to women [24,25], but due to few longitudinal datasets, this remains controversial. Little data exists on Asian elephants (*Elephas maximus*) outside zoos, where females appear to cease reproducing early (oldest breeding female in Europe was 36) [26] and have lower survival [27] compared to the wild. Such a reproductive pattern may result from captive breeding schedules that artificially accelerate reproductive senescence, when prolonged non-reproductive periods lead to increases in genital pathologies and accelerated follicular loss [28,29]. Moreover, while some field studies on African elephants (*Loxodonta africana*) suggest that females may cease reproduction at 40–50 years [23,24,30], others show females capable of reproducing until 60 [22,31,32].

Further complications result from large discrepancies between studies in the definitions of menopause and post-reproductive lifespan. First, human menopause is defined as a complete cessation of menstrual cycles and ovulation (operational menopause) [33]. However, investigating changes in ovulation frequency and hormonal profiles in wild populations is challenging, and the functional menopause preceding operational menopause is usually defined as the age at last reproduction [33]. Second, several definitions are in use for post-reproductive lifespan. In animals for which the existence of menopause is unknown and reproductive cessation cannot be directly observed, post-reproductive lifespan is often calculated simply as the population average of the difference between the ages at last reproduction and death [17,34,35]. To avoid problems such as including individuals which were actually able to reproduce but just died before that, the post-reproductive lifespan can also be calculated for only those individuals surviving long enough after last reproduction (based on mean inter-birth intervals and standard deviations after last birth) [9,17]. This method also takes better into account each individual's own reproductive history. However, this method excludes individuals with short but legitimate post-reproductive lifespan (the lower end of the distribution) [36]. Moreover, because inter-birth intervals increase markedly with age in many animals [17,19], these approaches may overestimate the incidence of post-reproductive lifespan when older individuals have greatly increased birth spacing. Finally, comparison of populations with these measures is problematic, because they are so correlated with overall longevity of a species [36]. Recently, Levitis and Lackey presented a new measure, post-reproductive representation, to allow easier comparisons between populations and species with differential longevities: the proportion of post-reproductive adults using life-table methods [36]. In line with this, they suggested that the main feature differentiating humans from other animals is not their ability to live beyond reproductive cessation and the subsequent post-reproductive lifespan, but the markedly greater proportion of mature individuals doing so [36,37]. Few studies exist however to compare such methods and their suitability for estimating reproductive cessation and post-reproductive lifespan.

Here we investigate changes in survival and reproduction with age and the prevalence and length of post-reproductive lifespan in humans and in Asian elephants. For elephants, we use the world's largest ($n = 1040$ ever reproduced females) multigenerational demographic dataset covering the life-history of several generations of logging elephants in 260 timber camps in Myanmar over a century [38-42]. Although these elephants are semi-captive and used in the timber industry subject to set workloads, they forage unsupervised in forests during nights and official rest periods and breed with captive and wild elephants. While basic veterinary care is available to treat injuries, survival rates reflect those of wild African elephants [27]. The elephants are not provisioned or aided in mating or calving and breeding rates are therefore unmanaged by humans. We compare female elephant reproductive and survival patterns with those of humans using as an example a comparable dataset on individual-based life-histories of historical women reproducing before the demographic transition (born <1850, $n = 5336$) to illustrate similarities/differences to humans with inevitable reproductive cessation (ie. menopause) and long subsequent post-reproductive lifespan. Specifically, we examine (a) age-specific changes in fertility, survival and birth intervals, and (b) reproductive cessation and subsequent post-reproductive survival in both species. We measure reproductive cessation by the observed age at last reproduction and by using a probabilistic model of the age-specific transition to a non-reproductive state. We determine post-reproductive lifespan by several measures that have been previously used in the literature including a recently published measure of post-reproductive representation that is particularly suited for comparisons between species [36].

## Results

### Age-specific changes in fertility, survival and birth intervals

#### *Age-specific fertility and survival*

First, we investigated age-specific changes in fertility and overall survival in elephant and human females. Reproductive Asian elephant females in Myanmar produced on average $2.6 \pm 1.76$ (SD) calves (range 1–11) during their lifetime (Table 1). In general, the age-specific female fertility showed a characteristic inverted U-shaped curve but compared to humans, elephant females had lower age-specific fertility as over 20% of women reproduced at the age of peak fertility, in contrast to only >10% of elephants (Figure 1a,b). Although at the population-level the fertility of female elephants did not show a well-defined peak but instead remained at its highest at 20–50 years, most offspring born into the population were produced by females aged 20–25 years (Figure 1c) and in humans a few years later (Figure 1d). The age-specific fertility decreased after age 50 also in elephants (Figure 1a), but the pattern did not show a similar abrupt total loss of fertility in old age as in women (Figure 1b). Consequently, the species differed significantly in their age-specific probability of reproducing after age 40, with elephants experiencing a less steep loss of fertility with old age ($n = 4989$ individuals, age*species –interaction: $F_{1,97526} = 144.84$, $P < 0.0001$, $\beta = 4.08 \pm 0.034$; age$^{2*}$ species –interaction: $F_{1,97526} = 179.38$, $P < 0.0001$, $\beta = -0.051 \pm 0.0038$ (S.E)).

The differences in the overall survival and lifespan between elephants and humans are also clear, although the maximum lifespan reaches almost 80 years in both species (Figure 1a,b, Table 1). The age-specific survival probability of reproductive females differed statistically between elephants and humans, with survival decreasing more rapidly after age 40 in elephants than in humans ($n = 4989$ individuals, age*species –interaction: $F_{1,120000} = 218.11$, $P < 0.0001$, $\beta = 0.085 \pm 0.0057$). For example, of all ever reproducing female elephants born into the population, 59% survived to age 50 in contrast to 75% in women. At older ages the difference becomes even more obvious as 9% of all reproductive female elephants survived to age 70, in contrast to 41% of all reproductive women (Figure 1a,b).

#### *Inter-birth intervals*

We then investigated how the age-specific changes in fertility at the population level reflected on individual-level variation in birth-intervals with age. Elephant females ($n = 1480$ intervals) produced calves on average every $5.99 \pm 2.99$ years (range 1.52-17.29). In elephants the inter-birth intervals decreased at older ages (age: $F_{1,842} = 0.19$, $P = 0.66$, $\beta = 0.0032 \pm 0.0073$; age$^2$: $F_{1,842} = 4.15$, $P = 0.042$, $\beta = -0.00024 \pm 0.00012$; Figure 2a; Additional file 1: Table S2), in contrast to many animals such as primates in which inter-birth intervals increase with maternal age [17,19]. The shortest inter-birth intervals were at old maternal ages (Figure 2a). The average inter-birth interval was 10.1 years for females younger than 10 years at the (first) reproduction, 6.0 years at ages 20–30 years, and only on average 4.7 years for females older than 40 years. The model controlled for significantly longer inter-birth intervals in wild-born as compared to captive-born mothers ($6.20 \pm 0.25$ vs. $5.44 \pm 0.23$ years), after firstborn compared to later-born calves ($5.98 \pm 0.24$ vs. $5.64 \pm 0.23$ years), for females with later age at last reproduction, as well as differences between birth cohorts and living areas (Additional file 1: Table S2). Inter-birth intervals did not differ after male and female births, with differing maternal lifespan and for censored or non-censored mothers (Additional file 1: Table S2). Age-related trends in birth-intervals might be confounded by higher infant mortality among youngest and oldest females, leading potentially to quicker replacement births

**Table 1 Descriptive statistics of Asian elephants and humans (ever reproduced females)**

| | n | Birth year | Mean afr ± S.D | Mean alr ± S.D | Mean fec ± S.D | Mean lifespan ± S.D | pr-lifespan ± S.D |
|---|---|---|---|---|---|---|---|
| Elephant females, all | 1040 | 1900-1993 | 19.85 ± 5.68 (5.30, 46.36) | 32.49 ± 10.61 (7.22, 64.9) | 2.62 ± 1.76 (1.0, 11.0) | 38.38 ± 11.58 (8.07, 79.64) | 5.89 ± 6.97 (0.00, 38.74) |
| Dead | 320 | 1900-1979 | 19.05 ± 7.05 (5.30, 46.36) | 32.52 ± 10.69 (7.22, 59.83) | 2.30 ± 1.53 (1.0, 10.0) | 40.40 ± 12.89 (8.07, 79.64) | 7.88 ± 7.06 (0.00, 38.74) |
| Alive | 720 | 1920-1993 | 20.10 ± 5.16 (8.00, 46.34) | 32.47 ± 10.58 (8.38, 64.92) | 2.77 ± 1.84 (1.0, 11.0) | 37.48 ± 10.83 (10.39, 79.09) | 5.00 ± 6.75 (0.00, 33.19) |
| Lifespan > =40 | 457 | 1911-1970 | 20.78 ± 6.86 (8.28, 46.36) | 40.35 ± 9.16 (13.07, 64.92) | 3.26 ± 2.02 (1.0, 11.0) | 48.83 ± 6.87 (40.03, 79.64) | 8.48 ± 8.33 (0.00, 38.74) |
| Lifespan > =40 + dead | 154 | 1911-1967 | 21.66 ± 9.36 (8.28, 46.36) | 39.77 ± 8.52 (18.92, 59.83) | 2.66 ± 1.73 (1.0, 10.0) | 51.09 ± 7.86 (40.03, 79.64) | 11.31 ± 7.84 (0.00, 38.74) |
| Captive born | 471 | 1936-1993 | 19.85 ± 5.68 (5.30, 46.36) | 29.88 ± 9.55 (7.22, 54.00) | 2.95 ± 1.93 (1.0, 10.0) | 35.02 ± 10.54 (8.07, 65.11) | 5.14 ± 6.68 (0.00, 34.31) |
| Wild born | 569 | 1900-1986 | n/a | 34.65 ± 10.96 (8.38, 64.92) | 2.36 ± 1.56 (1.0, 11.0) | 41.16 ± 11.67 (9.46, 79.64) | 6.51 ± 7.15 (0.00, 38.74) |
| Human females, all | 5336 | 1595-1849 | 26.22 ± 5.17 (15.00, 47.00) | 37.66 ± 6.27 (16.00, 52.30) | 5.20 ± 3.04 (1.0, 18.0) | 60.07 ± 16.50 (17.86, 100.31) | 22.42 ± 15.19 (0.00, 66.18) |
| Dead | 4529 | 1665-1849 | 26.18 ± 5.17 (15.00, 47.00) | 37.35 ± 6.30 (16.00, 52.30) | 5.11 ± 3.02 (1.0, 18.0) | 59.78 ± 17.28 (17.86, 100.31) | 22.43 ± 15.65 (0.00, 66.18) |
| Censored | 807 | 1595-1849 | 26.45 ± 5.17 (16.00, 46.20) | 39.37 ± 5.81 (17.00, 50.00) | 5.76 ± 3.14 (1.0, 17.0) | 61.71 ± 11.07 (20.00, 94.00) | 22.35 ± 12.28 (0.00, 66.01) |
| Lifespan > =42 | 4427 | 1595-1849 | 26.50 ± 5.32 (15.00, 47.00) | 38.98 ± 5.53 (17.00, 52.30) | 5.53 ± 3.05 (1.0, 18.0) | 65.54 ± 12.08 (42.00, 100.31) | 26.56 ± 13.24 (0.00, 66.18) |

Afr = age at first reproduction. Alr = age at last reproduction. Fec = number of births. Values of afr, alr, fec, lifespan (age at death or censoring) and pr-lifespan represent means with standard deviations (S.D.) and in brackets minimum and maximum values. Afr can only be calculated for captive-born elephants (sample size 38–471). Dead = individuals with known death date. Alive = individuals still alive (in elephants).

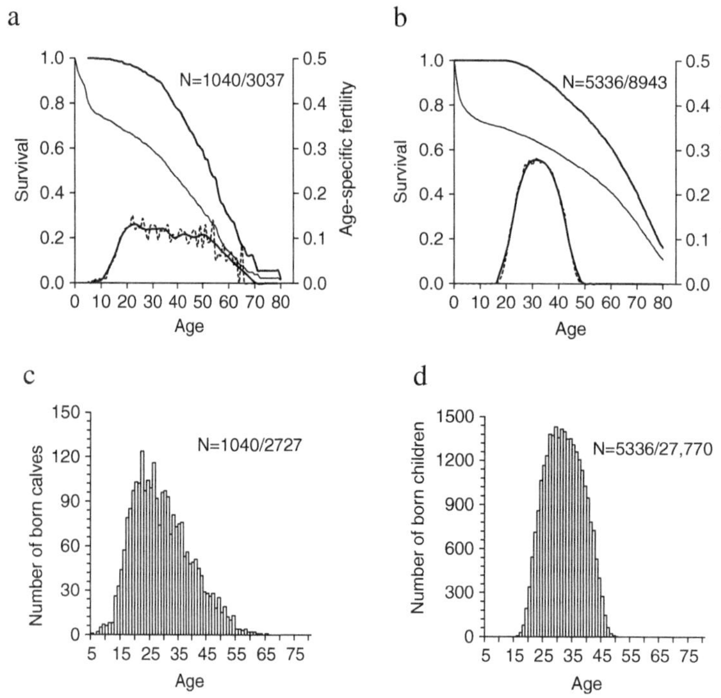


**Figure 1 Age-specific fertility, survival and offspring production until 80 years in Asian elephants and women.** **(a)** Elephant and **(b)** human age-specific fertility of ever reproducing females (dashed line) and 4-year average (solid line) shown along survival of all females born into population (elephants: $n = 3037$; humans: $n = 8943$, narrow line) and ever reproduced females only (elephants: $n = 1040$; humans: $n = 5336$, bold line). **(c)** Elephant and **(d)** human number of offspring born at each age (elephants: $n = 2727$ calves for 1040 females; humans: $n = 27{,}770$ offspring for 5336 women).

among such age-groups as compared to mothers with calves to care for. We therefore also re-ran the above analysis controlling for the survival of previous calf to age 1 ($n = 1222$). The birth interval was on average 0.7 years shorter when the previous calf died early (survived: 6.00 ± 0.20 vs. died: 5.33 ± 0.31; $F_{1,631} = 5.56$, $P = 0.019$, $\beta = -0.12 \pm 0.050$) but maternal age had a similar effect on inter-birth intervals irrespective of whether the previous calf died or survived (age*calf survival to age 1 –interaction: $F_{1,630} = 0.16$, $P = 0.69$, $\beta = -0.0023 \pm 0.0057$; age$^2$*calf survival to age 1 –interaction: $F_{1,629} = 1.30$, $P = 0.25$, $\beta = -0.00055 \pm 0.00048$).

In contrast to elephants, the average inter-birth interval during lifetime was 2.72 ± 1.23 years (range 0.75-9.02) for women and the intervals increased linearly with maternal age ($n = 21{,}033$, age: $F_{1,16592} = 262.59$, $P < 0.0001$, $\beta = 0.0095 \pm 0.00059$; Figure 2b, Additional file 1: Table S3). Such effects were controlled for significant variation in birth-interval length between parishes, birth cohorts, socio-economic classes, with poorest women having the longest intervals (rich: 2.65 ± 0.023; average: 2.79 ± 0.028, poor: 3.02 ± 0.039), birthorder, with shorter birth-intervals after firstborn than later-born (first: 2.73 ± 0.027, later: 2.91 ± 0.024), longer-living women having longer birth intervals and women with later age at last reproduction having shorter birth intervals (Additional file 1: Table S3). Birth-intervals did not differ after girl and boy births and according to whether the mother was censored or not (Additional file 1: Table S3). To control for potential effects of age-variation in replacement births, we again repeated the analysis fitting child survival to age 1 in the model ($n = 20{,}656$). Women whose child died under 1 had 1 year shorter following birth interval than women whose child survived beyond age 1 (1.97 ± 0.020 vs. 2.98 ± 0.024; $F_{1,16223} = 3293.91$, $P < 0.0001$, $\beta = -0.42 \pm 0.0073$). However, maternal age had a similar effect on birth intervals irrespective of whether the previous child died or not (age*child survival to age 1 –interaction: $F_{1,16222} = 0.32$, $P = 0.57$, $\beta = -0.00067 \pm 0.0012$).

### Reproductive cessation and subsequent post-reproductive survival

#### *Age at last reproduction*

Next, we investigated the patterns of reproductive cessation in elephants and humans. In captive-born female elephants ($n = 471$), with known reproductive history since birth, the mean age at first birth was 19.9 ± 5.7 years (range 5.3-46.4) and the mean age at last reproduction for all females ($n = 1040$) was 32.5 ± 10.6 years (range 7.2-64.9) (Table 1). Historical women ($n = 5336$) had their first birth on average at 26.2 ± 5.2 years (range 15.0-47.0) and last birth at 37.7 ± 6.3 (range 16.0-52.3). The age at last reproduction

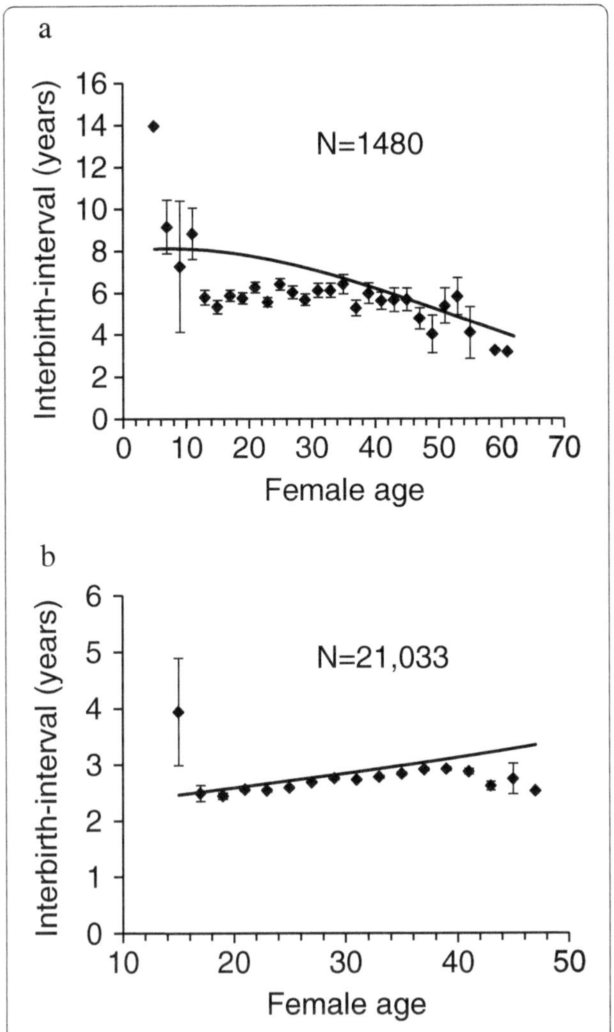


**Figure 2 Length of inter-birth interval with female age in elephants and women. (a)** Inter-birth intervals decreased with female age in Asian elephants ($n = 1480$), **(b)** but increased in women ($n = 21{,}033$) (Additional file 1: Tables S2 and S3). The figure also shows two year averages and standard errors of inter-birth intervals in raw data.

differed statistically between the species: elephants which were still alive at the end of the study (and thus more likely to have reached old age) had significantly later age at last reproduction than humans, whereas those elephant females that had already died (many so prematurely) had on average lower age at last birth (species*censoring interaction: $F_1 = 34.95$, $P < 0.0001$; species: $F_1 = 0.56$, $P = 0.45$; other significant variables: censoring: $F_1 = 120.59$, $P < 0.0001$; length of lifespan: $F_1 = 1201.80$, $P < 0.0001$; censoring*lifespan: $F_1 = 165.77$, $P < 0.0001$). This implies that survival to old age was a restricting factor for reproduction in elephants, whereas the age at last reproduction was similar in censored and non-censored humans. The variance in the age at last reproduction was significantly greater in elephants as compared to humans (Levene's test: $F_1 = 226.60$, $P < 0.0001$, $\beta = 1.95 \pm 0.13$). This is illustrated with 75% of elephant females in this sample having had last reproduced before 40.0 years and 99% before 57.4 (Figure 3a) whereas 75% of women gave birth to the last child before 42.2 and 99% before 47.0 years. The maximal age at last birth in elephants was 65 years compared to 52 years in humans. Hence, the main difference in the reproductive pattern between the species concerns a more rapid loss of fertility with age in women, characterized by the last quartile in women dividing into a much narrower scale than in elephant females (Figure 3a,b), and the total cessation of reproduction occurring about 10 years earlier in women than in elephants illustrated by both the 99% quartile and the maximal age at last reproduction.

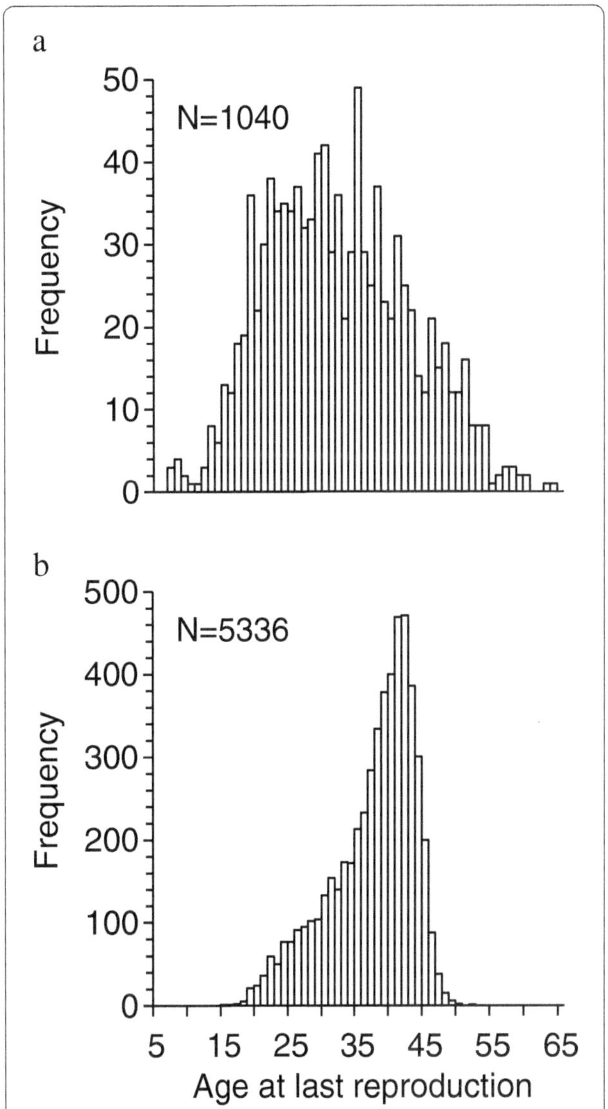


**Figure 3 Distribution of age at last reproduction. (a)** Asian elephants ($n = 1040$) and **(b)** women ($n = 5336$).

To confirm that the length of lifespan was restricting the potential to reproduce at later ages more in elephants than in humans, we also investigated reproductive cessation among only those elephants which survived beyond age 40 (calculated as the age by which 75% of females had given birth to their last offspring), and hence had an opportunity to breed at old age. These females had their last calf on average at 40.4 ± 9.2 years (range 13.1-64.9), and the age at last reproduction therefore increased by 7.9 years from the population average (Table 1). In humans, women who lived into old age (42 years; calculated as the age by which 75% of females had given birth to their last offspring) had their last child at 39.0 ± 5.5 years (range 17.0-52.3), which differs only by 1.3 years from the population average. This implies that survival rates to old age affected the reproductive pattern in elephants but not so in humans.

#### *Transition to a non-reproductive state with age*

We next investigated how age affected the probability of ceasing reproduction in both species. In elephants, the cubic term of age had a significant effect on the probability of ceasing reproduction and thus the probability of entering a non-reproductive state increased steadily from the earliest age at reproduction, 5 years, reaching 0.95 at age 38 and finally exceeding 0.99 at age 49 (Figure 4a, Additional file 1: Table S4). However, the probability did not reach 1 until at 65 years. When we restricted the sample to those females which had the possibility to continue reproducing at old age (living beyond age 40, see methods for definition), the probability of entering a non-reproductive state exceeded 0.99 almost ten years later, at age 56 (Figure 4a, $n = 457$ females; 10,559 observations), and again did not reach 1 at any age (age$^3$: $F_{1,10538} = 26.81$, $P < 0.0001$, $\beta = 0.00011 \pm 0.000022$). In human females, the cubic term of age had also a significant effect on the probability of entering a non-reproductive state but the probability increased much more abruptly compared to elephants, from low values at age 35 (0.25), reaching 0.95 at age 45, 0.99 at age 47 and finally 1 at age 56 (Figure 4b, Additional file 1: Table S5). When we restricted the sample to those females living until old age (42 years) and having the potential to reproduce at old age, the probability of entering the non-reproductive state exceeded 0.99 at age 48 and 1 at age 56 (age$^3$: $F_{1,149000} = 257.22$, $P < 0.0001$, $\beta = 0.00062 \pm 0.000039$), at the same ages as in the entire population (Figure 4b, $n = 4339$ females; 149,079 observations).

#### *Post-reproductive lifespan*

We then investigated how long females of each species survived beyond their last reproductive event. In the whole population of Asian elephants, the average length of female post-reproductive lifespan was 5.9 ± 7.0 years (range 0–38.7; Table 1). The mean length of post-reproductive lifespan for deceased females surviving beyond 40 (calculated as the age by which 75% of females had given birth to their last offspring) years was 11.3 ± 7.8 years (range 0.0-38.7). In comparison, the length of post-reproductive lifespan among all pre-industrial women was 22.4 ± 15.2 years (range 0.0-66.2; Table 1) and in women surviving beyond 42 years (calculated as the age by which 75% of females had given birth to their last offspring) 26.6 ± 13.2 years (range 0.0-66.2; Table 1). The difference between the species in post-reproductive lifespan is illustrated in Figure 1, with the gap between fertility and survival curves of ever reproducing females at old age much smaller in elephants than in women (Figure 1a,b).

#### *Post-reproductive representation*

Finally, in elephants ($n = 1040$), 12.8% (PrR = 0.128) of ever reproducing adult females in the population at any time point were post-reproductive and the post-reproductive phase covered one eighth of the adult lifespan. In women ($n = 5336$), 51.2% of adult females were post-reproductive (PrR = 0.512) and hence, the

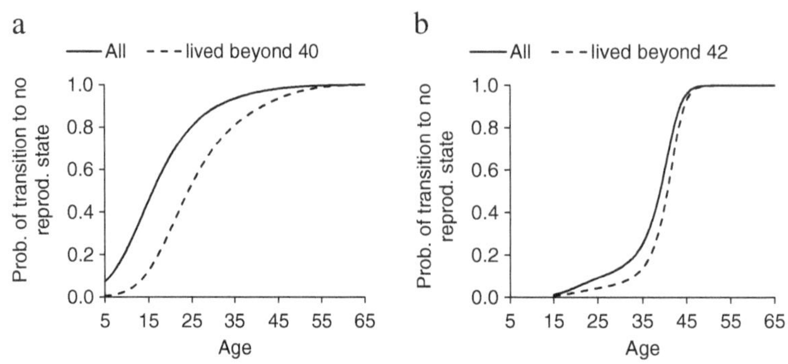


**Figure 4 Age-specific probability of entering a non-reproductive state. (a)** Asian elephants ($n = 16,369$ observations, 1019 females) and **(b)** women ($n = 157,039$ observations, 5176 women).

post-reproductive lifespan lasted half of female adult lifespan.

## Discussion

The ability for post-reproductive lifespan appears widespread in mammals [6], although only a small proportion of all females may exhibit such a trait [37]. Elephant lifespan represents the upper end recorded for terrestrial mammals along with humans, yet reliable information on their post-reproductive longevity or reproductive termination is lacking. We used several published methods to describe the end of fertility and the prevalence of post-reproductive lifespan in Asian elephants (e.g. Additional file 1, [17,36]) compared to humans. As in women, elephant fertility declined with age, but their pattern of reproductive cessation differed from women who terminated reproduction more abruptly and ~10 years earlier in relation to overall lifespan and had a markedly longer post-reproductive lifespan (Table 1, Figure 1). The percentage of adult females experiencing post-reproductive lifespan was 4-fold in humans as compared to elephants. These results add to the current controversy surrounding the prevalence and evolutionary origins of reproductive cessation and post-reproductive lifespan in species with different life-history and overall longevity.

We faced three difficulties whilst estimating the reproductive cessation and post-reproductive longevity in Asian elephants. First, as for most species, lack of direct data on ovulation frequency and hormonal profiles meant that we relied on age at last reproduction as a measure of functional menopause [33]. Second, because the majority of elephants were alive at the end of the study, the results may under-estimate reproductive and survival ability. Nevertheless, these estimates are to date the best available for any Asian elephant population and indeed avoid problems affecting most field studies on long-lived species where ages of all old cohorts are estimates rather than based on dates of birth. The exceptional record-keeping of the Myanma Timber Enterprise over a century provides accurate age data for all captive-born individuals, and our longevity data are thus likely more accurate than those published for wild elephants, where ages for most elephants are estimates. Moreover, if anything our finding that elephant females lack clear-cut menopause and generally maintain higher reproductive ability at old age than women is conservative, since the truncation of some elephant life-histories would produce the opposite result.

A third problem arises from the range of definitions available for post-reproductive lifespan, which all entail different problems and complicate generalisations. We tested several of such measures but overall, they all point to a lack of clear-cut menopause and wide-spread extended post-reproductive lifespan in Asian elephants. We also constructed a probabilistic model in both species to illustrate the age at which the probability of entering a non-reproductive state reached 1. In elephants, the probability increased steadily with age but did not reach 1 before age 65 years. Importantly, the probability increased more slowly in those elephants that survived to old age, which is in line with some individuals continuing to reproduce beyond 60 years. In contrast, in humans, the probability increased fast after age 35, reaching 1 at age 56 irrespective of the longevity of women. Such a difference between the species results from the restrictions that menopause poses on human women.

Using a recent alternative method for estimating post-reproductive survival [36], we found that 13% of ever reproducing adult females in the population at any time point were post-reproductive in elephants and 51% in humans. The post-reproductive lifespan covered only one eighth of adult lifespan in elephants in contrast to half of that in women. This method seems most appropriate for our study because it takes into account survival patterns of each species, is more comparable between populations and indicates the proportion of post-reproductive females in each population. Life expectancy at ages when 95% of fertility is realised gives a comparable value for post-reproductive lifespan, with 16.7 years after age 47 in elephants and 27.4 years after age 42 in humans. By any of the above method used, the length of post-reproductive lifespan in females living until old age (40/42, see methods) falls in elephants within a range of 11–17 years and in humans between 26–27 years. Consequently, the value for post-reproductive survival in elephants is greater than 10 years but less than 13% of females experience it, whereas the post-reproductive lifespan exceeds 25 years in humans and half of adult women experience it.

The PrR value of 0.13 calculated for Myanmar elephants is higher than those found in primates living in wild or semi-wild conditions (*Pan troglodytes* 0.018; *Papio hamadryas* 0.084; *Macaca fuscata* 0.055) (Figure 5) and corresponds to values found in many zoo populations of primates [36]. The value, 0.51, calculated for Finnish women correspond with patterns found in other historical or hunter-gatherer populations [36], where PrR ranges from 0.3 to 0.47 (in modern populations <0.76). The PrR-value for short-finned pilot whales (*Globicephala macrorhyncus*), known to also exhibit menopause, is 0.28 [36], which is over twice of that in Asian elephants, despite having shorter maximum longevity (Figure 5). The post-reproductive period of 14 years in short-finned pilot whales [43] is therefore potentially under stronger selection than the ~10 years in Asian elephants, because many more whales experience it compared to elephants.

Our understanding of elephant reproductive physiology and its changes with age has so far been largely based on zoo populations, which experience an absence of natural hormonal cycling [48] and high frequencies of obesity and

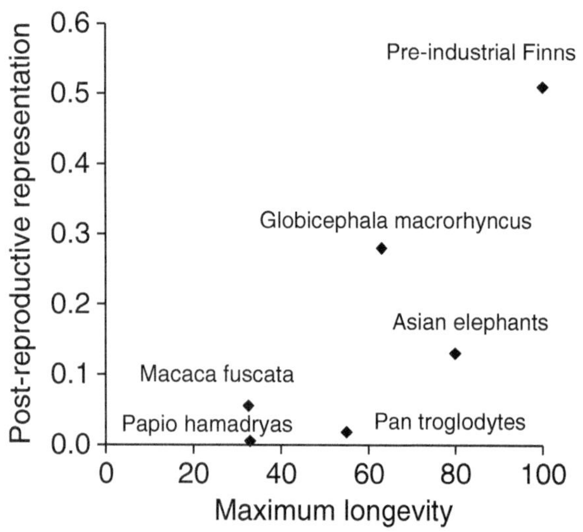


**Figure 5 Post-reproductive representation values plotted against maximum longevities in Asian elephants and humans compared to other long-lived mammals.** Post-reproductive representation values and maximum longevities in Asian elephants and humans from this study, other PrR-values according to Levitis and Lackey 2011 [36]. Maximum longevities in *Macaca fuscata* from [44], *Pan troglodytes* from [45], *Papio hamadryas* from [46] and *Globicephala macrorhyncus* from [47].

infertility attributed to reproductive track abnormalities, post-parturient problems and accelerated follicular loss arising from repetitive and continuous non-fertile oestrous cycles [28,29,48]. Such zoo populations also exhibit higher mortality rates (Asian and African zoo elephants: median lifespan 19 and 17 years, respectively) than those reported for either wild African elephants (median lifespan 56 years) or for the Myanmar working Asian elephants (median lifespan 42 years) [27]. The Myanmar timber elephant dataset provides detailed life-history data of known individuals that live in their natural habitat and breed with wild conspecifics. Nevertheless, workload may reduce their fertility as compared to wild counterparts. Importantly, however, workload may only reduce reproductive rate of prime-aged but not old females, given females enter workforce at 17 years and all females >54 years are retired from work, but their logbooks are maintained until death. Therefore, the potentially lower reproductive rate of the timber elephants should not prevent us from observing continued reproduction at old age (or menopause, should this exist).

One of the few other studies available on Asian elephants on a captive Indian population reported that females aged 50–55 years reproduced as frequently as the younger cows, although a decline occurred beyond this age [20]. Also, the African elephants in Amboseli showed a drop after 50 years despite that the majority of females estimated to be over 50 years have been seen in oestrus and several reproduced close to death, the eldest mother being potentially over 60 years [21,22]. The percentage of reproductively active females (pregnant and/or lactating) declined after age 50 also in Kruger National Park of South Africa, but females were capable of producing calves until close to an estimated age of 60 years [31,32]. These findings seem to mirror our results of fertility patterns in Asian elephants. We did not find any previously published estimates of post-reproductive period for either African or Asian elephants other than claims that female elephants can have a significant post-reproductive lifespan [6,49], and our study may thus provide the first quantitative data on this.

Overall, no comparable data exists to date on the fertility and survival patterns of very long-lived mammals with a similar lifespan to humans, except for two species of whales, which are known to have abrupt termination of reproduction (menopause) in relation to their long lifespan. Elephants seem to be clearly different from these species. For example, in killer whales, the fertility pattern with age resembles human pattern, peaking around age 20 and dropping sharply close to zero around age 50 [14]. Also the survival of killer whales at old age is remarkably similar to humans as >20% of females survive beyond 70 years, in line with 27% in humans (of all females born into the population). That inter-birth intervals decreased with female age in Asian elephants, whereas they increase in humans and killer whales [50] as well as in short-finned pilot whales [43] suggests that the elephant reproductive senescence patterns are ultimately different from those of the species exhibiting menopause.

## Conclusions

We conclude that whatever the method for estimating the duration and prevalence of post-reproductive survival, many older elephants are capable of reproducing despite fertility generally decreasing with age. The complete, irreversible cessation of reproductive capacity seen in some long-lived social animals such as pilot and killer whales [14,43] and humans around age 50 is not perceivable in elephants. Instead, similarly to many other animals including non-human primates [51], decline in elephant fertility occurs in parallel with declines in survivorship and overall health. Our study provides the first quantitative assessment of reproductive termination and post-reproductive lifespan in elephants. The results suggest that species with same longevities do not necessarily show similar ageing patterns [8], but instead, unusual kin-selected benefits may drive the ageing patterns both in humans [15,52] and in killer whales [16]. In these species, two types of indirect fitness benefit of reproductive cessation and post-reproductive longevity have been highlighted: adaptive benefits that old post-reproductive females bring to the success of their adult offspring [15,16] and benefits that menopause provides for avoiding reproductive competition between generations within multi-generational families [52]. Although elephants, too, have a family system where kin help in raising offspring is pronounced and females live in multi-generational family groups [20,21], our results show that in contrast to humans and killer whales, Asian elephant females retain the ability to reproduce until close to death. This suggests that different factors may have driven the evolution of their exceptional longevity.

## Methods

### Study populations

#### *Asian elephants*

The extensive demographic dataset available on semi-captive Asian elephants from Myanmar has been collated from elephant log-books and annual extraction reports archived and maintained by the Myanma Timber Enterprise. State ownership of thousands of elephants enables recording data of all registered elephants on: registration number and name; origin (wild-caught or captive-born); date and place of birth; mother's registration number and name; method, year and place of capture (if wild-captured); dates and identities of all calves born; date and cause of death or last known date alive. The log-books are maintained and updated by local veterinarians and extraction managers at least bi-monthly to check individual elephant health and working ability. Between-individual variation in workload or rest periods is limited by law: all state-owned elephants are subject to the same regulations set by central government for hours of work/week, working days/year, and tonnage to extract/elephant. The ages of captive-born elephants are known precisely from dates of birth, whereas wild-caught elephants are aged by comparing their height and body condition at capture with captive elephants of known age, as well as many other physical features varying with age in Asian elephants such as wrinkling, depigmentation and ear folding [53-56]. The extent of depigmentation (freckles) on trunk, face and temporal areas, and the degree of folding of the upper edge of the ear increase with age, while hairiness of the tail tuft and degree of corrugation or wrinkliness of the skin reduce with increasing age. The Myanmar elephant catchers and trainers take careful consideration of all physical features in estimating age of wild-caught elephants. The method is considered relatively accurate, and the error in these estimates is likely to be within one year for young animals (under 20), which form the majority (68%) of those captured from the wild [57].

The entire studbook currently includes 8759 elephants born and/or captured 1900–2000; data available for this study from 2000 onwards only includes updated calving/survival status information for 207 adolescent/adult elephants and 639 calves born 2000–2012. Of all 4742 females (of which many died before reaching reproductive age), 1463 reproduced by 2000/2012. We excluded wild-caught individuals captured before 1952 because only limited records were available prior then. We also excluded 18 individuals with erroneous age at first or last reproduction, lifespan or calving date. The remaining sample includes 1040 females (471 captive-born and 569 wild-born) of which 320 had already died before 2000/2012 whilst 720 were still alive in 2000/2012 (Table 1). These females delivered 2727 calves by 2000/2012. The maximum lifespan in the sample for captive-born females is 65 years, died in 2006. The oldest wild-born female died in 1995 at 79.6 years. These elephants come from 31 timber extraction areas within ten regions in Myanmar: Ayeyarwaddy, Bago, Chin, Kachin, Magway, Mandalay, Rakhine, Sagaing, Shan and Tanin.

#### *Historical Finnish people*

To compare the patterns in Asian elephants with those in human populations living without access to modern medical care and contraceptive methods, we calculated age-specific survival and reproductive patterns (see Table 1) for pre-industrial Finnish women whose life-history was collected using historical population registers. Although the used human population was an agriculture rather than hunter-gathering based economy, it serves as a suitable comparison to the elephant patterns, given that the obtained values reflect those published for other traditional human populations [36] and the large sample size available for this population allowed enough old individuals being included in the study. The Lutheran Church was obliged by law to submit census registers of all births, marriages and deaths in each parish since 1749 [58]. Our data contain 5336 ever reproduced women from eight farming/fishing communities (Hiittinen, Ikaalinen, Jaakkima, Kustavi, Pulkkila, Rautu, Rymättylä, Tyrvää) with complete life-history records ($n = 27{,}770$ offspring). These women were born 1595–1849, and the study period therefore ended before industrialism, healthcare improvements and modern birth-control methods influenced survival and reproduction in Finland [59]. Information on husband occupation (when children were born, e.g. landowner, craftsman, tenant farmer, servant etc.) allowed categorisation of the family socio-economic status (wealthy, average or poor), a correlate of resources available [15].

### Statistical analyses

We used the following measures to investigate reproductive output and survival in Asian elephants in comparison to humans: (a) age-specific changes in fertility, survival and birth intervals, and (b) reproductive cessation and subsequent post-reproductive survival. Reproductive cessation was measured as an age when reproduction ceased (last birth), the probability of entering a non-reproductive state with age and the following methods to estimate post-reproductive survival: length of post-reproductive lifespan, proportion of females terminating reproduction ([17], see Additional file 1) and post-reproductive representation. Table 1 shows descriptive statistics of both datasets. All analyses were conducted using SAS (SAS Institute, release 9.3, 2002–2010). All biologically interesting interactions were tested, but omitted if not statistically significant and if the model fit was not improved according to Akaike Information Criterion.

### Age-specific changes in fertility, survival and birth intervals

#### *Age-specific fertility and survival*

We included all reproductive females for quantification of population-level age-specific fertility ($n = 1040$ for elephants and $n = 5336$ for humans) and all females born into the population ($n = 3037$ for elephants and $n = 8943$ for humans) for survival functions. We calculated age-specific fertility as the total number of offspring born during each year divided by the number of ever reproducing females alive at the end of each year. For age-specific fertility function we included females captured from the wild the first time at their capture age and thus the total sample size was adjusted to the actual number of females in a population at each age. For survival functions, we included those females still alive at the end of the study and those females disappearing before death as censored observations (right censoring) and captured females entering the population at the age of capture (left truncation), whereas captive born females entered the analysis at birth. We formally compared the differences between elephants and humans in their probability of reproducing and surviving after age 40 with time event analysis (also known as event history analysis) [60], including the species, female age and interactions with these two. Our dataset included all ages for all individuals which survived beyond age 40, from age 40 (for those who were alive) until the age of 65 years (latest age at last reproduction in elephants). For each age we coded whether the individual reproduced or not (1/0) and whether the individual was still alive or dead. The comparison of the probability of reproducing between the species included only ages when each individual was alive, and contained 97,532 observations from 4989 individuals ($n = 457$ elephants; $n = 4532$ women). The comparison of the probability of surviving between the species included 119,845 observations from 4989 individuals ($n = 457$ elephants; $n = 4532$ women).

#### *Inter-birth intervals*

For both species, we calculated inter-birth interval length following each birth across a female's life, and maternal age at each birth in years (using information on year, month and date of birth). Lastborn and offspring without siblings were excluded from this analysis because their mothers did not have a following birth-interval, and stillbirths were excluded to focus on inter-birth intervals which are within the normal range. Elephant pregnancy lasts around 22 months and birth intervals <1.5 and >17.3 years (99% quartile of birth interval length) were therefore excluded as erroneous or outliers. Similarly, in humans, birth intervals <0.75 (9 months) and >9.03 (99% quartile) were excluded, resulting in sample sizes of 1480 births for 630 female elephants and 21,033 births for 4435 women. We investigated the effect of maternal age on inter-birth intervals using Glimmix procedure in SAS with gamma distribution and log-link. The elephant analysis included as fixed effects female age and age$^2$, living area (10 levels), origin (wild vs. captive-born), lifespan (exact or censored), age at last reproduction, whether the female was censored or not, as well as offspring sex, birthorder (firstborn vs. later-born), and birth cohort (1960–1969, 1970–1979, 1980–1989, 1990–1999, ≥2000) to control for maternal quality [61] and finally, female identity as a random term to adjust for inclusion of several birth-intervals from same mothers. The human analysis included as fixed effects female age and age$^2$, living parish (8 levels), socioeconomic status (rich, average, poor), and lifespan (exact or censored), age at last reproduction, whether the female was censored or not, as well as offspring sex, birthorder (firstborn vs. later-born), and birth cohort (<1760, 1760–1799, 1800–1829, 1830–1859, 1860–1899), and female identity as a random term. We also conducted analyses including the terms above and a categorical variable on whether the offspring survived to age 1 or not to control for potential variation in inter-birth intervals because of replacement births following early offspring death (sample sizes in these analyses: $n = 1222$ for elephants, $n = 20,656$ for humans). Final models were determined using backward elimination technique and AIC criterion for model fit to the data.

### Reproductive cessation and subsequent post-reproductive survival

#### *Age at last reproduction*

We used both all reproduced individuals ($n = 1040$ for elephants and $n = 5336$ for humans) and only those females survived until 'old age' ($n = 457$ for elephants and $n = 4427$ for humans) in both species to determine how the species-specific survival pattern modifies the age at last reproduction. We defined 'old age' as the age by which 75% of females had given birth to their last offspring (40 years in elephants, 42 in women). The females living until 'old age' are especially interesting because they can show the existence of menopause or, alternative, they can continue to reproduce at old age. We also analysed with the total sample (n = 6376, including 1040 elephants and 5336 humans) whether the species differed in their age at last reproduction with Glimmix procedure in SAS with negative binomial distribution and log-link. The analysis included as fixed effects the species (elephants vs. humans), lifespan (exact or censored), whether the female was censored or not and interactions between these variables. We tested with Levene's test whether the variances of age at last reproduction were equal between the species.

#### *Transition to a non-reproductive state with age*

For both elephants and humans, we investigated at which age the probability of stopping reproducing, i.e. of entering a non-reproductive state, reached 1. We used discrete time event analysis and thus a dataset including all ages for each individual from the earliest age at first reproduction (5 years in elephants, 15 in humans) until the age of 65 years (latest age at last reproduction in elephants). Before and at the age at last reproduction, the reproductive state was coded as '0' (still reproductive), and after the last reproduction, the reproductive state was coded as '1' (non-reproductive). The time since previous reproduction (in years) at each age was calculated based on the exact birth dates of each offspring. For each age, survival was coded as survived versus died (or already dead during previous years) or missing (when the individual was censored at an earlier age). In the analyses, we only included the ages when the individual was coded as alive. We performed the analyses with pseudo likelihood estimation technique in GLMM (GLIMMIX procedure in SAS) with binary error structure and logit link function to determine the effect of age on the probability of entering a non-reproductive state in both species. First we conducted analyses on the entire populations and second we included only those females surviving until old age (40 in elephants/ 42 in humans, see the definition above) to investigate whether the results change when we include only those individuals who have the potential to reproduce at old age.

The elephant analysis was carried out on 16,369 data points between ages 5–65 ($n = 1019$ females; excluding those females who produced only one calf and died within the same year). The model included the effects of age, age$^2$, age$^3$, living area (10 levels), whether the individual was censored or had an exact death date, origin (captive born vs. wild caught), birth cohort (1900–1949, 1950–1959, 1960–1969, 1970–1979, 1980–1993), lifespan (censored or exact) and time since previous birth (Additional file 1: Table S4). The human analysis was carried out on 157,039 data points between ages 15–65 ($n = 5176$ females; excluding those who produced only one child and died within the same year and those having missing information on socioeconomic status). The model included age, age$^2$, age$^3$, living parish (8 levels), whether the individual was censored or had an exact death date, socioeconomic status (rich, average, poor), lifespan (exact or censored), birth cohort (<1786, 1786–1813, 1813–1833, 1833–1849) and time since previous birth (Additional file 1: Table S5).

#### *Post-reproductive lifespan*

We calculated the mean length of post-reproductive lifespan by subtracting the age at last birth from the age at death/censoring [34,35] for all reproduced females, as well as for only those females that survived until old age (see above) and consequently with an opportunity to reproduce also at an advanced age.

#### *Post-reproductive representation*

Finally, we calculated post-reproductive representation (PrR) according to [36] to describe the proportion of adult years lived as post-reproductive and proportion of post-reproductive females. This value was calculated for the population of ever reproducing females in both species. The measure is independent of differences in infant and juvenile mortality between species and populations but dependent upon survival through the reproductive and post-reproductive periods. PrR is calculated based on life-table methods using functions evaluated at two ages: Age B, the beginning of reproductive lifespan or adulthood and Age M, the end of fecund lifespan. Ages B and M correspond to ages at which 5% and 95% of all births in a population have been realized independently of mortality, respectively. In elephants 5% and 95% fecundity is realized at ages 15 and 47, and in historical Finns at ages 21 and 42. The simplest population measure of survival beyond reproductive cessation is $e_M$, remaining life expectancy of those individuals who have survived to age M. In reproductive elephants $e_M = 16.9$ years and in humans $e_M = 27.4$ years. We also calculated values for life-expectancy at age when 5% of fecundity has been realized; in elephants $e_B = 39.5$ and in humans $e_B = 44.8$. Finally, we calculated $l_B$ and $l_M$ corresponding to numbers of individuals surviving to ages when 5% and 95% of fecundity has been realised, respectively. In elephants $l_B$ is 796 and in humans 5322, whereas $l_M$ for elephants is 238 and for humans 4463. We used these values to quantify the proportion of post-reproductive adults according to the following formula:

$$\mathrm{PrR} = \frac{l_M}{l_B} * \frac{e_M}{e_B}$$

## Additional file

**Additional file 1:** Includes the methods, results and discussion for calculating reproductive termination [17], the results of the GLMM of the factors affecting on the length of the following inter-birth interval length in Tables S2 and S3, and the results of the discrete time event models of the factors affecting on the probability of entering a non-reproductive state with age in Tables S4 and S5 in elephants and historical humans.

#### Competing interests

The authors declare that they have no competing interests.

### Authors' contributions

KUM collected the elephant data set, VL collected the human data set, ML and VL conceived and designed the paper. ML performed the analyses. ML, KUM and VL wrote the paper. All authors read and approved the final manuscript.

### Acknowledgements

We thank the Myanma Timber Enterprise, the Karelian database, Lasse Iso-Iivari, Kimmo Pokkinen, Aino Siitonen, Veli-Pekka Toropainen and Timo Verho for data collection and ERC (VL), Academy of Finland (ML), NERC (KUM, VL), Royal Society of London (VL), and Rufford Foundation (KUM) for funding. We also thank Hannah Mumby and the rest of the Myanmar Timber Elephant Project for help and support, and Simon Verhulst and anonymous referees for insightful comments.

### Author details

[1]Section of Ecology, Department of Biology, University of Turku, Turku FIN-20014, Finland. [2]Department of Animal and Plant Sciences, University of Sheffield, Sheffield S10 2TN, UK.

### References

1. Williams GC: **Pleiotropy, natural selection, and the evolution of senescence.** *Evolution* 1957, **11**:398–411.
2. Hamilton WD: **The moulding of senescence by natural selection.** *J Theor Biol* 1966, **12**:12–45.
3. Kirkwood TBL: **Evolution of ageing.** *Nature* 1977, **270**:301–304.
4. Medawar PB: *An unsolved problem of biology.* London: HK Lewis; 1952.
5. Bouwhuis S, Choquet R, Sheldon BC, Verhulst S: **The forms and fitness cost of senescence: age-specific recapture, survival, reproduction, and reproductive value in a wild bird population.** *Am Nat* 2012, **179**:E15–E27.
6. Cohen AA: **Female post-reproductive lifespan: a general mammalian trait.** *Biol Rev* 2004, **79**:733–750.
7. Holmes DJ, Ottinger MA: **Birds as long-lived animal models for the study of aging.** *Exp Gerontol* 2003, **38**:1365–1375.
8. Jones OR, Scheuerlein A, Salguero-Gómez R, Camarda CG, Schaible R, Casper BB, Dahlgren JP, Ehrlén J, Garcia MB, Menges ES, Quintana-Ascencio PF, Caswell H, Baudisch A, Vaupel JW: **Diversity of ageing across the tree of life.** *Nature* 2014, **505**:169–173.
9. Reznick D, Bryant M, Holmes D: **The evolution of senescence and post-reproductive lifespan in guppies (*Poecilia reticulata*).** *PLoS Biol* 2006, **4**:136–143.
10. Kidd NAC, Tozer DJ: **On the significance of post-reproductive life in aphids.** *Ecol Entomol* 1985, **10**:357–359.
11. Chen J, Senturk D, Wang JL, Müller HG, Carey JR, Caswell H, Caswell-Chen EP: **A demographic analysis of the fitness cost of extended longevity in *Caenorhabtidis elegans*.** *J Gerontol A Biol Sci Med Sci* 2007, **62**:126.
12. Peccei JS: **Menopause: adaptation or epiphenomenon?** *Evol Anthropol* 2001, **10**:43–57.
13. Marsh H, Kasuya T: **Evidence for reproductive senescence in female cetaceans.** *Rep Int Whal Commn* 1986, **8**(Special Issue):57–74.
14. Ward EJ, Parsons K, Holmes EE, Balcomb KC, Ford JKB: **The role of menopause and reproductive senescence in a long-lived social mammal.** *Front Zool* 2009, **6**:4.
15. Lahdenperä M, Lummaa V, Helle S, Tremblay M, Russell AF: **Fitness benefits of prolonged post-reproductive lifespan in women.** *Nature* 2004, **428**:178–181.
16. Foster EA, Franks DW, Mazzi S, Darden SK, Balcomb KC, Ford JKB, Croft DP: **Adaptive prolonged postreproductive life span in killer whales.** *Science* 2012, **337**:1313.
17. Caro TM, Sellen DW, Parish A, Frank R, Brown DM, Voland E, Borgerhoff Mulder M: **Termination of reproduction in nonhuman and human female primates.** *Int J Primatol* 1995, **16**(2):205–220.
18. Fedigan LM, Pavelka MSM: **Reproductive cessation in female primates. Comparisons of Japanese macaques and humans.** In *Primates in Perspective.* Edited by Campbell C, Fuentes A, MacKinnon KC, Panger M, Bearder S. USA: Oxford University Press; 2006:437–447.
19. Thompson ME, Jones JH, Pusey AE, Brewer-Marsden S, Goodall J, Marsden D, Matsuzawa T, Nishida T, Reynolds V, Sugiyama Y, Wrangham RW: **Aging and fertility patterns in wild chimpanzees provide insights into the evolution of menopause.** *Curr Biol* 2007, **17**:2150–2156.
20. Sukumar R: *The living elephants: evolutionary ecology, behaviour and conservation.* Oxford: Oxford University Press; 2003.
21. Moss CJ, Croze H, Lee PC: *The Amboseli elephants: A long-term perspectives on a long-lived mammal.* Chicago: The University of Chicago Press; 2011.
22. Moss CJ: **The demography of an African elephant (*Loxodonta africana*) population in Amboseli, Kenya.** *J Zool* 2001, **255**:145–156.
23. Laws RM, Parker ISC, Johnstone RCB: *Elephants and their habitats: The ecology of elephants in North Bunyoro, Uganda.* Oxford: Clarendon Press; 1975.
24. Lee PC: **Reproduction.** In *The illustrated encyclopedia of elephants.* Edited by Eltringham SK. London: Salamander Books Ltd; 1991:64–77.
25. Kuhle BX: **An evolutionary perspective on the origin and ontogeny of menopause.** *Maturitas* 2007, **57**:329–337.
26. Clubb R, Mason G: *A review of the welfare of zoo elephants in Europe.* Horsham, UK: RSPCA; 2002:203–205.
27. Clubb R, Rowcliffe M, Lee P, Mar KU, Moss C, Mason GJ: **Compromised survivorship in zoo elephants.** *Science* 2008, **322**:1649.
28. Hildebrandt T, Goritz F, Pratt NC, Brown JL, Montali RJ, Schmitt DL, Fritsch G, Hermes R: **Ultrasonagraphy of the urogenital tract in elephants (*Loxodonta africana* and *Elephas maximus*): An important tool for assessing female reproductive function.** *Zoo Biol* 2000, **19**:321–333.
29. Hermes R, Hildebrandt T, Goritz F: **Reproductive problems directly attributable to long-term captivity–asymmetric reproductive aging.** *Anim Reprod Sci* 2004, **82–83**:49–60.
30. Whitehouse AM, Hall-Martina AJ: **Elephants in Addo Elephant National Park, South Africa: reconstruction of the population's history.** *Oryx* 2000, **34**(1):46–55.
31. Smuts GL: **Reproduction and population characteristics of elephants in the Kruger National Park.** *J S Afr Wildl Mgmt Ass* 1975, **5**:1–10.
32. Freeman EW, Whyte I, Brown JL: **Reproductive evaluation of elephants culled in Kruger National Park, South Africa between 1975 and 1995.** *Afr J Ecol* 2008, **47**:192–201.
33. Atsalis S, Videan E: **Reproductive aging in captive and wild common chimpanzees: factors influencing the rate of follicular depletion.** *Am J Primatol* 2009, **71**:271–282.
34. Sommer V, Srivastava A, Borries C: **Cycles, sexuality and conception in free-ranging langurs (*Presbytis entellus*).** *Am J Primatol* 1992, **28**:1–27.
35. Paul A, Kuester J, Podzuweit D: **Reproductive senescence and terminal investment in female Barbary macaques (*Macaca sylvanus*) at Salem.** *Int J Primatol* 1993, **14**:105–124.
36. Levitis DA, Lackey LB: **A measure for describing and comparing postreproductive life span as a population trait.** *Methods Ecol Evol* 2011, **2**:446–453.
37. Levitis DA, Burger O, Lackey LA: **The human post-fertile lifespan in comparative evolutionary context.** *Evol Anthropol* 2013, **22**:66–79.
38. Hayward AD, Mar KU, Lahdenperä M, Lummaa V: **Early reproductive investment, senescence and lifetime reproductive success in female Asian elephants.** *J Evol Biol* 2014, **27**(4):772–783.
39. Mar KU, Lahdenperä M, Lummaa V: **Causes and correlates of calf mortality in captive Asian elephants.** *PLoS ONE* 2012, **7**(3):e32335.
40. Mumby HS, Courtiol A, Mar KU, Lummaa V: **Climatic variation and age-specific survival in Asian elephants from Myanmar.** *Ecology* 2013, **94**:1131–1141.
41. Mumby HS, Courtiol A, Mar KU, Lummaa V: **Birth seasonality and calf mortality in a large population of Asian elephants.** *Ecol Evol* 2013, **11**:3794–3803.
42. Robinson MR, Mar KU, Lummaa V: **Senescence and age-specific trade-offs between reproduction and survival in female Asian elephants.** *Ecol Lett* 2012, **15**:260–266.
43. Marsh H, Kasuya T: **Changes in the ovaries of the short-finned pilot whale, *Globicephala macrorhyncus*, with age and reproductive activity.** *Rep Int Whal Commn* 1984, **6**(Special issue):311–335.
44. Pavelka MSM, Fedigan LM: **Reproductive termination in female japanese monkeys: a comparative life history perspective.** *Am J Phys Anthropol* 1999, **109**:455–464.
45. Hill K, Boesch C, Goodall J, Pusey A, Williams J, Wrangham R: **Mortality rates among wild chimpanzees.** *J Hum Evol* 2001, **40**:437–450.
46. Bronikowski AM, Alberts SC, Altmann J, Packer C, Carey KD, Tatar M: **The aging baboon: Comparative demography in a non-human primate.** *Proc Natl Acad Sci U S A* 2002, **99**:9591–9595.
47. Kasuya T, Marsh H: **Life history and reproductive biology of the short-finned pilot whale, Globicephala macrorhynchus, off the Pacific coast of Japan.** *Rep Int Whaling Comm* 1984, **10**(Special Issue):259–310.

48. Hermes R, Saragusty J, Schaftenaar W, Goritz F, Schmitt DL, Hildebrandt TB: **Obstetrics in elephants.** *Theriogenology* 2008, **70:**131–144.
49. McComb K, Shannon G, Durant SM, Sayiale K, Slotow R, Poole J, Moss C: **Leadership in elephants: the adaptive value of age.** *Proc Biol Sci* 2011, **278:**3270–3276.
50. Olesiuk PF, Bigg MA, Ellis GM: **Life history and population dynamics of resident killer whales (*Ornicus orca*) in the coastal waters of British Columbia and Washington State.** *Rep Int Whal Commn* 1990, **12**(Special issue):209–243.
51. Alberts S, Altmann J, Brockman DK, Cords M, Fedigan LM, Pusey A, Stoinski TS, Strier KB, Morris WF, Bronikowski AM: **Reproductive aging patterns in primates reveal that humans are distinct.** *Proc Natl Acad Sci U S A* 2013, **110:**13440–13445.
52. Lahdenperä M, Gillespie DOS, Lummaa V, Russell AF: **Severe intergenerational reproductive conflict and the evolution of menopause.** *Ecol Lett* 2012, **15**(11):1283–1290.
53. Mar KU: *The demography and life-history strategies of timber elephants in Myanmar.* UK: PhD thesis. University College London; 2007.
54. Ilangakoon AD: **A preliminary study of the captive elephants in Sri Lanka.** *Gajah* 1993, **11:**29–42.
55. Kurt F, Kumarasinghe JC: **Remarks on body growth and phenotypes in Asian elephant *Elephas maximus*.** *Acta Theriol Suppl* 1998, **5:**135–153.
56. Goswami VR, Madhusudan MD, Karanth KU: **Application of photographic capture-recapture modelling to estimate demographic parameters for male Asian elephants.** *Anim Conserv* 2007, **10:**153–163.
57. Clubb R, Rowcliffe M, Lee P, Mar KU, Moss C, Mason GJ: **Fecundity and population viability in female zoo elephants: problems and possible solutions.** *Anim Welf* 2009, **18:**237–247.
58. Luther G: *Suomen tilastotoimen historia vuoteen 1970.* Helsinki: WSOY; 1993.
59. Soininen AM: *Old traditional agriculture in Finland in the 18$^{th}$ and 19$^{th}$ centuries.* Forssa: Forssan kirjapaino Oy; 1974.
60. Singer J, Willett JB: *Applied longitudinal data analysis: Modeling change and event occurrence.* Oxford: Oxford University Press; 2003.
61. van de Pol M, Verhulst S: **Age-dependent traits: a new statistical model to separate within- and between- individual effects.** *Am Nat* 2006, **167:**764–771.

# 5

# The role of juvenile hormone in dominance behavior, reproduction and cuticular pheromone signaling in the caste-flexible epiponine wasp, *Synoeca surinama*

Hans C Kelstrup[1,4*], Klaus Hartfelder[2], Fabio S Nascimento[3] and Lynn M Riddiford[1]

## Abstract

**Background:** The popular view on insect sociality is that of a harmonious division of labor among two morphologically distinct and functionally non-overlapping castes. But this is a highly derived state and not a prerequisite for a functional society. Rather, caste-flexibility is a central feature in many eusocial wasps, where adult females have the potential to become queens or workers, depending on the social environment. In non-swarming paper wasps (e.g., *Polistes*), prospective queens fight one another to assert their dominance, with losers becoming workers if they remain on the nest. This aggression is fueled by juvenile hormone (JH) and ecdysteroids, major factors involved in caste differentiation in most eusocial insects. We tested whether these hormones have conserved aggression-promoting functions in *Synoeca surinama*, a caste-flexible swarm-founding wasp (Epiponini) where reproductive competition is high and aggressive displays are common.

**Results:** We observed the behavioral interactions of *S. surinama* females in field nests before and after we had removed the egg-laying queen(s). We measured the ovarian reproductive status, hemolymph JH and ecdysteroid titers, ovarian ecdysteroid content, and analyzed the cuticular hydrocarbon (CHC) composition of females engaged in competitive interactions in both queenright and queenless contexts. These data, in combination with hormone manipulation experiments, revealed that neither JH nor ecdysteroids are necessary for the expression of dominance behaviors in *S. surinama*. Instead, we show that JH likely functions as a gonadotropin and directly modifies the cuticular hydrocarbon blend of young workers to match that of a reproductive. Hemolymph ecdysteroids, in contrast, are not different between queens and workers despite great differences in ovarian ecdysteroid content.

**Conclusions:** The endocrine profile of *S. surinama* shows surprising differences from those of other caste-flexible wasps, although a rise in JH titers in replacement queens is a common theme. Extensive remodeling of hormone functions is also evident in the highly eusocial bees, which has been attributed to the evolution of morphologically defined castes. Our results show that hormones which regulate caste-plasticity can lose these roles even while caste-plasticity is preserved.

**Keywords:** Cuticular hydrocarbons, Ecdysteroids, Endocrine, Epiponini, Juvenile hormone, Swarm founding, Wasps

* Correspondence: kelstrup@sun.ac.za
[1]Janelia Research Campus, Howard Hughes Medical Institute, Ashburn, VA 20147, USA
[4]Present address: Department of Botany and Zoology, Stellenbosch University, Private Bag XI, Matieland 7602, South Africa
Full list of author information is available at the end of the article

## Background

Division of labor, which is the fundamental condition for the ecological success of social insects, reaches its apex with the evolution of morphologically specialized queens and workers. These distinct phenotypes separate the highly advanced eusocial species from the 'primitively eusocial' ones, such as *Polistes* paper wasps, where caste is not determined until the adult stage. Yet caste totipotency is not always lost as a consequence of increased social specialization. For example, many of the Neotropical swarm founding wasps (Epiponini) – which are permanently social and polygynous, i.e. have multiple queens in their nests – have retained extreme caste flexibility [1-3]. As has been shown in *Metapolybia aztecoides* and *Synoeca surinama*, the social context in which a young adult female finds herself is the overwhelming factor in determining her caste fate [1-3]. So long as there are queens present and the colony does not split (i.e. form a reproductive swarm), young females will almost certainly become workers, a fate assured by direct suppression by older workers [2-5]. If a colony becomes queenless, the youngest cohort of females on the nest – which would have otherwise become workers – become the next group of queens [3]. The high frequency of sporadic nest destruction (e.g., by army ants or birds) or abandonment has placed a premium on being able to expeditiously adjust the ratio of egg layers to workers to avoid colony extinction [1-3,5,6]. Thus, caste identity is provisional, since queens may transform into workers when worker numbers are low, and young workers (e.g. builders) can become queens if queens disappear [1-3].

The question of caste determination in wasps has been best studied in temperate climate paper wasps of the genus *Polistes*, but these are not continuously social, as nests are independently founded by overwintered gynes (young queens that do not yet head a colony). These may either found a nest individually or as a group of females. In the latter case, a dominance hierarchy is established primarily through direct acts of aggression and differential oophagy, which typically results in a monogynic society [7-10]. Queens of some caste totipotent epiponines maintain their position through ritualized displays of dominance [11], namely abdomen-bending (or "bending"), a lateral hinging at the waist which bears a resemblance to a stinging posture [2,3] (Figure 1A). Queens display aggressively toward one another, often on the margins of the nest, but bending is also elicited by the approach of workers who engage in a ritualized act of their own: the spasmodic 'queen-dance', interpreted as an act of "arrested or inhibited aggression" [1] which, if released, can give way to attacks toward subordinate behaving queens [1-3]. Defeated queens, after passing through a period of idleness [1], may become workers until eventually only one queen remains (which has the effect of increasing relatedness within an otherwise polygynous society) [1,3,12]. Possible selective forces at work (e.g., mutualism, kin selection) in swarm-founding wasp societies have been discussed [1,3,13], but what remains to be investigated are the proximate mechanisms (i.e., behavioral and physiological modifiers) that underlie their extraordinary caste flexibility and performance.

In insects, the most frequently employed factor for orchestrating differentially expressed phenotypes (i.e., polyphenisms) is juvenile hormone (JH) [14], a sesquiterpenoid. In addition to its widely conserved roles in pre-imaginal development [15,16], JH is also a widespread gonadotropin [17,18]. In *Polistes* wasps, caste determination and maintenance is modulated by context- and nutrition-dependent effects of JH [19-22]. Among competing foundresses, JH drives ovarian growth and aggression [23-25], its hemolymph titers remain high in queens [21,26], and it is tightly linked to the production of chemical signatures of fertility in the cuticular hydrocarbon (CHC) profile [27]. Ecdysteroids, which in adult female insects are primarily produced by the ovaries and loaded into growing oocytes, may also be released into the hemolymph, as is the case in *Polistes.* They augment dominance in competing foundresses [24,25,28] but are not required to maintain dominance behaviors [29]. In *Polistes* females emerging in queenright conditions, methoprene (a stable JH mimic) treatments induced precocious worker behaviors, such as guarding or foraging [21,30]. Yet if these otherwise worker-destined females are removed from their environment and fed *ad libitum,* methoprene treatments resulted in enhanced oocyte development [19], and indeed, queenless workers that fight their way up the dominance hierarchy have especially high amounts of JH [26].

*Polistes* and epiponine wasps share a common caste-based ancestor [31] which was also likely to be caste-totipotent in the adult stage [32], setting the expectation for conserved functions for JH and the ecdysteroids in both caste transition and the subsequent modulation of caste physiology and behavior. Yet aside from the evidence that methoprene accelerates the onset of worker behaviors in the epiponine wasp *Polybia occidentalis* [33], the JH and ecdysteroid profiles in another caste-flexible wasp, *Polybia micans,* were remarkably different from that of *Polistes dominula* [34]. In *P. micans*, both JH and ecdysteroid titers were low in competing queens while JH increased only after all direct competitors were eliminated. Neither hormone thus appears to be essential for maintaining basic queen physiology, be it ovarian growth or chemical signaling. Despite these differences, JH titers were increased in potential reproductives following queen removal in both *P. micans* [34] and *P. dominula* [26], suggesting that JH fuels competitive ability and/or increases reproductive potential in challenging, unstable

**Figure 1 Nests of *Synoeca surinama*. (A)** Worker performing the "queen dance" and the queen responding with a pronounced bending display. **(B)** Aggressive defense response in a three compartment nest. The hole in the side was an observation window, and the opening on the bottom was incidental, leading to the emergence of colony defenders. **(C)** A single compartment nest with the envelope removed. Most nests were studied this way. **(D)** A massive 6 or 7 compartment nest discovered in Chapada Diamantina, Bahia, Brazil. Nest expansion proceeds upwards (yellow arrow). Red arrows indicate nest entrance/exit in B-D.

conditions, analogous to testosterone function in vertebrates [35,36].

The observed loss of hormone function with increasing level of sociality is paralleled in other lineages of Hymenoptera [14]. In the adults of primitively eusocial bumble bees, circulating JH and ecdysteroids are important for ovarian growth and reproductive behaviors [37-39]. Yet in swarm-founding, caste-dimorphic honey bees and stingless bees, these hormones have no obvious functions for reproduction or related behaviors [14,40]. Significant modification of JH function in eusocial Hymenoptera does not always accompany the switch to a swarming lifestyle, as JH has also relinquished gonadotropic functions in some ants [14,41]. The reason for convergent loss of hormone functions among these eusocial lineages remains a mystery, but the wasps are the best candidates for tracking the mechanisms of *how* these transitions occurred. Although there are no extant intermediate representatives between bumble bees and swarming-founding bees, there are many genera phylogenetically intermediate to *Polistes* and *Polybia* [31].

*Synoeca surinama*, whose social biology is very similar to the closely related *Metapolybia* [1], is a derived epiponine wasp, as is *Polybia* [31]. As opposed to *P. micans*, where queen selection involves, in part, an onslaught of physical attacks [34], in colonies of *Synoeca* queen succession is usually peaceful: even newly emerged females, which are relatively soft-bodied, can assume bending and gain acceptance as queens without a fight [1]. We show that the endocrinology of *S. surinama* is distinct from *Polistes* and *Polybia*, indicating that hormone functions in social wasps are liable to evolutionary change, even as caste-totipotency is conserved. In addition, the CHC components, including prospective queen signals, have diverged considerably between these two swarm founding wasps, possibly reflecting strong intragroup selection for the Epiponini as compared to some temperate wasps [42].

## Results

### Colony demographics and behavior

In total, 12 colonies were studied (Table 1). Queen number ranged from 1–32 while worker number ranged from approximately 65 to 300. All nests had produced at least one generation of females (i.e., were post-emergent colonies). Colony 6, the single pre-emergent colony studied, had founded their nest at least 4 weeks prior, indicating that all females were at least that old. Colonies 4–10 were first observed when in a queenright condition and, following queen removal, in a queenless state (Table 1).

In queenright nests where the envelope was removed but had not been manipulated otherwise (i.e. queen removal), no female was observed to work until at least two days after eclosion, and all females had begun building (envelope reconstruction) once they were four days old (N = 186 from 4 colonies). None of these females were observed to forage before building. Females that were not queens were never seen to lay an egg, but queens (i.e. egg-laying females that frequently clustered together, walked slowly and engaged in bending displays) were occasionally seen to participate in building activities. For instance, on three occasions fully established mated queens from colony 5 (Table 1) were observed to add a ball of pulp to the nest envelope.

When queens were removed, the next cohort of potential queens (i.e., females that engaged in bending displays) usually included the youngest adults present (idle females and workers) and/or those that had emerged after queen removal (Table 1). The number of females permitted to become queens is regulated in part by physical aggression from worker nestmates [1,3]. For example, those females which emerged subsequent to the cohort of accepted

**Table 1 Basic information for colonies used in this report**

| Colony | | | Original queens | | | Prospective queens | | Workers | Collection |
|---|---|---|---|---|---|---|---|---|---|
| ID | Colony phase | Envelope removed? | # | Ctlr. age[a] | Oocytes >70% LME[b] | ♀♀ present[c] | ♀♀ benders[d] | # | Month/Year |
| 1 | post-em' | No | n.d. | n.d. | n.d. | n.a. | n.a. | ~200 | Feb. 2010 |
| 2 | post-em' | No | n.d. | n.d. | n.d. | n.a. | n.a. | ~300 | Mar. 2010 |
| 4 | post-em' | Yes | 4 | 5 | 9 (5–12) | W | W | ~65 | July 2010 |
| 5 | post-em' | Yes | 32 | 1.5-4 | 3.4 (1–4) | P, W | P, W | ~300 | Nov. 2010 |
| 6 | pre-em' | Yes | 12 | 4.5-5 | 7.9 (4–10) | W | W | ~200 | May 2011 |
| 7 | post-em' | Yes | 7 | 4.5-5 | 6.4 (4–9) | P, I, W | P | ~90 | June 2011 |
| 8 | post-em' | Yes | 11 | 4.5-5 | 6.5 (5–8) | P, W | P | ~150 | Oct. 2011 |
| 9 | post-em' | Yes | 1 | 5 | 13 | P, I, W | P, I, W | ~70 | Nov. 2011 |
| 10 | post-em' | Yes | 4 | 4.5 | 14.8 (10–18) | P, I, W | P, I, W | ~250 | Nov. 2011 |
| 11 | post-em' | Yes | 1 | n.a. | n.d. | n.a. | n.a. | ~100 | Nov. 2011 |
| 12 | post-em' | Yes | 5 | n.a. | n.d. | n.a. | n.a. | ~200 | Nov. 2011 |

Colony phase indicates whether the nest had produced the first generation of adults [(pre- and post-emergent (em')]. Males were only seen in Colony 4. When all queens were removed from a nest, they could be replaced by: 1) females that were still pupae (P) at that time; 2) the youngest adults on the nest which were not observed to work and were idle (I); and/or 3) workers (W).

[a]Age of the wasps is represented as cuticular (Ctlr.) age which was estimated on a scale of 1–5 according to the darkness of the cuticle. Scale shown in Additional file 9: Figure S1.

[b]Average number of oocytes >70% length of mature egg (LME) is given along with the range within a given colony.

[c]♀♀ present indicates the types of females present when the queen was removed.

[d]♀♀ benders indicates which females became the next cohort of prospective queens.

*n.a.*, Not applicable; *n.d.*, Not determined.

benders were attacked (e.g., were mounted and mandibulated without causing ostensible injury) by older nest mates (2 colonies; see below). Even in queenright colonies with established queens, young idle females were observed to receive aggression when walking across the comb (N = 8 from 3 colonies). Finally, in a nest made monogynic by the removal of three queens (which presumably led to an abrupt drop in queen-derived signals in the colony), all young idle females which initiated bending (0–2 days old; N = 10; colony 10) were chased and attacked by the remaining queen and several workers.

### Ovarian development in young females from queenright and queenless colonies

Oocyte length (given as % length of mature egg; LME) was assessed in young queenright (worker-destined) and queenless (queen-destined) females from colonies 5 and 7 (Figure 2). In both colonies, the ovaries of queenright workers showed slight maturation over the first few days, before fully regressing as early as day 7 following eclosion (oocytes <8% LME were considered filamentous and not measurable). Oocytes of queenless, unsuppressed benders began to differ from those of queenright females around the adult age of days 2 or 3 (colony 5) (Figure 2A).

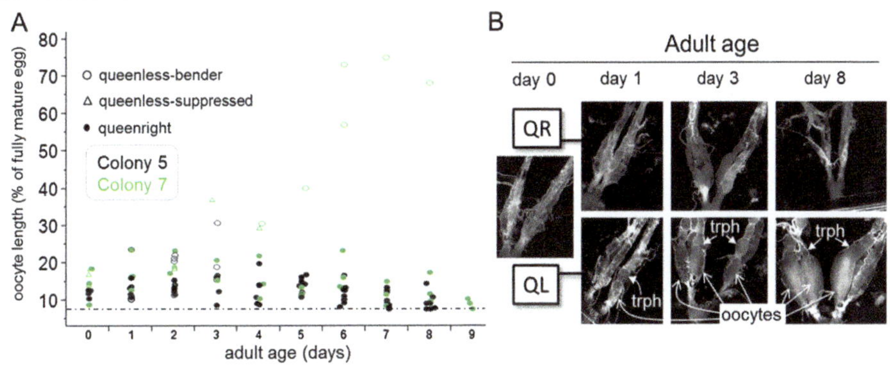


**Figure 2 Ovarian development in young females. (A)** Time course of ovarian maturation in young queenright (QR) and queenless (QL) females from two nests. The average of the longest two primary oocytes is indicated, so that each point represents an individual ovary. Dotted line indicates filamentous ovaries lacking measurable oocytes (<8% LME). **(B)** Typical ovaries of a newly emerged wasp (day 0) and those of older queenright and queenless females. Oocytes and trophocytes ("trph") are indicated.

## JH titers

### *Correlation with caste*

Queens consistently showed higher JH titers than other female types assayed alongside (Additional file 1: Figure S2 (blue axes)). In a pooled analysis including data from all queenright colony conditions, queens had significantly higher JH titers than newly emerged females (<24 hours since eclosion), 1–3 day old pre-workers and workers, workers ≥4 days old, and young attacked benders from colony 10 (see Figure 3A). To determine whether JH titers correlated with reproductive dominance, we recorded oviposition events over an 8-day period (8 h/day) in colony 5 when cells became vacant because of adult emergence in addition to egg removal by the observer. Queens (N = 28), some of which were bled after the observations concluded, were observed to oviposit 0–8 times. There was no difference in JH titer between queens with low (0–2 ovipostion events; N = 6) and high (6–8 oviposition events; N = 5) short-term fecundity (two tailed *t*-test, t = 0.49; DF = 9; P = 0.62; $Z_r = 0.076$) (Additional file 1: Figure S2C). Also, there was no significant correlation between ovary size (number of oocytes >70% LME) and JH titers among queens sampled (colony 4: N = 4, Pearson's $r = 0.2$, P = 0.80, Zr = 0.08; colony 5: N = 11, $r = -0.45$, P = 0.18, Zr = 0.28; colony 6: N = 12, $r = 0.57$, P = 0.08, Zr = 0.64; colony 7: N = 7, $r = -0.46$, P = 0.30, Zr = 0.11; colony 10: N = 4, $r = 0.8$, P = 0.2, Zr = 0.09).

Queenright females aged 1–3 adult days, which include pre-workers and young workers, had higher JH titers than workers over the age of 3 (Figure 3A), although this pattern was not seen in all colonies (e.g., Additional file 1: Figure S2). Among workers from colonies 4–10 where most of the envelope was removed (e.g., Figure 1C), JH titers were very low (Figure 3 and Additional file 1: Figure S2). In colonies where collected foragers had a 'cuticular age' score higher than builders (two-tailed *t*-test on cuticular age: colony 9: t = 3.00, DF = 6, P = 0.024; colony 10: t = 4.87, DF = 13, P = 0.0003) or did not (colony 4), JH titers were nearly identical between the builders and foragers (means within 1 pg/μl). Yet in colonies with only a small piece of the envelope removed (e.g., Figure 1B), a wider range of JH titers was observed, with nectar foragers having the highest of all (Additional file 2: Figure S3).

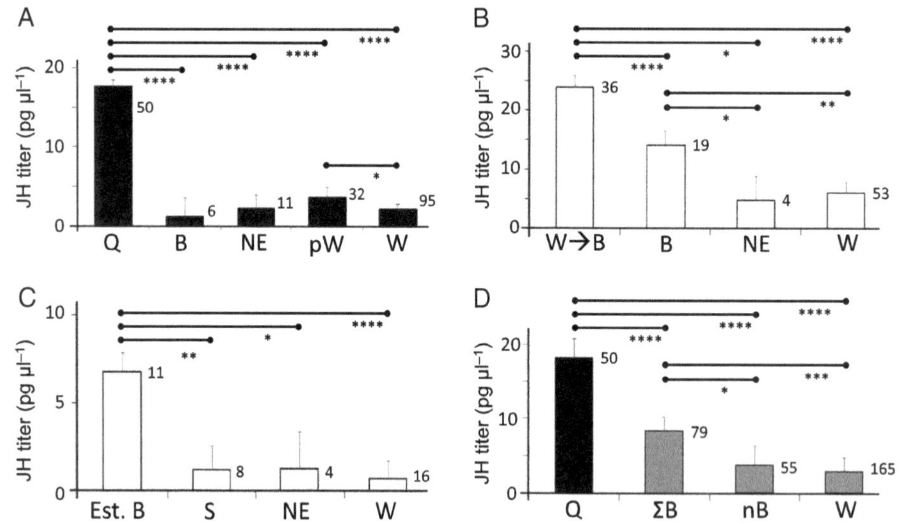


**Figure 3 JH titers according to female status.** A Mixed Model analysis of samples taken from colonies 4–10, using up to two fixed factors (Female Status and Colony Condition) and a random one (Colony). Black, white and gray bars indicate samples from queenright (QR) conditions, queenless (QL) conditions and a combination of samples from QR and QL conditions, respectively. Least Squared Means (±SE) are shown along with sample size. **(A)** In QR conditions, JH differed between the female groups (F4,186=63.29, P<0.0001). Queens (Q) had higher titers than newly emerged (NE) females (t=−8.34), new benders without history of working (B) (t=−6.52), 1–3 day old pre-working and working females (pW) (t=−9.85), and workers >3 days old (W) (t=−15.35)). Also, pW females had higher JH titers than workers (t=−2.14). **(B)** In QL conditions, 1–3 days after queen removal, JH levels differed between the female groups (F3,109=20.32, P<0.0001). Workers that transitioned into benders (W→B) had higher JH titers than NE females (t=−4.09), bending females which were never observed to work (t=−2.61), and workers (t=−7.53). The latter bending females also had higher JH titer than NE females (t=−2.13) and workers (t=−2.75). **(C)** In QL conditions, 7–8 days after queen removal, JH differed between the female groups (F3,36=5.73, P=0.003). Established benders (Est B) had higher JH titers than all female groups (vs. NE females (t=−2.42); vs. suppressed (S) females (i.e., attacked benders and huddlers) (t=−3.23); vs. workers (t=−3.85). **(D)** When QR and QL groups were pooled, with females classified as either queens, non-queen benders of types (ΣB), 0–3 day old non-benders (nB) and workers, JH differed according to Female Status (F3,341=65.83, P<0.0001), Colony Condition (F1,341=16.18, P<0.0001) and their interaction (F2,341=11.33, P<0.0001). Queens had higher JH titers than all other groups, and Benders had higher JH titers than nB females and workers. For all models, Colony had no effect.

### Effects of queen removal

Females that engaged in bending displays for 1–3 days following queen removal (referred to as 'new benders') consistently had higher JH titers than comparably aged workers before queen removal (Additional file 1: Figure S2 and Additional file 3: Figure S4A) as well as workers that continued to work in the absence of queens (Additional file 1: Figure S2 (red axes)). As mentioned, new benders could be comprised of former workers, younger idle females, and/or females that emerged in the absence of queen (Table 1). In a pooled analysis including data from queenless nests, benders that had worked prior to queen removal had higher JH titers than benders without a history of working, and both types of benders had higher JH titers than newly eclosed wasps and workers (Figure 3B). Finally, statistical analyses revealed no significant correlation between JH titer and oocyte length among these new benders (e.g., colony 5: N = 15, Pearson's $r = 0.15$, $P = 0.59$, $Z_r = 0.09$; colony 6: N = 20, Spearman's $\rho = -0.05$, $P = 0.83$, $Z_r = 0.14$).

In colonies 6 and 7, we allowed a set of 'new benders' to become established for one week (Additional file 1: Figure S2D-E (green x-axes)). In both colonies, the JH titers of these 'established benders' were highly variable and showed no statistical correlation with oocyte length (colony 6: N = 5, Pearson's $r = -0.41$, $P = 0.49$, $Z_r = 0.064$; colony 7: N = 6, Pearson's $r = 0.47$, $P = 0.35$, $Z_r = 0.34$). In colony 6, one bender was being chased and attacked by a worker, and a former bender was observed building. These 'defeated benders' had baseline JH titers. In colony 7 (Additional file 1: Figure S2E), the established benders were comprised of 4–8 day old adults that had eclosed after queen elimination. Females aged 1–4 days did not bend, frequently huddled (i.e., remained idle) instead of working and were sometimes attacked. These suppressed 'huddlers' had very low JH titers (Additional file 1: Figures S2E and Additional file 3: S4B). In a pooled analysis that included data from both of these colonies, the established benders had higher JH titers than the suppressed females (the attacked bender of colony 6 and huddlers of colony 7), newly emerged females and their worker counterparts (Figure 3C).

When data for all females from queenright and queenless conditions were combined, both queens and benders (encompassing all pre-reproductive bending subtypes) had higher JH titers than 0–3 day old non-benders and workers ≥3 days of adult age (Figure 3D).

In colony 10, the removal of 3 of 4 queens (single queen reign is indicated by the light blue x-axis bar in Additional file 1: Figure S2G) led to the emergence of young benders in queenright conditions (see above). Among the benders, attacking workers and queen, only the latter had an elevated JH titer (Additional file 1: Figure S2G and Additional file 4: Figure S5). Once the last queen was removed, JH titers were clearly elevated in the three queenless benders (2–5 day old adults) on the following day (Additional file 1: Figure S2G (red axis) and Additional file 4: Figure S5).

### Correlation with oocyte length

JH titers were positively correlated with oocyte length in all six colonies where ovaries were measured (Spearman's ρ for each nest: $P < 0.004$) (Additional file 5: Figure S6). This correlation was counteracted to some extent by the higher than expected JH titers (based on oocyte length) of new benders and the lower than expected titers of suppressed 'hopeful reproductives' (e.g., huddling and/or attacked non-workers). These patterns, along with the lack of statistical correlation between oocyte length and JH titer within bending groups (see above), suggest that JH titers are more associated with the activation of the ovaries than with basal oocyte size.

### Ecdysteroids

Ovarian ecdysteroids were correlated with queen ovary size (number of oocytes >70% LME) (Figure 4A). Among females where oviposition was not observed, ovarian ecdysteroids clearly correlated with oocyte length in pooled workers and benders from colony 4 (Figure 4B).

Hemolymph ecdysteroids were measured in two colonies. Queens and workers of colony 8 showed no difference in ecdysteroid hemolymph titers (Figure 4C). Also, only 3 of 13 samples from colony 4 had detectable amounts, none of which came from queens (N = 4).

### Effects of exogenous JH

In experiments where we were interested to see whether JH may affect behavior we used JH III, the natural JH of Hymenoptera [43-45], and methoprene, a JH mimic. In most insects, including bees [46], the natural JH is rapidly metabolized, and so if a change in behavior (e.g., aggression) was not observed within 24 hours of JH III application, we repeated the treatment for a total of three consecutive days. By contrast, in experiments where we were interested to find out whether JH may have a gonadotropic effect or provoke an alteration in the cuticular hydrocarbon pattern in these wasps we used methoprene. This JH mimic is a more stable molecule than JH III as it is not degraded as efficiently by JH esterase [47]. It is therefore better suited for investigating longer lasting effects and requires fewer applications.

On a nest containing brood of all ages, all queens and workers <3 weeks of adult age were removed (colony 8). Forty-six of the ~55 remaining workers received a dose of 10 μg of JH III (N = 23) in 1 μl of cyclohexane or just 1 μl of cyclohexane (N = 23) per day over three consecutive days starting with the day of removal of the last queen. The nest produced no adults during the first two

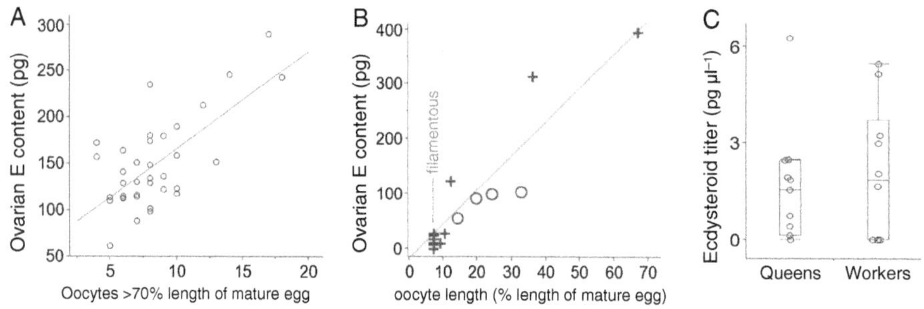


**Figure 4 Ecdysteroids levels in ovaries and hemolymph.** Ovarian ecdysteroid contents correlate positively with: **(A)** ovary size (number of oocytes >70% length of mature egg) in queens from colonies 6–10 (N = 36, Spearman's ρ = 0.51, P = 0.0014), and **(B)** oocyte length in workers (+) and benders (o) from colony 4 (N = 15; Spearman's ρ = 0.87; P <0.0001). **(C)** Hemolymph ecdysteroid titers did not statistically differ between queens and workers in colony 8 (Kruskal-Wallis test: N = 21, $H_1$ = 0.18, P = 0.86, $Z_r$ = 0.075). Box plots show the median and the middle two quartiles; the whiskers indicate the 1.5 interquartile range.

days of treatment, and two newly emerged females were removed when they appeared on the third and fourth days. No bending or queen-like behavior was observed among any of the treated or non-treated workers.

After finding no behavioral effect upon JH III treatment of old workers of colony 8, treatments were redirected toward newly emerged queenless females (one treatment per day for the first three days after eclosion). The first six females to eclose on the queenless nest assumed bending and were not treated. The following 15 females to emerge joined the cohort of future queens, irrespective of whether or not they had received JH III. All females to emerge thereafter were attacked and became workers, and JH III treatment had no ostensible effect on their bending behavior (Figure 5A). Thus, JH III treatment failed to induce the dominance behaviors frequently displayed by females with uncertain reproductive fates.

### Methoprene effects on behavior and ovarian development

To determine the effect of exogenous JH on ovarian development, young workers from queenright colonies 11 and 12 were treated twice (at adult age of 3 and 6 days) with methoprene or cyclohexane as a solvent control. Colony 11 also included non-treated wasps that were anesthetized alongside the treated females. Each female had begun building before the initial treatment. By the adult age of 4 days on colony 11, all methoprene-treated females began to receive one-on-one biting attacks from older workers that contacted them. The methoprene-treated females were mounted and bitten over large portions of their body (e.g., head, thorax, abdomen and/or appendages) including the face which induced an offering of liquid to the attacker or at least a gesture of it. The methoprene-treated female, having endured the attacks motionless in a crouched or slightly rolled over to the side position, would sometimes dash away when the aggression waned or the attacker moved on. This interaction strongly resembles the stereotyped, suppressive caste-determining attacks often received by young idle females in nests of *Synoeca* (and was observed in colony 11 when the nest envelope was first removed) and potential reproductives of related genera [2,4]. By days 7–9, the attacks had increased in severity and the attackers were, apparently, no longer appeased by subordinate gestures, compelling the attacked methoprene-treated wasps to flee towards isolated areas on or off the nest; by day 10 the first of the methoprene-treated females had disappeared (4 of 4). Ovarian measurements and cuticular hydrocarbon extractions (see below) were therefore done on females at the adult age of 7 days. Young workers from colonies 11 and 12 were monitored and treated at the same time. Observations were restricted to colony 11 (1–2 hours/day for 17 of 18 days), although attacks were also observed in colony 12 (N = 4) when females were collected for treatment. In both nests, only methoprene-treated wasps were observed to be attacked; in colony 11, 100% of the methoprene-treated wasps (N = 17) were attacked.

For this experiment, all six primary oocytes in the two ovaries (three ovarioles each) were measured to provide an in-depth description of the ovarian status. When evaluating separately the length of each primary oocyte (i.e., each set of ovaries potentially contributing six data points), methoprene-treated females had significantly longer oocytes than cyclohexane and non-treated controls for monogynic colony 11 and, to a lesser but also significant extent, polygynic colony 12 (Figure 5B), thus implicating JH as a gonadotropin.

### Cuticular hydrocarbons (CHCs)

An analysis of worker CHCs identified 22 compounds (Additional file 6: Table S1 and Additional file 7: Table S2) which appeared in every sample at detectable amounts. Identified compounds were either alkenes or linear alkanes.

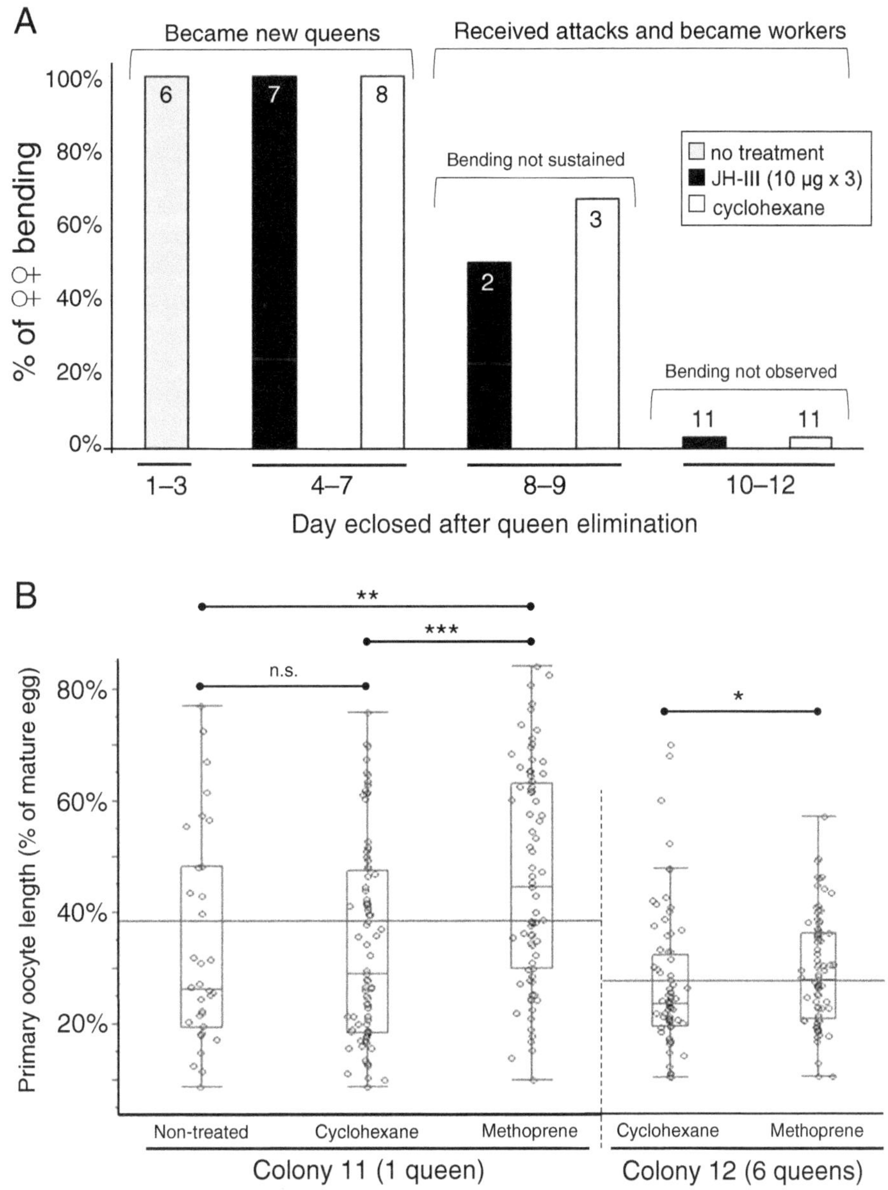


**Figure 5 Hormone manipulation experiments. (A)** JH III treatment of young queenless workers. JH III treated females showed no difference in bending propensity compared to cyclohexane-treated controls. **(B)** Effect of methoprene treatment on oocyte development in young queenright females from colonies 11 and 12. Each data point represents 1 of a possible 6 measurable primary oocytes from treated and non-treated females (the horizontal line represents the grand mean shown for each nest). In colony 11, methoprene-treated females had longer oocytes than both control types (Steel-Dwass all pairs: vs. non-treated: N = 111, $Z$ = 3.25, P = 0.003 = **; vs. cyclohexane: N = 166, $Z$ = 4, P = 0.0002 = ***). In colony 12, where a comparison was made with only a solvent control, the difference was less obvious but still significant (Mann-Whitney U test: N = 165, $Z$ = 2.12, P = 0.03 = *).

**Queens vs. workers**

The CHC profiles of queens (N = 67) and workers over the age of 4 days (N = 95) from colonies 4–8 were compared. A representative chromatogram for each phenotype is shown in Figure 6A. Adult females of 4 days and younger were excluded from this analysis because younger females experience a major age-related change in their CHC profile (see below). A stepwise discriminant function analysis (DA) based on all 22 shared compounds significantly separated individuals according to their status. A Principal Components Analysis (PCA) reduced the variable (i.e., CHC) number to 16, and the DA based on these compounds produced similarly robust results (see Figure 6B for statistics). The C25:1 monoene, 9-Pentacosane (CHC ID number 6 in Figure 6), was the only compound that was significantly higher in queens

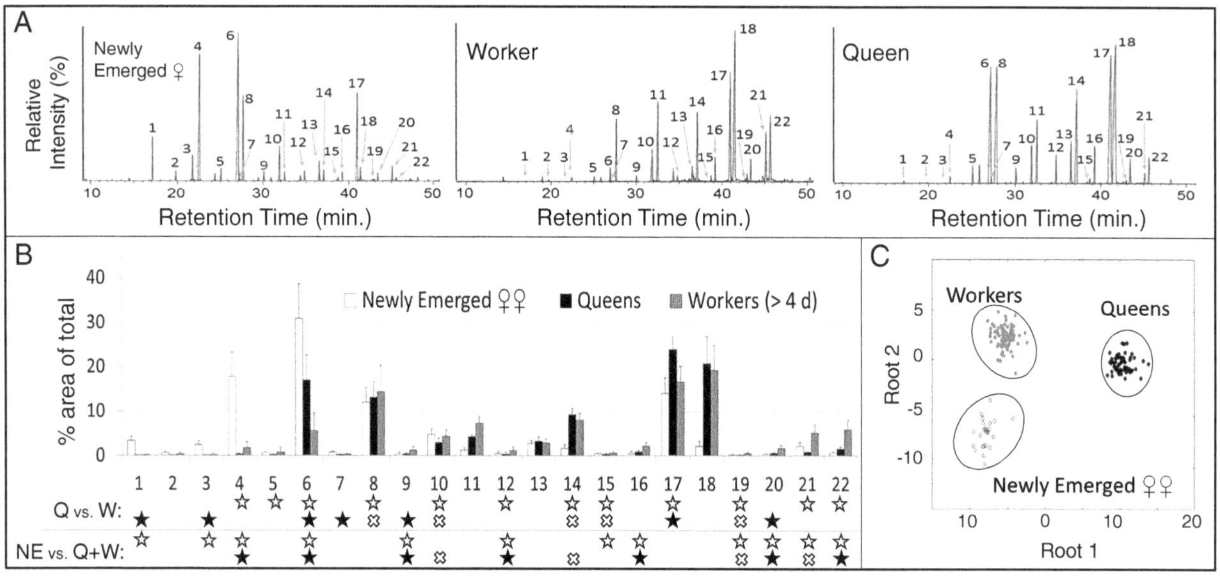


**Figure 6 Cuticular hydrocarbon (CHC) profiles of *Synoeca surinama*. (A)** Representative chromatograms of whole-body extracts for main female types. For compound numbering and identification see Additional file 6: Table S1. **(B)** CHC profiles of newly emerged females (NE), queens (Q) and workers (>4 days since eclosion) (W) from colonies 4–8 (mean % ± SD). Discriminant analyses were performed for queens vs. workers (w/out PCA: Global Wilks' $\lambda = 0.01489$; $F_{16,145} = 599.68$; $p < 0.001$; w/ PCA: Global Wilks' $\lambda = 0.03855$; $F_{11,150} = 340.06$; $p < 0.001$), and newly emerged females vs. pooled queens + workers (w/out PCA: Global Wilks' $\lambda = 0.00148$; $F_{34,334} = 245.96$; $P < 0.001$; w/ PCA: Global Wilks' $\lambda = 0.15546$; $F_{15,170} = 61.570$; $p < 0.001$). White and black stars indicate the main contributors of separation prior to and following PCA, respectively. "X" denotes a compound removed based on the PCA analysis. **(C)** Canonical scatterplot based on a discriminatory analysis for all three groups.

(but not in newly emerged females; see below) across all colonies (the C31:1 monoene, CHC ID number 17, was higher in queens in most but not all colonies examined). Workers tended to have a higher proportion of longer hydrocarbons (e.g., C32, C33:1, and C33).

**Newly emerged vs. older females**

Newly emerged females (<24 h since eclosion) had a very different CHC profile than older females (Figure 6A-C). A DA based on all compounds shared between newly emerged females (N = 24) and queens + workers (N = 162) significantly separated individuals according to their age. The PCA reduced the variable (i.e., CHC) number to 17, and the DA based on these compounds also separated the groups, although the difference was less marked (see Figure 6B for statistics). A canonical plot of a DA performed for all queens, workers and newly emerged females is shown in Figure 6C.

**The emergence of caste-specific profiles**

The CHC profile of newly emerged females became more worker- or queen-like depending on the social context (Figure 7A). For all maturing females, shorter CHCs (e.g., C21 and C23 alkanes, and C23:1 monoene) decreased in proportion while longer ones (C27, C29, C31:1, C31) increased. In queenright females, C25:1 decreased while larger compounds, such as C32, C33:1 and C33, increased in proportion. Maturing queenless benders showed just the opposite pattern, marking the emergence of a queen-like CHC profile. In support of this, a DA for colony 7, where both 4–9 day old queenright workers and queenless benders were sampled, showed that benders have a CHC profile intermediate between workers and queens (Additional file 8: Figure S7A).

The development of a queen-like CHC profile was also evident in old workers that made the transition to bending (colony 6). After a day of displaying, these females did not show a distinct CHC profile from their worker counterparts sacrificed alongside or on the days prior (during queenright condition) (Additional file 8: Figure S7B and Additional file 6: Table S1). A week after queen removal, both workers and established benders showed a change in their CHC profiles, with benders showing a queen-directed shift (Additional file 8: Figure S7B). The two failed benders from this colony (see above) grouped according to their behavior at the time of collection: a former bender that was observed working had a queenless worker profile whereas the active bender that was chased and attacked grouped with the cohort of established, accepted benders (Additional file 8: Figure S7B).

For colony 8, established queens and, subsequently, young replacements queens (10–12 days old) were sacrificed. Unlike benders sampled from other colonies, these benders had begun laying eggs, and the majority of them were observed to fly off the nest near dusk,

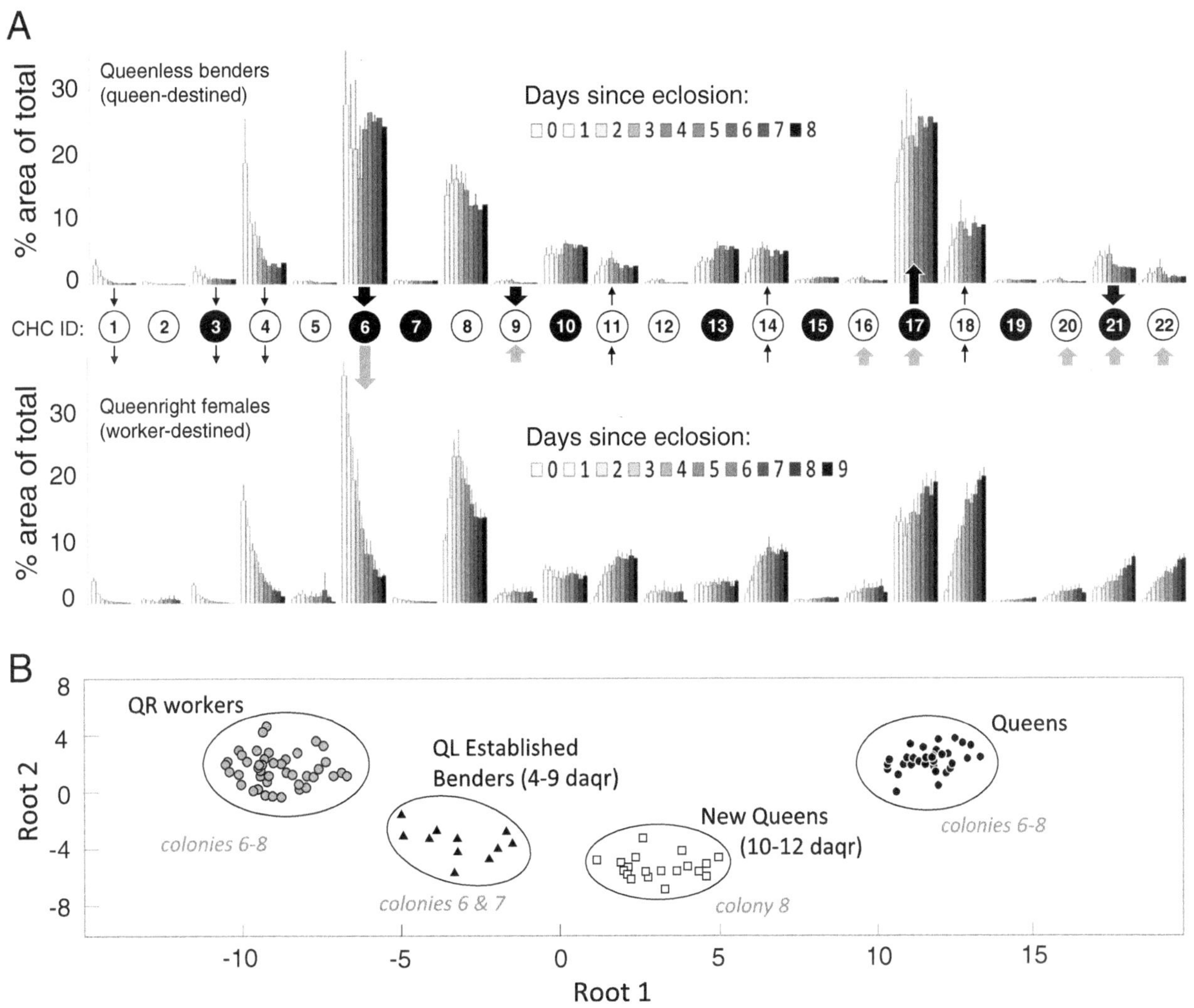


**Figure 7 Age and status related changes in cuticular hydrocarbon (CHC) profiles of *Synoeca surinama*.** **(A)** Relative percentages of CHC compounds in females in queenless (top) and queenright (bottom) conditions from colonies 5 and 7 (mean % ± SD). Linear alkanes are shown in white circles; alkenes are shown in black circles (for identities, see Additional file 6: Table S1). The surrounding arrows indicate evident changes in a compound's proportional representation over the first 8 or 9 days of adult life and also show a difference between queen and workers (see Figure 6). Small arrows specify an age-specific shift, thick black arrows indicate a queen-like shift and thick gray arrows signify a worker-like shift. **(B)** Canonical scatterplot based on a discriminant analysis of CHC profiles for females belonging to colonies 6–8. daqr = days after queen removal.

perhaps in search of mating opportunities. A DA based on all 22 compounds (PCA identified all compounds as contributors) was performed for queenright workers (>4 days old), queens and established benders from colonies 6–8. As expected, these new queens exhibited a CHC profile that was intermediate to established benders (i.e., pre-queens) and established queens (Figure 7B).

### CHC profile of methoprene-treated queenright females

The CHC profiles of the attacked methoprene-treated workers from colonies 11 and 12 (see above) were distinct from their cyclohexane and non-treated counterparts (Figure 8 and Additional file 7: Table S2). Methoprene-treated females possessed an established bender-like CHC profile, while non-treated females were nested within the cyclohexane-treated workers (Figure 8B). Importantly, when females from each nest were split into two groups according to oocyte length, irrespective of treatment, the differences, although significant, were notably less so (colony 11: Global Wilks' $\lambda = 0.650$; $F_{5,37} = 4.0$; $P < 0.001$; 81% classification; colony 12: Global Wilks' $\lambda = 0.182$; $F = 25.4$; $P < 0.001$; 95% classification). Therefore, methoprene application appears to affect the CHC profiles of young females independently of its gonadotropic role.

## Discussion

To our knowledge, this is the first study to couple endogenous endocrine measurements and hormone

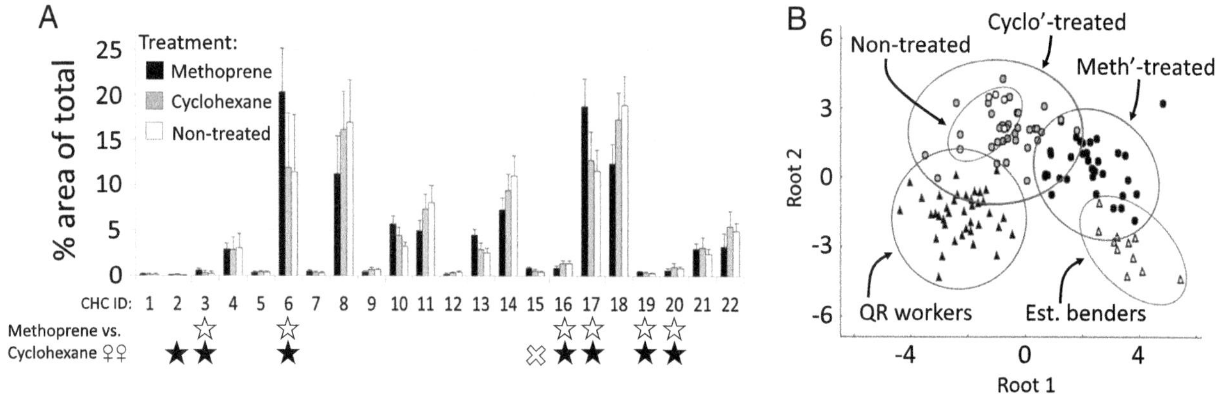


**Figure 8 Effect of the methoprene on the cuticular hydrocarbon (CHC) profile of *Synoeca surinama.* (A)** Methoprene application induced a CHC profile similar to that of incipient reproductives. The CHC profile of methoprene-treated females was distinct from cyclohexane- and non-treated females (pooled from colonies 11 and 12). For compound identification, see Additional file 6: Table S1. White and black stars indicate the main contributors of separation prior to and following PCA, respectively. "X" denotes a compound removed based on the PCA analysis. **(B)** Canonical scatterplot based on a discriminatory analysis for all treatment groups from colonies 11 and 12, including established benders and queenright workers from colonies 6 and 7 showing a clearly significant difference, done with and without PCA (for both: Global Wilks' $\lambda = 0.185$; $F_{13,51} = 17.3$; $p < 0.001$; 100% predicted classification). The non-treated controls group within the cyclohexane-treated wasps whereas methoprene-treated females overlap with the established benders, but not the queenright workers.

manipulation experiments for a permanently eusocial wasp. For *Synoeca surinama*, we show that JH titers are consistently higher in incipient and actual queens than in workers, suppressed 'hopeful reproductives' and newly emerged females (Figure 3 and Additional file 1: Figure S2). This pattern suggests that JH has conserved gonadotropic, behavioral and/or chemical signaling functions that these wasps share with their non-swarming primitively eusocial relatives (e.g., *Polistes*) [27,48,49]. Yet in further hormone assays combined with subsequent JH and methoprene application experiments for *S. surinama*, it became evident that JH is not required for the expression of aggressive displays and failed to influence the caste trajectory of females with uncertain fates. Instead, JH appears to both promote ovarian development and direct a shift in the CHC profile toward that of a replacement queen. This is in contrast to another caste-flexible epiponine wasp, *Polybia micans*, where high JH titers are not typically sustained in maturing reproductives [34]. One shared aspect in the endocrinology of *S. surinama* and *P. micans* is the apparent lack of ecdysteroid function in the hemolymph, which in *Polistes* is important for establishing dominance [25,28,29].

### JH and behavior in *S. surinama*

In *S. surinama*, queens from each colony had higher JH titers than other types of females (Figure 3A,D) while accepted queenless benders had more than their worker or idle counterparts that did not display (Figure 3B,C and S2). Nonetheless, JH titers were highly variable in these cohorts of accepted queenless benders, with some females having no detectable amounts. Without exception, all queenright and queenless benders which were suppressed through physical aggression had very low JH titers (Figure 3A,C). What remains to be explained is why queenless females which transitioned from working to bending had higher JH titers than bending cohorts which had not previously worked (Figure 3B).

These data show that elevated JH levels are not required for the expression of aggressive displaying in *S. surinama*, but these assays could not test whether JH plays a role in augmenting aggression. To address this, we showed that JH-III treatment failed to induce bending in old workers on a nest lacking both queens and young females. Despite their putative totipotency (cf. old workers-turned-benders in colony 6), all workers, treated or not, forewent the opportunity for direct reproduction on a nest where new adults were about to emerge. JH-III treatment also had no effect on young females that had eclosed in a period where an individual's caste fate was less restricted due to lack of queen(s) (Figure 5A). One caveat to these results, which showed that there was no obvious behavioral response in treated females, is that the manipulations lacked a positive control, i.e. a treatment with a JH mimic that is more stable than JH III. Thus, we cannot rule out the possibility that exogenous JH-III were counteracted to some extent by JH-degrading enzymes, most probably a JH esterase (JHE), as shown in honey bees [50], where JH-III treatment strongly induced JHE expression. That said, we were deliberately testing for a trigger effect of JH-III on aggressive behaviors, and it is likely that the exogenous JH-III was present for at least a few hours to a day before it could be fully degraded. Thus, we conclude that JH is not a direct activator of aggressive displays in females with open fates in *S. surinama*, a result completely at odds with other social wasps, where a boost in JH is thought to enhance competitive ability in unstable circumstances. For example, in the fellow epiponine *P. micans*, JH levels were seen to spike in potential reproductives that may become queens [34]. In *Polistes*, where physical aggression is the common expression of dominance [51], JH levels are higher in winners among competing reproductives [48,49,52], and are particularly high in workers that compete to replace a fallen queen [26].

A link between dominance and JH titers might not become established until incipient reproductives mate and become queens, where competition is high, egg-guarding vigils initiate and ritualized aggression is common [1,2,6]. Although no dominance hierarchy appears to be operational in *S. surinama*, inequalities among queens are likely to exist. Nonetheless, among queens, JH titers did not significantly correlate with large oocyte number in any colony, and when comparing queens with relatively high and low oviposition rates, there was no difference in JH titers (Additional file 1: Figure S2C), suggesting that JH is not directly related to dominance or fecundity within a cohort of queens. Yet because there were always a few brood cells empty in the aforementioned nest, this may have encouraged egg-laying among queens while reducing dominance expressed by way of differential oophagy, a behavior highly correlated with dominance in wasps [10]. In terms of direct displays of dominance, it is noteworthy that the one queen observed to physically attack other benders had the highest JH titer in the colony (Additional file 1: Figure S2G), suggesting that in some cases JH may elevate during periods of intense competition. Overall, however, *S. surinama* stands alone among the currently studied social wasps in providing evidence that JH is possibly only of secondary importance for behavioral signals of dominance (i.e., aggression) in the context of reproductive competition.

Other candidate drivers of aggression are the ecdysteroids. In *Polistes*, the ovaries release ecdysteroids into the hemolymph [28], and injection of 20E increases a foundress' chance of becoming dominant (similar to JH) [25]. Yet in *S. surinama* the hemolymph ecdysteroid levels were low and indistinguishable between queens and workers (Figure 4C). Therefore, an unknown factor must incites aggressive displays in *S. surinama*. Biogenic amines, functioning as neurohormones, are good candidates given their role in reproduction [53,54,3], aggression [55,56] and arousal [57] in other eusocial Hymenoptera.

Although worker behavior in *S. surinama* was not studied herein in detail, queenright females in envelope-damaged nests clearly built before they transitioned to foraging, as observed by West-Eberhard [1]. In colonies where only pieces of the nest were removed, some old foragers had elevated titers (Additional file 2: Figure S3). The scatter of JH titer values may, thus, reflect a change in JH titers associated with task transition, as has been suggested in other eusocial wasps [26]. Yet we found that females that built and/or foraged in envelope-removed nests generally had low JH titers, and only appeared to be slightly elevated in 1–3 day old queenright females (Figure 3A), an age span concomitant (in this study) with the onset of building and slight ovarian growth (Figure 2). Also, methoprene treatment to young queenright builders of *S. surinama* led to an increase in ovarian growth and received aggression (see below) instead of precocious worker activity, contrary to reports of other caste-totipotent polistine wasps: *Polistes canadensis* [21], *P. dominula* [30], and in the epiponine *Polybia occidentalis* [33].

**JH acts to coordinate ovarian maturation and incipient queen-like CHC profile in *S. surinama***

Although JH was not directly linked to reproductive dominance, there are multiple lines of evidence suggesting that JH functions as a gonadotropin in *S. surinama*: 1) JH levels were increased in benders that were not physically challenged by nestmates (Figure 3B); 2) JH titers were correlated with oocyte length in all colonies (Additional file 5: Figure S6); 3) JH was higher in queenright 1–3 day old females than established workers (Figure 3A) as their oocytes increased in length in the few days following eclosion (e.g., Figure 2); and 4) methoprene application led to an increase in oocyte length in queenright workers (Figure 5B). In the latter experiment, methoprene application had a much stronger effect on workers in a single-queen nest, suggesting that chemical signals from queens can modulate the gonadotropic effects of exogenous methoprene.

Methoprene application to the aforementioned queenright workers also induced a queen-directed shift in their CHC profiles, consistent with the positive relationship between JH titers and the queen-like CHC profiles of queenless benders. The putative fertility information conveyed in their CHC profile probably led to their attack by nestmates, although it is possible that other chemical signals were co-affected (e.g., pheromones from the head [2]). The intermittent suppressive attacks were occasionally observed on nests with young females, indicating that the methoprene-treated wasps were not attacked based on a foreign odor, especially since treated females were not observed to be attacked on the day of initial treatment. As has been suggested in other wasps [2,4,9,20,58] and honey bees [59], the objective of such stereotyped aggressive interactions is likely to curb reproductive growth and/or stimulate working in prospective reproductives. In methoprene-treated workers of *S. surinama*, the CHC profile continued to change toward a reproductive-like profile despite the initial attacks. This persistence likely resulted in the unrestrained aggression which eventually led to their permanent eviction from the colony.

Does JH directly affect the CHC profile of *S. surinama*, or might JH stimulate ovarian development which in turn modifies the CHCs? Some solvent-treated females that were never attacked had ovaries that were larger than those of some of the methoprene-treated females that were always attacked. This suggests that signaling through an altered CHC profile is more important in this context than oocyte size. Also, grouping females according to treatment (methoprene vs. solvent) resulted in a much better separation by CHCs than by oocyte length in both colonies studied. And as mentioned, biting attacks were observed a day after treatment, scarcely affording sufficient time for significant ovarian growth in the attackees. These results suggest a direct role of JH in modulating the CHC profile.

### Age- and social context-related changes in cuticular hydrocarbon profiles

In social insects, CHCs convey important information regarding age, reproductive status, sex, nest membership and more [60]. In *S. surinama*, newly emerged females possessed high amounts of relatively small hydrocarbons that became drastically reduced over the first few days of adult life (Figure 7A), a pattern comparable to *P. micans* [34]. The CHC profile of these aging females also becomes more worker- or queen-like depending on the social context of the nest (Figure 7A). For example, a proportional increase in long chain hydrocarbons was only observed in females that became workers (Figure 7A). Of particular interest is the dynamic of 9-Pentacosene (C25:1), which showed both age and caste related changes. In the CHC profile of newly eclosed females, this alkene represents 40% of the total CHC blend. In workers it drops precipitously to ~5%, whereas in incipient queens (i.e., females which emerged in queenless conditions) it stays above 15% (Figure 6B; compound number 6 in Figure 7). It is possible, then, that the aggressive approaches of worker and queens toward newly emerged females and other queens are elicited by the high representation of C25:1, and whether a physical attack results or not may depend on both the colony state (e.g., queenright or queenless nest) and the condition of approached female (e.g., expressed through behavioral and/or other chemical signals). Nevertheless, the fact that no hydrocarbon was found to be consistently higher in queens than in both newly emerged females and workers suggest that the composite of the CHC profile of a queen is more important in keeping workers sterile than any given compound. In other eusocial hymenopterans, including vespine wasps, single saturated hydrocarbons – expressed at relatively high levels in queens for a given species – are sufficient to reduce oocyte length in workers [61]. Yet in *S. surinama* those saturated hydrocarbons that showed caste-related differences were always higher in workers (Additional file 6: Table S1, Figures 6, 7), and so it unlikely that this class of pheromones is involved in controlling nestmate reproduction in this species.

In both *S. surinama* and *P. micans*, prospective reproductives with well-developed ovaries exhibited a rise in hydrocarbons associated with queen status, as would be expected, although the queen-associated compounds differed between the species. Indeed, whereas a methyl-branched alkane appears to be important for *Polybia* [34,62], likely the most derived genus of Epiponini [31,32], they were not consistently detected in either *S. surinama* (Figure 6 and Additional file 6: Table S1) or another epiponine, *Parachartergus aztecus* [63], which are characterized by a high composition of alkenes. Indeed, the diversity of hydrocarbon signaling within the Epiponini – in apparent contrast to temperate vespine wasps [42,61] – suggests these putative pheromones may be subject to strong social selection between workers and queens [1,42].

### Concluding remarks

With respect to social evolution in wasps, *Polistes*, *Synoeca* and *Polybia* are likely to share a eusocial ancestor that was caste-totipotent [31,32], and so it was expected that a common endocrine ground plan may underlie caste plasticity in these wasps, with JH and the ecdysteroids playing important roles in competition and reproduction, as shown in *Polistes* [14,22,24,48]. Yet when summarizing the major findings for these three wasp genera (Table 2), the idea of a grand unifying pattern of hormonal regulation in social wasp adults appears untenable, for aside from the observed rise in JH titers in potential queens (and some evidence that JH is involved in the behavior transitions of workers), the endocrine profiles of these caste-flexible wasps have little in common. Indeed, social wasp endocrinology appears to be as labile as the plasticity it was thought to underlie. And whereas the endocrine profile of caste-flexible *Polybia micans* appears to align more closely with that of highly eusocial bees than other wasps studied to date [34], JH titers for *S. surinama* appear more bumble bee-like than either *P. micans* or *Polistes*. For example, JH is likely gonadotropic in both bumble bees [64] and *S. surinama*, and although JH titers are generally associated with reproductive dominance [64], JH or methoprene treatment does not increase dominance in queenright and queenless conditions in either species [65,66]. To our knowledge, a role for JH in modulating the CHC profile of bumble bees has not been explored. In addition to the species listed in Table 2, there is evidence that JH drives age-related changes in behavior in

**Table 2 Summary of differences in the endocrinology of three genera of wasps**

| Hormone & effect | *Synoeca surinama* | *Polybia micans* | *Polistes sp.* |
|---|---|---|---|
| JH titers rise in reproductive competitors in a queenless colony | Yes (Figure 3B; Additional file 1: Figure S2) | Yes [34] | Yes [26] |
| JH titers remain elevated in potential reproductives as they become queens | Yes (Figure 3; Additional file 1: Figure S2) | No [34] | Yes [26] |
| JH higher in queens than workers | Yes (Figure 3A) | No [34] | Yes [21] |
| JH higher in lone (vs. multiple) queens | Maybe (Additional file 1: Figure S2F,G) | Yes [34] | n/a |
| JH and/or methoprene induces aggressive behaviors in potential reproductives | No (Figure 5A) | Not tested | Yes [23-25] (but see [48]) |
| High JH sustains ovarian development | Yes (Figure 5B) | No [34] | Yes [22,23,49] |
| JH involved in regulating the CHC profile of maturing reproductives | Yes (Figure 8) | Maybe [34] (does not sustain) | Yes [27] (correlated in *P. dominula*) |
| JH involved in age-related changes of behavior of workers | Maybe (Additional file 2: Figure S3) | Maybe (H.Kelstrup, unpubl.) | Probably [21,30] |
| Ecdysteroids present in the ovaries | Yes (Figure 4A,B) | Yes [34] | Yes [28] |
| Ecdysteroids present in the hemolymph and augment aggressive behaviors | No (Figure 4C) | Doubtful [34] | Yes [25,28] |

*Polybia occidentalis* [33] and functions as a gonadotropin in the primitively eusocial *Ropalidia marginata* [67], but hormone levels have not been measured in either case.

If caste-plasticity is conserved among a group of eusocial animals, how and why would the original regulatory factors controlling caste fate change? The diversity and apparent susceptibility to modification of wasp endocrinology is a treasure trove for such evolutionary and developmental research, and the ability to study behavior and endocrinology *in situ* in the field is a major advantage [68]. Fortunately, there are over 225 species from roughly 20 genera of Epiponini, facilitating the phylogenetic tracking of CHC signaling evolution and its effectors. Finally, this physiological work should keep an eye on future genomic studies, which in wasps [69], as well as ants and bees [70], have revealed lineage-specific changes not just in gene composition but in the regions where hormones, their receptors and co-activators operate: the regulatory elements [70].

## Methods

### Wasps

All behavioral and physiological data were collected from *Synoeca surinama* [Hymenoptera: Vespidae: Epiponini] nests found on the campus of Universidade de Federal Sergipe (UFS), São Cristóvão, Sergipe, Brazil. All nests of *S. surinama* were studied *in situ*, anywhere from 1 to 12 meters off the ground. Although nests can grow quite large (Figure 1D), for queen removal colonies we focused on relatively small nests where all females could be marked according to their activity within a couple days. Voucher specimens are preserved at Universidade de São Paulo (USP), Ribeirão Preto.

### Field studies

The paper nests of *Synoeca* consist of a vertical planar comb of cells and a brittle corrugated envelope (Figure 1). Envelopes were partially or completely removed to facilitate observations inside the nest [3,71]. The destruction of the nest envelope unavoidably induces a drastic change in task allocation toward restoring the envelope (e.g., building and pulp foraging). Thus, on some nests, only small pieces of envelope were removed (Figure 1B). On most nests, all females were marked with oil-based Sharpie pens based on their behavior (e.g., bending, building, etc.) or day of emergence from the brood cell. Queens and workers are easily distinguished based on behavior [3]. Queens, in addition to rarely working, walk much more slowly than workers, spend most of their time huddled with other queens in the corners of the nest, exhibit bending behavior toward both workers and fellow queens and receive the 'queen dance' from workers. In this study, all sampled females (N = 71) having this repertoire turned out to have large ovaries and opaque spermatheca, evidence of having mated. Young queenright adults, which are idle for several days following eclosion, also huddle in corners of the nest but do not exhibit bending behavior (except in rare cases as reported here), and after several days they transition to building in queenright colonies. Each colony was observed for at least 4 days to identify the general status of each female and the adult age of newly emerged ones. In colonies where all queens are removed, the above queen-repertoire is taken up by a subset of females that are referred to as "new benders". Queenless benders that have displayed for a full week are called "established benders", and their ascension to queenhood is considered complete when they begin laying eggs. When queen fecundity was assessed (colony 5), eggs were periodically removed from cells.

To avoid stress-related endocrine responses, no females were sacrificed until two days after the envelope was first removed. To limit the effect of possible circadian endocrine changes [72], all wasps were collected for processing from 13:00–17:00, placed in clean glass vials and buried in ice (0°C) within 15 s of removal. Collection almost always included multiple types of females, with the relevant pair determined by the objective of the assay (e.g., queens vs. workers).

Relative age of workers and queens was estimated by scoring (blind) the degree of apodeme and cuticular darkening on the $5^{th}$ gastral sternite [6,7,73]. A pictorial representation of this feature and the scores assigned are shown in Additional file 9: Figure S1 and is referred to as 'cuticular age'.

### Collecting hemolymph for hormone measurements, cuticular hydrocarbon wash, and ovary measurements

Wasps were transferred on ice to the laboratory for processing; for details see [34]. All wasps were bled within 2 h since longer periods of cold anesthesia significantly affect JH titers in honeybees [74]. Two to 10 μl of hemolymph was withdrawn from between the anterior-most segments of the gaster with a microcapillary (Drummond Scientific Company, Broomall, PA, USA). Samples destined for JH measurement by radioimmunoassay (RIA) were transferred to 500 μl of acetonitrile. Samples destined for ecdysteroid measurement were preserved in 500 μl of methanol.

Subsequent to bleeding, cuticular hydrocarbons (CHCs) were extracted from females by placing them in 2 ml of hexane for 2–2.2 min. Hormone and CHC samples were kept at −20°C. The ovaries of these wasps were carefully removed in cold E & B Ringer solution (7.5 g NaCl and 0.35 g KCl/1 L distilled water) and photographed with a Leica EZ4D Microscope Camera. Ovaries were then placed in 500 μl of methanol for ecdysteroid measurement.

Queen ovaries were quantified by counting the number of oocytes >70% the length of a mature egg (LME). The two largest oocytes of females that were not queens were measured using ImageJ (NIH, Bethesda, MD, USA) and averaged. Oocyte length is given as a percentage of a LME, and oocytes <8% LME were considered filamentous.

### Hemolymph juvenile hormone titer analysis by radioimmunoassay (RIA)

JH was extracted from the haemolymph sample in acetonitrile following a liquid-phase separation protocol developed for honeybees [75]. For the radioimmunoassay, we used [10-$^3$H (N)]-JH III (spec. activity 19.4 Ci/nmol, Perkin Elmer Life Sciences, Waltham, MA, USA), JH-III (Fluka, Munich, Germany), and a JH-specific antiserum [76] as detailed by [77] and [34]. JH titers of the samples were calculated by non-linear four-parameter regression on standard curve values (ImmunoAssay Calculations spreadsheet, Bachem, Bubendorf. Switzerland) and are expressed as JH-III equivalents (pg/μl hemolymph).

### Haemolymph titer and ovarian ecdysteroid content analysis by radioimmunoassay (RIA)

Haemolymph samples in 500 μl in methanol were cold centrifuged (4°C), and the supernatant transferred to RIA glass vials and dried by vacuum centrifugation. For the quantification of ovarian ecdysteroid content, interfering lipids were removed from the methanolic extract by reversed phase chromatography [38].

Ecdysteroids were quantified by RIA, as previously described [77,78] using an antiserum prepared against a hemisuccinate derivative of ecdysone [79,80] [23,24-$^3$H (N)] ecdysone (Perkin Elmer) (NEN, spec. act. 102 Ci/mmol), and 20-hydroxyecdysone (20E; Sigma St. Louis, MO, USA). Details of this RIA protocol are in [77]. Results are expressed as 20E equivalents, calculated by the same regression analysis used for JH titers (see above), reported as pg/μl for the hemolymph samples, or as pg/ovary for the ovary samples.

### Cuticular Hydrocarbon (CHC) analysis

After evaporation of the wash hexane, the extract was re-suspended in 50 μl of hexane and 1 μl was injected into a combined gas chromatography-mass spectrometer (GCMS) (model QP2010, Shimadzu, Kyoto, Japan). Separation was achieved on a DB-5MS column of 30 m, with a helium gas carrier at 1.0 ml $min^{-1}$. Oven temperature was initially set to 150°C, and ramped up 3°C $min^{-1}$ until it reached 280°C and held for 20 min. Analyses were performed in the splitless mode. The mass spectra were obtained by 70 eV ionization. The chromatographs were analyzed with GCMS solutions (Shimadzu). CHCs were identified using authentic standards (linear alkanes) and/or by their molecular diagnostics ion. The positions of unsaturations in alkenes were identified according to the dimethyl disulfide derivatization technique [81] and analyzed with the same GCMS system mentioned above.

### Exogenous JH III and methoprene application

Females were treated with either JH III (Sigma) or methoprene (Zoecon, Palo Alto, CA). Both JH III and methoprene were dissolved in cyclohexane (HPLC grade, Sigma-Aldrich) at final concentrations of 10 μg/μl each and 1 μl was topically applied to the gaster of cold-anesthetized wasps. Control wasps received a solvent application, and in some cases, non-treated wasp controls were also included. To test for immediate behavioral responses, 10 μg of JH III was applied on three consecutive days to old (>3 weeks) queenless workers and, in another experiment, to 1–3 day old adult females who emerged as the colony was transitioning from being queenless to queenright. To assess JH effects on ovarian growth and chemical signaling, 10 μg of methoprene, which has longer-lasting effects than JH III [47], was applied to working wasps at 3 and 6 days following eclosion. All wasps were kept in isolation for 45–60 min before they were returned to the nest to reduce the spread of the topically applied hormone to nestmates. Unless otherwise noted, all nests containing treated females were observed for 1–2 hours each day before treatment or capture.

### Statistical analyses

The range of JH titers for a given nest was not always the same, attributable to the fact that the samples were collected and analyzed over a two year period. Also, low sample size for certain female types precluded statistical tests within individual colonies. Therefore, we looked for differences in pooled data sets by performing a Mixed Model analysis, the Restricted Maximum Likelihood Method, on software SPSS 21 (IBM, Armonk, NY, USA) and JMP 10.0 (SAS Corporation, Cary, NC, USA). Fixed factors included Status and, in a grand analysis, Colony Condition (i.e., queenright vs. queenless nest). Colony was designated as a Random factor. To look for relationships between variables within a colony or from samples run on the same RIA, a Pearson's $r$ (normal data) or Spearman's $\rho$ (non-normal data) test implemented in JMP10.0 was used for significance. When comparing two groups, a two-tailed $t$-test (normal data) or a Kruskal-Wallis rank sums test (non-normal data) was used. In the case of multiple comparisons, the Steel-Dwass method for all pairs was employed using JMP 10.0. In cases where no relationship was found between two variables, a power analysis was employed and reported as Fisher's Refined Z (Zr) using JMP 10.0 and Statistica 12.0 (StatSoft, Tulsa, OK, USA).

### Cuticular hydrocarbon data

Principal components analysis (PCA) was used to define the main components to be compared, such that compounds missing in most individuals of a group, as well as compounds contributing less than 5% to the first two factors in PCA were excluded from the statistical analysis. The relative concentrations of the compounds used in the discriminant analysis were readjusted to 100% and each peak was transformed according to [82]. Following this, a stepwise discriminant function analysis was done to see if combinations of variables could be useful in predicting group. Wilks' λ values were used to verify the individual contribution of each variable to the model. For details of the CHC profile analyses, see [34]. The statistical analyses were performed using the software Statistica 10.0.

## Additional files

**Additional file 1: Figure S2.** JH titers according to female status. JH titers across colonies 4–10. *Blue axis* indicates queenright (QR) conditions; *red axis* indicates queenless (QL) conditions within 3 days of queen removal; *green axis* indicates QL conditions 7–8 days after queen removal; *light blue axis* indicates QR conditions with 3 of 4 queens removed. Queens (Q) with relatively high (Q↑) and low (Q↓) oviposition rates (see text) are distinguished in colony 5 (C), as is the remaining lone queen (LQ) for colony 10 (G). Newly emerged (NE) females were collected <24 after eclosion. Workers (including worker-destined females) were split into two groups: pre-workers and new workers 1–3 days since eclosion (dse) (pW) and older workers (W). New benders (B), which had been observed to bend from 1–3 days, are separated according to their history: workers-turned-benders (W➔B), idle females-turned-benders (I➔B), females that were within their pupal case when the last queen was removed (P➔B) and newly emergent benders (NE/B). In some nests, workers attacked new benders (W➔Atk). As benders became established for 4 days or more (Est B), suppressed 'hopeful reproductive', such as defeated benders (Def B) or idle huddlers (I), were also present. Asterisks in C and E indicate groups that are shown in more detail in Additional file 2: Figure S3. Box plots show the median and the inner two quartiles; the whiskers indicate the 1.5 interquartile range. Sample size is also indicated.

**Additional file 2: Figure S3.** JH titers in workers from envelope-intact colonies. Relative age and task performed by workers from colonies 1 and 2. The cuticular age is based on the score of apodeme and cuticular tanning of the 5th sternite (see Additional file 9: Figure S1).

**Additional file 3: Figure S4.** JH titers in young queenright (QR) and queenless (QL) females. (A) Colony 5: QL females appear to have higher JH titers, although QR females showed a rise during the first few days following eclosion. QL females of adult age 1–3 were observed to bend, and when compared to 1–3 day old QR females, QL benders had higher JH titers than QR workers (two tailed $t$-test, $t = 4.81$, $DF = 25$, $P = 0.0004$). (B) Colony 7: QR females had very low to undetectable JH levels, whereas QL benders, fated to become the next queens, showed a scatter from low to relatively high titers. Younger QL females that emerged after the establishment of accepted benders were suppressed and had low JH titers.

**Additional file 4: Figure S5.** In Colony 10, queens and queenless benders had significantly higher JH titers than pooled QR + QL workers (vs. Queens: $Z = -3.06$ $P = 0.012$; vs. QL benders: $Z = -2.7$). Queens had significantly more JH than QR benders ($Z = -2.76$) while QL benders tended to have higher JH titers than QR ones ($Z = -2.45$) (Steel-Dwass all pairs, $P < 0.05$). Box plots show the median and the inner two quartiles; the whiskers indicate the 1.5 interquartile range. Sample size is also indicated.

**Additional file 5: Figure S6.** Correlation analysis of JH titers versus oocyte length. Among all females types, JH levels were positively correlated (Spearman's $\rho$) with primary oocyte length (% length of fully mature egg) in (A) colony 4 ($N = 35$, $\rho = 0.54$, $P = 0.0009$), (B) colony 5 ($N = 58$, $\rho = 0.43$, $P = 0.0008$), (C) colony 6 ($N = 80$, $\rho = 0.58$, $P < 0.0001$), (D) colony 7 ($N = 70$, $\rho = 0.48$, $P < 0.0001$), (E) colony 9 ($N = 53$, $\rho = 0.73$, $P < 0.0001$), and (F) colony 10 ($N = 43$, $\rho = 0.45$, $P = 0.004$). Behavioral category along with queen state (QR = queenright, QL = queenless) is indicated in color. "New Benders" includes all females observed to bend within 3 days of queen removal; the "Suppressed ♀♀" category includes huddlers or bending females that received aggression.

**Additional file 6: Table S1.** Mean percentage of composition and standard deviation (SD) of cuticular hydrocarbons in various types of females of *Synoeca surinama*. Colony (s) of origin noted on top. Ret. Time = Retention Time; dse = days since eclosion; QR = queenright; QL = queenless; D = days after queen removal. (PDF 396 kb)

**Additional file 7: Table S2.** Mean percentage of composition and standard deviation (SD) of cuticular hydrocarbons in various types of methoprene-, cyclohexane- (cyclo) non-treated females from queenright colonies 11 and 12. Ret. Time = Retention Time; dse = days since eclosion.

**Additional file 8: Figure S7** Intracolonial alterations in the cuticular hydrocarbon (CHC) profile of emerging reproductives. The canonical scatterplots are based on discriminant analyses of the CHC profiles. (A) From colony 7: queens, queenright (QR) workers 4–9 days since eclosion (dse) and queenless (QL) established (Est.) benders of the same age. (B) From colony 6: queens, QR workers; QL workers and new benders 1–3 days after queen removal (daqr); and QL workers, benders and two defeated (Def.) benders 7 days after queen removal. All females from colony 6 had eclosed over a month prior. Circles encompass all profiles for a given group.

**Additional file 9: Figure S1.** Subjective measure of relative age of individuals based on the analysis of apodeme and other cuticular darkening on the anterior portion of 5$^{th}$ abdominal sternite. Representative samples: 1 = newly eclosed female; 2 = 7-day old builder; 3 = 12-day old worker; 4 = 20-day old forager; 5 = >35 day old forager.

### Abbreviations

CHC: Cuticular hydrocarbons; DA: Discriminate analysis; JH: Juvenile hormone; LME: Length of mature egg.

### Competing interests

The authors declare that they have no competing interests.

### Authors' contributions

HCK conceived the study, participated in its designed, performed the field work, collected the physiological samples, carried out the radioimmunoassays (RIAs), performed the hormone manipulation experiments, participated in the processing of the cuticular hydrocarbon (CHC) samples, quantified the CHC profiles, participated in the statistical analyses and drafted the manuscript. KH participated in the design of the experiments, oversaw and assisted with the RIAs, helped with statistical analyses and participated in the drafting of the manuscript. FSN participated in the processing of the CHC samples as well as the quantification and statistical analyses of the data. LMR oversaw the research and participated in the design of the study. All authors read and approved the final manuscript.

### Acknowledgements

This work was made possible by the generosity of the Laboratório de Entomologia at the Universidade Federal de Sergipe (UFS). Biologists Leandro S. Souto, Yana T. Reis, Ana Paula Marques Costa, Bianca Ambrogi and others provided lab space, equipment and continuous support with the procurement of materials required for this research. Discussions with Mary Jane West-Eberhard provided the early inspiration for this work. Michel Precisão and Adriano Aquino from the UFS Laboratório de Análise de Compostos Orgânicos Poluentes supplied the solvents. For assistance in the field, we thank Eduardo Nascimento, Beryl Jones and Lucas Oliveira. Carlos Alberto dos Santos and Dr. Fernando Noll (São José do Rio Preto) showed us how to work with this aggressive wasp. We thank Dave Mellert and Natasha Mothapo for assistance with the statistical analyses. Finally, Danielle Caroline and Isabel Cristina Turatti helped with the processing of CHC samples at the Universidade de São Paulo.

### Author details

[1]Janelia Research Campus, Howard Hughes Medical Institute, Ashburn, VA 20147, USA. [2]Faculdade de Medicina de Ribeirão Preto, Universidade de São Paulo, Universidade de São Paul, Av. Bandeirantes 3900, Ribeirão Preto 14049-900, SP, Brazil. [3]Departamento de Biologia da Faculdade de Filosofia, Ciências e Letras de Ribeirão Preto, Universidade de São Paulo, Av. Bandeirantes 3900, Ribeirão Preto 14040-900, SP, Brazil. [4]Present address: Department of Botany and Zoology, Stellenbosch University, Private Bag XI, Matieland 7602, South Africa.

### References

1. West-Eberhard MJ: **Intragroup selection and the evolution of insect societies.** In *Natural Selection and Social Behavior.* Edited by Alexander RD, Tinkle DW. New York: Chiron Press; 1981:3–17.
2. West-Eberhard MJ: *The Establishment of Dominance of the Queen in Social Wasp Colonies.* Netherlands: Wageningen; 1977:223–227.
3. West-Eberhard MJ: **Temporary queens in *Metapolybia* wasps - nonreproductive helpers without altruism.** *Science* 1978, **200**:441–443.
4. Platt TG, Queller DC, Strassmann JE: **Aggression and worker control of caste fate in a multiple-queen wasp, *Parachartergus colobopterus*.** *Anim Behav* 2004, **67**:1–10.
5. West-Eberhard MJ: **The nature and evolution of swarming in tropical social wasps (Vespidae, Polistinae, Polybiini).** In *Social Insects in the Tropics.* Edited by Jaisson P. Paris: University of Paris Press; 1982:97–128.
6. Forsyth AB: *Studies on the Behavioral Ecology of Polygynous Social Wasps.* Cambridge, MA: Harvard Universtiy; 1978.
7. West MJ: *The Social Biology of Polistine Wasps*, Ph.D. Thesis. Ann Harbor, Michigan: University of Michigan; 1969.
8. Pardi L: **Dominance order in *Polistes* wasps.** *Physiol Zool* 1948, **21**:1–13.
9. Reeve H: **Polistes.** In *The Social Biology of Wasps.* Edited by Ross K, Matthews R. Ithaca, NY: Cornell University Press; 1991.
10. Spradberry J: **Evolution of queen number and queen control.** In *The Social Biology of Wasps.* Edited by Ross K, Matthews R. Ithaca, NY: Cornell University Press; 1991:366–388.
11. West-Eberhard M: **Sexual selection, social competition, and evolution.** *Proc Am Phil Soc* 1979, **123**:222–234.
12. Hughes CR, Queller DC, Negrón-Stomayor JA, Strassmann JE, Solis C, Gastreich KR: **The maintenance of high genetic relatedness in multi-queen colonies of social wasps.** In *Queen Number and Sociality in Insects.* Edited by Keller L. Oxford: Oxford University Press; 1993:153–170.
13. Strassman J, Sullender B, Queller DC: **Caste totipotency and conflict in a large-colony social insect.** *Proc R Soc Lond B Biol Sci* 2001, **269**:263–270.
14. Hartfelder K, Emlen D: **Endocrine control of insect polyphenism.** In *Insect Endocrinology.* Edited by Gilbert LI. San Diego: Academic Press; 2012:464–522.
15. Riddiford LM: **Cellular and molecular actions of juvenile hormone. 1. general considerations and premetamorphic actions.** *Adv Insect Physiol* 1994, **24**:213–274.
16. Jindra M, Palli SR, Riddiford LM: **The juvenile hormone signaling pathway in insect development.** *Annu Rev Entomol* 2013, **58**:181–204.
17. Raikhel A, Brown M, Bellés X: **Hormonal control of reproductive processes.** In *Comprehensive Molecular Insect Science.* Edited by Gilbert L, Iatrou K, Gill S. Oxford: Elsevier; 2005:433–491.
18. Wyatt GR, Davey KG: **Cellular and molecular actions of juvenile hormone. 2. roles of juvenile hormone in adult insects.** *Adv Insect Physiol* 1996, **26**:1–155.
19. Tibbetts EA, Levy S, Donajkowski K: **Reproductive plasticity in *Polistes* paper wasp workers and the evolutionary origins of sociality.** *J Insect Physiol* 2011, **57**:995–999.
20. West-Eberhard MJ: **Wasp societies as microcosms for the study of development and evolution.** In *Natural History and Evolution of Paper Wasps.* Edited by Turillazzi S, West-Eberhard MJ. London, England: Oxford University Press; 1996:290–317.
21. Giray T, Giovanetti M, West-Eberhard MJ: **Juvenile hormone, reproduction, and worker behavior in the neotropical social wasp *Polistes canadensis*.** *Proc Natl Acad Sci U S A* 2005, **102**:3330–3335.
22. Tibbetts EA, Izzo AS: **Endocrine mediated phenotypic plasticity: condition-dependent effects of juvenile hormone on dominance and fertility of wasp queens.** *Horm Behav* 2009, **56**:527–531.
23. Barth RH, Lester LJ, Sroka P, Kessler T, Hearn R: **Juvenile-hormone promotes dominance behavior and ovarian development in social wasps (*Polistes annularis*).** *Experientia* 1975, **31**:691–692.
24. Röseler PF: **Reproductive competition during colony establishment.** In *The Social Biology of Wasps.* Edited by Ross K, Matthews R. Cornell, Ithaca: Cornell University Press; 1991:309–335.
25. Strambi A: **Physiology and reproduction in social wasps.** In *Social Insects - an Evolutionary Approach to Castes and Reproduction.* Edited by Engels W. Heidelberg: Spring; 1990:59–75.
26. Tibbetts EA, Huang ZY: **The challenge hypothesis in an insect: juvenile hormone increases during reproductive conflict following queen loss in *Polistes* wasps.** *Am Nat* 2010, **176**:123–130.
27. Izzo A, Wells M, Huang Z, Tibbetts E: **Cuticular hydrocarbons correlate with fertility, not dominance, in a paper wasp, *Polistes dominulus*.** *Behav Ecol Sociobiol* 2010, **64**:857–864.
28. Röseler PF, Röseler I, Strambi A: **Role of ovaries and ecdysteroids in dominance hierarchy establishment among foundresses of the primitively social wasp, *Polistes gallicus*.** *Behav Ecol Sociobiol* 1985, **18**:9–13.
29. Röseler PF, Röseler I: **Dominance of ovariectomized foundresses of the paper wasp, *Polistes gallicus*.** *Insectes Sociaux* 1989, **36**:219–234.

30. Shorter JR, Tibbetts EA: **The effect of juvenile hormone on temporal polyethism in the paper wasp *Polistes dominulus*.** *Insectes Sociaux* 2009, **56**:7–13.
31. Pickett KM, Carpenter JM: **Simultaneous analysis and the origin of eusociality in the Vespidae (Insecta: Hymenoptera).** *Arthropod Syst Phylogeny* 2010, **68**:3–33.
32. Noll FB, Wenzel JW: **Caste in the swarming wasps: 'queenless' societies in highly social insects.** *Biol J Linn Soc* 2008, **93**:509–522.
33. O'Donnell S, Jeanne RL: **Methoprene accelerates age polyethism in workers of a social wasp (*Polybia occidentalis*).** *Physiol Entomol* 1993, **18**:189–194.
34. Kelstrup HCP, Hartfelder K, Nascimento FS, Riddiford LM: **Reproductive status, endocrine physiology and chemical signaling in the Neotropical, swarm-founding eusocial wasp, *Polybia micans* Ducke (Vespidae: Epiponini).** *J Exp Biol* 2014, **217**:2399–2410.
35. Wingfield JC, Hegner RE, Dufty AM, Ball GF: **The challenge hypothesis - theoretical implications for patterns of testosterone secretion, mating systems, and breeding strategies.** *Am Nat* 1990, **136**:829–846.
36. Tibbetts EA, Crocker KC: **The challenge hypothesis across taxa: social modulation of hormone titres in vertebrates and insects.** *Anim Behav* 2014, 281–290.
37. Shpigler H, Amsalem E, Huang ZY, Cohen M, Siegel AJ, Hefetz A, Bloch G: **Gonadotropic and physiological functions of juvenile hormone in bumblebee (Bombus terrestris) workers.** *PLoS One* 2014, 9(6):e100650.
38. Geva S, Hartfelder K, Bloch G: **Reproductive division of labor, dominance, and ecdysteroid levels in hemolymph and ovary of the bumble bee *Bombus terrestris*.** *J Insect Physiol* 2005, **51**:811–823.
39. Bloch G, Wheeler D, Robinson G: **Endocrine influences on the organization of insect societies.** In *Hormones, Brains and Behavior.* Edited by Pfaff D, Arnold A, Etgen A, Fahrbach S, Moss R, *et al.* New York, NY: Academic Press; 2002.
40. Hartfelder K, Bitondi MMG, Santana WC, Simoes ZLP: **Ecdysteroid titer and reproduction in queens and workers of the honey bee and of a stingless bee: loss of ecdysteroid function at increasing levels of sociality?** *Insect Biochem Mol Biol* 2002, **32**:211–216.
41. Penick CA, Liebig J, Brent CS: **Reproduction, dominance, and caste: endocrine profiles of queens and workers of the ant *Harpegnathos saltator*.** *J Comp Physiol A* 2011, **197**:1063–1071.
42. van Zweden JS, Bonckaert W, Wenseleers T, d'Ettorre P: **Queen signaling in social wasps.** *Evolution* 2013. doi:10.1111/evo.12314.
43. Hagenguth H, Rembold H: **Identification of juvenile hormone 3 as the only juvenile hormone homolog in all developmental stages of the honey bee.** *Zeitschrift für Naturforschung* 1978, **33C**:847–850.
44. Sommer K, Hölldobler B, Rembold H: **Behavioral and physiological-aspects of reproductive control in a *Diacamma* species from Malaysia (Formicidae, Ponerinae).** *Ethol* 1993, **94**:162–170.
45. Burns SN, Teal PEA, Meer RKV, Nation JL, Vogt JT: **Identification and action of juvenile hormone III from sexually mature alate females of the red imported fire ant, *Solenopsis invicta*.** *J Insect Physiol* 2002, **48**:357–365.
46. Mackert A, Hartfelder K, Bitondi MMG, Simões ZLP: **The juvenile hormone (JH) epoxide hydrolase gene in the honey bee (*Apis mellifera*) genome encodes a protein which has negligible participation in JH degradation.** *J Insect Physiol* 2010, **56**:1139–1146.
47. Staal GB: **Insect growth-regulators with juvenile-hormone activity.** *Annu Rev Entomol* 1975, **20**:417–460.
48. Röseler P: **Endocrine basis of dominance and reproduction in polistine paper wasps.** *Fortschr Zool* 1985, **31**:259–272.
49. Tibbetts EA, Izzo A, Huang ZY: **Behavioral and physiological factors associated with juvenile hormone in *Polistes* wasp foundresses.** *Behav Ecol Sociobiol* 2011, **65**:1123–1131.
50. Mackert A, Do Nascimento AM, Bitondi MMG, Hartfelder K, Simões ZLP: **Identification of a juvenile hormone esterase-like gene in the honey bee, Apis mellifera L - expression analysis and functional studies.** *Comp Biochem Physiol B* 2008, **150**:33–44.
51. West-Eberhard MJ: **Dominance relations in Polistes-canadensis (L), a tropical social wasp.** *Monit Zoologico Italiano-Italian J Zool* 1986, **20**:263–281.
52. Röseler PF, Röseler I, Strambi A: **The activity of corpora-allata in dominant and subordinated females of the wasp Polistes gallicus.** *Insectes Sociaux* 1980, **27**:97–107.
53. Sasaki K, Yamasaki K, Nagao T: **Neuro-endocrine correlates of ovarian development and egg-laying behaviors in the primitively eusocial wasp (*Polistes chinensis*).** *J Insect Physiol* 2007, **53**:940–949.
54. Sasaki K, Yamasaki K, Tsuchida K, Nagao T: **Gonadotropic effects of dopamine in isolated workers of the primitively eusocial wasp, *Polistes chinensis*.** *Naturwissenschaften* 2009, **96**:625–629.
55. Bloch G, Simon T, Robinson GE, Hefetz A: **Brain biogenic amines and reproductive dominance in bumble bees (*Bombus terrestris*).** *J Comp Physiol A* 2000, **186**:261–268.
56. Cuvillier-Hot V, Lenoir A: **Biogenic amine levels, reproduction and social dominance in the queenless ant *Streblognathus peetersi*.** *Naturwissenschaften* 2006, **93**:149–153.
57. Robinson GE, Heuser LM, Le Conte Y, Lenquette F, Hollingworth RM: **Neurochemicals aid bee nestmate recognition.** *Nature* 1999, **399**:534–535.
58. O'Donnell S: **Polybia wasp biting interactions recruit foragers following experimental worker removals.** *Anim Behav* 2006, **71**:709–715.
59. Visscher PK, Dukas R: **Honey-bees recognize development of nestmates ovaries.** *Anim Behav* 1995, **49**:542–544.
60. Blomquist G, Bagnères A: **Introduction: history and overview of insect hydrocarbons.** In *Insect Hydrocarbons.* Edited by Blomquist G, Bagnères A. Cambridge: Cambridge University Press; 2010:3–18.
61. van Oystaeyen A, Oliveira RC, Holman L, van Zweden JS, Romero C, Oi CA, d'Ettorre P, Khalesi M, Billen J, Wäckers F, Millar JG, Wenseleers T: **Conserved class of queen pheromones stops social insect workers from reproducing.** *Science* 2014, **343**:287–290.
62. Singer T, Espelie K, Gamboa G: **Nest and nestmate discrimination in independent-founding paper wasps.** In *Pheromone Communication in Social Insects.* Edited by Vander Meer R, Breed M, Winston M, Espelie K. Boulder, CO: Westview Press; 1998:104–125.
63. Espelie KE, Hermann HR: **Congruent cuticular hydrocarbons: biochemical convergence of a social wasp, an ant and a host plant.** *Biochem Syst Ecol* 1988, **16**:505–508.
64. Bloch G, Borst DW, Huang ZY, Robinson GE, Cnaani J, Hefetz A: **Juvenile hormone titers, juvenile hormone biosynthesis, ovarian development and social environment in *Bombus terrestris*.** *J Insect Physiol* 2000, **46**:47–57.
65. Van Doorn A: **Factors influencing dominance behavior in queenless bumblebee workers *Bombus terrestris*.** *Physiol Entomol* 1989, **14**:211–222.
66. van Doorn A: **Investigations into the regulation of dominance behavior and of the division of labor in bumblebee colonies *Bombus terrestris*.** *Neth J Zool* 1987, **37**:255–276.
67. Agrahari M, Gadagkar R: **Juvenile hormone accelerates ovarian development and does not affect age polyethism in the primitively eusocial wasp, *Ropalidia marginata*.** *J Insect Physiol* 2003, **49**:217–222.
68. Zera AJ, Zhao Z, Kaliseck K: **Hormones in the field: evolutionary endocrinology of juvenile hormone and ecdysteroids in field populations of the wing-dimorphic cricket *Gryllus firmus*.** *Physiol Biochem Zool* 2007, **80**:592–606.
69. Ferreira PG, Patalano S, Chauhan R, Ffrench-Constant R, Gabaldon T, Guigó R, Sumner S: **Transcriptome analyses of primitively eusocial wasps reveal novel insights into the evolution of sociality and the origin of alternative phenotypes.** *Genome Biol* 2013, **14**:R20.
70. Simola DF, Wissler L, Donahue G, Waterhouse RM, Helmkampf M, Roux J, Nygaard S, Glastad K, Hagen DE, Viljakainen L, Reese JT, Hunt BG, Graur D, Elhaik E, Kriventseva E, Wen J, Parker BJ, Cash E, Privman E, Childers CP, Munos-Torres MC, Boomsma JJ, Bornberg-Bauer E, Currie C, Elsik CG, Suen G, Goodisman MAD, Keller L, Liebig J, Rawls A, *et al*: **Social insect genomes exhibit dramatic evolution in gene composition and regulation while preserving regulatory features linked to sociality.** *Genome Res* 2013, **23**:1235–1247.
71. Nascimento FS, Tannure-Nascimento IC, Zucchi R: **Behavioral mediators of cyclical oligogyny in the Amazonian swarmfounding wasp *Asteloeca ujhelyii* (Vespidae, Polistinae, Epiponini).** *Insectes Sociaux* 2004, **51**:17–23.
72. Zera AJ: **Endocrine analysis in evolutionary-developmental studies of insect polymorphism: hormone manipulation versus direct measurement of hormonal regulators.** *Evol Dev* 2007, **9**:499–513.
73. Noda SCM, Shima SN, Noll FB: **Morphological and physiological caste differences in *Synoeca cyanea* (Hymenoptera, Vespidae, Epiponini) according to the ontogenetic development of the colonies.** *Sociobiol* 2003, **41**:547–570.
74. Lin HR, Dusset C, Huang ZY: **Short-term changes in juvenile hormone titers in honey bee workers due to stress.** *Apidologie* 2004, **35**:319–327.
75. Huang ZY, Robinson GE, Borst DW: **Physiological correlates of division of labor among similarly aged honey bees.** *J Comp Physiol A* 1994, **174**:731–739.

76. Goodman WG, Coy DC, Baker FC, Xu L, Toong YC: **Development and application of a radioimmunoassay for the juvenile hormones.** *Insect Biochem* 1990, **20**:357–364.
77. Hartfelder K, Bitondi M, Brent C, Guidugli-Lazzarini K, Simões Z, Stabentheiner A, Tanaka ED, Wang Y: **Standard methods for physiology and biochemistry research in *Apis mellifera*.** *J Apicultural Res* 2013, **52**:1–47.
78. Hartfelder K, Feldlaufer MF: **Relationship of the neutral sterols and ecdysteroids of the parasitic mite, *Varroa jacobsoni* to those of the honey bee, *Apis mellifera*.** *J Insect Physiol* 1997, **43**:541–545.
79. Bollenbacher WE, O'Brien MA, Katahira EJ, Gilbert LI: **A kinetic analysis of the action of the insect prothoracicotropic hormone.** *Mol Cell Endocrinol* 1983, **32**:27–46.
80. Warren JT, Gilbert LI: **Ecdysone metabolism and distribution during the pupal-adult development of *Manduca sexta*.** *Insect Biochem* 1986, **16**:65–82.
81. Carlson DA, Roan CS, Yost RA, Hector J: **Dimethyl disulfide derivatives of long chain alkenes, alkadienes, and alkatrienes for gas chromatography/mass spectrometry.** *Anal Chem* 1989, **61**:1564–1571.
82. Aitchison J: **The statistical analysis of compositional data.** *J R Stat Soc B* 1982, **44**:139–177.

# 6

# The evolution of metazoan α-carbonic anhydrases and their roles in calcium carbonate biomineralization

Nathalie Le Roy[1], Daniel J Jackson[2*], Benjamin Marie[3], Paula Ramos-Silva[4,5] and Frédéric Marin[5]

**Abstract**

The carbonic anhydrase (CA; EC 4.2.1.1) superfamily is a class of ubiquitous metallo-enzymes that catalyse the reversible hydration of carbon dioxide. The α-CA family, present in all metazoan clades, is a key enzyme involved in a wide range of physiological functions including pH regulation, respiration, photosynthesis, and biocalcification. This paper reviews the evolution of the α-CA family, with an emphasis on metazoan α-CA members involved in biocalcification. Phylogenetic analyses reveal a complex evolutionary history of α-CAs, and suggest α-CA was independently co-opted into a variety of skeleton forming roles (*e.g.* as a provider of $HCO_3^-$ ions, a structural protein, a nucleation activator, *etc.*) in multiple metazoan lineages. This evolutionary history is most likely the result of multiple gene duplications coupled with the insertion of repetitive or non-repetitive low-complexity domains (RLCDs/LCDs). These domains, of largely unknown function, appear to be lineage-specific, and provide further support for the hypothesis of independent recruitment of α-CAs to diverse metazoan biocalcification processes. An analysis of α-CA sequences associated with biocalcification processes indicates that the domains involved in the activity and conformation of the active site are extremely conserved among metazoans.

**Keywords:** α-Carbonic anhydrase, Metazoa, Biocalcification, Biomineralization, Molecular evolution, Low complexity domains (LCDs), Repetitive low complexity domains (RLCDs)

## Introduction

Biocalcification refers to the physiological and molecular process by which living systems produce mineralized structures based on calcium salts, mainly calcium carbonate. From an evolutionary perspective, biocalcified structures represent a true innovation in the evolutionary history of metazoans. Biocalcitic structures fulfil a wide range of functions [1] such as: support and protection (*e.g.* sponge spicules, coral and bryozoan exoskeletons, mollusc and brachiopod shells, mineralized carapaces of crustaceans, calcified tubes of annelids, and the test, spines and teeth of echinoderms); organs for spatial equilibration (*e.g.* statoliths in jellyfishes, otoliths in fishes); calcium storage structures (*e.g.* gastroliths in crustaceans); organs for reproduction (*e.g.* gastropod love darts); desiccation prevention (*e.g.* avian eggshells); detoxification (*e.g.* calcium granules in molluscs). Despite the diversity and abundance of metazoan biocalcified structures, and the relevance of their evolutionary origins to our understanding of how metazoan life diversified, the molecular processes that underlie their formation remain generally poorly understood. Nevertheless, many recent molecular studies focused on a broad phylogenetic range of biomineralising animals have revealed that carbonic anhydrase (CA) is a key conserved enzyme involved in the process of biocalcification [2-22].

* Correspondence: djackso@uni-goettingen.de
[2]Courant Research Centre Geobiology, Georg-August-University of Göttingen, Goldschmidtstr 3, 37077 Göttingen, Germany
Full list of author information is available at the end of the article

### The carbonic anhydrase super-family

Carbonic anhydrases constitute a group of metallo-enzymes of the carbon-oxygen lyase sub-class and of the hydro-lyase infra-class. They catalyse the reversible hydration of $CO_2$ to form one bicarbonate ion and one proton: $H_2O + CO_2 \Leftrightarrow H_2CO_3 \Leftrightarrow HCO_3^- + H^+$. In most cases, the active site of these enzymes is composed of one Zinc(II) ion coordinated by three histidine residues. CAs occur in all three domains of life (Eubacteria,

Archaea and Eukaryota), and play roles in diverse metabolic pathways. The CA superfamily is subdivided into five families: α-CA, β-CA, γ-CA, δ-CA and ζ-CA [23,24]. Interestingly, large-scale phylogenetic analyses reveal no homology between these five families [25], and they are therefore thought to be the result of convergent evolution, *i.e.* there was no common ancestor from which these five families evolved. This viewpoint is supported by the fact that CA conformations differ between families: while monomers and dimers are the active forms of α-CAs [26], γ-CAs are active in homotrimeric form [27], and β-CAs can be active in dimeric, tetrameric, hexameric or octameric forms. The amino acids involved in the coordination of the zinc ion are three histidine residues, or two cysteines and one histidine residues in the case of β-CAs (*e.g.* dicotyledonous plant *Pisum sativum*; [28]).

Despite their similar enzymatic properties and distinct evolutionary histories, the presence of a given CA family in a certain lineage does not exclude the presence of another. For example, α-CAs are present in bacteria, algae, in the cytoplasm of green plants and fungi, and in metazoans [29-31], while the β-CAs are present in bacteria, fungi, algae and in the chloroplasts of monocotyledones and dicotyledones [32,33]; γ-CAs are found in archaea and bacteria [34-36], suggesting that they are the oldest class, while the δ-CAs and the ζ-CAs are present in some marine diatoms [37-39]. All CA families are characterized by the presence of a zinc ion, the exception being the ζ family, for which cadmium replaces zinc, for example in marine diatoms such as *Thalassiosira weissflogii* [37,40]. Because α-CAs have most often been associated with biocalcification roles in various metazoans, here we will focus our attention on this family.

### The α-carbonic anhydrases family

In 1933 Meldrum and Roughton [41], and simultaneously Stadie and O'Brien [42], discovered the first α-CA in vertebrate erythrocytes *via* its enzymatic activity. Today, with the advances in molecular techniques and the development of high-throughput sequencing methods, hundreds of α-CAs in many isoforms have been described for vertebrate and invertebrate metazoans.

Different α-CA isozymes are produced by distinct tissues. In mammals, they have diverse subcellular localizations: cytosolic (isozymes CAI, II, III, VII and XIII), membrane-bound (isozyme CAIV – GPI-link), transmembrane (isozymes CAIX, XII and XIV), mitochondrial (isozymes CAVA and VB) or extracellular (isozyme CAVI). They are involved in many biological functions including: acid–base balance, $CO_2$ transport, urea cycling, gluconeogenesis, fatty acids/amino acids synthesis and calcification/decalcification [43]. It is very likely that the primitive function of the α-CA was the intracellular regulation of acid–base balance, and/or the metabolism of $CO_2$ [44]. However, during evolution, the function of the enzyme was coopted for other physiological processes, such as biomineralization. In some metazoan groups, α-CAs have been described as a calcium-dissolving enzyme, as in the bones of vertebrates and the spicules of calcareous sponges such as *Sycon raphanus* [45]. Recently, eight CAs in murine, produced by ameloblasts, were identified to be involved in biomineralization during teeth development [46]. In corals, α-CA isoforms interact in the carbon cycle of the endosymbiotic zooxanthellae playing an indirect role in skeletal formation [19]. In the calcifying species listed in Table 1, we have summarized the presence of α-CAs potentially involved in the deposition of $CaCO_3$ biominerals. For some species, such as brachiopods, data are currently missing; for others, such as sponges, crustaceans or cnidarians, they are widely incomplete [7,9,14,47]. In molluscs, sequence data is relatively abundant, but are focused on a limited number of species, such as the pearl oyster (*Pinctada* sp.) or the giant owl limpet (*Lottia gigantea*) [11,21]. Within the deuterostomes, most of the available α-CA data in echinoderms are related to the genome sequencing of the sea urchin *Strongylocentrotus purpuratus* [48]. The urochordates and vertebrates also possess $CaCO_3$ structures (*e.g.* spicules, otoliths, eggshell). Indeed, α-CAs were identified in the calcifying tissues that produce those structures [3,4,6,49].

The α-CA family contains numerous isozymes, corresponding to the expression of a set of paralogous genes that are the product of gene duplication and speciation events that occurred throughout the Phanerozoic [65-67]. All genes encoding α-CAs are likely derived from a single ancestral gene, and are therefore homologous [35].

To date, very few molecular phylogenetic studies of α-CAs have been conducted in relation to the biocalcification processes [7,9,10,18,19]. These few previous studies were performed with CA sequences of interest from poriferan, cnidarian and molluscan representatives in addition to various invertebrate/vertebrate datasets (Figure 1). The first phylogenetic study of α-CAs [35] revealed a clear dichotomy between membrane-associated/secreted and cytosolic/mitochondrial α-CAs, and this pattern is generally found in subsequent studies (Figure 1) [7,9,10,18]. In all of these studies poriferan α-CAs form a monophyletic clade, and occupy an early branching position (Figure 1). In other respects α-CA tree topology can differ significantly, for instance the position of vertebrate GPI-linked α-CAs as a sister group to the poriferan α-CAs [68], or the early-branching position of the molluscan nacreins [10]. One significant difficulty that all phylogenetic analyses of the α-CAs face, is the lack of a full genomic complement of α-CAs from a diverse taxonomic range of metazoan representatives. Due to the ever-growing availability of whole genomes and transcriptomes from non-model

**Table 1 Biomineralization types and the presence of α-CA in various metazoan phyla**

| | Envir. | $CaCO_3$ structure | Predominant form of mineral | Organic phase | Calcifying cells/ tissue | α-CAs | Technique | Tissue localization | Ref. |
|---|---|---|---|---|---|---|---|---|---|
| **PORIFERA** (appearance in Ediacaran) | | | | | | | | | |
| Hexactinellida | DS | Spicules | Silica, aragonite | nd | Sclerocytes | nd | - | - | |
| Demospongia | M, FW | | | "Spongine", silicatein, silintaphin-1, chitin | | ✓ | RT-PCR & EST lib. | Adult whole organism, embryos | [7] |
| Calcarea | M | | Calcite, aragonite | nd | | ✓ | EST lib. | Adult whole organism | [45] |
| **CNIDARIA** (appearance in Ediacaran) | | | | | | | | | |
| Octocorallia | M | Sclerites, spicules, skeleton | High-Mg calcite | Glycoproteins, lipids, polysaccharides… | Calicoblastic cells | ✓ | biochemistry, RT-PCR, EST lib. | Spicules, tentacules, polyps, larva, embryos | [50] |
| Hexacorallia | | Hard skeleton | Aragonite | Glycoproteins, lipids, polysaccharides… | | ✓ | EST lib., RT-PCR, genome | Calicoblastic cells | [9,14,51-53] |
| Hydrozoa | M, FW | Granules, skeleton | Aragonite | nd | | ✓ | EST lib., genome | Polyps | [54] |
| **BRACHIOPODA** (appearance in Cambrian) | | | | | | | | | |
| Craniiformea | M | Shell | Low-Mg calcite | nd | Mantle cells | nd | - | - | |
| Linguliformea | | | $Ca_3(PO_4)_2$ | nd | | nd | - | - | |
| Rynchonelliformea | | | Low-Mg calcite | Glycoproteins + polysaccharides… | | nd | - | - | |
| **MOLLUSCA** (appearance in Cambrian) | | | | | | | | | |
| Solenogastra & Caudofoveata | M | Spicules | Calcite, aragonite | nd | Mantle cells | nd | - | - | |
| Polyplachophora | | | | | | nd | - | - | |
| Bivalvia | M, FW | Shell & granules | | Periostracum/OM | | ✓ | RT-PCR, EST lib. | Mantle, gills, adductor muscles, larva | [2,11,13,15] |
| Gastropoda | M, FW, T | | | | | ✓ | RT-PCR, EST lib., genomes | Mantle, nervous system, whole body | [16,21] |
| Cephalopoda | M | Internal shell | | Nautilin 63 | | ✓ | - | - | [55,56] |
| **ANNELIDA** (appearance in Cambrian) | | | | | | | | | |
| Pogonophora | M | Tube & soft granules | Calcite, aragonite | Mucopoly-saccharides | Epidermal glands | ✓ | RT-PCR | Branchial plume | [57] |
| **ARTHROPODA** (appearance in Cambrian) | | | | | | | | | |
| Hexapoda | A, T | Exoskeleton | Calcite | Chitin, proteinaceous organic matrix | Cuticle epidermis | ✓ | RT-PCR, EST lib., genomes | Whole organism | [54] |
| Pancrustacea | A, M, FW, T | Exoskeleton, gastrolithes, sternal plates, caecal concretions | Calcite, vaterite, amorphous $CaCO_3$ | | Cuticle epidermis, stomacal disc | ✓ | RT-PCR, EST lib. | Larvae, oocytes, gills, epidotes, brain, antennal glands… | [47,58-60] |

**Table 1 Biomineralization types and the presence of α-CA in various metazoan phyla** *(Continued)*

| | | | | | | | | | |
|---|---|---|---|---|---|---|---|---|---|
| **ECHINODERMATA** (appearance in Cambrian) | | | | | | | | | |
| Echinoidea | M | Test, teeth, spicules, spines | High-Mg calcite | OM | Syncitium/ "fenestrated" mineral | ✓ | EST lib., genomes | Pluteus ams, calcifying organic matrix | [48,61] |
| Crinoidea | | Articulated ossicules | nd | nd | nd | nd | - | - | |
| Ophiuroidea | | | nd | nd | nd | nd | - | - | |
| Asteroidea | | Spicules | nd | nd | nd | ✓ | Biochemistry | - | [62] |
| Holothuridea | | Tiny spicules or ossicules | nd | nd | nd | nd | - | - | |
| **UROCHORDATA** (appearance in Cambrian) | | | | | | | | | |
| Ascidiacea | M | Spicules | Calcite | OM | Sclerocytes | ✓ | EST lib. | nd | [54] |
| Appendicularia | | nd | nd | nd | nd | ✓ | Genome | nd | [63] |
| **VERTEBRATA** (appearance in Silurian) | | | | | | | | | |
| Actinopterygii | M, FW | Otoliths | Aragonite, vaterite | Endolymphatic constituants | Sacculus | ✓ | RT-PCR | Retina, kidney, erythrocytes, gills… | [49,64] |
| Aves | T | Eggshell | Calcite | Uterine fluid | Distal oviduct | ✓ | Genome | Testis, retina… | [54] |

For each taxon we collected several lines of information: the environment in which each taxon can be found (A: amphibious; DS: deep-sea; FW: freshwater; M: marine; T: terrestrial); the general calcium carbonate structure; the predominant form of the mineral; the general nature of the organic phase within the mineral (OM: organic matrix); the source of the calcifying tissue/cells; whether carbonic anhydrases have been specifically detected within these cells; the method of identification (EST lib.: EST library); tissues where CAs were detected; references to studies from which this information was collated. nd: not determined.

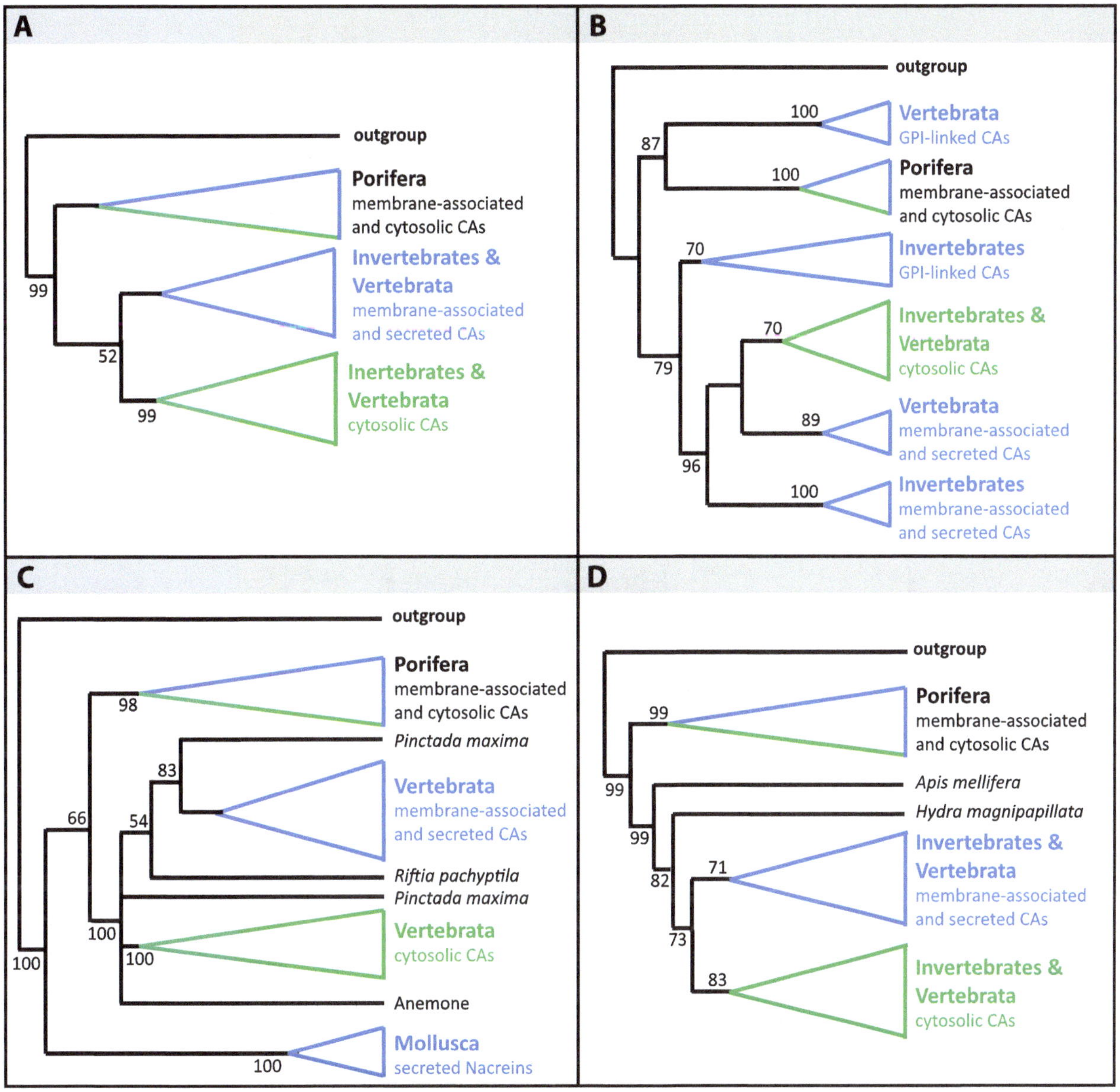


**Figure 1 Simplified phylogenetic trees from four studies of α-CAs derived from metazoan taxa.** These trees were modified from Jackson et al., 2007 **(A)**, Moya et al., 2008 **(B)**, Jackson et al., 2010 **(C)** and Moya et al., 2012 **(D)** [7,9,10,18]. Membrane-associated/secreted CAs are indicated in blue and cytosolic/mitochondrial CAs are indicated in green. Bootstrap or posterior probability values are indicated.

organisms, this issue is being addressed and will allow more complete analyses to be performed.

### Evolution of metazoan α-CAs

We explored the evolutionary relationships of metazoan α-CAs using two datasets compiled from publicly available genomic, EST and RNASeq databases (*i.e.* JGI, SpBase and NCBI [48,54,69]) derived from several metazoan species. In compiling these datasets we endeavoured to employ only those species with complete, or likely close to complete, genomic complements of α-CA. We explicitly excluded α-CA sequences from incomplete genomes and incomplete and/or highly divergent α-CA sequences that were insufficiently conserved for alignment. We performed two phylogenetic analyses using Maximum likelihood [70,71] and Bayesian methods [72,73] (Additional file 1) that generated largely congruent tree topologies. The first analysis was conducted using 138 α-CA sequences (6 α-CAs from non-metazoans as outgroups, and 132 α-CAs from metazoans, Figure 2, Additional files 2 and 3: Figure S1 and S2), and the second, from 98 α-CA sequences (3 α-CAs from protists and 94 α-CAs from metazoans with a focus on molluscan α-CAs likely to be involved in biocalcification, Figure 3, Additional file 4: Figure S3).

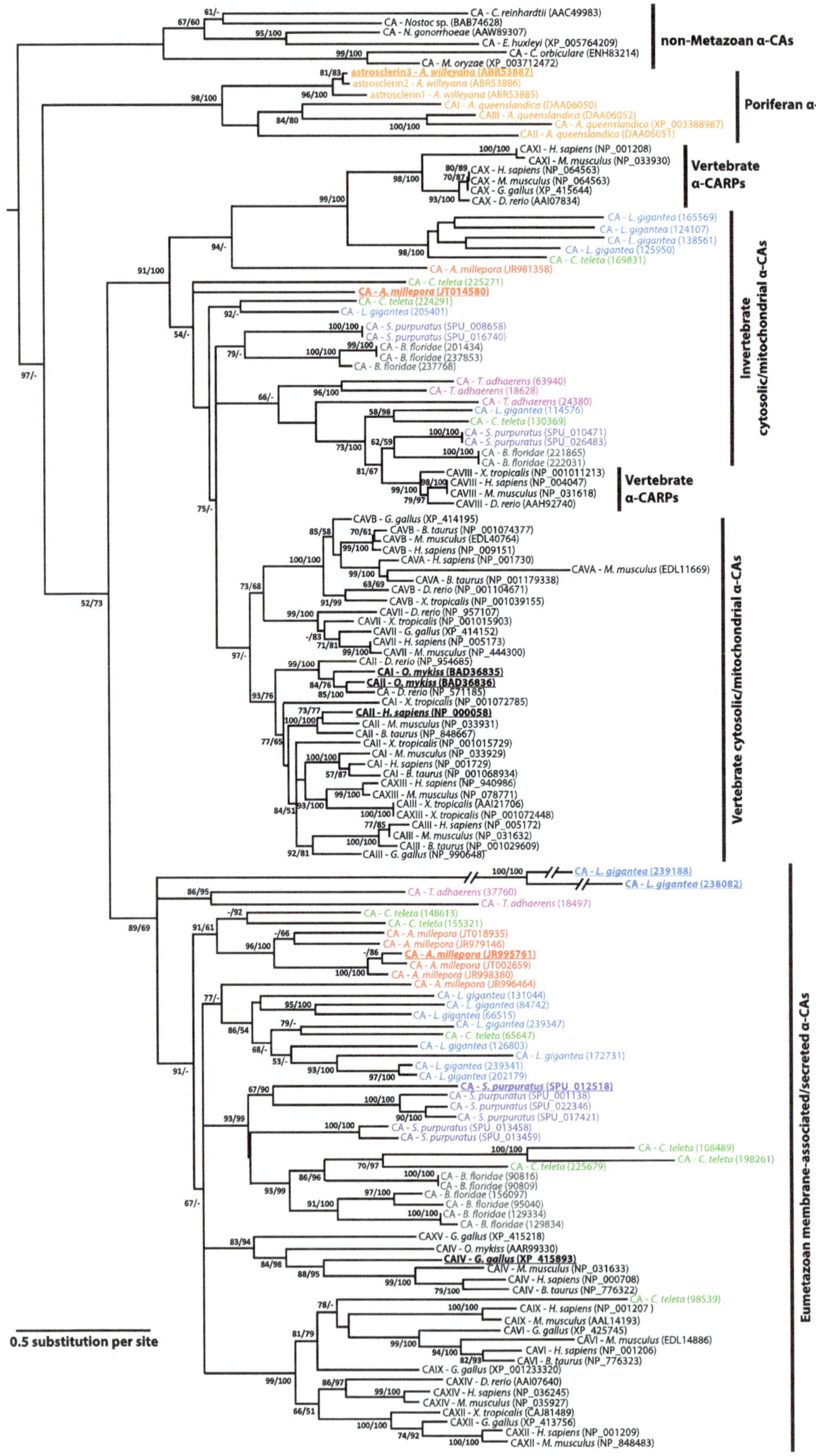


**Figure 2** (See legend on next page.)

(See figure on previous page.)

**Figure 2 Phylogenetic reconstruction of metazoan α-CAs relationships.** The phylogeny was reconstructed using PhyML and Bayesian inference methods, the topology presented is that resulting from the PhyML method (see Additional file 2: Figure S1 for the Bayesian topology). Node support values indicate aLRT values, and Bayesian posterior probabilities expressed as a percentage respectively. Only values above 50 are indicated. Sequence data was derived from 4 vertebrate species (*Homo sapiens, Mus musculus, Bos taurus, Gallus gallus, Oncorhynchus mykiss and Xenopus tropicalis*, in black); 1 cephalochordate (*Branchiostoma floridae*; in grey); 1 sea urchin (*Strongylocentrotus purpuratus*, in purple); 1 annelid (*Capitella teleta*, in green); 1 mollusc (*Lottia gigantea*, in blue); 1 coral (*Acropora millepora*, in red); 1 placozoan (*Trichoplax adhaerens*, in pink); 2 sponges (*Amphimedon queenslandica* and *Astrosclera willeyana*, in orange); 1 fungi (*Colletotrichum orbiculare*, in black); 2 chromalveolates (*Emiliania huxleyi* and *Vaucheria litorea*, in black); 1 chlorophyte (*Chlamydomonas reinhardtii*, in black); and 2 bacteria (*Nostoc* sp. *and Neisseria gonorrhoeae*, in black). The α-CAs involved in biocalcification are in bold/underlined. CARPs: Carbonic Anhydrase Related-Proteins.

In these trees, as previously mentioned, α-CA isozymes can be located in one of four broad positions: cytosolic, membrane-associated (*i.e.* transmembrane, membrane-bound, GPI-linked), secreted and mitochondrial. The location of a given α-CA seems to be married to its function; for example the majority of molluscan α-CAs known to be involved in shell formation are located extracellularly [2,15,21], while CAV members are thought to be involved in insulin regulation in the pancreas [74]. The fact that the ancient (>550 MYA) relationships of α-CA family members are largely reflected in where they are located highlights the important role that gene duplication played in

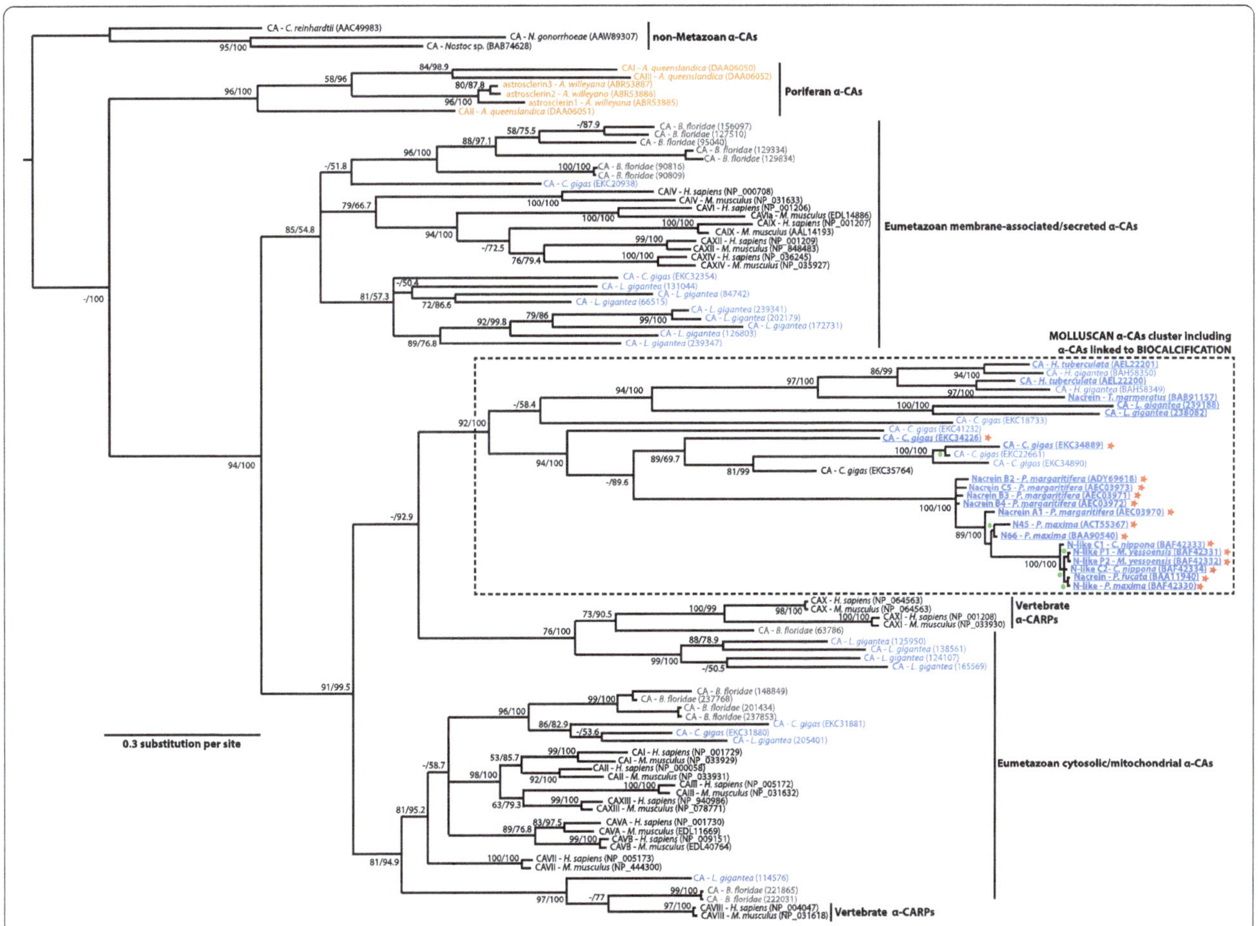


**Figure 3 Phylogenetic reconstruction of metazoan α-CAs with a focus on molluscan sequences.** The phylogeny was reconstructed using PhyML and Bayesian inference methods, the topology presented is that resulting from the Bayesian method (see Additional file 4: Figure S3 for the PhyML topology). Node support values indicate aLRT values, and Bayesian posterior probabilities expressed as a percentage respectively. Only values above 50 are indicated. Sequence data was derived from 2 vertebrate species (*Homo sapiens* and *Mus musculus*, in black); 1 cephalochordate (*Branchiostoma floridae*, in grey); 6 bivalve molluscs (*Crassostrea gigas, Crassostrea nippona, Patinopecten yessoensis, Pinctada fucata, Pinctada margaritifera* and *Pinctada maxima*, in blue); 4 gastropod molluscs (*Haliotis gigantea, Haliotis tuberculata, Lottia gigantea, Turbo marmoratus*, in blue); 2 sponges (*Amphimedon queenslandica* and *Astrosclera willeyana*, in orange); and 3 unicellular organisms (*Nostoc* sp., *Chlamydomonas reinhardtii* and *Neisseria gonorrhoeae*, in black). The α-CAs involved in biocalcification are in bold/underlined. Nacreins and nacrein-likes are indicated by a red star. Well supported aLRT/bootstrap values (>90) are indicated by a green point due to lack of space. CARPs: Carbonic Anhydrase Related-Proteins.

allowing this class of enzymes to diversify and acquire novel functions.

In the tree of Figure 2, we observe that the seven poriferan α-CAs share a monophyletic relationship, and that they occupy an early-branching position. This finding is congruent with earlier publications [7,10,18] (see Figure 1 A, C and D), and confirms a recent analysis reported by Moya et al. [9]. While we do not observe a monophyletic clade of secreted and membrane-bound α-CAs, we do recover a highly supported clade of cytosolic and mitochondrial α-CAs. Puzzling is also the fact that the poriferan α-CAs are excluded from the membrane-associated/secreted cluster in the different phylogenetic reconstructions [7,9,10,18]. Among them, six are membrane-associated or secreted while one is cytosolic (CAII *Amphimedon queenslandica*, GenBank:DAA06051) [7]. The presence of membrane-associated/secreted α-CAs both in the poriferan and in eumetazoan α-CAs, suggests that the ancestral metazoan α-CA was membrane-associated or secreted. It must be noted however, that there is a putative cytosolic α-CA in the poriferan cluster, which belongs to the sponge *A. queenslandica* (CAII, GenBank:DAA06051). This may be explained by either the internalization of an ancestral membrane-associated/secreted enzyme, or the early evolution of a membrane-associated/secreted CA in the poriferan lineage. Nevertheless, it is obvious that the dearth of α-CA sequence data from the phylum Porifera is a critical point for any interpretation of the evolution of metazoan α-CA.

### The distribution of metazoan α-CAs in relation to biocalcification

Our second phylogenetic analysis (Figure 3) includes 44 α-CAs from molluscan species, and sheds light on the evolution of A-CAs involved in molluscan biocalcification. One key-feature of this analysis is that molluscan mantle-secreted α-CAs and nacrein/nacrein-like sequences clearly group together. In their 2010 paper, Jackson et al. [10] described a basal position of molluscan nacreins relative to cytosolic/mitochondrial and secreted/membrane-associated α-CAs. Our phylogenetic reconstruction uses a more complete α-CA dataset, and generates a nacrein-type α-CA clade in a more derived position (Figure 3). Together with other molluscan α-CAs (*Lottia gigantea*, JGI:239188, 238082, *Crassostrea gigas*, GenBank:EKC18733, EKC41232, EKC34226, EKC34889, EKC22661, EKC34890, *Haliotis tuberculata*, GenBank:AEL22201, AEL22200, *Haliotis gigantea*, GenBank:BAH58350, BAH58349, *Turbo marmoratus*, GenBank:BAB91157), the nacrein/nacrein-like cluster is the sister group of a clade containing vertebrate related-protein α-CAs (CARPs: CAX and CAXI). Nacreins are clustered with other molluscan α-CAs likely to be involved in biocalcification (*e.g. Lottia gigantea*, JGI:239188, 238082).

With regard to its complement of 17 α-CAs, the giant owl limpet, *L. gigantea*, is unusual. Currently, this represents the largest known repertoire of α-CAs for an invertebrate. Sixteen of these sequences were used in our phylogenetic reconstruction, with 8 of these located within the membrane-associated/secreted clade, 6 within the vertebrate cytosolic/mitochondrial α-CAs and CARPs, and 2 (JGI:239188, 238082]) within the molluscan cluster containing nacreins (Figure 3). Two recent proteomic analyses confirmed the presence of these 2 α-CAs (JGI:239188, 238082) in the shell of *L. gigantea* [17,21]. Nonetheless, among the 8 sequences that are distributed in the membrane-associated/secreted cluster, 3 have a signal peptide (JGI:239341, 202179, 172731) while 5 are incomplete sequences with missing N-termini. We also note the presence of a putatively cytosolic α-CA (JGI:239347) in both trees (Figures 2 and 3), within the membrane-associated/secreted α-CAs. Even so, if these sequences are genuine cytosolic α-CAs, this may be explained by the evolutionary internalisation of a formerly secreted α-CA, as may be the case for the poriferan DAA06051 sequence (see above). Alternatively, these proteins may be secreted by non-canonical secretory pathways, or represent incomplete sequencing/assembly/or gene predictions of the N-terminal region of the protein. Moreover, the location of the majority of the invertebrate α-CAs are yet to be experimentally validated, and therefore the groupings that we observe in our phylogenetic analysis must be considered in that light. Despite these uncertainties, the overall pattern of molluscan α-CA evolution appears to be one that took place largely independently of vertebrate α-CA evolution (see dashed square in Figure 3).

Similar to the *L. gigantea*, the 10 α-CAs from the purple sea urchin *Strongylocentrotus purpuratus*, and the 8 from the stony coral *Acropora millepora*, are spread across the membrane-associated/secreted cluster (6 *A. millepora* α-CAs and 6 *S. purpuratus* α-CAs), and in the cytosolic/mitochondrial cluster (2 *A. millepora* α-CAs and 4 *S. purpuratus* α-CAs; Figure 2). One of the *S. purpuratus* α-CAs (JGI:SPU_012518, also called Sp-CAra7LA) has been identified in the extracellular matrix of the adult urchin test [12]. In our phylogenetic tree, Sp-Cara7LA is localized in the same cluster that contains the 2 α-CAs involved in the shell biocalcification of *L. gigantea* (JGI:239188, 238082) [17,69] and the 2 others involved in the skeleton precipitation of *A. millepora* (GenBank:JR995761 and JT014580) [18]. Despite the large phylogenetic distance between these three organisms they share a common α-CA ancestor that could be independently recruited in the purple sea urchin, in the giant owl limpet and in the stony coral, to be included in their specific biocalcification processes.

The dispersion of α-CAs involved in biocalcification in cnidarians, molluscs and echinoderms (Figures 2 and 3) supports the hypothesis of the independent evolution of calcifying matrices in these different metazoan lineages. This is also true for vertebrate α-CAs taking part in the calcification processes (*e.g.* otolithes, eggshell). Our finding is congruent with the proteomic analyses on skeletal matrices that show that calcifying secretory repertoires are highly divergent in different metazoan phyla [22,75], even within one clade such as molluscs [10,15,21,68,76].

The position of α-CAs involved in (1) the calcification of eggshell of the red junglefowl *Gallus gallus* (GenBank:XP_415893, [77]), of otoliths of the rainbow trout *Oncorhynchus mykiss* (GenBank:BAD36835, BAD36836, [49,64]) and (2) the decalcification of bones of the human *Homo sapiens* (GenBank:NP_000058, [78]), still remains inside vertebrate α-CA clades (Figure 2). Nevertheless, they are not monophyletic. Among these α-CAs playing a role in the biocalcification, CAII of *H. sapiens* and CAI and II of *O. mykiss* are clustered with cytosolic/mitochondrial α-CAs while CAIV of *G. gallus* is included within membrane-associated/secreted α-CAs. The use of different α-CA isoforms in these species could be explained by an independent recruitment of α-CAs for biocalcification functions inside the vertebrate subphylum.

The phylogenetic distribution of α-CAs involved in metazoan biocalcification (Figure 2) reveals a complex evolutionary history that requires more whole-genome data in order to clarify how the different α-CA isoforms evolved. Obviously, broader taxon sampling from groups such as sponges would serve to shed light on the evolutionary history of these biomineral forming mechanisms.

### Conserved domains in α-CAs involved in the metazoan biocalcification processes

The catalytic domain of all α-CAs is highly conserved. In order to identify the most conserved domains in α-CAs involved in biocalcification, we calculated the substitution rate per site using the methodology described in Petit et al. (2006) [79] and Martin et al. (2009) [80]. Briefly, the topology of the tree provided by the Maximum Likelihood method with MEGA 5.1 (using an alignment of 34 sequences for the alignment of α-CAs involved in the calcification process and another alignment of 33 sequences from cytosolic/mitochondrial α-CAs in eumetazoans, Additional file 5: Table S1) was considered as the user tree to run the parsimony program Protpars included in the PHYLIP v3.69 package [81]. A moving average (window size = 5) was calculated for each substitution number per site, and then plotted in parallel with the structural domains of α-CAs (Figure 4). The major reversed peaks correspond to the catalytic sites that contain histidine residues, but also to conserved domains that maintain the conformation of the catalytic site for a correct activity. Seven short conserved α-CA domains (from 4 to 8 residues) spread along the sequence were identified: QSPI, LHVH, GSEH, EAHL, FVVVGVFL, GSLTTP and ESVLW. These conserved regions represent about 25% of the overall sequences used for the analysis. These domains contain residues essential for the regeneration of the active site [51,82-86], and an amino acid involved in the proton shuffling process during the catalytic activity. Two of the poriferan α-CAs (Astrosclerin 1 and 2: GenBank: ABR53885, ABR53886) are characterized by the replacement of one of the three zinc coordinating histidine residues by a glutamine. In accordance with this Astrosclerin 1 was reported to be catalytically inactive, however recombinant Astrosclerin 2 could not be generated [7]. Such mutations in the active site of α-CA are likely to cause a weakening or elimination of zinc-binding capacity, and therefore of catalytic function [87].

The substitution rate is two times less in the α-CAs involved in the biocalcification (1.6 to 8.3; graph A Figure 4) than in the cytosolic/mitochondrial α-CAs (4 to 16.8; graph B Figure 4). In the cytosolic/mitochondrial α-CAs the substitution rate is higher in regions that are not conserved in comparison with the substitution rate in the first graph (A, Figure 4). In general it seems that α-CAs involved in the biocalcification process are highly conserved when compared to the cytosolic/mitochondrial ones. We observe that the independent recruitment of these α-CAs to the biocalcification process seems to be followed by a higher conservation of amino acid sequences than in α-CAs involved in other functions. This pressure is probably link to the biocalcification function, which needs a high conserved α-CA activity as we can observe in the cluster of molluscan α-CAs linked to the formation of the shell (*e.g.* Figure 3). The catalytic activity could appear at different levels of this process. Indeed, CAs can provide $HCO_3^-$ ions and/or $CO_2$ depending of the direction of the reaction. The catalytic hydration of $CO_2$ produces protons in addition to $HCO_3^-$ and these protons have been suggested to promote the uptake of $Ca^{2+}$ during larval shell formation in the freshwater gastropod *Lymnaea stagnalis* [88]. α-CAs may also play structural roles in biomineralization; the enzyme is occluded in the extracellular organic matrix of the biomineral structure as in some coral, molluscan and urchin species (*e.g. A. millepora*, *U. pictorum*, *L. gigantea*, *S. purpuratus*).

### Low Complexity Domains (LCDs) in α-CAs related to $CaCO_3$ biomineralization

*In silico* sequence analysis (Additional file 1) focused on the identification and characterization of LCD domains present in α-CA sequences. We selected 33 α-CAs among which 33 have been demonstrated to be present in calcifying tissues or in the $CaCO_3$ biomineral itself (*e.g.* nacreins in the shell of pearl oysters; [2]). Thus, we

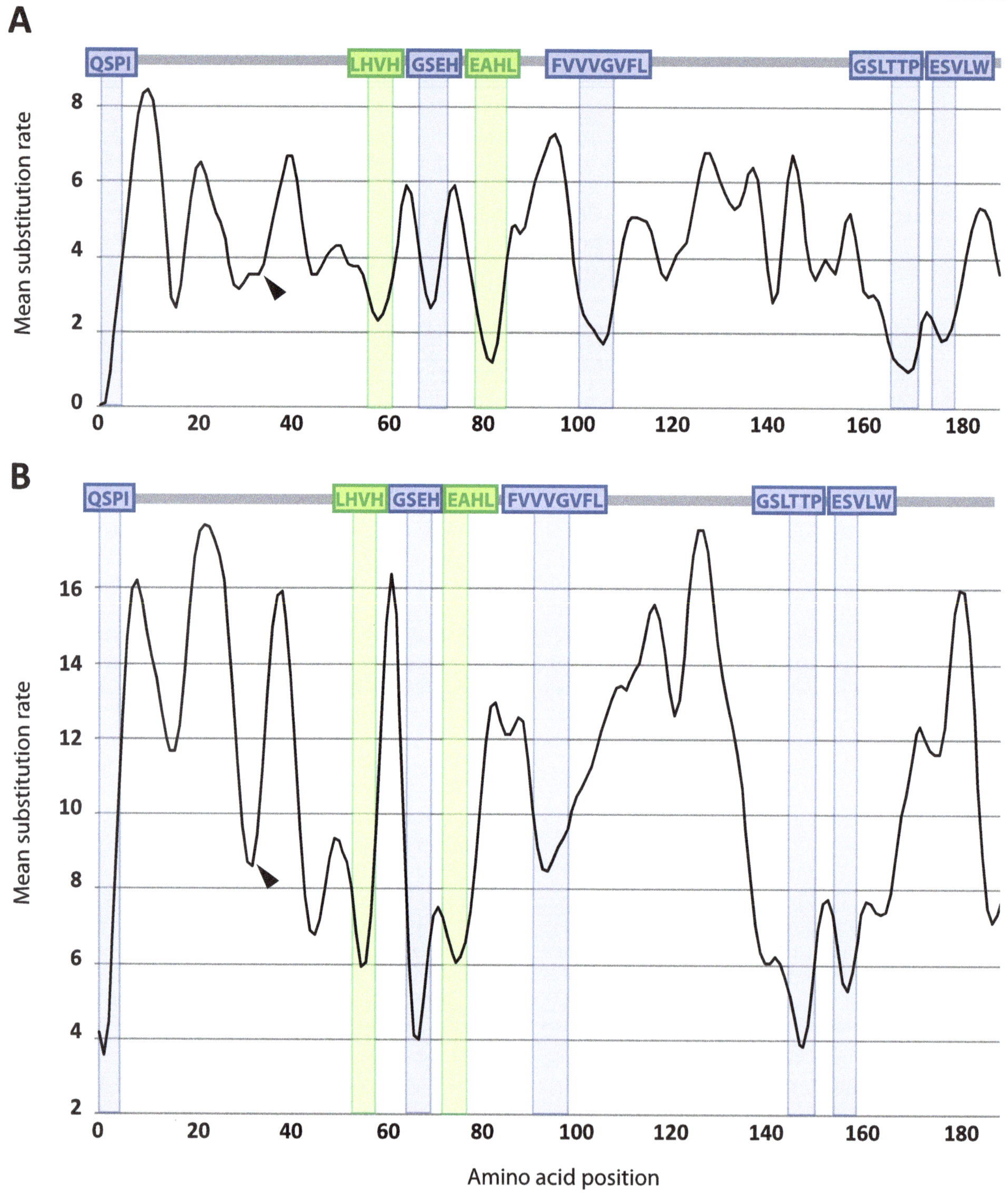


**Figure 4 Substitution rates within metazoan α-CAs.** The number of substitutions per site between poriferan and vertebrate sequences was calculated for an alignment of amino acids derived from 34 α-CA sequences involved in biocalcification **(A)**, and an alignment of 33 cytosolic/mitochondrial α-CAs **(B)**. Sequences used for the two analyses are listed in the Additional file 5: Table S1. The first position begins with the first α-CA conserved peptide (QSPI). Blue peptides are involved in the CA conformation and green peptides contain histidine residues that coordinate the $Zn^{2+}$ ion in the active site. The black arrowhead indicates the substitution rate of the proton shuffling amino acid.

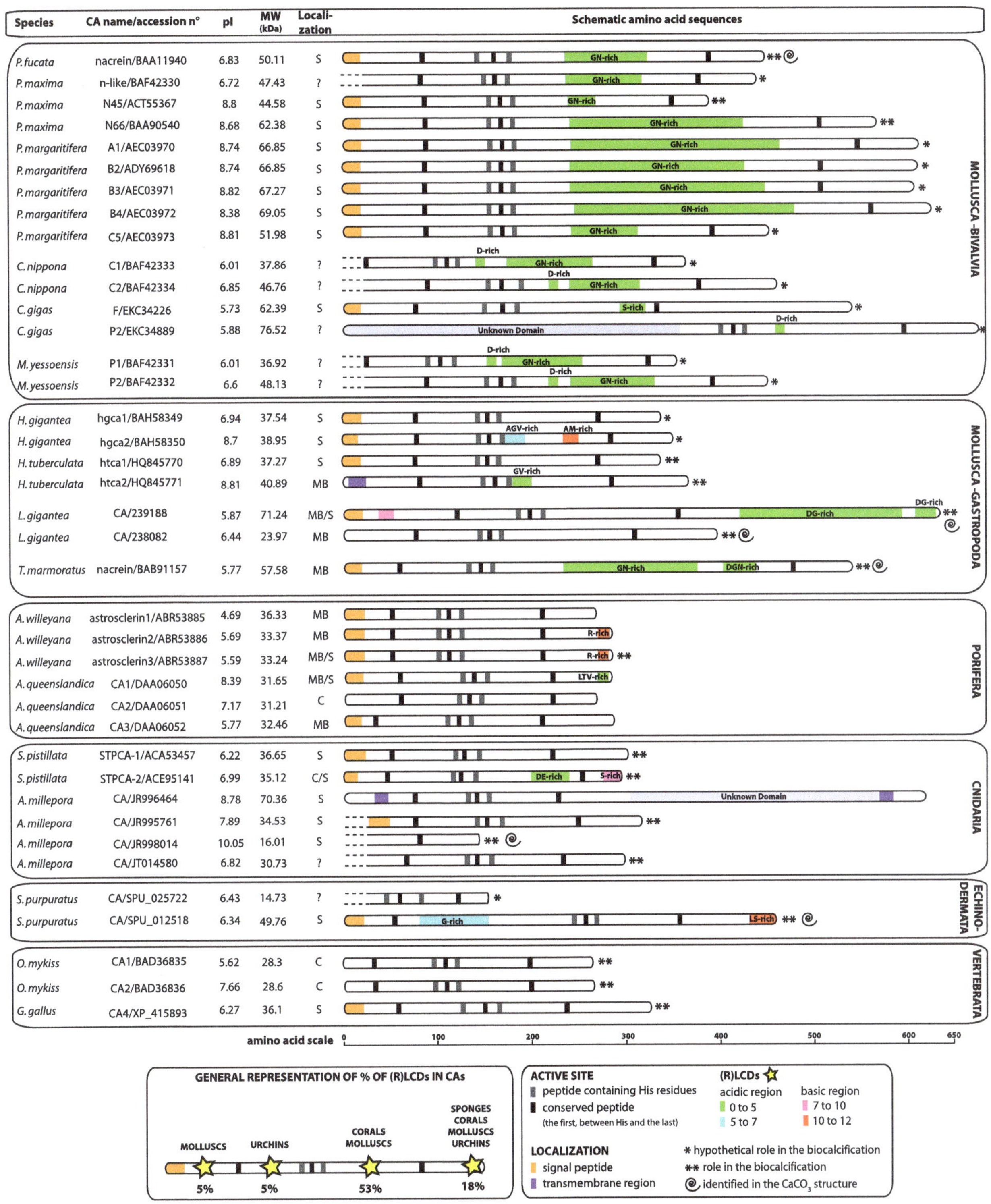


**Figure 5 LCDs in α-CAs, which are presumably associated with $CaCO_3$ biominerals.** Schematic representations of 38 α-CA sequences that are membrane-associated (MA), secreted (S) or cytosolic (C).

assume that these α-CAs are directly involved in the biocalcification process.

The full-length sequences of these biomineral-associated α-CAs varies from 278 (*A. willeyana*, GenBank:ABR53885) to 678 amino acids (*C. gigas*, GenBank:ECK34889). This disparity in size is primarily due to the presence/absence of LCDs of variable length. Among the selected 39 biomineralization-associated α-CA sequences, 25 of them contain LCDs, which can make up to half the length of the sequence. Figure 5 presents a schematic representation of the primary structure of these 39 CAs.

With three exceptions (*C. gigas*, GenBank:EKC34889; *L. gigantea*, JGI:239188 and *S. purpuratus*, JGI:SPU_012518), all of these LCDs are located in the C-terminal half of the protein. The homology of these domains is difficult to discern as they differ in length, and are composed of a variety of amino acid sequences (Figure 5). Moreover, our phylogenetic analyses indicate that α-CAs containing these LCDs are not necessarily orthologous, but were likely acquired independently from lineage to lineage. The most common LCD is the GXN repeat found in molluscan nacreins [2,89]. Within the bivalves, the homology of such GXN domains is hinted by the degree of nucleic acid similarity (Additional file 6: Table S2). This similarity does not extend to the GN repeat found in the nacrein of the gastropod *Turbo marmoratus* (Figure 6). Beside issues of LCD homology is the question of origin. In bivalves, the repetition of the GXN motif may be the result of a partial gene duplication followed by a series of unequal recombination events, as it has been suggested for other proteins such as mucins [90-92]. Such events result in the successive, tandem addition of short motifs. Transposable elements may also have generated such RLCDs in molluscan biomineralizing proteins. To test this idea we searched for signatures of transposases and the presence of inverted repeats in RLCD containing α-CA using the RepeatMasker software [93]. We could not find evidence of either in these sequences.

There is strong proteomic evidence that some nacreins are occluded in the calcified molluscan shell, and consequently belong to the so-called "calcifying matrix" [15,94-97]. It is known that the GXN repeat may display different functions related to calcification: (1) it may interact with calcium ions in solution, by behaving as a low affinity, high capacity calcium-binding domain, and could therefore be involved in $CaCO_3$ nucleation [98,99]. This suggests that nacreins work as both as enzymes, converting $CO_2$ into bicarbonate, and also as $CaCO_3$ nucleating polymers. (2) The GXN repeat domain is also known to

| | Species | CA name - accession n° | LCD & RLCD details |
|---|---|---|---|
| PORIFERA | *A. willeyana* | astrosclerin2 - ABR53886 | ...GEDKKRKRRNNHRGRHHHG... |
| | *A. willeyana* | astrosclerin3 - ABR53887 | ...GEDKKRKRRNNRHGRDHHG... |
| | *A. queenslandica* | CA1 - DAA06050 | ...LMSTTTALLAFLLIAVLA... |
| CNIDARIA | *S. pistillata* | STPCA-2 - ACE95141 | ...DEDEDDEDIAGDDPDAVEEEEVDAKKKIK... |
| | *A. millepora* | CA - JR996464 | ...RSFQRPSSPCSKVSDTLMECSVLIGNKVVFYAAEVNMCTLPATVTLQVSQPDIGYKFNKMIYSDIHSTVQIGNLDYGFTLELHSHVQETRPSTLHITINHANPLSSGIQKE VISADFEEKACQLGSCQKTKCNKLWCSGKAFGFTKQDLLYIGFSSTFKDLSTSSPSVHVNFYRNESLKWHLAESATFVNNAERKVHLLGNLVHASSSAKHKRGTLQSSQW EYVTFKAKITKTTNKVLYQLLMIPANGLYYRIFKDEITLD... |
| MOLLUSCA | *P. fucata* | nacrein - BAA11940 | ...DNNENGNGDNGNNGYNGDNGNN GDNGNNSYNGDNGNN GVNGNNGYNGDNGNN GDNGNNGYNGDNGNN GDN GNNGEN GNNGEN GNNGENG... |
| | *P. maxima* | nacrein-like - BAF42330 | ...DNNENGNGDNGNNGYNGDNGNN GDNGNNGYNGDNGNN GDNGNNGYNGDNGNN GDN GNNGEN GNNGEN GNNGDN GNNGENG... |
| | *P. maxima* | N45 - ACT55367 | ...NNGDNGNNCDNGNGDNGYNGDNGNSDG... |
| | *P. maxima* | N66 - BAA90540 | ...NNGDNGNNCDNGNNGNNDNNGNNGNNGNGNNGYNGNNGDNGNNGNGNGNNGYNGNNGYN GNNGNNGNGNNDNNGNDNNGNNGGN GNNGNN GNGNNGNNGNGNNGNNGGN GNNGNNGNSNNGNNGNGNNGNNGGN GNNGNNGNGNNENNGNGSNGNNGGN GNNGNN GDNGNGDNGYNGDNGNSDG... |
| | *C. nippona* | C1 - BAF42333 | ...DNNENGNGDNGNNGYNGDNGNN GDNGNNGYNGDNGNN GVNGNNGYNGDNGNN GDNGNNGYNGDNGNN GDN GNNGEN GNNGEN GNNGENG... |
| | *P. margaritifera* | A1 - AEC03970 | ...NNGDNGYNCDNGNNGNNGNNGNGNNDYNGNNGDNGNGNNGYNGNNGYNGNNGNGNGYN GDNGNGNNGNNGGNGNNGNNGNGNNGNNGNGNN GNNGGNSNNGNSGYGNGNNGNGNNGNN GGNDNNGNNGNSNNGNNGNGNNGNNGGNGNNGNSGYGNGNNGNGNNGNNGGNGNNGNNGNGNNGNNGN GDNGSNGNNGGNGNNGNN GDNGNGDNGYNGDNGNSDG... |
| | *P. margaritifera* | C5 - AEC03973 | ...DNNGDNGNNGNGNNGNNGGNGNNGNGNNGSNGNGDHGSNGNNGGNGNNGNN GDNGNGDNGYNGDNGNSDG... |
| | *H. gigantea* | hgCA2 - BAH58350 | ...GGNASGSVAGQAVVAV... |
| | *H. tuberculata* | htCA2 - AEL22201 | ...GGNGSGSVAGLAVVAVFLS... |
| | *L. gigantea* | CA - 239188 | ...GNGGNGMMKNLLK... ...RREEDDERGDGRHDLRDDDDNYDDDDYYNDDYSNDDYYDDDYYYDDYDDDTDDDHKDD GRRDRGGGDDKGGRGKGDDRGGRDNGDNRTGRGNRNDRGRRGNGDDSGGRGNGNNRDGR GNGDSRDRNNGNGNGRENGGVRGNGNDRDGRRDNGNGGDNGTRRGNGDDRGGRRNEDRG ENRRG... ...DGRRRRRRFNGRRRRRGRGDD KGDD |
| | *T. marmoratus* | nacrein - BAB91157 | ...YGDGDNGNGNGNGNGNGNGKDNGNGNGNDNGYGNGNGNGNGNGNGNGNGNGNGY GYGNGNGNGNGNGNGNGNGNGNGNGNGNGNGYGYGNGNGNGNGNGNGNGNGNGNG NGNGNGNGNGNGNGNGNGNGNGNGKDNGDN... ...GYNGENGNGDGNGNGGNGEI GDNGDYGYCDDDQCPDD... |
| ECHINO. | *S. purpuratus* | CA - SPU_012518 | ...VGAGGNQAGQGGGQGGQGGSGFNWGGAGAGGGGGGVGGVGGGGGWNWNSWWNGNGWGGNNGGGQNAVGNQ... ...SANNSANLISVNLFLLLSAIMAF VCRSL |

**Figure 6 Details of 17 LCDs of α-CAs/nacreins from invertebrate species.** Three categories sharing likely homology are indicated (black boxes): C-term basic LCDs in the sponge *A. willeyana* (highlighted in pink), acidic GXN repeated domains in bivalves (highlighted in blue) and acidic LCDs in the two species of gastropods *Haliotis* (highlighted in green).

interact with $CaCO_3$ *in vitro*. Miyamoto et al. [5], conducted *in vitro* calcification assays with wild type and truncated recombinant nacrein proteins. According to that work, the GXN domain inhibits the activity of nacrein and could act as a negative regulator of the $CaCO_3$ precipitation in the shell formation. The reason for such an α-CA inhibitory role is as yet unknown.

Our *in silico* analysis also reveals the occurrence of domains with similar acidic domains to the nacrein GXN repeat in the longer α-CA of *L. gigantea* (JGI:239188) and in the spicule specific α-CA [12] of *S. purpuratus* (JGI: SPU_012518). In *L. gigantea*, an acidic DG-rich peptide is located at the C-terminal end, and another more basic peptide is present just downstream of the signal peptide. In the purple sea urchin, the G-rich domain is less acidic than the nacrein GXN-rich domain. The position of this domain is also uncommon because it is upstream of the active α-CA site, in the N-terminal half of the protein.

Other short LCDs are strikingly different from the nacrein GXN repeat. Some are basic, *i.e.*, positively charged in standard pH conditions and are represented by the C-terminal R-rich, S-rich and LS-rich ends in four α-CAs (GenBank:ABR53886, GenBank:ABR53887, GenBank:ACE95141, JGI:SPU_012518). In *A. willeyana* the basic K-rich domain exhibits the KKRKRR motif containing also several arginine residues similarly to the C-termini of two family of proteins present in the prism matrix of the shell of the pearl oyster *Pinctada fucata*: shematrins and K-rich mantle proteins [100-102]. The presence of such basic domains suggests two possible functions: (1) an interaction with negatively charged bicarbonate ions. If so, the basic domain may concentrate bicarbonate ions at the vicinity of the nucleation site. (2) Electrostatic interaction with acidic (D/E-rich) domains of other proteins of the calcifying matrix. In this case, the basic domains may then anchor the acidic-soluble proteins to the three dimensional framework of the matrix. Here again, these basic domains indicate that the corresponding α-CAs display at least two functions, the enzymatic activity and the interaction with bicarbonate and/or acidic macromolecules.

Although the way by which the "mosaic" α-CAs are evolutionarily constructed is not known, exon shuffling can be envisaged as one possible mechanism [103]. Such a mechanism can generate new functions through the combination of different pre-existing functional domains into one protein. This "genetic tinkering" is well documented for proteins present in vertebrate extracellular matrices [104,105]. It is believed that this mechanism is subject to elevated rates of evolution, in comparison to those observed for intracellular proteins [106].

## Conclusion

The α-CA family is a key family for understanding how the molecular mechanisms of metazoan biocalcification strategies evolved. Independent lines of evidence (phylogenetic relationships of biomineral associated α-CAs and patterns of LCD insertions) suggest that the recruitment of α-CAs into biocalcification roles likely occurred independently in different metazoan lineages. A major limitation of this work is the difficulty in reconstructing a robust phylogeny of metazoan α-CAs. The primary causes of this difficulty are likely to include the deep phylogenetic history of the α-CA family, the fact that many independent duplications have generated a variety of isoforms, and that these isoforms have subsequently been recruited to a wide variety of physiological roles. We remain optimistic that both broader and deeper taxon sampling (i.e. complete genome sequences) may bring more resolution to these relationships, and we anticipate that in the coming years such data will make these analyses more accurate. Such complete genome data may also provide additional lines of evidence for assigning CA orthology *via* gene synteny. In addition, physiological, biochemical and functional studies aimed at understanding the role of specific α-CAs in biomineralization processes will provide insight into the evolutionary history of mineral-relevant α-CA isozymes. A rich field of research regarding the origins of the LCDs associated with mineralizing α-CAs also awaits further investigation. Understanding how such LCDs are generated, how and why they come to be associated with mineralizing α-CAs, and what their functions are remain open and intriguing questions.

## Additional files

**Additional file 1: Description of methods used for Phylogenetic analyses (Figures 2 and 3) and *in silico* analyses (Figure 5).**

**Additional file 2: Figure S1.** Topology of the tree presented in Figure 2 resulting from the Bayesian method. Bayesian posterior probabilities expressed as a percentage above 50 are indicated. For more details of the figure see legend of Figure 2.

**Additional file 3: Figure S2.** PhyML phylogenetic reconstruction with percentage bootstrap values, using the dataset of Figure 2.

**Additional file 4: Figure S3.** Topology of the tree presented in Figure 3 resulting from the PhyML method. The aLRT values expressed as a percentage above 50 are indicated. For more details of the figure see legend of Figure 3.

**Additional file 5: Table S1.** List of α-CA sequences used for the substitution rate analysis (Figure 4). For each α-CA, the name of species and the GenBank or JGI ID are provided.

**Additional file 6: Table S2.** Percentage of identity between nucleotide sequences encoding GN repeat domains in bivalve nacreins using nBLAST in NCBI.

### Abbreviations

CA: Carbonic anhydrase; $CaCO_3$: Calcium carbonate; EST: Expressed sequence tag; GPI: GlycoPhosphatidylInositol; LCD: Low complexity domain; RLCD: Repetitive LCD; CARP: Carbonic anhydrase related-protein.

### Competing interests

The authors declare that they have no competing interests.

### Authors' contributions

NLR collected the dataset, carried out alignments and phylogenetic analyses, wrote and drafted the manuscript. DJ participated in the phylogenetic analyses and drafted the manuscript. FM conceived the review, and participated in the design, coordination and drafting of manuscript. All authors read, approved and corrected the final manuscript.

### Acknowledgements

The authors thank Dr. Daniel Petit from UMR INRA 1061 UGMA of the University of Limoges (France) for his kind help and advice in phylogenetic analyses and interpretations. This work was supported by a PhD Fellowship from the Ministère de l'Education Nationale et de la Recherche (contract number 27264-2007), an ANR project (ACCRO-Earth, ref. BLAN06-2_159971, coordinator Gilles Ramstein, LSCE, Gif/Yvette, France) for years 2007–2011. Additional financial support was provided by INSU (INTERRVIE project, year 2010), by the COST network TD0903 "Biomineralix" (Chair: D. Medakovic/F. Marin) and by the ECTS (internship grant, 2012). We would like to thank two anonymous reviewers that greatly improved the quality of this manuscript.

### Author details

[1]Centre Scientifique de Monaco, 8 quai Antoine Ier, Monte-Carlo 98000, Monaco. [2]Courant Research Centre Geobiology, Georg-August-University of Göttingen, Goldschmidtstr 3, 37077 Göttingen, Germany. [3]UMR CNRS 7245 MCAM, Muséum National d'Histoire Naturelle, 12 rue Buffon, Paris, France. [4]Instituto Gulbenkian d e Ciência, Rua da Quinta Grande 6, Apartado 14, Oeiras P-2781-901, Portugal. [5]UMR CNRS 6282 Biogéosciences, Université de Bourgogne, 21000 Dijon, France.

### References

1. Lowenstam HA, Weiner S: *On Biomineralization.* New York: Oxford University Press; 1989.
2. Miyamoto H, Miyashita T, Okushima M, Nakano S, Morita T, Matsushiro A: **A carbonic anhydrase from the nacreous layer in oyster pearls.** *Proc Natl Acad Sci U S A* 1996, **93:**9657–9660.
3. Tohse H, Mugiya Y: **Effect of enzyme and anion transport inhibitors on in vitro incorporation of inorganic carbon and calcium into endolymph and otoliths in salmon *Oncorhynchus masou*.** *Comp Biochem Physiol A* 2001, **128:**177–184.
4. Beier M, Anken RH, Rahmann H: **Susceptibility to abnormal (kinetotic) swimming fish correlates with inner ear carbonic anhydrase-reactivity.** *Neurosci Lett* 2002, **335:**17–20.
5. Miyamoto H, Miyoshi F, Kohno J: **The carbonic anhydrase domain protein nacrein is expressed in the epithelial cells of the mantle and acts as a negative regulator in calcification in the mollusc *Pinctada fucata*.** *Zool Sci (Tokyo)* 2005, **22:**311–315.
6. Beier M, Anken R: **On the role of carbonic anhydrase in the early phase of fish otolith mineralization.** *Adv Space Res* 2006, **38:**1119–1122.
7. Jackson DJ, Macis L, Reitner J, Degnan BM, Wörheide G: **Sponge paleogenomics reveals an ancient role for carbonic anhydrase in skeletogenesis.** *Science* 2007, **216:**1893–1895.
8. Marie B, Luquet G, Bédouet L, Milet C, Guichard N, Medakovic D, Marin F: **Nacre calcification in the freshwater mussel *Unio pictorum*: carbonic anhydrase activity and purification of a 95 kDa calcium-binding glycoprotein.** *ChemBioChem* 2008, **9:**2515–2523.
9. Moya A, Tambutté S, Bertucci A, Tambutté E, Lotto L, Vullo D, Supuran CT, Allemand D, Zoccola D: **Carbonic anhydrase in the scleractinian coral *Stylophora pistillata*.** *J Biol Chem* 2008, **283:**25475–25484.
10. Jackson DJ, McDougall C, Woodcroft B, Moase P, Rose RA, Kube M, Reinhardt R, Rokhsar DS, Montagnani C, Joubert C, Piquemal D, Degnan BM: **Parallel evolution of nacre building gene sets in molluscs.** *Mol Biol Evol* 2010, **27:**591–608.
11. Joubert C, Piquemal D, Marie B, Manchon L, Pierrat F, Zanella-Cléon I, Cochennec-Laureau N, Gueguen Y, Montagnani C: **Transcriptome and proteome analysis of *Pinctada margaritifera* calcifying mantle and shell: focus on biomineralization.** *BMC Genomics* 2010, **11:**613.
12. Mann K, Poustka A, Mann M: **Phosphoproteomes of *Strongylocentrotus purpuratus* shell and tooth matrix: identification of a major acidic sea urchin tooth phosphoprotein, phosphodontin.** *Proteome Sci* 2010, **8:**6.
13. Marie B, Zanella-Cléon I, Le Roy N, Becchi M, Luquet G, Marin F: **Proteomic analysis of the acid-soluble nacre matrix of the bivalve *Unio pictorum*: detection of novel carbonic anhydrase and putative protease inhibitor proteins.** *ChemBioChem* 2010, **11:**2138–2147.
14. Bertucci A, Tambutté S, Supuran CT, Allemand D, Zoccola D: **A new coral carbonic anhydrase in *Stylophora pistillata*.** *Mar Biotechnol* 2011, **13:**992–1002.
15. Marie B, Le Roy N, Zanella-Cléon I, Becchi M, Marin F: **Molecular evolution of mollusc shell proteins: insights from proteomic analysis of the edible mussel *Mytilus*.** *J Mol Evol* 2011, **72:**531–546.
16. Le Roy N, Marie B, Gaume B, Guichard N, Delgado S, Zanella-Cléon I, Becchi M, Auzoux-Bordenave S, Sire JY, Marin F: **Identification of two carbonic anhydrases in the shell-forming mantle of the European abalone *Haliotis tuberculata* (Gastropoda, Haliotidae): phylogenetic implications.** *J Exp Zool Part B* 2012, **318:**353–367.
17. Mann K, Edsinger-Gonzales E, Mann M: **In-depth proteomic analysis of a mollusc shell: acid-soluble and acid-insoluble matrix of the limpet *Lottia gigantea*.** *Proteome Sci* 2012, **10:**28.
18. Moya A, Huisman L, Ball EE, Hayward DC, Grasso LC, Chua CM, Woo HN, Gattuso JP, Forêt S, Miller DJ: **Whole transcriptome analysis of the coral *Acropora millepora* reveals complex responses to CO2-driven acidification during the initiation of calcification.** *Mol Ecol* 2012, **21:**2440–2454.
19. Bertucci A, Moya A, Tambutté S, Allemand D, Supuran CT, Zoccola D: **Carbonic anhydrase in anthozoan corals-a review.** *Bioorg Med Chem* 2013, **21:**1437–1450.
20. Drake J, Mass T, Haramaty L, Zelzion E, Bhattacharya D, Falkowski PG: **Proteomic analysis of skeletal organic matrix from the stony coral *Stylophora pistillata*.** *Proc Natl Acad Sci U S A* 2013, **110:**3788–3793.
21. Marie B, Jackson DJ, Ramos-Silva P, Zanell-Cléon I, Guichard N, Marin F: **The shell-formaing proteome of *Lottia gigantea* reveals both deep conservations and lineage-specific novelties.** *FEBS J* 2013, **280:**214–232.
22. Ramos-Silva P, Kaandorp J, Huissman L, Marie B, Zanemmé-Cléon I, Guichard N, Miller DJ, Marin F: **The skeletal proteome of the coral *Acropora millepora*: the evolution of calcification by co-option and domain shuffling.** *Mol Biol Evol* 2013, **30:**2099–2112.
23. Supuran CT: **Carbonic anhydrase: novel therapeutic applications for inhibitors and activators.** *Nature Rev Drug Discov* 2008, **7:**168–181.
24. Elleuche S, Pöggeler S: **Evolution of carbonic anhydrase in fungi.** *Curr Genet* 2009, **55:**211–222.
25. Capasso C, Supuran CT: **An overview of the alpha-, beta- and gamma-carbonic anhydrases from bacteria: can bacterial carbonic anhydrases shed new light on evolution of bacteria?** *J Enzyme Inhib Med Chem.* doi:10.3109/14756366.2014.910202.
26. Alterio V, Di Fiore A, D'Ambrosio K, Supuran CT, De Simone G: **Multiple binding modes of inhibitors to carbonic anhydrases: how to design specific drugs targeting 15 different isoforms?** *Chem Rev* 2012, **112:**4421–4468.
27. Ferry JG: **The γ class of carbonic anhydrase.** *Biochim Biophys Acta* 2009, **1804:**374–381.
28. Kimber MS, Pai EF: **The active site architecture of *Pisum sativum* β-carbonic anhydrase is a mirror image of that of α-carbonic anhydrases.** *EMBO J* 2000, **19:**1407–1418.
29. Moroney JV, Bartlett SG, Samuelsson G: **Carbonic anhydrase in plants and algae.** *Plant Cell Physiol* 2001, **24:**141–153.
30. Tripp BC, Smith K, Ferry JG: **Carbonic anhydrase: new insights for an ancient enzyme.** *J Biol Chem* 2001, **276:**48615–48618.
31. Bahn YS, Mühlschlegel FA: **$CO_2$ sensing in fungi and beyond.** *Curr Opin Microbiol* 2006, **9:**572–578.
32. Fawcett TW, Browse JA, Volokita M, Bartlett SG: **Spinach carbonic anhydrase primary structure deduced from the sequence of a cDNA clone.** *J Biol Chem* 1990, **265:**5414–5417.
33. Burnell JN, Suzuki I, Sugiyama T: **Light induction and the effect of nitrogen status upon the activity of carbonic anhydrase in maize leaves.** *Plant Physiol* 1990, **94:**384–387.
34. Alber BE, Ferry JG: **Characterization of heterologously produced carbonic anhydrase from *Methanosarcina thermophila*.** *J Bacteriol* 1994, **178:**3270–3274.
35. Hewett-Emmett D, Tashian RE: **Functional diversity, conservation, and convergence in the evolution of the α-, β-, and γ-carbonic anhydrase gene families.** *Mol Phylogenet Evol* 1996, **5:**50–77.

36. Smith KS, Ferry JG: **A plant-type (β-class) carbonic anhydrase in the thermophilic methanoarchaeon *Methanobacterium thermoautotrophicum*.** *J Bacteriol* 2000, **181**:6247–6253.
37. Lane TW, Saito MA, George GN, Pickering IJ, Prince RC, Morel FM: **A cadmium enzyme from a marine diatom.** *Nature* 2005, **435**:42.
38. Park H, Song B, Morel FMM: **Diversity of the cadmium-containing carbonic anhydrase in marine diatoms and natural waters.** *Environ Microbiol* 2007, **9**:403–413.
39. Vullo D, Del Prete S, Osman SM, De Luca V, Scozzafava A, AlOthman Z, Supuran CT, Capasso C: **Sulfonamide inhibition studies of the d-carbonic anhydrase from the diatom *Thalassiosira weissflogii*.** *Bioorg Med Chem Lett* 2014, **24**:275–279.
40. Xu Y, Feng L, Jeffrey PD, Shi YG, Morel FMM: **Structure and metal exchange in the cadmium carbonic anhydrase of marine diatoms.** *Nature* 2008, **452**:56–61.
41. Meldrum N, Roughton FJW: **Carbonic anhydrase: its preparation and properties.** *J Physiol* 1933, **80**:113–142.
42. Stadie W, O'Brien H: **The catalysis of the hydration of carbon dioxide and dehydration of carbonic acid by an enzyme isolated from red blood cells.** *J Biol Chem* 1933, **103**:521–529.
43. Tashian RE: **The carbonic anhydrase: widening perspectives on their evolution, expression and function.** *Bioessays* 1989, **10**:186–192.
44. Henry RP: **Multiple roles of carbonic anhydrase in cellular transport and metabolism.** *Annu Rev Physiol* 1996, **58**:523–538.
45. Müller WEG, Wang X, Grebenjuk VA, Korzhev M, Wiens M, Schlossmacher U, Schröder HC: **Common genetic denominators for $Ca^{2+}$ –based skeleton in metazoa: role of osteoclast-stimulating factor and of carbonic anhydrase in a calcareous sponge.** *PLoS One* 2012, **7**:14.
46. Reibring C-G, El Shahawy M, Hallberg K, Kannius-Janson M, Nilsson J, Parkkila S, Sly WS, Waheed A, Linde A, Gritli-Linde A: **Expression patterns and subcellular localization of carbonic anhydrases are developmentally regulated during tooth formation.** *PLoS One* 2014, **9**:e96007.
47. Pongsomboon S, Udomlertpreecha S, Amparyup P, Wuthisuthimethavee S, Tassanakajon A: **Gene expression and activity of carbonic anhydrase in salinity stressed *Penaeus monodon*.** *Comp Biochem Physiol A* 2009, **152**:225–233.
48. **SpBase Strongylocentrotus purpuratus.** In [http://www.spbase.org]
49. Tohse H, Murayama E, Ohira T, Takagi Y, Nagasawa H: **Localization and diurnal variations of carbonic anhydrase mRNA expression in the inner ear of the rainbow trout *Oncorhynchus mykiss*.** *Comp Biochem Physiol B* 2006, **145**:257–264.
50. Rahman M, Isa Y, Takemura A, Uehara T: **Analysis of proteinaceous components of the organic matrix of endoskeletal sclerites from the alcyonarian *Lobophytum crassum*.** *Calcif Tissue Int* 2006, **78**:178–185.
51. Silverman D, Lindskog S: **The catalytic mechanism of carbonic anhydrase implications of a rate-limiting protolysis of water.** *Accounts Chem Res* 1988, **21**:30–36.
52. De Boer M, Krupp DA, Weis VM: **Two atypical carbonic anhydrases homologs from the planula larva of the scleractinian coral *Fungia scutaria*.** *Biol Bull* 2006, **211**:18–30.
53. Grasso L, Maindonald J, Rudd S, Hayward DC, Saint R, Miller DJ, Ball EE: **Microarray analysis identifies candidate genes for key roles in coral development.** *BMC Genomics* 2008, **9**:540.
54. **National Center for Biotechnology Information.** In [http://www.ncbi.nlm.nih.gov]
55. Hirohashi N, Alvarez L, Shiba K, Fujiwara E, Iwata Y, Mohri T, Inaba K, Chiba K, Ochi H, Supuran CT, Kotzur N, Kakiuchi Y, Kaupp UB, Baba SA: **Sperm from sneaker male squids exhibit chemotactic swarming to $CO_2$.** *Curr Biol* 2013, **23**:775–781.
56. Hu MY, Lee J-R, Lin L-Y, Shih T-H, Stumpp M, Lee M-F, Hwang P-P, Tseng Y-C: **Development in a naturally acidified environment: $Na^+/H^+$-exchanger 3-based proton secretion leads to $CO_2$ tolerance in cephalopod embryos.** *Front Zool* 2013, **10**:51.
57. De Cian M, Andersen AC, Bailly X, Lallier FH: **Expression and localization of carbonic anhydrase and ATPases in the symbiotic tubeworm *Riftia pachyptila*.** *J Exp Biol* 2003, **206**:399–409.
58. Leu J-H, Chang C-C, Wu J-L, Hsu C-W, Hirono I, Aoki T, Juan H-F, Lo C-F, Kou G-H, Huang H-C: **Comparative analysis of differentially expressed genes in normal and white spot syndrome virus infected *Penaeus monodon*.** *BMC Genomics* 2007, **8**:120.
59. Towle DW, Smith CM: **Gene discovery in *Carcinus maenas* and *Homarus americanus* via expressed sequence tags.** *Integr Comp Biol* 2006, **46**(6):912–918.
60. Serrano L, Halanych KM, Henry RP: **Salinity-stimulated changes in expression and activity of two carbonic anhydrase isozymes in the blue crab *Callinectes sapidus*.** *J Exp Biol* 2007, **210**:2320–2332.
61. Love AC, Andrews ME, Raff RA: **Gene expression patterns in a novel animal appendage: the sea urchin pluteus arm.** *Evol Dev* 2007, **9**:51–68.
62. Morino Y, Koga H, Tachibana K, Shoguchi E, Kiyomoto M, Wada H: **Heterochronic activation of VEGF signaling and the evolution of the skeleton in echinoderm pluteus larvae.** *Evol Dev* 2012, **14**(5):428–436.
63. Denoeud F, Henriet S, Mungpakdee S, Aury JM, Da Silva C, Brinkmann H, Mikhaleva J, Olsen LC, Jubin C, Canestro C, Bouquet JM, Danks G, Poulain J, Campsteijn C, Adamski M, Cross I, Yadetie F, Muffato M, Louis A, Butcher S, Tsagkogeorga G, Konrad A, Singh S, Jensen MF, Cong EH, Eikeseth-Otteraa H, Noel B, Anthouard V, Porcel BM, Kachouri-Lafond R, *et al*: **Plasticity of animal genome architecture unmasked by rapid evolution of a pelagic tunicate.** *Science* 2010, **330**:1381–1385.
64. Esbaugh AJ, Perry SF, Bayaa M, Georgalis T, Nickerson J, Tufts BL, Gilmour KM: **Cytoplasmic carbonic anhydrase isozymes in rainbow trout *Oncorhynchus mykiss*: comparative physiology and molecular evolution.** *J Exp Biol* 2005, **208**:1951–1961.
65. Eicher EM, Stern RH, Womack JE, Davisson MT, Roderick TH, Reynolds SC: **Evolution of mammalian carbonic anhydrase loci by tandem duplication: close linkage of Car-1 and Car-2 to the centromere region of chromosome of the mouse.** *Biochem Genet* 1976, **14**:651–660.
66. Hewett-Emmett D, Hopkins PJ, Tashian RE, Czelusniak J: **Origins and molecular evolution of the carbonic anhydrase isozymes.** In *Biology and Chemistry of the Carbonic Anhydrase*, Volume 429. Edited by Tashian R, Hewett-Emmett D. New York: Annals of the New York Academy of Sciences; 1984:338–358.
67. Peterson RE, Tu C, Linser PJ: **Isolation and characterization of a carbonic anhydrase homologue from the zebrafish (*Danio rerio*).** *J Mol Evol* 1997, **44**:432–439.
68. Jackson DJ, McDougall C, Green K, Simpson F, Wörheide G, Degnan BM: **A rapidly evolving secretome builds and patterns a sea shell.** *BMC Biol* 2006, **4**:40.
69. **Doe Joint Genome Istitute.** In [http://www.jgi.doe.gov]
70. **Méthodes et Algorithmes pour la Bioinformatique LIRMM.** In [www.phylogeny.fr]
71. Guindon S, Gascuel O: **A simple, fast, and accurate algorithm to estimate large phylogenies by maximum likelihood.** *Syst Biol* 2003, **52**:696–704.
72. Huelsenbeck JP, Ronquist F: **MRBAYES: Bayesian inference of phylogeny.** *Bioinformatics* 2001, **17**:754–755.
73. Ronquist F, Huelsenbeck JP: **MRBAYES 3: Bayesian phylogenetic inference under mixed models.** *Bioinformatics* 2003, **19**:1572–1574.
74. Parkkila AK, Scarim AL, Parkkila S, Waheed A, Corbett JA, Sly WS: **Expression of carbonic anhydrase V in pancreatic beta cells suggests role for mitochondrial carbonic anhydrase in insulin secretio.** *J Biol Chem* 1998, **273**:24620–24623.
75. Livingston BT, Killian CE, Wilt F, Cameron A, Landrum MJ, Ermolaeva O, Sapojnikov V, Maglott DR, Buchanan AM, Ettensohn CA: **A genorne-wide analysis of biomineralization-related proteins in the sea urchin *Strongylocentrotus purpuratus*.** *Dev Biol* 2006, **300**:335–348.
76. Marie B, Le Roy N, Marie A, Dubost L, Milet C, Bédouet L, Becchi M, Zanella-Cléon I, Jackson D, Degnan B, Luquet G, Marin F: **Nacre evolution: A proteomic approach.** In *MRS Symposium Proceedings; 14–17 April 2009*. Edited by Kisailus DEL, Gupta H, Landis W, Zavattieri P. San Fransisco: Materials Research Society; 2009:69–75.
77. Nys Y, De Laage X: **Effects of suppression of eggshell calcification and of 1,25 (OH)2D3 on $Mg^{2+}$, $Ca^{2+}$ and $Mg^{2+}$ $HCO_3^-$ ATPase, alkaline-phosphatase, carbonic-anhydrase and CABP levels. 1: the laying hen uterus.** *Comp Biochem Physiol A* 1984, **78**:833–838.
78. Lehenkari P, Hentunen TA, Laitala-Leinonen T, Tuukkanen J, Väänänen K: **Carbonic anhydrase II plays a major role in osteoclast differentiation and bone resorption by effecting the steady state intracellular pH and $Ca^{2+}$.** *Exp Cell Res* 1998, **242**:128–137.
79. Petit D, Maftah A, Julien R, Petit JM: **En bloc duplications, mutation rates, and densities of amino acid changes clarify the evolution of vertebrate alpha-1,3/4-fucosyltransferase.** *J Mol Evol* 2006, **63**:353–364.

80. Martin R, Gallet PF, Rocha D, Petit D: **Polymorphism of the prion protein in mammals: a phylogenetic approach.** *Recent Pat DNA Gene Seq* 2009, **3:**63–71.
81. Felsenstein J: *PHYLIP (Phylogeny Inference Package) version 3.6. Distribution by the author.* Seattle: Department of Genome Sciences, University of Washington; 2004.
82. Vallee B, Auld DS: **Active-site zinc ligands and activated $H_2O$ of zinc enzymes.** *Proc Natl Acad Sci U S A* 1990, **87:**220–224.
83. Stams T, Christianson DW: **X-ray crystallographic studies of mammalian carbonic anhydrase isozymes.** *EXS* 2000, **90:**159–174.
84. Briganti F, Mangani S, Orioli P, Scozzafava A, Vernagoline G, Supuran CT: **Carbonic anhydrase activators: X-ray crystallographic and spectroscopic investigations for the interaction of isozymes I and II with histamine.** *Biochemistry* 1997, **36:**10384–10392.
85. Di Fiore A, Monti SM, Hilvo M, Parkkila S, Romano V, Scaloni A, Pedone C, Scozzafava A, Supuran CT, De Simone G: **Crystal structure of human carbonic anhydrase XIII an dits complex with the inhibitor acetazolamide.** *Proteins* 2008, **74:**164–175.
86. Toba S, Colombo G, Merz KM Jr: **Solvent dynamics and mechanism of proton transfer in human carbonic anhydrase II.** *J Am Chem Soc* 1999, **121:**2290–2302.
87. Muhammad E, Leventhal N, Parvari G, Hanukoglu A, Hanukoglu I, Chalifa-Caspi V, Feinstein Y, Weinbrand J, Jacoby H, Manor E, Nagar T, Beck JC, Sheffield VC, Hershkovitz E, Parvari R: **Autosomal recessive hyponatremia due to isolated salt wasting in sweat associated with a mutation in the active site of carbonic anhydrase 12.** *Hum Genet* 2011, **129:**397–405.
88. Ebanks SC, O'Donnell MJ, Grosell M: **Characterization of mechanisms for $Ca^{2+}$ and $HCO_3^-/CO_3^{2-}$ acquisition for shell formation in embryos of the freshwater common pond snail *Lymnea stagnalis*.** *J Exp Biol* 2010, **213:**4092–4098.
89. Miyamoto H, Yano M, Miyashita T: **Similarities in the structure of nacrein, the shell-matrix protein, in a bivalve and a gastropod.** *J Molluscan Stud* 2003, **69:**87–89.
90. Gum JR: **Mucin genes and the proteins they encode-structure, diversity, and regulation.** *Am J Respir Cell Mol Biol* 1992, **7:**557–564.
91. Desseyn J, Aubert JP, Porchet N, Laine A: **Evolution of the large secreted gel-forming mucins.** *Mol Biol Evol* 2000, **17:**1175–1184.
92. Lane T, Hansson GC, Samuelsson T: **Gelforming mucins appeared early in metazoan evolution.** *Proc Natl Acad Sci U S A* 2007, **104:**16209–16214.
93. **Current Version: open-4.0.3 (RMLib: 20130422 & Dfam: 1.2).** In [http://www.repeatmasker.org/cgi-bin/WEBRepeatMasker]
94. Marin F, Luquet G: **Molluscan shell proteins.** *C R Palevol* 2004, **3:**469–492.
95. Gaume B, Fouchereau-Peron M, Badou A, Helléouet M-N, Huchette S, Auzoux-Bordenave S: **Biomineralization markers during early shell formation in the European abalone *Haliotis tuberculata*, Linnaeus.** *Mar Biotechnol* 2011, **158:**341–353.
96. Marin F, Le Roy N, Marie B: **He formation and mineralization of mollusk shell.** *Front Biosci* 2012, **S4:**1099–1125.
97. Marin F, Marie B, Ben Hamada S, Silva P, Le Roy N, Guichard N, Wolf S, Montagnani C, Joubert C, Piquemal D, Saulnier D, Gueguen Y: **"Shellome": proteins involved in mollusc shell biomineralization - diversity, functions.** In *Recent Advances in Pearl Research.* Edited by Watabe S, Maeyama K, Nagasawa H. Tokyo: Terrapub; 2013:149–168.
98. Shiba K, Minamisawa T: **A synthesis approach to understanding repeated peptides conserved in mineralization proteins.** *Biomacromolecules* 2007, **8:**2659–2664.
99. Isowa Y, Sarashina I, Setiamarga DHE, Endo K: **A comparative study of the shell matrix protein aspein in pterioid bivalves.** *J Mol Evol* 2012, **75:**11–18.
100. Yano M, Nagai K, Morimoto K, Miyamoto H: **Shematrin: a family of glycin-rich structural proteins in the shell of the pearl oyster.** *Comp Biochem Physiol B* 2006, **144:**254–262.
101. Zhang C, Xie L, Huang J, Liu X, Zhang R: **A novel matrix protein family participating in the prismatic layer formation of pearl oyster, *Pinctada fucata*.** *Biochem Biophys Res Commun* 2006, **344:**735–740.
102. McDougall C, Aguilera F, Degnan BM: **Rapid evolution of pearl oyster shell matrix proteins with repetitive, low complexity domains.** *J R Soc Interface* 2013, **10:**20130041.
103. Bork P: **Shuffled domains in extracellular proteins.** *FEBS L* 1991, **286:**47–54.
104. Patthy L: **Genome evolution and the evolution of exon-shuffling – a review.** *Gene* 1999, **238:**103–114.
105. Patthy L: **Modular assembly of genes and the evolution of new functions.** *Genetica* 2003, **118:**217–231.
106. Julenius K, Pedersen AG: **Protein evolution is faster outside the cell.** *Mol Biol Evol* 2006, **23:**2039–2048.

# 7

# Biodiversity, ecology, and behavior of the recently discovered insect order Mantophasmatodea

Steffen Roth[1*], Jorge Molina[2] and Reinhard Predel[3]

**Abstract**

The spectacular discovery of the new insect order Mantophasmatodea in 2002 was immediately followed by detailed studies on morphology and scattered information on different aspects of its behavior and general biology. A distinct feature of these predatory insects is the development of large arolia, which are typically held upright; hence, their common name is heelwalkers. The first mantophasmatodean species were described based on two museum specimens originally collected in Tanzania and Namibia. To date, these insects have been observed at surprising levels of diversity and abundance in Namibia and South Africa. For our studies on the phylogenetic relationships within Mantophasmatodea, we collected and analyzed numerous populations that belong to all known mantophasmatodean lineages, including East African populations. These collections not only provided a comprehensive biogeographical overview but also facilitated a comparative analysis of behavior, which was mainly analyzed under laboratory conditions. Here, we review and discuss the published data, as well as provide additional information on Mantophasmatodea distribution, evolutionary lineages, morphology, and biology, with a specific focus on reproductive biology.

**Keywords:** Behavior, Biodiversity, Evolutionary lineages, Life history, Mantophasmatodea, Polyneoptera

## Introduction

The insect order Mantophasmatodea [1,2] was the first new order described since 1914 [3], and several taxa were immediately characterized (e.g. [4-6]). These descriptions were accompanied by a comprehensive description of the morphological features, such as genitalia of both sexes [4], the structure of sperm [7], eggs [8,9], head [10,11], circulatory organs [12], arolium and tarsus [13-16], and antennae [4,17,18], as well as studies on reproductive biology [19], communication [20-22], peptide hormones [23,24], physiology [25], and development [17,26,27]. In addition to molecular studies [4,28-31], these studies provided the initial information for reconstructing the Mantophasmatodea phylogeny. Although the interordinal relationship has been disputed with Mantophasmatodea either to Phasmatodea [28] or Grylloblattodea [13,29,31,32], more convincing morphological and also molecular data support Mantophasmatodea as the closest relatives of Grylloblattodea. Considering the short period since Mantophasmatodea was first described, an extensive body of literature has been generated, which includes behavior and morphology as well as systematic and phylogenetic studies. However, certain general aspects of their biology and ecology are less well-known. Scattered reports have been published as short notes either in diverse scientific (see references above) or more popular publications (e.g. [33-36]). In this manuscript, we summarize the available information on Mantophasmatodea distribution, ecology, life history, reproduction, and feeding behaviors. Most data are based on our field studies and laboratory observations (see Additional file 1 for methods), for which we sampled specimens from all mantophasmatodean lineages (see [31]) in Southern Africa (Figure 1) and East Africa. In addition to presenting our data, we review the previously available information on ecology and behavior from the scientific literature and other public resources.

## Review

### General remarks and intraordinal relationships

Mantophasmatodeans superficially resemble mantises and stick insects. Different common names are used in the literature, including gladiator [37,38], rock crawler

* Correspondence: steffen.roth@um.uib.no
[1]University Museum of Bergen, The Natural History Collections, P.O. Box 7800, N-5020 Bergen, Norway
Full list of author information is available at the end of the article

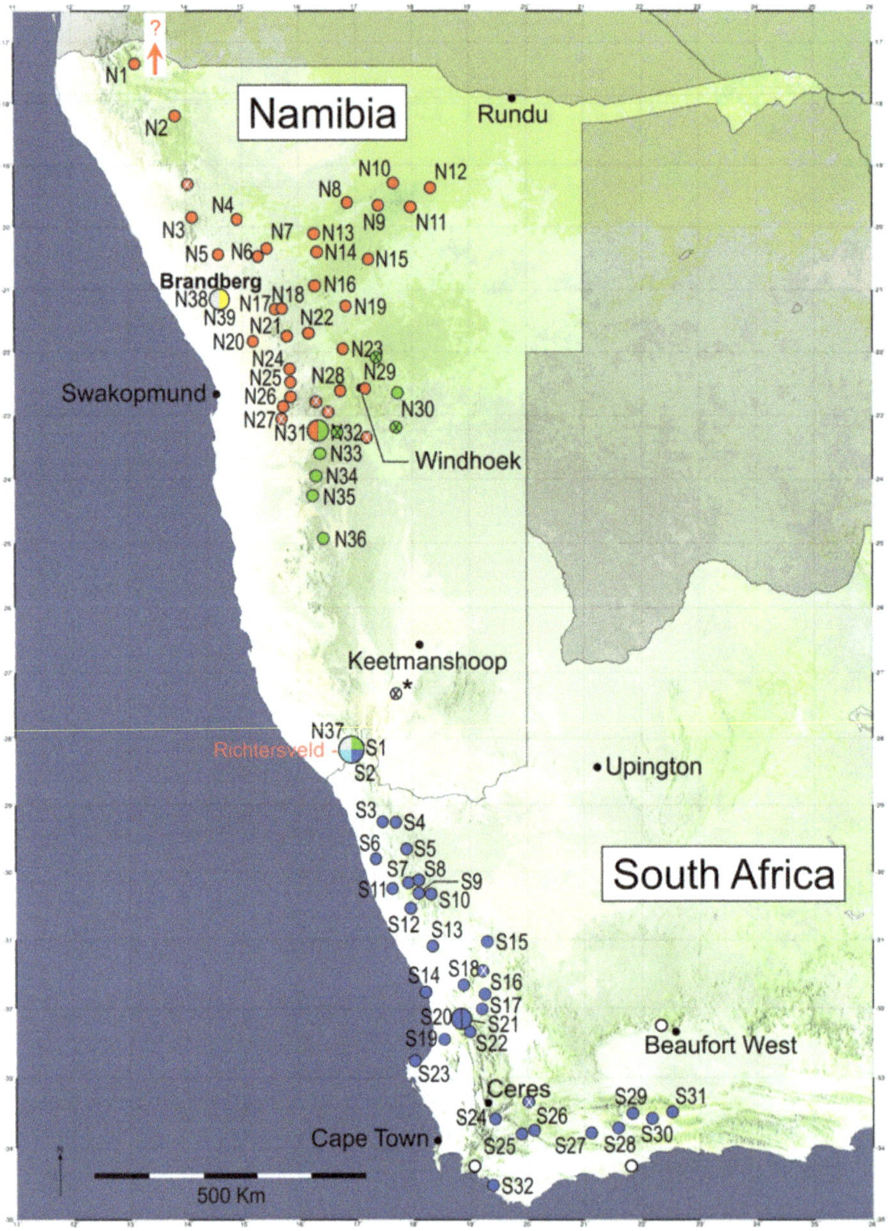


**Figure 1 Distribution for Mantophasmatodea in Southern Africa (modified from [31]).** The numbered sample spots refer to populations that were analyzed using peptidomics [33]. South African Austrophasmatidae taxa for which the mtDNA sequences were analyzed by [30]) have been collected within the same distributional range in the Western and Northern Cape provinces, except for the populations of Austrophasmatidae marked with white circles (see [30]). The circles marked with an X refer to populations that were recently collected and are not yet included in the phylogenetic studies. The large circles represent areas where several taxa co-occur (genus level). Whereas the respective taxa from Brandberg (N 39, greenish *Pachyphasma*; N38 grey brownish *Tyrannophasma*), and Clanwilliam (S21, greenish *Viridiphasma*; S20, grey brownish *Karoophasma*) inhabit different microhabitats, different taxa in the Gamsberg area (N31, greenish *Mantophasma*; N32, greenish *Striatophasma*) were occasionally identified in the same bushes. The Richtersveld (arrowhead), which is part of the most northern extension of the winter rainfall region in South Africa, is inhabited by at least four distantly related taxa (*Praedatophasma*, *Namaquaphasma*, and two additional undescribed taxa) and, thus, shows exceptionally high diversity at the genus level. A presumptive *Praedatophasma* population has been reported from the Fish River catchment area (asterisk), which is approximately 100 km north of the established range (H. Vollbrecht, EduVentures Windhoek; pers. comm.). Red circles, *Mantophasma*; green circles, *Striatophasma*; blue circles, Austrophasmatidae sensu [4]); grey circles, *Tyrannophasma/Praedatophasma*; yellow, *Pachyphasma*; light blue/light green, undescribed taxa from Richtersveld.

[39], and heelwalker (e.g., [25,30,40,41]). Because 'heelwalker' reflects the typical Mantophasmatodea phenotype, we prefer this designation. Both sexes of all known species, including fossil species [38,39,42] are apterous, which yields a nymph-like appearance. This appearance might be one reason that the Mantophasmatodea specimens were

not recognized as members of a new insect order for a long time. Mantophasmatodean fossils were discovered in Baltic amber [38,39,43] which indicates a wide distribution in the past. Jurassic fossil findings (165 mya) in China confirm this hypothesis [42]. Currently, Mantophasmatodea have a relict global distribution; most of the 18 described species are concentrated in western Southern Africa areas, except for *Tanzaniophasma subsolanum* (Tanzaniophasmatidae) in East Africa [1,4,5,30,44,45].

The extant Mantophasmatodea species have been provisionally grouped into three families: Tanzaniophasmatidae (which comprise only the described Tanzanian species of Tanzaniophasma), Austrophasmatidae (9 described species from African genera *Namaquaphasma, Karoophasma, Hemilobophasma, Lobatophasma, Austrophasma, Viridiphasma*), and the Namibian genus *Striatophasma*, Mantophasmatidae (8 described species from Namibian genera *Tyrannophasma, Praedatophasma, Mantophasma, Sclerophasma, Pachyphasma*). Another family-level group has been proposed for accommodating Praedatophasma and Tyrannophasma [30]. In particular, the species identity within the Mantophasmatidae and the placement of *Striatophasma* in Mantophasmatidae or Austrophasmatidae must be clarified in future studies. A recent phylogenetic analysis of peptide hormone sequences suggests at least eight distinct lineages in Mantophasmatodea, including *Tyrannophasma, Praedatophasma, Pachyphasma, Mantophasma* (incl. *Sclerophasma*), *Tanzaniophasma, Striatophasma*, Austrophasmatidae sensu [4], and an undescribed taxon from the Richtersveld (see [31]). A second undescribed taxon from the Richtersveld (Wipfler & Predel; in prep.) may link *Striatophasma* from Namibia and the South African Austrophasmatidae. The monotypic *Tyrannophasma, Praedatophasma*, and *Pachyphasma*, as well as the two undescribed taxa from the Richtersveld, are relict taxa with a limited distribution range. In contrast, *Mantophasma, Striatophasma*, the South African Austrophasmatidae sensu [4], and most likely the poorly known *Tanzaniophasma* clade are currently successful and widespread taxa. Although a number of *Mantophasma* species have been described [1,4,46], recent analyses of numerous populations of the genus *Mantophasma* did not reveal conclusive lineage sorting within this taxon [31]. Even the relationship between the *Mantophasma* lineage and *Sclerophasma paresisense* is poorly resolved [16] and requires further analyses to resolve the taxonomical status of this clade. The relationships within the family Austrophasmatidae, which was originally proposed to encompass taxa from the Western and Northern South African Cape Provinces, are well-supported at the genus and species levels. Ambiguities exist on an intrafamilial level for analyses of mitochondrial DNA sequences [30] and peptide hormone sequences [31]. However, these analyses clearly indicate that the family Austrophasmatidae derived from Namibian Mantophasmatodea species. Species from this family may have evolved in a rapidly speciating lineage, which produced a monophyletic group of mostly allopatric species [31].

### Distribution patterns and habitats

All records of extant Mantophasmatodea are from Africa, south of the equator. In East Africa, in the mountainous regions of Malawi (Dedza Mountains, Nyika Plateau), we identified slightly distinguishable populations based on their general appearance and habitus, which most likely belong to the *Tanzaniophasma* clade (Figure 2). These localities are near the South Tanzanian region, where the only scientifically described (museum) *Tanzaniophasma* specimen was collected in 1950 (see [1]). In Malawi, the specimens were located in areas characterized by a mixture of grasses and dense bushes (open shrubland and savanna); no specimens were observed in forest biomes. The Mantophasmatodea distribution in East Africa, at least in Malawi, may be negatively affected by extensive vegetation burn off, which is common and widespread in Malawi [47,48]. We observed heelwalkers that were

**Figure 2** *Tanzaniophasma* **spec. female from the Dedza Mts. (Malawi) with an adult fleshfly as her prey (photo taken in laboratory).**

restricted to dense bushes of different size on rocky hills that were not recently burned (Figures 3 and 4). Hockman et al., [17] noted putative *Tanzaniophasma* specimens from Cape Maclear, which is on the southern bank of Lake Malawi. In this area we observed nymphs in bush edging an open patch of unused dense shrubland along a creek. The data on Mantophasmatodea distribution in Namibia and South Africa have been discussed in detail by [31]. In short, the southern African distribution appears largely confined to the westernmost Great Escarpment upfoldings; the range of the genus *Mantophasma* likely extends into southwest Angola (see Figure 1 with multiple new localities). In the Western Cape (South Africa), a few species also inhabit the coastal region, including Sand Fynbos vegetation in the Strandveld, and the Dune Thicket (Fynbos); no specimens have been observed on the Cape Peninsula. The distribution map (Figure 1) suggests that a contemporary dispersal barrier (between 25 and 27 degrees south latitude) exists between the species in the winter rainfall area of Southern Africa and species living in the summer rainfall region of Namibia. A similar distributional discontinuity between Namibia and Namaqualand has been reported for several faunal elements, including other insects, birds, frogs, and mammals [49] (and references therein), and for plant species [50]. Our phylogenetic analyses and their recent distribution suggest that *Striatophasma*-like species have crossed this region in the past and established a founder population that yielded the present Austrophasmatidae sensu [4] (see [31]).

The spatial patterns of Mantophasmatodea in Southern Africa have been influenced by range fragmentation on a large scale (Namibian versus South African species) and range-edge speciation on a smaller scale. The latter form of speciation was likely driven by geological uplift during the Miocene to Pliocene [51], serial climatic oscillations during the Pleistocene [52], and alternating dry and wet

**Figure 3** **Characteristic habitat for *Tanzaniophasma* in Malawi (Nyika Plateau).**

**Figure 4** **Preferred type of vegetation for *Tanzaniophasma* at Nyika Plateau (Malawi).**

climatic fluctuations. Thus, Mantophasmatodea speciation in Southern Africa was most likely allopatric through vicariance, which has been assumed for other insects [53] and arthropods (Diplopoda; [54]) in the region. Remarkably, the core regions of their recent distribution, such as the Brandberg in Namibia and Succulent Karoo-Fynbos biomes [55-57] typically show exceptional levels of biodiversity and endemism. In these regions, Mantophasmatodea are part of a very unique insect fauna [58,59]. Notably in the context of Mantophasmatodea distribution, no additional mantophasmatodean taxa have been identified within the large *Mantophasma* distributional range, whereas two distinct monotypic genera are sympatric in different ecological niches on the Brandberg Massif immediately outside the *Mantophasma* range. Both the genera *Tyrannophasma* and *Pachyphasma* appear restricted to the Brandberg Massif and have not been observed in other areas. The Brandberg Massif (the highest mountain in Namibia) has volcanic origins and is isolated from the longitudinal Namibian escarpment (*Mantophasma* habitat); this isolation has generally ensured that it is a significant relict habitat for numerous endemic organisms (see [53,55]). On the Brandberg Massif plateau, *Pachyphasma brandbergense* was found in the same biotope as *Tyrannophasma gladiator*. The large and impressive-looking grey-brown *T. gladiator* is mainly found in grass tussocks (see also [33]); however, under dry conditions, nymphs have also been found in dense bushes (altitude 1400-2400 m). The much smaller and greenish *P. brandbergense* is the only mantophasmatodean taxon primarily collected from flowering composite bushes; the ecological relevance of this finding is unknown. The Brandberg Plateau vegetation is characterized by plant species confined to the upper regions of mountains (higher than 1500 m); Asteraceae is the most speciose family [60]. The different *Mantophasma* populations (including *Sclerophasma paresisense*) inhabit well-vegetated mountains in central Namibia up to the Otavi Mountains in the northeast; they also live in mountainous mopane savanna (Kaokoveld) in the northwest. Throughout most of their range, these insects are common in grasses, shrubs, and even trees given sufficient rainfall (see [31]). The density of individuals in a population can be surprisingly high, which is somewhat unexpected for an insect group that has been overlooked for such a long time period. The genus *Striatophasma*, which was mainly collected south/southeast of the *Mantophasma* distribution, typically inhabits dry regions with scattered vegetation, and specimens have often been collected from dwarf bushes. In the Gamsberg/Hakos Mountains, *Mantophasma* and *Striatophasma* are sympatric. Austrophasmatidae, which live in the winter-rain South African region, were typically found in habitats with sparse to scattered vegetation. Eberhard et al., [44] reported a preference of *Viridiphasma clanwilliamense* to a small tree (*Euclea recemosa*), but for most species, we did not observe a preference for hiding on a specific plant species; clearly, Austrophasmatidae select dense bushes or grass tussocks, which provide camouflage and a sufficient prey spectrum. The Kraalbos *Galenia africana*, a bush that is poisonous to sheep and is common in overgrazed or otherwise disturbed Western/ Northern Cape areas, appears to be particularly attractive to mantophasmatodeans. The occurrence of heelwalkers in strongly disturbed areas suggests that these insects are less sensitive to environmental damage as long as prey is available. At least for the *Austrophasma* and *Hemilobophasma* populations, we found that nymphs prefer bushes, whereas adults were typically collected from large grass tussocks. Copulating heelwalkers have regularly been observed in these grasses. Most Austrophasmatidae clades were collected from low altitudes to mountainous locations up to 2000 m, which indicates that different altitudes with changing vegetation types are not effective dispersal barriers for heelwalkers.

Only few *Praedatophasma maraisi* specimens have been detected in the Richtersveld/Ai-Ais region in the Orange River/Fish River vicinity. The first specimen was collected in 1990 by E. Marais (Windhoek, Namibia) on the Namibian side of the Orange River when it approached a light trap (see [38]); two additional specimens were collected by the authors nearby on the South African Orange River site (Richtersveld). However, this species' distribution appears larger because scholars from Namibia detected a putative *Praedatophasma* specimen in 2005 in the dry Gaub/Fish River region, which is approximately 90 km NW of the confirmed localities (H. Vollbrecht, EduVentures Namibia; pers. communication). It is currently unclear whether these large heelwalkers (Figure 5) hide in vegetation similar to the other mantophasmatodeans. Extensive collection activities in the Richtersveld have yielded three additional and distantly related taxa (see Figure 1) with limited ranges. The succulent Karoo in the Richtersveld/Ai-Ais region, where these four mantophasmatodean taxa have been identified, is only sparsely covered with vegetation and represents the most desert-like habitat in which mantophasmatodeans have been found to date. The Richtersveld is the only recognized arid biodiversity hotspot [61] and is located in the summer- and winter-rain transition zone.

**Figure 5 Mating pair of *Praedatophasma* spec. collected during an EduVentures expedition with scholars from Windhoek in the Fish River catchment region.** Photo courtesy of Telané Greyling, Namibia Horse Safari Company, Aus, Namibia.

## General habits

Klass et al., [4] Wipfler et al., [45] Wipfler [62] have provided detailed descriptions of mantophasmatodean morphology; here, we focus on certain characteristics related to the ecological and behavioral aspects of their biology. Adults in all Mantophasmatodea species are in the 13-26 mm range, except for the larger species *P. maraisi* and *T. gladiator* (up to 35 mm). In all species, mature females with developed eggs are substantially larger than males (see below). Both sexes are secondarily wingless and highly flexible along the longitudinal axis. This flexibility enables the species to clean their external genitalia with their mouthparts (Figure 6), but it is also used to attack large prey (Figure 7). The known species' ground color is brownish, grey, or greenish, which indicates camouflage. The *Mantophasma, Tanzaniophasma, Pachyphasma, Viridiphasma, Striatophasma,* and *Austrophasma/Lobatophasma* species are typically green; however, the *Austrophasma gansbaaiense* and *Lobatophasma redelinghuysense* specimens have been observed with different ground colors (green or brownish) in the same local population, and the *Mantophasma* specimens have regularly been identified with uniform pale-brown, pale-yellow, or whitish coloring (see also [4,33]). The type of ground color is consistent during nymphal development and cannot be changed, but no detailed information is available regarding the heredity of color frequencies at the population level. Under laboratory conditions without vegetation, many specimens (particularly *Mantophasma*) became paler, and it is advantageous

**Figure 6 Abdomen grooming by a female *T. gladiator* (N 38, Brandberg), which shows the flexible body axis.** Note the use of a single arolium to adhere the heavy body to the wall of the plastic container (photo taken in the laboratory).

to document the color of specimens in the field. In addition to these general color types, the insect's body might be covered by many spots and median stripes (for a detailed description, see [4]). A number of greenish species exhibit color polymorphism within a population and vary in color and markings. Brownish dorso-median stripes are typical of a few species (Figure 8). In *Striatophasma*, only males show the brownish stripe; the coloration is from brown pigments in the dorsal diaphragm. Other species with brownish stripes (e.g., *Austrophasma* and *Lobatophasma*) belong to Austrophasmatidae, and both sexes have these stripes. In general, whereas there appears to be no specific link to a particular plant species, for most species, the color matches the habitat structure. Green species occupy the more mesic fynbos vegetation in the Cape region, savanna, and hilly grasslands in Namibia and East Africa, and the grey or drab brown morphs occupy more arid habitats in the succulent Karoo, such as many Austrophasmatidae and *P. maraisi*. A good example for color adaptation are the Western Cape's green *Viridiphasma clanwilliamense* and the sympatric *Karoophasma biedouwense* which were mainly collected from green bushes or low-growing grey and brown karooid shrubs; respectively. In *Lobatophasma redelinghuysense*, brown and grey striped morphs dominate in dry sandplain fynbos vegetation, while green striped nymphs dominate in more mesic mountain fynbos vegetation [44]. The only greyish/brownish species north of the winter rainfall South African region is *T. gladiator*, which has a restricted distribution in the Brandberg Massif.

**Figure 7 Green morph of a non-pregnant *Austrophasma gansbaaiense* female (S32, Gansbaai) attacking a pregnant female from an undescribed Austrophasmatid species (S18; Vanrynsdorp).** The sequence demonstrates the use of the flexible body axis and the arolia in a vigorous attack, as well as the ability to strike extremely rapidly (image from a video recorded in the laboratory).

Apart from hunting, when Mantophasmatodea stalk or walk quickly over very short distances to catch their prey, they predominantly sit and hide, and they rarely walk in vegetation. Most species, at least in the family Austrophasmatidae, hide in the centers of bushes or grass tussocks during the day. Night observations verified the presence of Mantophasmatodea in the periphery of bushes or grasses during feeding or mating. Whereas most specimens that were detected with an electric torch tried to hide themselves (Figure 9), one *H. montaguense* male was observed drumming on a grass stalk (Figure 10). Heelwalkers that were caught and held in hand rarely showed fast escape behaviors. However, in certain cases, we observed a strange "curling" behavior when the specimens were knocked out of a bush. This behavior was particularly conspicuous because, after dropping, Mantophasmatodea typically do not immediately move away. Heelwalkers sometimes jump for small distances, such as a few centimeters (Figure 11; see Additional file 2: Videos 1A, Additional file 3: Video 1B, Additional file 4: Video 1C, Additional file 5: Video 1D), however, we did rarely observe 10 cm distances as described in [5] for *Mantophasma*.

Occasionally, Mantophasmatodea actively drop in the vegetation. In most cases, they land upright on their legs ("cat-like") on the ground or in the correct direction in small branches of shrubs and grasses. Experiments using high-speed cameras suggest that heelwalkers actively influence the landing position due to their highly flexible body axis (see Additional file 6: Video 2A and Additional

**Figure 8 Development of brownish dorso-median stripes in different Mantophasmatodea taxa. a)** Female *Austrophasma gansbaaiense* (S32, Gansbaai); the brownish stripe with white margins. **b)** *Striatophasma naukluftense;* the brownish stripe is only developed in males; females with less distinct green stripe. Western population (Gamsberg) with posteriorly broadened stripe. The inset shows a male from an eastern population (70 km E of Rehoboth) with distinctly outlined abdominal stripe. **c)** An undescribed species from the Richtersveld with weak male dorsal stripe development.

file 7: Video 2B). Similarly sized unwinged mantises and phasmids were unable to behave similarly; however, the mantises attempted to perform a 'loop,' but required much more distance between the start and end points (Predel, unpublished). In the field and in captivity, healthy heelwalkers are reluctant to immediately walk away after dropping. When these insects did not land on their legs but on their backs, they changed their positions with a sudden movement immediately after landing and stopped again (see Additional file 8: Videos 3A and Additional file 9: Video 3B). Due to this behavior, which may be maintained for 30 sec or even longer, it is sometimes difficult to detect heelwalkers in detritus after beating a bush; many arthropods, such as ants, beetles, cockroaches, and spiders, run around actively under these circumstances. In nature, this strategy likely prevents detection by insect-hunting mammals or birds.

**Figure 9 Mantophasmatodea females attempting to avoid detection under sudden light exposure. a)** *Mantophasma* spec. (N15, Waterberg) on a grass stalk at 11 p.m. (field observation); **b)** *T. gladiator* (N 38, Brandberg) (photo taken in laboratory).

Both sexes possess very large arolia on the pretarsi of all legs (see [1,13-15]), which are, however, mostly held in an upright position; this highly characteristic position ("heelwalker") is quite unique among insects. Studies have shown that arolia are rarely used for walking, if at all, and are not necessary to produce, detect, or respond

**Figure 10 *H. montaguense* male (S29, Calitzdorp) in the field drumming at night (11 p.m.) on a grass stalk.**

**Figure 11** ***T. gladiator*** **male (N 38, Brandberg) forced to jump from a glass slide.** In contrast to grasshoppers, for example, Mantophasmatodea are typically reluctant to jump. (images from a video recorded in the laboratory).

to vibrational communication signals [15]. We observed arolia use when walking or adhering to smooth surfaces. For example, when cleaning body parts or eating large prey, the animals sometimes hung upside-down (Figures 6 and 12) and often used only a single hindleg arolium to adhere to smooth plastic surfaces. We also observed the insects using arolia to stabilize their position after landing or when a large prey was attacked (see Figure 7). Effective adhesion is supported by secretions produced in large glands in the arolia [15]. Secretion production is typically energy-consuming, which may be one reason to maintain the large arolia in an upward position when not in use. All species that have been maintained in our laboratory frequently clean their arolia but also antennae and genital segments using their mouthparts (see Additional file 10: Video 4; also [5] for. *T. gladiator*)); tarsi were not used to clean any body part.

**Figure 12** ***T. gladiator*** **female (N 38, Brandberg) using hind leg arolia to adhere to the wall of a plastic container while eating a cricket.** The fore- and hind legs were used to fix the prey (photo taken in the laboratory).

The Mantophasmatodea antennae are long, filiform, and have a flagellum well separated in basi- and distiflagellum; unique feature among insects [4,17,18]. While walking or hiding, these insects continuously wave their antennae in a manner similar to that of cockroaches (i.e., both antennae alternatively move up and down) (see Additional file 11: Video 5). During these motions, the flagellomeres are typically bent downwards. Antenna-flickering has been observed upon contact with a sexual partner and prey. This behavior indicates intensive chemical signal use. Potential glandular structures have been identified in two of the distal flagellomeres [4,18,36]. Occasionally, males walk with their abdomens upwards, similarly to scorpions (e.g., when approaching a sexual partner).

A diurnal habit has been observed in *Mantophasma* specimens (see also [5]). Occasionally, we found these insects sitting in the upper parts of grasses and outer shrub branches during the hottest portion of the day (11 a.m. - 5 p.m.); however, the *Mantophasma* specimens were more abundant in these locations after nightfall. In the laboratory, *Mantophasma* specimens showed greater activity during the day compared with Austrophasmatidae species, including foraging and sexual behaviors (see [33] and below). For all species, mating and drumming behaviors were predominantly observed at night in the laboratory, including dusk and dawn.

#### Life cycle, growth, and life history

Mantophasmatodea are annual and univoltine species. Females oviposit in pods of foam that form a cocoon, which hardens due to incorporated sand and contains eggs from both ovaries (in *T. gladiator* 26-30, in *Mantophasma* females 20-24; Figures 13 and 14). Tojo et al., [19] reported 10-12 eggs in pods from *K. biedouwense* (Austrophasmatidae), which indicates that these insects may occasionally oviposit eggs from a single ovary (the egg number per ovary is similar to that of *Mantophasma*). These pods are laid superficially (0.5 - 2 cm deep) in sandy soil (Figure 15; see also [33]). In the field, we identified two empty pods

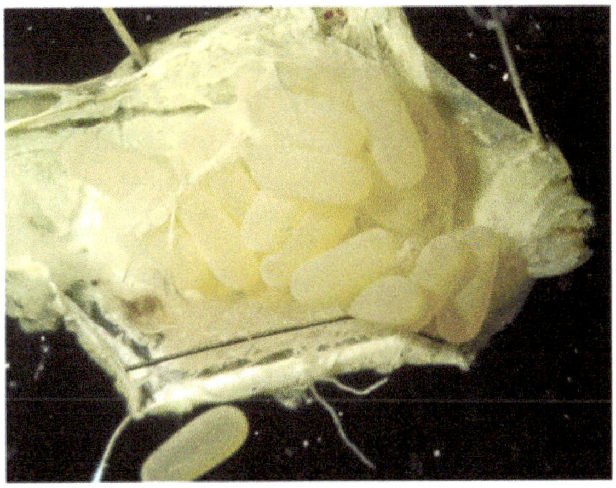

**Figure 13 Female *Mantophasma kudubergense* (N21, Erongo Mts.) abdominal body cavity with 20 large eggs from both ovaries.**

**Figure 15 *Mantophasma kudubergense* female (N21, Erongo Mts.) laying an egg pod under laboratory conditions.** The soil is not from the sample location.

close together, which suggests that females do not hesitate to lay eggs at approximately the same spot. The pods are formed from sand and a secretory liquid to protect the eggs. In the pods, the eggs exhibit a regular vertical arrangement similar to that of bottles in a beer crate (see Figure 14). We did not observe sterile egg pod deposition, which is commonly observed in praying mantises [63].

Mantises and cockroaches, which were often collected at the same location as the heelwalkers, laid eggs/egg cases even under simplified laboratory conditions (e.g., in plastic boxes without a soil cover), but for Mantophasmatodea, oviposition was not observed under these simplified conditions. This reluctance to oviposit was evident. However, Mantophasmatodea specimens readily copulated in the laboratory if provided the opportunity (see below). Pregnant *Mantophasmatodea* females laid eggs almost immediately if sufficient soil was provided, and we obtained more than 30 pods with fertile eggs from 20 *T. gladiator* females after copulation in the laboratory. Uchifune [64] obtained multiple egg pods from *Karoophasma biedouwense* (Austrophasmatidae); individual females laid up to four egg pods over a short period, i.e., a few days, and 50 - 100 eggs total during their adulthood [19]. In this species, oviposition was observed in September and typically occurred in the morning when the temperature was low and the humidity was high [64]. Under artificial laboratory conditions, Mantophasmatodea breeding was unsuccessful in most cases. We tested different dormancy scenarios using *T. gladiator* egg pods (G. Köhler, unpubl. data). Neither a dry and warm period over several months nor dry conditions followed by a cooling (4°C) period over several months initiated hatching. In all scenarios, a water supply was used to terminate dormancy, which simulated the rainy season and appeared to initiate hatching in the field. The egg fertilization rates varied from 48-95% in our experiments, and the embryos reached the anatrepsis phase. From these observations, we assumed a diapause. However, successful rearing in *Sclerophasma paresisense* and *Praedatophasma maraisi* (in fact, *Tyrannophasma gladiator*, see locality) was reported by [17], who used increased moisture and rising temperatures following a cold period.

**Figure 14 *T. gladiator* (N39, Brandberg) egg pod with 29 eggs laid into sand provided in the laboratory.** Note the regular and vertical arrangement of the eggs, which became visible after the removal of the lower sand layer.

In nature, the egg stages survive at least eight months of a dry summer (Austrophasmatidae) or winter (Mantophasmatidae, Tanzaniophasmatidae) within the resistant egg pods, and the first instar nymphs hatch the following year during the rainy season (i.e., in mid-summer for Namibia/East Africa and autumn/winter (May or later) for South Africa) (see also [17,19]). In the Naukluft Mountains (Namibia), we identified multiple (n > 20) first instar *Striatophasma naukluftense* nymphs [45] in the morning

14 days after a strong summer rain; scattered along a section of 200 m. The same section was searched the previous evening without success. This finding suggests highly synchronized egg hatching from different pods during the night (see also [17] for observations on *Karoophasma*). Analyzing the occurrence of nymphs and adults of different Austrophasmatidae species in South African sample sites, we found that development varied considerably at the local scale throughout two subsequent years (Figure 16), depending on the local rainfall patterns and latitude. These observations indicate highly variable life history patterns on both species and population levels as an adaptation to local climate heterogeneity. In Namibia at the end of March 2014, we collected *Striatophasma* (Hakos Mountains) and *Mantophasma* (Emeritus farm) adults and first instar nymphs in the same habitat. For the Hakos locality, detailed precipitation recordings exist and verified strong rainfall at the end of December 2013 and again in mid-March (see Figure 16). Therefore, facultative bivoltine behavior may occur during a long rainy season but this scenario needs to be verified experimentally. Other than the direct influence of rainfall on egg development, there is no information on which additional factors induce the end of dormancy in eggs after the dry period. Under dry experimental conditions and room temperature, most *T. gladiator* embryos remained alive after two years. Similarly prolonged egg dormancy is also known from other Polyneoptera inhabiting semi-deserts; in Southern Africa eggs of the armoured ground cricket *Acanthoplus bechuanus* (Orthoptera) are able to undergo dormancy at least for two years [65].

*T. gladiator* nymphs that hatched reached adulthood after 3.5-4 months [5]; the adults survived at least two months in our laboratory. First instar *Striatophasma naukluftense* nymphs were collected and reared; the oldest adult died at 136 days (4.5 months). For the winter-

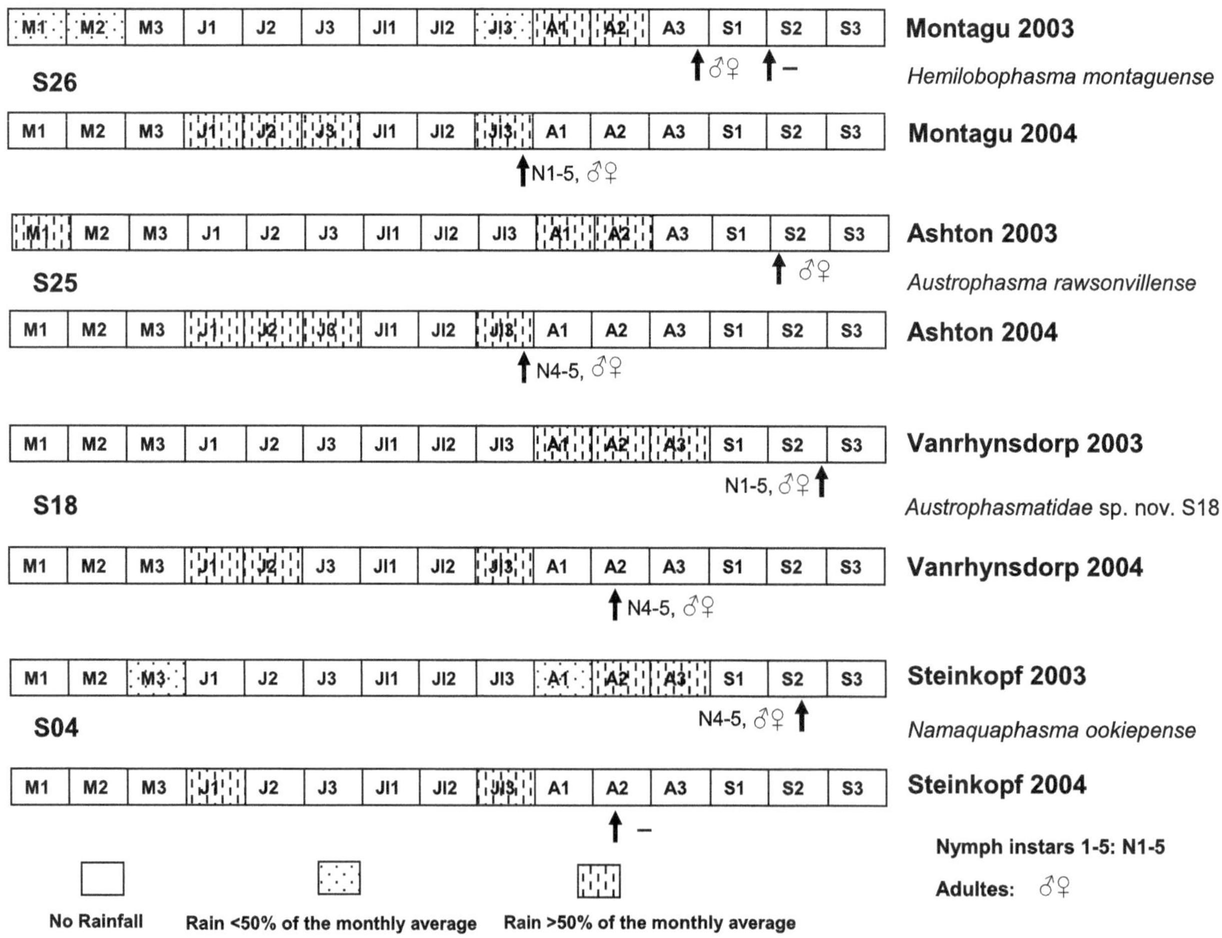


**Figure 16 Mantophasmatodea phenology at four sample sites as a function of rainfall patterns in 2003 and 2004.** Time scale is in periods of ten days from May through September. For each year and sampling site (sample code see [31], the rainfall per period is ranked into 3 categories: no rain, <50%, or ≥50% of the monthly average rainfall (1962-1990; South African Weather Service [http://www.weathersa.co.za/]). The sampling days are provided for each locality (arrow) and the detection of adults and nymphs is given, respectively. The data indicate a life cycle that responds to the period and amount of rain during spring.

active Austrophasmatidae species, the developmental period appears to be shorter, but the unpredictable weather conditions strongly influence their life spans. Drilling and Klass [17] indicated approximately 55 days (excluding the first instar phase) in *L. redelinghuysense*; our laboratory experiments (21-23°C) have shown that third instar Austrophasmatidae larvae (localities: S04, S15, S18, S26, S32; see Figure 1) reached adulthood within less than a month. In the laboratory, we observed a shorter life span in males compared with fertile females. Typically, Mantophasmatodea molted during the night and ingested their exuviae; therefore, counting the number of instar stages for these insects was difficult. In only a few cases, the fragile Mantophasmatodea exuviae were found hanging on the top or wall of the rearing container. Hockman et al., [17] used antenna development (i.e., the increasing number of basiflagellum annuli) as the criterion for assessing instars and reported 5 nymph instars in *L. redelinghuysense*. The first instar nymphs possessed four annuli in the basiflagellum and seven hairy annuli in the distiflagellum. For each instar, two additional annuli were derived from the most basal annulus (meriston); therefore, the number of annuli in the adult *L. redelinghuysense* basiflagellum was 14. Because consistent antennal development was also observed for other mantophasmatodean taxa [17], the assumed 8-10 instars in *T. gladiator* [5] is likely an overestimate.

There is no available information on mortality factors, including predation and parasitism, in the field. Camouflage appears to be important for these small predators, which are vulnerable to attack by birds, lizards, and predaceous insects that are present in the same microhabitat. Escape behavior, i.e., jumping away from a bush or grass tussock, was rarely observed. However, heelwalkers can jump for a short distance (see above). In particular, the summer-active Mantophasmatidae species regularly co-occur with mantises, mainly the genus *Miomantis*, which were collected using the same collection method. However, on the microhabitat scale, we found that a high heelwalker density was primarily linked to relatively low mantis abundance, even if mantises were more frequent elsewhere in the immediate neighborhood. Although we did not perform quantitative studies, this observation indicates extermination or possible avoidance of intra-guild predation between these insects.

### Prey and feeding behavior

Mantophasmatodea are carnivorous insects. Similarly to many other predatory insects, they may be generalists regarding their prey [66] because they consume different types of arthropods, mainly spiders and insects. Mantophasmatodea clearly coincide with microhabitats that have a high arthropod density. Co-occurrence with other predators, as mantises, was rare in the winter rainfall range (Austrophasmatidae), but common in Namibia; however, co-occurrence was not observed in the same microhabitat (see above). We did not observe heelwalkers in bushes with high ant abundance. However, this phenomenon does not exclude ants as prey. We investigated the gut content of 14 specimens from *Mantophasma* spec. (N15, Waterberg) at different nymphal stages and in adults collected at 5:30 p.m. and 11:00 p.m.; we primarily found remains of ants, other adult Hymenoptera, Diptera, caterpillars, and moths, as well as different size spiders. In a single sample, oribatid mite remains were detected. However, phytophagous locusts, bugs, or beetle remains were not observed. Four of the 14 *Mantophasma* specimens did not contain food remains in their gut (3 × 11:00 p.m., 1 × 5:30 p.m.). In an investigation under laboratory conditions, fasting *Mantophasma* sp. males from Waterberg (N15; N = 5) were fed small crickets and dissected at 1 h intervals. After two hours the mid-gut contained considerable quantities of partially digested cricket (Figure 17). However, the foregut stored food remains for many hours. Therefore, food in the foregut does not necessarily reflect recent intake. From these experimental data, we concluded that at least two out of eight (25%) *Mantophasma* specimens (N15) collected at 11 p.m. consumed a meal a few hours in advance. In contrast, the six specimens collected at 5:30 p.m. did not show signs of a recent meal. One specimen from the same location was collected at 7 a.m., and both the foregut and mid-gut were entirely filled with food particles. These experiments indicate hunting activity during the evening and at night.

To further survey the Mantophasmatodea food spectrum, we investigated the potential prey from a single bush (*Suaeda fruticosa*, Amaranthaceae) at 8 a.m. in the dry Richtersveld (South Africa), which yielded four adult specimens from an undescribed Mantophasmatodea taxon (Figure 18). In total, we collected 12 mites (2 species), 2 pseudoscorpions, 32 spiders (6 species; most individuals

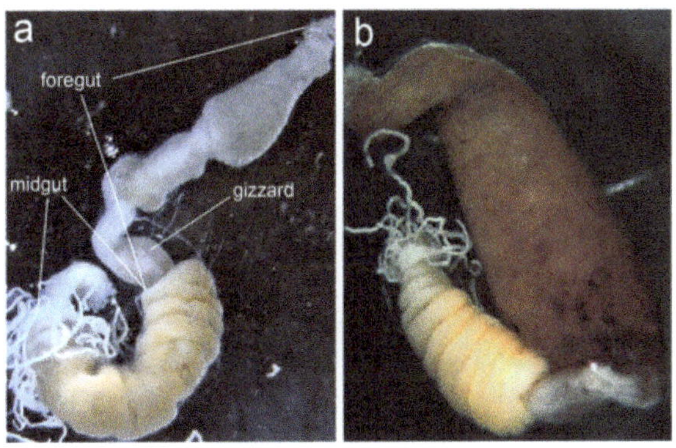


**Figure 17 Dissected adult *Mantophasma* spec. (N15; Waterberg) gut tissues before (a) and 2 h after (b) feeding with crickets.** In the latter specimen, the gizzard and mid-gut contained food particles 2 h after the meal.

**Figure 18 A *Suaeda fruticosa* (arrow) sample site with four adult Mantophasmatodea specimens (8 a.m.).** In addition to the Mantophasmatodea, we collected the remaining arthropods from this bush to survey the potential prey spectrum. Subsequently, the gut content of the four Mantophasmatodea specimens and the identity of the collected arthropods were analyzed in the laboratory.

less than 3 mm), a single termite, 7 adult Psocoptera, 19 small, adult Heteroptera (likely Geocoridae), 3 Homoptera nymphs, 1 myrmicine ant, 1 caterpillar, 1 plume moth, 9 adult Coleoptera (7 Curculionidae, 1 Nitidulidae, and 1 Latridiidae), 2 phytophagous beetle larvae, and 4 winged Diptera (3 Brachycera and 1 Nematocera); a few Diptera escaped. The prey remains in the guts of the four Mantophasmatodea collected from the same bush indicated that each specimen consumed a meal the night before, but the gut contents did not entirely reflect the potential prey spectrum (the prey spectrum that was present in the morning). In the gut lumen of these Mantophasmatodea, we found brachyceran fly remains; additional remains were assigned to a caterpillar, a spider, and a few small adult insects (likely Holometabola) that were not identical to the species listed above.

The literature includes only a few observations on how Mantophasmatodea find and handle their prey in the field [5,33]. A hide and wait strategy (in combination with slow stalking) is very likely and is concordant with the nocturnal activity of most species. In the laboratory, Mantophasmatodea nymphs and adult Austrophasmatidae were mainly fed different *Drosophila* species. Larger Mantophasmatidae and Tanzaniophasmatidae specimens were fed house and flesh flies, as well as different size *Gryllus* nymphs. Additional prey that were successfully provided in captivity include mosquitos, plant hoppers, booklice, moths (including larvae), termites, bush crickets, antlion lacewings, may flies, and small earthworms. Adis et al., [33] also fed the insects dead mealworms (*Tenebrio*). In captivity, crickets were refused as prey by many Austrophasmatidae, but not Mantophasmatidae. For *T. gladiator,* [5] suggested that cockroaches and moths were prey in the Brandberg area. In a single experiment, we placed two late instar *Mantophasma* spec. nymphs (N21, Erongo Mountains) with a mantis nymph (*Miomantis* sp.) of the same size and from the same locality in a plastic box (100x100x40). The mantis was the prey. Cannibalism among nymphs and adults (also between females) has been regularly observed in captivity (see also [19,33]). In certain instances, females have successfully attacked males after mating or without mating contacts in both small pots and larger cages.

When hunting *Drosophila*, Mantophasmatodea waited or sometimes walked a few millimeters to the flies. With a sudden motion, they moved forward and caught the prey with their mandibles (see Additional file 12: Videos 6A, Additional file 13: Videos 6B, Additional file 14: Videos 6C and Additional file 15: Videos 6D). Thus, small prey, such as fruit flies, can be caught without using the legs. Occasionally, we observed that Mantophasmatodea used the tarsi of a foreleg to catch a fruit fly and bring it to their mouthparts. The insects embrace (face-to-face) larger prey using their spiny forelegs and, often, their mid-legs (Figure 12; see also [5] for *T. gladiator*); the prey is immediately consumed. In a single case, we removed an *A. gansbaaiense* female's victim (a male of the same species) immediately after the first attack to the neck and found the ventral nerve cord severed. This behavior was not consistent, but [5] also reported initial neck bites with larger flies. Meal consumption requires a short time and does not necessarily exclude body appendages (see Additional file 16: Video 7). Adis et al., [33] observed that Mantophasmatodea occasionally bite their prey's head off (e.g., *Musca*). We occasionally observed this behavior when we fed the Austrophasmatidae specimens using *Drosophila*. In few cases, the heelwalkers only used the heads for eating/feeding and neglected the remaining bodies from up to ten fruit flies.

## Sexual behavior

### *General remarks*

All Mantophasmatodea species exhibit strong sexual dimorphism in the external genital structure [4], and males are, on average, smaller and more slender compared with females. The male reproductive system includes testes, a deferent duct, seminal vesicles, and accessory glands [7] and is proliferated in the abdomen. Depending on the species, pregnant females have 8-15 eggs per ovary, which are surrounded by a thin cover of tissue and typically oviposited in a single pod (but see also [19], which includes a description of pods with 10-12 eggs). The female's weight increases remarkably during egg maturation (wet weight in g) (e.g., *Mantophasma kudubergense* (0.14-0.22) [5] and *Austrophasmatidae* sp. n. (0.063-0.152; S18, Vanrhynsdorp)). For *Karoophasma biedouwense* and *Hemilobophasma montaguense*, 50-100 eggs total per lifetime have been reported [19]. Under laboratory conditions,

Mantophasmatodea mate readily; therefore, we observed different aspects of their sexual behavior, such as sexual communication, partner finding, courtship, and copulation, in several species and compared these data with previous descriptions in [20].

***Drumming behavior***

Both sexes use single vibrational calls, which are produced by tapping the abdomen onto the substrate (see Additional file 17: Videos 8A and Additional file 18: Video 8B), for mate location and recognition [20]. The male subgenital plate has a semicircular lip-like process of taxon-specific extension ([1,4]; Figure 19), which temporarily contacts the surface during drumming (='drumstick'). Females do not develop this abdominal process and tap with the middle part of their ventral abdomen. Female calls consist of single repeated pulses, whereas male calls are more complex pulse trains that likely contain coding for species recognition. Eberhard and Picker [20] detailed a behavior associated with duet drumming in *K. biedouwense*. The males randomly emitted drumming calls to test their environment for receptive females, whereas the females typically drummed in response to the male calls. After stimulation by a male, the females stopped walking, began drumming, and rapidly moved their antennae. In response, males initiated rapid antennal movement, rubbed their abdomen on vertical stems before drumming, and, finally, began their partner-searching behavior. Rubbing the abdomen appears facultative because we did not observe this behavior before drumming in our laboratory experiments. Substrate vibration is likely detected by leg scolopidial organs [21].

An analysis of two sympatric Austrophasmatidae (*Viridiphasma clanwilliamense* and *Karoophasma biedouwense* [20,44] species revealed different call parameters

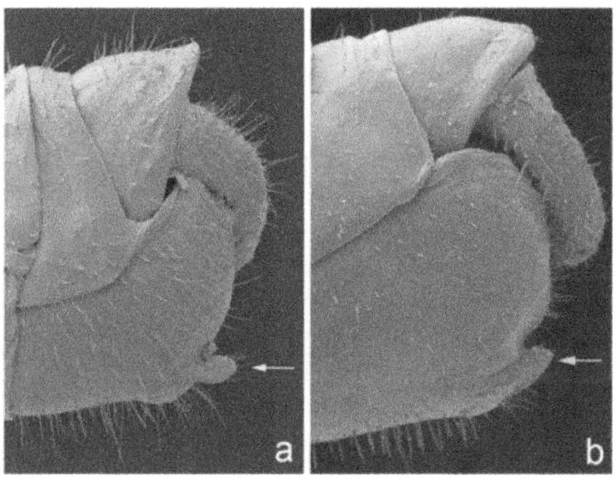


**Figure 19 SEM micrographs showing taxon-specific development of male subgenital plates (arrow) from *Pachyphasma brandbergense* and *Striatophasma naukluftense*.** (modified from [45]).

in both species and a clear preference for conspecific calls. However, most Mantophasmatodea species are not sympatric [31], and species recognition via species-specific call parameters may be relatively unimportant. This assumption was partially confirmed by [22], who analyzed vibrational communication in 13 Mantophasmatodea species, which included 11 Austrophasmatidae and two Mantophasmatidae species from Namibia. Nevertheless, slightly different call parameters were reported for all species studied. For females, a greater pulse repetition time in Austrophasmatidae was observed compared with species from Namibia [22]. For our report, we studied vibrograms and temporal call characteristics for male *Striatophasma naukluftense* from Namibia (Figure 20, Table 1). We studied the following five parameters for the male calls: 1) pulse repetition time (the time between single beats), 2) number of pulses per pulse train, 3) pulse train duration, 4) inter-pulse train intervals (the time between the end of one pulse train and the beginning of the next pulse train), and 5) pulse train repetition time (the time between initiating two pulse trains). For reasons explained below, only the first three parameters are presented here (see Table 1). In addition, we analyzed six males from the *Mantophasma/Sclerophasma* clade, which currently consists of *Sclerophasma paresisense* and several *Mantophasma* species. In contrast to the specimens studied in [22], our six *Mantophasma/Sclerophasma* specimens originated from different localities/populations (see Figure 1). The calls from these males were similar in structure, but the temporal patterns of all measured characteristics (see Table 1) differed significantly (Kruskal-Wallis test results are not shown). When comparing coefficients of variation (CV) values of the measured calls with those reported for other Mantophasmatodea species (CVs for all parameters <15%, see [22]), we found similar CV values for the duration of pulse trains and the number of pulses/train (CVs: 0.4-17%, see Table 1). In contrast, the pulse repetition time had much higher CVs on inter-individual level (38% and 57% for *Striatophasma* and *Sclerophasma/Mantophasma*, respectively) in our analysis compared with the study by [22]. Because the intra-individual comparisons did not demonstrate this variation, we excluded a systematic error. The number of individuals investigated was low and further specimens have to be investigated, but the obtained intra-specific variability in *Striatophasma* and *Mantophasma* already suggests variation at the population level. A direct comparison of call parameters obtained in our analyses with those reported in [22] is complicated by incorrect assignments (and GPS data) for their Namibian species; the calls assigned to specimens from the Brandberg should be from *Tyrannophasma gladiator* instead of *Sclerophasma paresisense*, and the calls assigned to specimens from Paresisberge should be from *Sclerophasma paresisense* instead of *Praedatophasma maraisi*. A notable

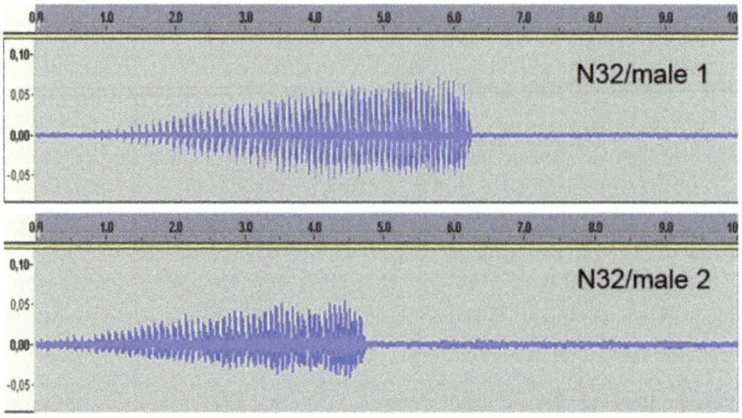


**Figure 20 Vibrogram of substrate vibrational signals of two male *Striatophasma naukluftense*.** The specimens were caught at Gamsberg Pass, Namibia (N32); the signals were recorded at ca. 20°C in the laboratory. The male call consists of repeated pulse trains and one pulse train for each individual is shown. The duration of pulse train is about 6000 ms (male 1) and 4500 ms (male 2), respectively. Note the increasing amplitude in the first segment and the more constant amplitude in the second segment of the pulse train.

difference between our analyses and [20,22] is that their recordings were collected during the day, and the insects were stimulated. In our recordings, the Austrophasmatidae species (not shown) drummed spontaneously, but only during the evening, at night, and, occasionally, in the morning. In contrast, both sexes in *Mantophasma* were also observed drumming during the day (see also [5]); however, the main drumming activity occurred at night without a detectable activity window. In general, the intra- and inter-individual variability in call parameters were

**Table 1 Temporal characteristics of vibrational signals for three males of *Striatophasma naukluftense* and six males from the *Sclerophasma/Mantophasma* clade**

| Taxa | Pulse repetition time (ms)/CV(%) | Pulse train duration (ms)/CV(%) | Number of pulses per pulse train/CV(%) |
|---|---|---|---|
| ***Striatophasma* (N32)** | 19.46 ± 2.16/11% | 5365.4 ± 409.6/7% | 57.4 ± 5.38/9% |
| | n = 70 | n = 7 | n = 7 |
| ***Striatophasma* (N32)** | 19.2 ± 2.41/125% | 4395.7 ± 221.7/5% | 50.2 ± 3.3/7% |
| | n = 40 | n = 4 | n = 4 |
| ***Striatophasma* (N33)** | 35.6 ± 2.43/6% | 2537.5 ± 323.3/13% | 28.8 ± 2.93/10% |
| | n = 60 | n = 6 | n = 6 |
| Average (N32,N32,N33) | **24.75 ± 9.39/38%** | **4099.53 ± 1437.02/35%** | **45.46 ± 14.8/32.5%** |
| ***Sclerophasma* (N14)** | 21.03 ± 3.77/17% | 1300 ± 122.5/9% | 14.89 ± 1.17/7% |
| | n = 90 | n = 9 | n = 9 |
| ***Mantophasma* (N9)** | not analyzed | 1939.2 ± 266.7/13% | 20.7 ± 2.6/12.5% |
| | | n = 11 | n = 11 |
| ***Mantophasma* (N10)** | 24.05 ± 7.5/31% | 2059.5 ± 277.8/13% | 24 ± 3.0/12.5% |
| | n = 60 | n = 6 | n = 6 |
| ***Mantophasma* (N15)** | 39.03 ± 3.65/9% | 1689 ± 83.3/4% | 19 ± 0.9/5% |
| | n = 60 | n = 6 | n = 6 |
| ***Mantophasma* (N20)** | 74.08 ± 4.7/6% | 1628.6 ± 111.3/7% | 12.1 ± 0.9/7% |
| | n = 70 | n = 7 | n = 7 |
| ***Mantophasma* (N31)** | 78.5 ± 2.12/2% | 3205 ± 15/0.4% | 13 ± 0 |
| | n = 20 | n = 2 | n = 2 |
| Average (N9,N10, N14, N15, N20, N31) | **47.34 ± 27.33/57%** | **1970.22 ± 660.1/33.5%** | **17.33 ± 4.79/28%** |

The data are the mean ± standard deviation; n = number of measured values analyzed per individual. For pulse repetition times, each pulse train was measured 10 times and then averaged. CV = coefficient of variance (i.e., standard deviation/mean (rounded)); localities (N9-N33) in accordance with [31]).

expressed as CVs that ranged from 10-40% ([22]; this study) and were in the variation range for communication in other insects (23.6% average CV; [67]).

Based on the studies conducted using Mantophasmatodea, the following questions remain. 1) It is not clear whether repetitive pulse trains constitute a single call, as proposed in [20] and [22]. Alternatively, repetitive pulse trains could compose consecutive calls. 2) Certain parameters studied ([20,22], and this study) are inter-correlated. The number of pulses per pulse train might depend on the pulse train duration. As another example, the pulse train repetition time is the sum of the inter-pulse train interval and pulse train duration. It is difficult to determine which parameter is most important for females to discriminate among male calls from different species because of these strong inter-correlations. 3) In insect communication, call performance also depends on the specimen condition and ambient temperature (Boumans, pers. communication). Stonefly studies have demonstrated that temperature has an important impact on the frequency but not the number of pulses [68]. However, physical constitution affects the number of pulses per pulse train (Boumans, in prep.). After submitting the manuscript, the authors collected in the Richterveld (South Africa) males of two different species *(Namaquaphasma ookiepense; Austrophasmatidae* gen. n. sp. n. S01) in the same bush and placed them in separate plastic boxes in the field. The *N. ookiepense* males produced short pulse trains of about a second (see also [22]). In striking contrast to these short signals were the pulse trains of the second species which lasted 4-6 sec.

### *Courtship and mating behavior*

If maintained in spacious cages, mating in Austrophasmatidae species primarily begins in the evening or at night. In contrast, mating of several *Mantophasma* specimens was observed at any time of day. Eberhard and Picker [20] hypothesized that males were attracted chemically through female pheromones over a long range and through continuous vibrational communication in a median range, such as where both partners are in the same bush. However, even if vibrational communication is part of a heelwalker's sexual behavior, it is not an obligate behavior for locating a sexual partner in the laboratory (see also [20]). If males and females were placed at different edges of a cage (200 × 200 × 300, with a grass tussock structure), they found each other without drumming. Sexual partners seemed to notice each other from approximately 8-10 cm. As they neared each other, rapid antennation (i.e., flickering of their antennae) occurred (see [20]), which indicates odor-mediated sexual communication. This approach was often interrupted by phases without movement. In most cases, the antennae did not directly contact the partner (see also [5]), but occasionally, contact was made. We frequently observed the males moving their abdomens up and down, while the females approached the males over a short range or continuously beat their abdomens on the surface/substrate. In one case, we observed a *Hemilobophasma montaguense* female that walked back and forth 3 cm in front of a male; suddenly, the female stopped and moved her abdomen up, and immediately, the male moved closer and mounted. Males mount rapidly. If the male's head was initially directed towards the tip of the female abdomen, the male quickly rotated until their heads were aligned (see Additional file 19: Video 9). Copulation occurs through a false male-above mating superposition, and males hold the females with their fore- and midlegs. Eberhard and Picker [20] reported arolia use during copulation, but this behavior was only observed (if ever) during the first second(s) of mounting to stabilize their position. Shortly after mounting, the males hold their arolia upwards again. For copulation, males extend their abdomen to the right side of the females, forming an "s" shape, while females project their abdomen upwards. Fixing the genitalia on the left side of the body was not observed (see also [20]). The male cerci are used as claspers to bring the genitalia of sexes close together. The large male genitalia evaginate and insert into the vagina (morphological description in [4]). During prolonged mating (see below), males remain mounted regardless of the female's behavior. During copula, we observed females walking, stalking, catching prey (see Additional file 20: Video 10), and excreting (see also [20]). Even if a female dropped down from the grass, the male remained sitting on the female's abdomen afterwards. Occasionally, we found males mounted on dead females. Both partners groomed during copula. When cleaning the forelegs, males occasionally stood with only their midlegs on the females, while the hind legs were held upwards. Under these circumstances, we observed that male midleg arolia contacted the female body. Except for grooming, the males did not move during copula; even the antennae did not flicker (see Additional file 11: Video 5). This observation might be interpreted as a behavioral adaptation to minimize the energy loss during the extended copulation time. The position of males during copulation was typical of Mantophasmatodea (Figure 21) and included upholding the hind legs, as well as resting from antenna flickering.

The males generally appear to be at risk of attack by the female, as indicated by their rapid mounting. Zompro et al., [5] reported that *Mantophasma kudubergense* males retreated by jumping away if the females appeared agitated after mounting. In *Austrophasmatidae* sp. n. (S18, Vanrhynsdorp), we observed that a female defended herself against a mounted male by both leg kicking and biting (see Additional file 21: Video 11). The elements of sexual behavior, such as drumming, antennae flickering for chemical cues, and antennae contact, can be interpreted

**Figure 21 Typical mating position of Mantophasmatodea. (a)** *Mantophasma* spec. (N15, Waterberg); righthand view. **(b)** *Mantophasma kudubergense* (N21, Erongo Mts.) with different color morphs; lefthand view. **(c)** *Striatophasma naukluftense* (N35, Naukluft Mts.; from [45]). **(d)** *Austrophasmatidae* sp. n. (S18, Vanrhynsdorp), with a conspicuous semi-liquid and gelatinous substance (arrow) between the external genital structures. **(e)** *T. gladiator* (N 38, Brandberg). Note the uniform position of the hind legs and antennae in males.

as species recognition behavior [20] or courtship behavior to reduce female attacks; the elements might be involved in both behaviors. As described for *Mantophasma kudubergense* by [33], we observed intense pumping movements at the tip of the abdomen for Austrophasmatidae males; thereafter, spherical eversion of the phallic lobes was observed. Eberhard and Picker [20] considered this action an indication of direct sperm transfer from males to females. This behavior was observed during the first 1-2 min of copulation and at least three times during the first hour of copulation. In our observations we did not identify similar time patterns for pumping in Austrophasmatidae. Pumping occurred at different periods during prolonged copulation. In *Austrophasmatidae* sp. n. (S18, Vanrhynsdorp) we observed a semi-liquid, gelatinous substance between the external genital structures of males and females during the last phase of copulation (Figure 21C). We assume that this substance was released by the male and that it was a type of mating plug, not a spermatophore.

#### *Copulation frequency and time*

Heelwalkers typically exhibited extended copulation times ([20,33]; our observations are summarized in Table 2). However, we do not know exactly when the sperm transfer occurred. Thus, we do not know if the prolonged mating time should be considered as mate guarding and which role prolonged copulation plays in the male and female Mantophasmatodea sexual strategies. Currently, there is no evidence that the copulation time depends on the sexual partners' mating status. Adis et al., [33] reported remating for *Mantophasma kudubergense*; we observed remating in other species, such as *A. gansbaaiense*, *K. botterkloofense*, and *Austrophasmatidae* sp. n. (S18, Vanrhynsdorp). Remating 24 - 48 h after the first mating has been observed both with different partners and the same partner, but no partner choice experiments have been conducted. *Austrophasmatidae* sp. n. (S18, Vanrhynsdorp) males occasionally began drumming 4 h after mating. In September 2003, in Little Karoo, we observed that 20.8% of all *H. montaguense* individuals were in copula during the day, which included 10 (i.e., five pairs) out of 48 (25 males, 23 females) individuals.

#### *Mating costs*

In captivity, female heelwalkers in copula were regularly observed hunting and feeding on insects. Therefore, the extra costs from mating may be low for females. In contrast, males do not eat during copula. Thus, males lose a significant amount of weight during copulation. We assume that a short starvation period is a natural phenomenon in Mantophasmatodea, similarly to other carnivorous arthropods (e.g., [69-71]). In addition to starvation and dehydration, transferring substantial amounts of protein and liquids to the females must also be considered. We attempted to assess the costs to males by comparing weight loss during copula versus weight loss during a similar starvation/dehydration period. The weight lost was considered an indicator of the cost to a male during prolonged mating. Weight loss during mating that exceeds a similar starvation period indicates remarkable protein transfer. Therefore, we compared the costs during mating with other biological periods of energy depletion to assess the consequences of prolonged mating for the male energy budget and fitness. In our experiments, *Austrophasmatidae* sp. n. (S18, Vanrhynsdorp) male weight loss was measured after starvation/dehydration for one group of males (starvation group, N = 6) and after mating for a

**Table 2 Copulation times for several Mantophasmatodea species in the laboratory**

| Species | Observed copulation times in hours | Number of observed copulations |
|---|---|---|
| *Karoophasma botterkloofense* (S16) | > 29 | 1 |
| *Karoophasma botterkloofense** | 68.5 (S.E. ±4.56) | 7 |
| *Austrophasmatidae* sp. n. (S18) | 7.5-31.5 | 17 |
| *Hemilobophasma montaguense* (S26) | 20-54 | 4 |
| *Austrophasma gansbaaiense* (S32) | 24 [R], 24 [R], 64 | 3 |
| *Mantophasma spec.* (N21) | 18-28 | 6 |
| *Mantophasma kudubergense*** | 12-72 | 12 |
| *Hemilobophasma sp.n.*[1] x *H. cf. montaguense*[2] | < 24 | 1 |
| Mantophasmatodea in general*** | Up to 96 | - |

*Data from [19].
**Data from [5], species assignment according to [46].
***Data from [36].
[R]Data from a remating pair of *A. gansbaaiense* (S32, Gansbaai).
[1]*Hemilobophasma sp. n.* - male (S31, De Rust).
[2]*Hemilobophasma cf. montaguense* - female (S29, Calitzdorp).

second group (mating group, N = 7). Starvation and mating were conducted under the same experimental conditions (temperature 22-24°C, 12 h light). Preliminary experiments showed 1) that males can lose up to 50% of their initial body mass within 24 h of starvation and dehydration, and 2) a significant positive correlation between initial body mass and percent of weight lost was identified (Spearman rank test: r = 0.89, N = 10, p = 0.01). Therefore, in the following experiments, we only used males with the same body class and feeding status (i.e., both groups did not differ significantly in their average initial weight) (starvation group: 0.033 g, s.d. 0.004/mating group: 0.0325 g, s.d. 0.003; t-Test: t = 0.226, df = 11, p = 0.82). Mating was not artificially interrupted, and we assumed copulation with sperm (and protein) transfer in all cases. As a result, the copulation times varied among tested males and were in the 11-21 h range (average 14.7 h), whereas the starvation time was 24 h for all individuals in the starvation group. The starving males lost body weight in the range of 2.6-29.6% (average: 13.29, s.d. 11.37), and the males in copulation lost body weight in the range of 3.5-9.0% (average: 6.38, s.d. 0.81). An analysis of variance confirmed that 1) the initial body weight had a positive effect on the percentage of body weight lost (i.e., the heavier individuals lost relatively more weight), 2) the weight lost between both experimental groups (starving/desiccating males versus mating males) did not differ, and 3) the treatment time (starvation or copulation time) did not affect the results (Table 3). Altogether, our data show that body weight lost during mating does not differ significantly from loss during a similar starvation and desiccation timeframe. Considering that males do not feed (and most likely do not take in water) during prolonged copulation, the data suggest that proteins are not excessively transferred from males to females.

## Conclusions

Only a few years after the first Mantophasmatodea was described, the literature includes an impressive scientific knowledge on its diversity and biology. In contrast to the well-known distribution range in Southern Africa, however, the Mantophasmatodea range and diversity in East Africa have not been explored as extensively. In addition, the assignments of the known genera to different erected families must be clarified in a thorough revision. Future novel discoveries regarding general aspects of their biology will largely depend on successful breeding in captivity. One particular aspect of their biology is their tendency toward broad species distribution with many local populations. This phenomenon is, in part, due to low dispersal abilities (winglessness, low mobility, and low chance of dispersal for all stages, including eggs). Using mass fingerprints of peptide hormones to map the different populations, it appears possible that intraspecific migration/replacement/hybridization can be followed over a long time period, which is another unique possibility provided by these fascinating insects.

**Table 3 Analysis of variance to determine the effects of 1) initial body weight (in g), 2) treatment duration (in hours), and 3) treatments (starvation/desiccating vs. mating) on the percentage of body weight lost in *Austrophasmatidae sp. n.* males (S18, Vanrhynsdorp)**

| Source of Variation | Sum of square | DF | Mean square | F | P |
|---|---|---|---|---|---|
| Corrected Model | 554.466[a] | 3 | 184.822 | 5.964 | 0.016 |
| Intercept | 276.439 | 1 | 276.439 | 8.921 | 0.015 |
| Main Effects | | | | | |
| Initial body weight | 385.618 | 1 | 385.618 | 12.444 | 0.006 |
| Duration of treatment | 129.356 | 1 | 129.356 | 4.174 | 0.071 |
| Treatment | 016.256 | 1 | 016.256 | 0.525 | 0.487 |
| (Starvation/mating) | 278.885 | 9 | 030.987 | | |
| Error | 2037.005 | 13 | | | |
| Total | 833.351 | 12 | | | |
| Corrected total | | | | | |

[a]$R^2$ = 0.665 (adjusted $R^2$ = 0.554).

## Sampling and export permissions

The insects were captured and exported with permission from the Western and Northern Cape Nature Conservation Boards (no. 2297/2003, 0697/2004, 0554/2004), the Ministry of Environment and Tourism of Namibia (research/collecting permits 891/2005, 1041/2006), and Malawi (Government Document: Department of Forestry Specimens Collection Permit 21/01/2011/no.1).

## Additional files

**Additional file 1: Methods.**

**Additional file 2: Video 1A - Jumping male; *Tyrannophasma gladiator*.** Note: movement of antennae is artificially affected by airflow using a blow-dryer.

**Additional file 3: Video 1B - Jumping male; *Tyrannophasma gladiator*.** Note: movement of antennae is artificially affected by airflow using a blow-dryer.

**Additional file 4: Video 1C - Jumping male; *Tyrannophasma gladiator*.** Note: movement of antennae is artificially affected by airflow using a blow-dryer.

**Additional file 5: Video 1D- Female is jumping over a male which tries to mount and copulate afterwards; *Karoophasma botterkloofense*.**

**Additional file 6: Video 2A- Flexible body whilst jumping; *Tyrannophasma gladiator*.**

**Additional file 7: Video 2B- Flexible body whilst jumping; *Tyrannophasma gladiator*.**

**Additional file 8: Video 3A - Landing behavior after dropping; *Tyrannophasma gladiator*.**

**Additional file 9: Video 3B - Landing behavior after dropping; *Tyrannophasma gladiator*.**

**Additional file 10: Video 4 - Grooming behavior; *Austrophasmatidae* sp. nov. (S18, Vanrhynsdorp).**

**Additional file 11: Video 5 - Antenna-flickering, *Austrophasmatidae* sp. nov. (S18, Vanrhynsdorp).**

**Additional file 12: Video 6A - Foraging behavior; *Karoophasma botterkloofense*.**

**Additional file 13: Video 6B - Foraging behavior; *Austrophasmatidae* sp. nov. (S18, Vanrhynsdorp).**

**Additional file 14: Video 6C - Foraging behavior; *Austrophasmatidae* sp. nov. (S18, Vanrhynsdorp).**

**Additional file 15: Video 6D - Foraging behavior; *Austrophasmatidae* sp. nov. (S18, Vanrhynsdorp).**

**Additional file 16: Video 7 - Foraging behavior (details); *Pachyphasma brandbergense*.**

**Additional file 17: Video 8a, Drumming behavior; *Austrophasmatidae* sp. nov. (S18, Vanrhynsdorp).**

**Additional file 18: Video 8B - Drumming behavior; *Austrophasmatidae* sp. nov. (S18, Vanrhynsdorp).**

**Additional file 19: Video 9B - Mating behavior; *Austrophasmatidae* sp. nov. (S18, Vanrhynsdorp).**

**Additional file 20: Video 10 - Female foraging whilst in copula; *Austrophasmatidae* sp. nov. (S18, Vanrhynsdorp).**

**Additional file 21: Video 11 - Female defense against male; *Austrophasmatidae* sp. nov. (S18, Vanrhynsdorp).**

**Competing interests**

The authors declare that they have no competing interests.

**Authors' contributions**

SR and RP conducted field samplings and observations. Rearing, laboratory observations including video and vibrational signal tapping were performed by RP and partly by SR. The mating cost experiment was carried out and analyzed by SR. JM analyzed the vibrational signals. Data interpretation and preparation of the manuscript was done by SR, RP and JM. The final manuscript was written by RP and SR. All authors read, commented on and approved the final manuscript.

**Acknowledgments**

We wish to thank the following individuals for support during the numerous field trips and for their help in the laboratory work with Mantophasmatodea: Susanne Neupert (Köln), Wolf Hütteroth (Konstanz), Mike Picker (Cape Town), Adrian Scheidt (Groningen), Martin Scheidt (Vienna), Moritz and Marie Predel (Jena), Rene Köhler (Köln), Alexander König (Göttingen), Udo Neugebauer (Saalburg), Thomas Schmalenberg (Leipzig), Holger Vollbrecht (Windhoek), Martin Fischer (Jena) and Benson Muramba (Windhoek). We thank Louis Boumans (Oslo) for useful comments on our vibrational studies. We also acknowledge comments from Mike Picker on a former version of the manuscript and Dirk U. Bellstedt (Stellenbosch) for plant determination. This study was supported by travel grants from the Boehringer Ingelheim Stiftung, the British Ecological Society, and the Orthoperists' Society of America. Remains of orbatids in intestines were confirmed by Torstein Solhøy, Roy A. Norton, and Arne Fjellberg (Bergen; in litt.).

**Author details**

[1]University Museum of Bergen, The Natural History Collections, P.O. Box 7800, N-5020 Bergen, Norway. [2]Departamento de Ciencias Biologicas, Universidad de Los Andes, Apartado Aerea 4976, Bogota, Colombia. [3]University of Cologne, Institute for Zoology, Zuelpicher Str. 47b, 50674 Cologne, Germany.

**References**

1. Klass K-D, Zompro O, Kristensen NP, Adis J: **Mantophasmatodea: A new insect order with extant members in the Afrotropics.** *Science* 2002, **296:**1456–1459.
2. Picker MD, Colville JF, van Noort S: **Mantophasmatodea, Now in South Africa.** *Science* 2002, **297:**1475.
3. Walker EM: **A new species of Orthoptera, forming a new genus and new family.** *Can Entomologist* 1914, **45:**93–99.
4. Klass K-D, Picker MD, Damgaard J, van Noort S, Tojo K: **The taxonomy, genitalic morphology, and phylogentic relationships of southern African Mantophasmatodea (Insecta).** *Entomologische Abhandlungen* 2003, **61:**3–67.
5. Zompro O, Adis J, Bragg PE, Naskrecki P, Meakin K, Wittneben M, Saxe V: **A new genus and species of Mantophasmatidae. Insecta: Mantophasmatodea) from Brandberg Massif, Namibia, with notes on behaviour.** *Cimbebasia* 2003, **19:**13–24.
6. Zompro O: **Inter- and intra-ordinal relationships of the Matophasmatodea, with comments on the phylogeny of polyneopteran orders (Insecta: Polyneoptera).** *Mitteilungen des Geologisch* 2005, **89:**85–114.
7. Dallai R, Frati F, Lupetti P, Adis J: **Sperm ultrastructure of *Mantophasma zephyra* (Insecta, Mantophasmatodea).** *Zoomorphology* 2003, **122:**67–76.
8. Tsutsumi T, Machida R, Tojo K, Uchifune T, Klass KD, Picker MD: **Transmission electron microscopic observations of the egg membranes of a South African heel-walker *Karoophasma biedouwensis* (Insecta: Mantophasmatodea).** *Proc Arthropodan Embryol Soc Jpn* 2004, **39:**23–29.
9. Uchifune T, Machida R, Tsutsumi T, Tojo K: **Chorion of a South African heel-walker, *Karoophasma biedouwensis* Klass et al. SEM observations (Insecta: Mantophasmatodea).** *Proc Arthropodan Embryol Soc Jpn* 2006, **41:**29–35.
10. Baum E, Dressler C, Beutel RG: **Head structures of Karoophasma sp. (Hexapoda, Mantophasmatodea) with phylogenetic implications.** *J Zool Syst Evol Res* 2007, **45:**104–119.
11. Klass K-D, Eulitz U: **The tentorium and anterior head sulci in Dictyoptera and Mantophasmatodea (Insecta).** *Zool Anz* 2007, **246:**205–234.
12. Pass G, Gereben-Krenn B-A, Merl M, Plant J, Szucsich NU, Tögel M: **Phylogenetic relationships of the orders of Hexapoda: contributions from**

the circulatory organs for a morphological data matrix. *Arthropod Syst Phylogeny* 2006, **64**:165–203.
13. Beutel RG, Gorb SN: **A revised interpretation of the evolution of attachment structures in Hexapoda with special emphasis on Mantophasmatodea.** *Arthropod Syst Phylogeny* 2006, **64**:3–25.
14. Beutel RG, Gorb SN: **Evolutionary scenarios for unusual attachment devices of Phasmatodea and Mantophasmatodea (Insecta).** *Syst Entomol* 2008, **33**:501–510.
15. Eberhard MJB, Pass G, Picker MD, Beutel R, Predel R, Gorb SN: **Structure and function of the Arolium of Mantophasmatodea (Insecta).** *J Morphol* 2009, **270**:1247–1261.
16. Buder G, Klass K-D: **The morphology of tarsal processes in Mantophasmatodea.** *Deutsche Entomologische Zeitschrift* 2013, **60**:5–23.
17. Hockman D, Picker MD, Klass K-D, Pretorius L: **Postembryonic development of the unique antenna of Mantophasmatodea (Insecta).** *Arthropod Struct Dev* 2009, **38**:125–133.
18. Drilling K, Klass KD: **Surface structures of the antenna of Mantophasmatodea (Insecta).** *Zool Anz* 2010, **249**:121–137.
19. Tojo K, Machida R, Klass K-D, Picker MD: **Biology of South African heel-walkers, with special reference to reproductive biology (Insecta: Mantophasmatodea).** *Proc Arthropodan Embryol Soc Jpn* 2004, **39**:15–21.
20. Eberhard MJB, Picker MD: **Vibrational communication in two sympatric species of Mantophasmatodea (Heelwalkers).** *J Insect Behav* 2008, **2008** (21):240–257.
21. Eberhard MJB, Lang D, Metscher B, Pass G, Picker MD, Wolf H: **Structure and sensory physiology of the leg scolopidial organs in Mantophasmatodea and their role in vibrational communication.** *Arthropod Struct Dev* 2010, **39**:230–241.
22. Eberhard MJB, Eberhard SH: **Evolution and Diversity of Vibrational Signals in Mantophasmatodea (Insecta).** *J Insect Behav* 2012, **26**:352–370.
23. Predel R, Roth S, Neupert S, Picker M: **New insect order Mantophasmatodea: Species differentiation by mass fingerprints of peptide hormones?** *J Zool Syst Evol Res* 2005, **43**:149–156.
24. Gäde G, Marco HG, Simek P, Marais E: **The newly discovered insect order Mantophasmatodea contains a novel member of the adipokenetic hormone family of peptides.** *Biochem Biophys Res Commun* 2005, **330**:598–603.
25. Chown SL, Marais E, Picker MD, Terblanche JS: **Gas exchange characteristics, metabolic rate and water loss of the Heelwalker, *Karoophasma biedouwensis* (Mantophasmatodea: Austrophasmatidae).** *J Insect Physiol* 2006, **52**:442–449.
26. Machida R, Tojo K, Tsutsumi T, Uchifune T, Klass KD, Picker MD, Pretorius L: **Embryonic development of heel-walkers: Reference to some preevolutionary stages (Insecta: Mantophasmatodea).** *Proc Arthropodan Embryol Soc Jpn* 2004, **39**:31–39.
27. Tsutsumi T, Tojo K, Uchifune T, Machida R: **Ovarian structure and oogenesis of the South African heel-walker *Karoophasma biedouwensis* (Insecta: Mantophasmatodea).** *Proc Arthropodan Embryol Soc Jpn* 2005, **40**:15–22.
28. Cameron SL, Barker SC, Whiting MF: **Mitochondrial genomics and the new insect order Mantophasmatodea.** *Mol Phylogenet Evol* 2006, **38**:274–279.
29. Terry MT, Whiting MF: **Mantophasmatodea and phylogeny of the lower neopterous insects.** *Cladistics* 2005, **21**:240–257.
30. Damgaard J, Klass K-D, Picker MD, Buder G: **Phylogeny of the Heelwalkers (Insecta: Mantophasmatodea) based on mtDNA sequences, with evidence for additional taxa in South Africa.** *Mol Phylogenet Evol* 2008, **47**:443–462.
31. Predel R, Neupert S, Huetteroth W, Kahnt J, Waidelich D, Roth S: **Peptidomics-based phylogeny and biogeography of Mantophasmatodea (Hexapoda).** *Syst Biol* 2012, **61**:609–629.
32. Kjer KM, Carle FL, Litman J, Ware J: **A molecular phylogeny of Hexapoda.** *Arthropod Systematics & Phylogeny* 2006, **2006**(64):35–44.
33. Adis J, Marais E, Moombolah-Goagoses E, Zompro O: **Gladiatoren: Gespenstische Räuber.** *Spektrum Wissensch* 2003, **2003**:64–69.
34. Walker JA: **Mantophasmatodea - a new order of insects.** *Bull Amateur Entomologists' Soc* 2003, **62**:72–78.
35. Zompro O: **Mantophasmatodea - Gladiatoren im Insektenreich.** *Arthropoda* 2008, **16**:4–25.
36. Klass K-D: **Mantophasmatodea, Die zuletzt entdeckte Insektenordnung.** *Nat Mus* 2009, **139**:218–227.
37. Adis J, Zompro O: **Gladiators: A new order of Insects.** *Sci Am* 2002, **287**:60–67.
38. Zompro O, Adis J, Weitschat W: **A review of the order Mantophasmatodea (Insecta).** *Zoologischer Anzeiger* 2002, **241**:269–279.
39. Arillo A, Engel MS: **Rock Crawlers in Baltic Amber (Notoptera: Mantophasmatodea).** *Am Mus Novit* 2006, **3539**:1–10.
40. Machida R, Tojo K: **Heel walkers, a new insect order Mantophasmatodea. (In Japanese).** *Kontyu to Shizen* 2003, **38**:26–31.
41. Picker M, Griffiths C, Weaving A: *Field guide to insects of South Africa.* Cape Town: Struik Publishers; 2004.
42. Huang DY, Nel A, Zompro O, Waller A: **Mantophasmatodea now in the Jurassic.** *Naturwissenschaften* 2008, **10**:947–952.
43. Arillo A, Ortuno VM, Nel A: **Description of an enigmatic insect from Baltic amber.** *Bulletin de la Société entomologique de France* 1997, **102**:11–14.
44. Eberhard MJB, Picker MD, Klass K-D: **Sympatry in Mantophasmatodea, with the description of a new species and phylogenetic considerations.** *Organisms Diversity & Evolution* 2011, **11**:43–59.
45. Wipfler B, Pohl H, Predel R: **Two new genera and species of Mantophasmatodea (Insecta, Polyneoptera) from Namibia.** *Zookeys* 2012, **166**:75–98.
46. Zompro O, Adis J: **Notes on Namibian Mantophasma Zombro, Klass Kristensen & Adis 2002, with descriptions of three new species (Insecta: Mantophasmatodea: Mantophasmatidae: Mantophasmatini).** *Russ Entomol J* 2006, **15**:21–24.
47. Mloza-Banda HR: **Development and application of conservation agriculture in Malawi's smallholder subsistence and commercial farming systems.** In *Proceedings of Workshop on Conservation Farming for Sustainable Agriculture: 20-24 October 2002.* Edited by Mloza-Banda HR, Kumwenda WF, Manda M, Bwalya M. Lilongwe, Malawi: Land Resources Conservation Department; 2003.
48. Williams J: **Adoption of conservation agriculture in Malawi.** In *Master thesis.* Durham, North Carolina: Nicholas School of Environment of Duke University; 2008.
49. Holm E: **Notes on faunas bordering on the Namib Desert.** In *Namib Ecology: 25 years of Namib Research.* Edited by Seely MK. Pretoria: Transvaal Museum Monographs; 1990:55–60.
50. Jürgens N: **Floristic biodiversity and history of African arid regions.** *Biodivers Conserv* 1997, **6**:495–514.
51. Partridge TC: **Evolution of landscapes.** In *Vegetation of Southern Africa.* Edited by Cowling RM, Richardson DM, Pierce SM. Cambridge: Cambridge University Press Cambridge; 1997:5–20.
52. Haughton SH: *Geological history of Southern Africa.* The Geological Society of South Africa: Cape Town; 1969.
53. Simmons RE, Griffin M, Griffin RE, Marais E, Kolberg H: **Endemism in Namibia: patterns, processes and predictions.** *Biodivers Conserv* 1989, **7**:513–530.
54. Redman GT, Hamer ML: **The distribution of southern African Harpagophoridae Attems, 1909 (Diplopoda: Spirostreptida).** *Afr Invertebr* 2003, **44**:213–226.
55. Kirk-Spriggs AH, Marais E: **Dâures-biodiversity of the Brandberg Massif.** *Namibia Cimbebasia Memoir* 2000, **9**:1–389.
56. Cowling R: *The ecology of Fynbos. Nutrition, fire and diversity.* Cape Town: Oxford University Press; 1992.
57. Lovegrove B: *The living deserts of Southern Africa.* Vlaeberg: Fernwood Press; 1993.
58. Picker MD, Samways MJ: **Faunal diversity and endemicity of the Cape Peninsula, South Africa–a first assessment.** *Biodiversity & Conservation* 1996, **5**:591–606.
59. Huber BA, Sinclair BJ, Lampe K-H (Eds): *African Biodiversity: Molecules, Organisms, Ecosystems. Proceedings of the 5th International Symposium on Tropical Biology, Museum Alexander Koenig, Bonn.* New York: Springer; 2005.
60. Craven P, Craven D: **The flora of the Brandberg, Namibia.** *Cimbebasia Memoir* 2000, **9**:49–67.
61. Myers N, Mittermeier RA, Mittermeier CG, da Fonseca GA, Kent J: **Biodiversity hotspots for conservation priorities.** *Nature* 2000, **403**:853–858.
62. Wipfler B: **Mantophasmatodea.** In *Insect Morphology and Phylogeny.* Edited by Beutel RG, Friedrich F, Ge S-Q, Yang X-K. Berlin: De Gruyter; 2014:272–277.
63. Cukier M, Guerrero GA, Maggese MC: **Parthenogenesis in *Coptopteryx viridis*, Giglio Tos (1915) (Dyctioptera, Mantidae).** *Biol Bull* 1979, **157**:445–452.

64. Uchifune T: **Collection and rearing of a South African heel-walker, Karoophasma biedouwensis Klass et al. (Insecta: Mantophasmatodea).** *Science Report of the Yokosuka City Museum* 2008, **55**:23–28.
65. Power JH: **On the biology of *Acanthoplus bechuanus* Per (Orthoptera: Tettigoniidae).** *J Entomological Soc South Africa* 1958, **21**:376–381.
66. New TR: *Insects as predators.* Sydney: New South Wales University Press; 1991.
67. Reinhold K: **Variation of acoustic courtship signals in insects and amphibians: No evidence for bimodality, but identical dependence on duration.** *Ethology* 2009, **115**:131–140.
68. Rupprecht R: **Drumming signals of Danish Plecoptera.** *Aquat Insects* 1992, **4**:93–103.
69. Dixon AFG, Russel RJ: **The effectiveness of *Anthocoris nemorum* and *A. confusus* (Hemiptera: Anthocoridae) as predators of the sycamore aphid, *Drepanophisum platanoides*.** *Entomologia Experimentales et Applicata* 1972, **13**:194–207.
70. Wise DH: *Spiders in ecological webs.* Cambridge: University Press; 1993.
71. Moran MD, Hurd LE: **Relieving food limitations reduces survivorship of a generalist predator.** *Ecology* 1997, **78**:1266–1270.

# Environmental change and long-term body mass declines in an alpine mammal

Tom HE Mason[1*], Marco Apollonio[2], Roberta Chirichella[2], Stephen G Willis[1] and Philip A Stephens[1]

**Abstract**

**Introduction:** Climate and environmental change have driven widespread changes in body size, particularly declines, across a range of taxonomic groups in recent decades. Size declines could substantially impact on the functioning of ecosystems. To date, most studies suggest that temporal trends in size have resulted indirectly from climate change modifying resource availability and quality, affecting the ability of individuals to acquire resources and grow.

**Results:** Here, we investigate striking long-term body mass declines in juvenile Alpine chamois (*Rupicapra rupicapra*), within three neighbouring populations in the Italian Alps. We find strong evidence that increasing population density and warming temperatures during spring and summer are linked to the mass declines. We find no evidence that the timing or productivity of resources have been altered during this period.

**Conclusions:** We conclude that it is unlikely that environmental change has driven body size change indirectly via effects on resource productivity or phenology. Instead, we propose that environmental change has limited the ability of individuals to acquire resources. This could be due to increases in the intensity of competition and decreases in time spent foraging, owing to high temperatures. Our findings add weight to a growing body of evidence for long-term body size reductions and provide considerable insight into the potential drivers of such trends. Furthermore, we highlight the potential for appropriate management, for instance increases in harvest size, to counteract the impacts of climate change on body mass.

**Keywords:** Body size, Body mass, Chamois, Climate change, Environmental change, Hunting, NDVI, Population density, Temperature, Ungulate

## Introduction

Climatic and other environmental change has impacted species in a variety of ways, from altering their spatial distributions (e.g. [1]) to changing the timing of their annual events (e.g. [2]). Recently, focus has broadened to include the impacts of climate change on life-history traits, body condition and population processes (e.g. [3-6]). One emergent generality is that responses to climate change include widespread body size changes, particularly declines, which could have pronounced negative impacts on the functioning and productivity of ecosystems [7,8]. Body size declines driven by climate change have been recorded in the past, for instance in large ungulates during the Pleistocene, and are thought to have led to the extinction of some species [9].

A variety of climatic drivers of recent body size declines have been proposed. In ectotherms, higher metabolic rates are predicted in warmer environments [10], so climate change might lead to decreased body size, unless individuals can increase their rate of food intake [8]. In endotherms, in line with Bergmann's rule [11], it could be beneficial to be smaller (and thus have a larger surface area to volume ratio) in warmer environments, due to a reduced need for heat conservation and a greater need for heat loss [12]. In mammals and birds, intra-specific trends between body size and both latitude [13,14] and temperature [15,16] have been observed, providing some support for this theory. However, the most frequently cited climatic driver of body mass declines is the indirect link via climatic impacts on resource availability, which has been implicated considerably more

* Correspondence: tom.h.e.mason@gmail.com
[1]School of Biological and Biomedical Sciences, Durham University, South Road, Durham DH1 3LE, UK
Full list of author information is available at the end of the article

frequently than any other mechanism (reviewed in [7]). Climate and environmental change can alter the timing, quality and quantity of resources, affecting the rate at which individuals acquire resources to invest in growth and energy storage, and ultimately body size (e.g. [6,17]). Non-climatic processes can also drive body size change and could be playing a role alongside climate change. For example, in harvested animal populations, a preference for larger bodied individuals by hunters can drive selection for decreased body size [18,19].

In ungulates, body mass is an important indicator of fitness [20,21] and can respond rapidly to environmental change [6]. High body mass is commonly related to high reproductive success (e.g. [22]) and survival (e.g. [23]). As such, changes in body mass can have important effects on population dynamics [6,21]. The indirect link between climate, resources and ungulate body mass is well studied and could be strongly influenced by climate change [24,25]. However, environmental change could also directly affect the ability of ungulates to acquire resources, particularly in seasonal environments, which could lead to temporal body mass change. For instance, changes in population density can alter the intensity of competition for resources, affecting the ability of individuals to forage and grow [26,27]. There is some evidence from other taxa of links between density and temporal body size change [28]. In recent decades, warming climates have led to changing snow cover and depth in some areas [29], altering the costs of locomotion and foraging for some species [30]. In alpine species, high temperatures in spring and summer can cause heat-stress, reducing the time individuals can spend foraging [31]. As a result, we propose that, in alpine areas, which have been strongly affected by climate warming [32], reductions in time spent foraging due to higher thermoregulatory costs could lead to reduced body size. Climate change is predicted to drive similar changes in temperature-dependent activity budgets in other taxa [33,34]. Furthermore, temperature induced changes in foraging behaviour have been detected in experimental studies [35].

Here, we explore long-term variation in the body mass of juveniles in three neighbouring populations of Alpine chamois (*Rupicapra rupicapra*) in the Italian Alps. The body condition of juvenile ungulates is particularly responsive to environmental variation [25,26,36]. This is largely because juveniles invest highly in growth, and not in reproduction, meaning that their body condition is very sensitive to the availability of resources. Experiencing poor environmental conditions in youth can suppress growth rates and result in smaller body size at maturity [6], which could have important implications for population dynamics. We begin by investigating whether there have been consistent, long-term changes in chamois body mass across sexes and populations. Having identified the existence of temporal trends in body mass, we seek to tease apart the different drivers of these trends. For reasons discussed below (see Methods), we dismiss the role of artificial selection driven by hunting. Consequently, motivated by the findings reviewed above, we seek to evaluate the evidence for three plausible drivers of body size change:

1. Climate-mediated changes in vegetation productivity or phenology, altering resource quality and availability.
2. Climate-mediated changes in behaviour, altering access to resources.
3. Changes in population density, altering per-capita resource availability.

## Methods

### Study area

The study area is located in Trento Province in the Central-Eastern Italian Alps (46°02'N, 10°38'E), across three chamois hunting districts: Adamello (area = 373 km$^2$), Presanella (146 km$^2$) and Brenta (263 km$^2$). The area is forested up to the tree-line at about 2,000 m, above which it consists of Alpine meadows, rocky outcrops, scree fields and open rock faces. The average altitude varies among the districts, though with considerable overlap (mean altitude ± SD: Adamello, 1,901 ± 616 m; Presanella, 2,098 ± 540 m; Brenta, 1,594 ± 603 m). Adamello and Presanella are characterised by nutrient-poor siliceous vegetation whilst Brenta is characterised by nutrient-rich calcareous vegetation [37]. Typically, meadows in Adamello and Presanella are dominated by *Festuca scabriculmis* and *Carex curvula*, whilst those in Brenta are composed of *Sesleria albicans* and *Carex firma*. Throughout the study area, meadows are grazed by small herds of livestock (sheep, goats and cows) during summer, a practice that has been maintained at consistent levels throughout the study period. Several potential predators of chamois were present during the study, including a small, stable population of brown bear (*Ursus arctos*) in Brenta, a very small number of Eurasian lynx (*Lynx lynx*) and the golden eagle (*Aquila chrysaetos*). However, predation on chamois is very rare here (personal communication, Adamello Brenta Nature Park, Trento Province, Italy).

### Body mass data

Chamois are hunted every year between mid-September and late-December. Data were collected on the eviscerated body mass and day of shooting of 10,455 yearling (≈1.5 year olds; hereafter juveniles) Alpine chamois (5,762 males and 4,693 females), hunted between 1979 and 2010 (see Additional file 1 for annual breakdowns of sample size). Hunting is heavily regulated and there is little potential for artificial selection by hunters, as chamois can easily detect hunters in the predominantly open habitat and will flee from hunters at particularly large distances [38,39]. Moreover, there is no evidence of hunters preferentially harvesting larger bodied age-classes in these populations [38]. Hunting pressure on yearlings varies among sites (mean proportion of yearlings hunted in census years: Adamello males, 0.40 ± 0.01; Adamello females, 0.32 ± 0.01; Presanella males, 0.32 ± 0.01; Presanella females, 0.24 ± 0.02; Brenta males, 0.37 ± 0.01; Brenta females, 0.31 ± 0.02). In order to account for intra-seasonal variation in body mass, which is not the focus of this study, a published model of seasonal body mass change [38], which considers inter-annual mass variation, was used to estimate juvenile mass standardised to a specific day of the year. Annual estimates (n = 32) of mean juvenile body mass were produced for each sex, within each site, standardised to day 300 of the year (27$^{th}$ October) (see Figure 1). Body mass was estimated after the vegetation growing season (hereafter 'growing season') because body condition at that time will have been influenced by the spring and summer environment, which is thought to have a strong influence on ungulate body mass [40].

There were clear negative temporal body mass trends in all sexes and sites (Figure 1). In order to examine drivers of deviations from the long-term trends (i.e. years in which mean body mass was particularly high or low, even given the trend), the body mass time series were detrended by fitting linear models and calculating residuals. However, detrending can remove long-term fluctuations related to environmental trends [41,42], which are of primary interest to us. As a result, we modelled body mass data, to examine drivers of long-term trends, and also modelled body mass residuals, to examine drivers of deviations from the trends.

### Environmental and demographic data

A range of climatic and non-climatic factors might be expected to influence chamois body mass. Negative effects of population density on mass are common in ungulates [26,27]. In the absence of natural predation, these effects generally operate through increased intra-specific competition at higher population densities, resulting in lower per-capita food intake, particularly during periods when food is scarce [43-45]. To investigate density-dependence in these chamois populations, site-specific population density estimates were used from total population censuses performed in September every year between 1981 and 2009 (with the exception of 1990 and 1991; data from these years were excluded from the analysis). Each year, a set of simultaneous censuses was performed from vantage points across different blocks of each hunting district. It was assumed that density estimates from this time of year would reflect the population density over the previous growing season.

To investigate a possible direct thermoregulatory link between climate and body mass, we calculated yearly site-specific estimates of mean daily growing season temperature between 1982 and 2007 from high-altitude meteorological stations located in each of the three study sites (Data provided by The Forecasts and Organization Office, Civil Protection and Infrastructures Department, Trento Province, Italy). Differences in the elevation of weather stations among sites contributed to inter-site temperature differences (see Figure 2a). However, this did not affect our analysis since the drivers of body mass

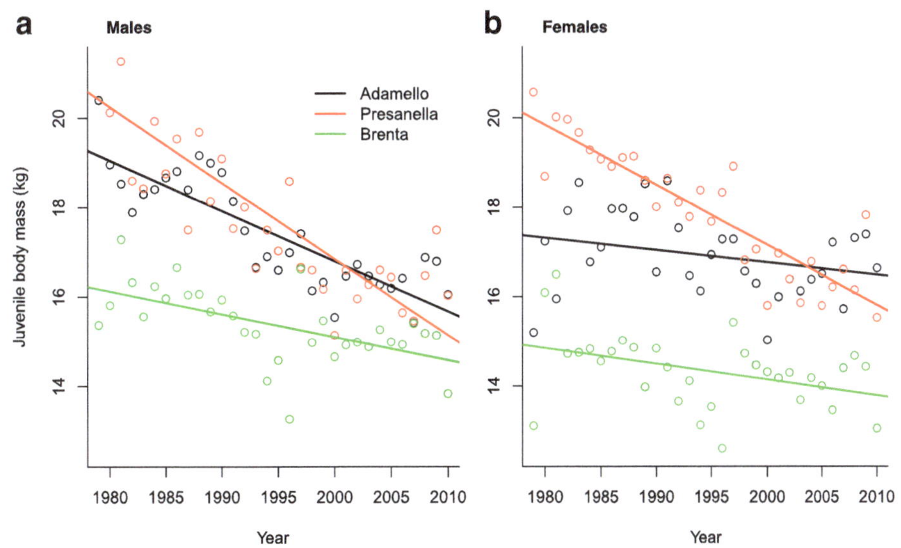


**Figure 1 Temporal juvenile body mass trends.** Long-term temporal trends in body masses of **a)** male and **b)** female juvenile chamois in the three study sites between 1979 and 2010. Points are annual mass estimates standardised to day 300 and straight lines are fitted trends.

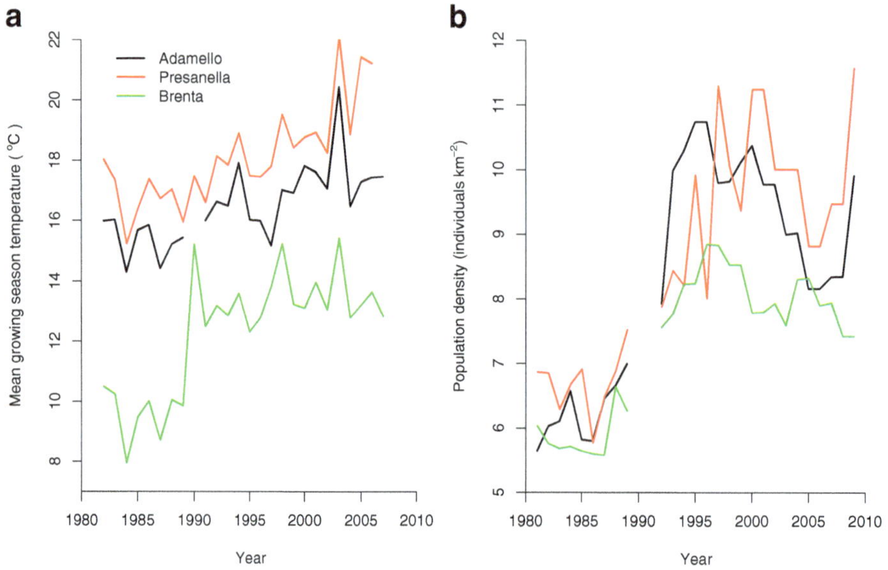


**Figure 2 Temporal trends in population density and mean growing season temperature.** Long-term variation in **a)** mean growing season daily maximum temperature, between 1982 and 2007, and **b)** population density, between 1981 and 2009, in Adamello (black), Presanella (red) and Brenta (green). Gaps show years with missing data. Whilst the three study areas do differ in their climate, some of the observed inter-site differences in temperature in a) are due to variation in the elevation of weather stations among areas.

trends were examined separately in each site and, additionally, temperatures were standardised within sites, along with all other environmental predictors (see 'Modelling variation in mass and mass residuals'). The bounds of the growing season were estimated using snow cover data, also from meteorological stations located within each site. The growing season was defined as the period between the snow-melt in spring, when snow cover was reduced to 0% (which generally occurs between late March and early May), and the first significant snowfall in winter that results in new snow settling on the ground (which generally occurs between early November and late December).

To investigate the effect of vegetation productivity and phenology on mass, NDVI (normalised difference vegetation index) data were used as a measure of vegetation productivity, processed by the Global Inventory Modelling and Mapping Studies group (GIMMS; [46,47]). These data are global at a 0.07 degree resolution (approximately 8 km by 8 km) and are available at fortnightly intervals between 1982 and 2006 (thus slightly restricting the yearly data range for analyses). In order to focus on vegetation types utilised by chamois for foraging, such as alpine meadows and sparsely vegetated areas, only NDVI pixels dominated by such vegetation types were considered. To do this the Corine land-cover 2006 data-set at a 100 m resolution [48] was used to select only NDVI pixels within each site containing less than 25% coniferous woodland. As each of the three sites encompassed a number of these NDVI pixels, mean NDVI from these pixels was calculated for each fortnightly time period, within each site. Previous studies have implicated a number of metrics relating to annual NDVI variation as being important to ungulate body condition (e.g. [24,25]). Here, we seek to derive NDVI metrics in a standardised fashion, despite inherent noise in NDVI estimates caused by factors such as cloud cover, water, snow or shadow [47]. As in previous studies [49], we used a smoothed function to characterise variation in NDVI with time in a given year. The following function was used (see Additional file 2, for an illustration of the functional form):

$$\bar{p}(s,y,t) = \alpha_{s,y} + \left(\beta_{s,y} - \alpha_{s,y}\right) \exp\left(-\left|\frac{t-t^*}{\sigma_{s,y}}\right|^{z_{s,y}}\right).$$

Here, $\bar{p}(s,y,t)$ is predicted NDVI at time-period $t$ in site $s$ and year $y$, $\alpha_{s,y}$ and $\beta_{s,y}$ are minimum and maximum NDVI respectively in site s and year y, $\sigma_{s,y}$ is a parameter related to the width of the function and $z_{s,y}$ is a parameter describing the shape of the function. Variation in NDVI data, $p(s,y,t)$, about the predicted mean was beta distributed. Thus, the likelihood of the model parameters, $\theta_{s,y}$, given the data, parameterised by $\bar{p}(s,y,t)$ and the dispersion coefficient $\phi_{s,y}$, is

$$L(\theta_{s,y}) = \prod_t \frac{\Gamma(a_{s,y,t} + b_{s,y,t})}{\Gamma(a_{s,y,t})\Gamma(b_{s,y,t})} p_{s,y,t}^{a_{s,y,t}-1} \left(1-p_{s,y,t}\right)^{b_{s,y,t}-1},$$

where $\Gamma(x)$ is the gamma function, $a_{s,y,t} = \bar{p}_{s,y,t}/\phi_{s,y}$ and $b_{s,y,t} = \left(1-\bar{p}_{s,y,t}/\phi_{s,y}\right)$. The most parsimonious fit was identified using Akaike's Information Criterion (AIC) [50,51].

The most parsimonious fitted relationships between mean NDVI and time in each year and site were calculated. Using these relationships, four NDVI metrics, described below, were calculated relating to vegetation productivity and phenology. All four of the metrics selected have been highlighted as important either to juvenile chamois specifically or to other ungulate species [47,52]. Previously, climate-induced changes in spring growing conditions [53] have been linked to higher juvenile body mass in ungulates, including chamois [52], due to longer growing-seasons and higher vegetation quality [24]. However, warmer springs have also been linked to negative impacts on body mass as higher temperatures lead to faster rates of vegetation 'green-up' and, thus, a shorter period of access to nutrient-rich emergent vegetation associated with early spring [25,54]. Here, the following four metrics were used: maximum rate of spring green-up, growing season duration, maximum NDVI and total growing season NDVI. Maximum rate of spring green-up ($NDVI_{rate}$) was calculated as the maximum first derivative of $\bar{p}_{s,y}(t)$ (i.e. the maximum rate of NDVI increase). The duration of the growing season ($NDVI_{dur}$) was calculated as the length of time between the maximum second derivative of $\bar{p}_{s,y}(t)$ (the start date of the growing season; when the rate of NDVI increase is increasing at its maximum rate) and the minimum second derivative of $\bar{p}_{s,y}(t)$ (the end date of the growing season; when NDVI is decreasing most rapidly). Maximum NDVI ($NDVI_{max}$) was calculated as the maximum value of $\bar{p}_{s,y}(t)$ and total growing-season NDVI (INDVI) as the integral of $\bar{p}_{s,y}(t)$ within the bounds of the growing-season. An illustration of the calculation of these metrics can be seen in Additional file 2.

### Modelling variation in mass and mass residuals

Environmental predictors were standardised by z-transformation within each site, as follows: $z_{s,y} = (x_{s,y} - \bar{x}_s)/\sigma$, where $z_{s,y}$ is a z-transformed predictor in year $y$ and site $s$, $x_{s,y}$ is the untransformed predictor, $\bar{x}_s$ is the site-specific mean of that predictor and $\sigma_s$ the site-specific standard deviation. General linear models were fitted to examine variation in body mass and body mass residuals using *R* version 2.12.0 [55]. Juvenile phenotypic quality is thought to be strongly influenced by lagged environmental effects [40]. As such, lagged environmental factors [56,57], affecting the availability and accessibility of resources to mothers, as well as population density [36,58], affecting the intensity of competition for resources, can be strong predictors of juvenile body mass. To account for lagged environmental effects on juvenile mass, density, temperature and NDVI data were used not only from year $y$ (the year a cohort was shot), but also from year $y$-1 (the year of a cohort's birth). Models were fitted with all possible biologically meaningful combinations of population density, $NDVI_{rate}$, $NDVI_{dur}$, $NDVI_{max}$, INDVI and temperature, from years $y$ and $y$-1. Models were considered with either the same intercept or different intercepts for males and females. In each site, several of the predictors were highly correlated (Pearson correlation coefficients ≥ 0.6), for instance $NDVI_{dur}$ with INDVI, and $density_y$ with $density_{y-1}$. To avoid problems of multicollinearity, highly correlated predictors were not permitted in the same model (but were permitted separately). Given the temporal range of the predictors, and to use the same temporal range of data in each model, data were used for the 19 years between 1983 and 2006, excluding 1990, 1991 and 1992, years for which estimates of $density_y$ and/or $density_{y-1}$ are not available (population censuses were not performed in 1990 and 1991). In order to identify the most parsimonious models, we used the two-step model selection process suggested by Richards [51]. First, all models having an AIC within six units of the smallest AIC calculated were selected (i.e. $\Delta AIC \leq 6$). Second, in order to remove overly complex models, we disregarded those that had a higher AIC than any simpler nested model. To investigate the potential for sex-specific environmental effects, we tested for sex interaction terms with all predictors within models in each site's top model set. To assess the relative importance of different predictors, Akaike model averaged coefficients were calculated from all models in each site's top model set [50]. Since we wanted to compare how the relative importance of predictors varied among sites, model averaged coefficients were z-transformed within each site. This standardised coefficients from different sites to the same scale, allowing the most important effects to be readily compared among sites.

## Results

Juvenile chamois body masses decreased strongly between 1979 and 2010 in all three populations (Figure 1). The extent of this decrease varied considerably among sexes and sites but decreases in male mass have been more pronounced than decreases in female mass in all sites (slopes of temporal mass trends ± SE: Adamello males, −0.11 ± 0.01; Adamello females, −0.03 ± 0.02; Presanella males, −0.17 ± 0.02; Presanella females, −0.13 ± 0.01; Brenta males, −0.05 ± 0.01; Brenta females, −0.04 ± 0.01). Decreases have been less pronounced in Brenta than in the other two sites. There have been striking increases in growing season temperatures in all sites between 1982 and 2007 (Figure 2a). During the same period, all three populations increased in density substantially, peaking in the mid-1990s before declining slightly in recent years (Figure 2b). This growth coincides with the implementation of stricter controls on hunting in the area (including increases in the number of rangers and a more strictly enforced quota system). In contrast, there have been no pronounced long-term trends in the four NDVI metrics between 1982 and 2006, although growing seasons have tended to be longer (Figure 3b) and more productive (Figure 3d) between 2004 and 2006.

The most parsimonious body mass models fitted the observed data well (Table 1, Figure 4) ($R^2$: Adamello, 0.78; Presanella, 0.83; Brenta, 0.69). In comparison, the fits of body mass residual models were poor ($R^2$: Adamello, 0.44; Presanella, 0.13; Brenta, 0.36; see Additional file 3); thus, our inferences focus on models that describe longer-term changes in body mass, rather than those focused on explaining variation around the long-term trend. Temperature, density and NDVI all appear to play a role in describing long-term variation in juvenile body mass (Table 1, Figure 5). Within each site, the top set of body mass models contains a number of closely competing models (Table 1) but clear and consistent patterns across sites are illustrated by model averaging (Figure 5). Temperature and/or density have the strongest negative effects on mass in all sites (Figures 5 and 6). Furthermore, strong negative effects of temperature and density in the current year are present in all of the top models within Adamello and Presanella (Table 1), providing good evidence for these effects. A slightly weaker negative effect of lagged temperature appears in most of the top models for these sites. In Brenta, there is some evidence for a negative effect of temperature, which appears in the top three models in this site but there is no evidence of a density effect. Despite mass declines in males being consistently stronger than in females (Figure 1), we found no support for any interaction terms between sex and environmental predictors, suggesting that there are not strong differences in the magnitude of environmental effects on body mass between males and females. The lack of evidence for sex-specific effects could stem from the slightly restricted temporal range of data used in the analysis.

The modelled effects of NDVI are much weaker than the effects of density and temperature, and are generally positive (Table 1, Figure 5). Only in Brenta is there strong evidence for an influence of vegetation productivity and phenology; a strong, lagged, positive effect of INDVI is present in all top models and there is some evidence for a lagged, positive effect of maximum NDVI.

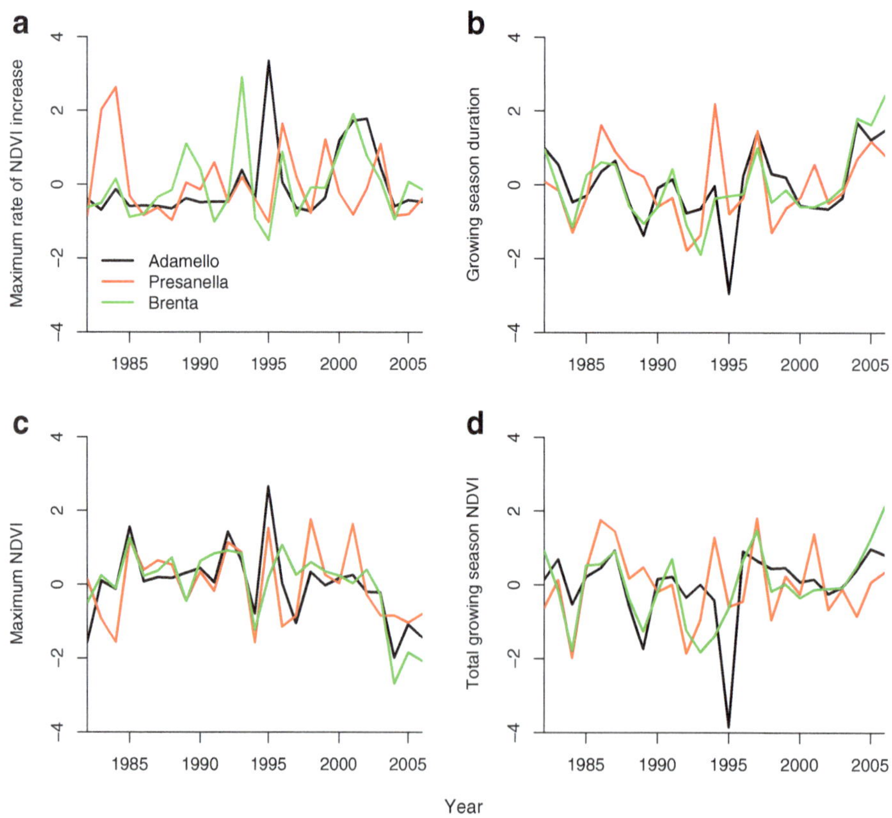


**Figure 3 Temporal variation in standardised NDVI metrics.** Long-term variation in standardised **a)** maximum rate of NDVI increase ($NDVI_{rate}$), **b)** growing season duration ($NDVI_{dur}$), **c)** maximum NDVI ($NDVI_{max}$) and **d)** total growing season NDVI (INDVI) between 1982 and 2006 in Adamello (black), Presanella (red) and Brenta (green).

**Table 1 Set of most parsimonious body mass models**

| Site | $Temp_y$ | $Temp_{y-1}$ | $Dens_y$ | $Dens_{y-1}$ | $NDVI_{dur,y}$ | $NDVI_{dur,y-1}$ | $INDVI_y$ | $INDVI_{y-1}$ | $NDVI_{max,y}$ | $NDVI_{max,y-1}$ | $NDVI_{rate,y}$ | $NDVI_{rate,y-1}$ | Sex | K | $R^2$ | ΔAIC | Weight |
|---|---|---|---|---|---|---|---|---|---|---|---|---|---|---|---|---|---|
| **Adamello** | **−0.42** | **−0.21** | **−0.61** | | | | **−0.18** | | | | | **0.22** | **0.32** | **8** | **0.78** | **0.0** | **0.24** |
| | −0.35 | −0.19 | −0.60 | | | −0.21 | −0.13 | | | | | | 0.32 | 8 | 0.77 | 0.7 | 0.17 |
| | −0.39 | −0.18 | −0.57 | | | −0.21 | | | | | | | 0.32 | 7 | 0.76 | 1.8 | 0.10 |
| | −0.42 | −0.21 | −0.61 | | | | −0.18 | | | | | 0.22 | | 7 | 0.75 | 2.4 | 0.07 |
| | −0.35 | −0.19 | −0.60 | | | −0.21 | −0.13 | | | | | | | 7 | 0.75 | 3.0 | 0.05 |
| | −0.39 | | −0.68 | | | −0.22 | −0.12 | | | | | | 0.32 | 7 | 0.75 | 3.3 | 0.05 |
| | −0.47 | | −0.69 | | | | −0.17 | | | | | 0.21 | 0.32 | 7 | 0.75 | 3.6 | 0.04 |
| | −0.39 | −0.16 | −0.56 | | | | | −0.18 | | | | | 0.32 | 7 | 0.75 | 3.6 | 0.04 |
| | −0.46 | −0.20 | −0.55 | | | | | | | | | 0.18 | 0.32 | 7 | 0.75 | 3.7 | 0.04 |
| | −0.38 | | −0.67 | | | | −0.13 | −0.20 | | | | | 0.32 | 7 | 0.74 | 3.7 | 0.04 |
| | −0.39 | −0.18 | −0.57 | | | −0.21 | | | | | | | | 6 | 0.73 | 3.9 | 0.03 |
| | −0.42 | | −0.65 | | | −0.22 | | | | | | | 0.32 | 6 | 0.73 | 3.9 | 0.03 |
| | −0.42 | | −0.64 | | | | | −0.20 | | | | | 0.32 | 6 | 0.73 | 4.7 | 0.02 |
| | −0.39 | | −0.68 | | | −0.22 | −0.12 | | | | | | | 6 | 0.72 | 5.3 | 0.02 |
| | −0.47 | | −0.69 | | | | −0.17 | | | | | 0.21 | | 6 | 0.72 | 5.5 | 0.02 |
| | −0.39 | −0.16 | −0.56 | | | | | −0.18 | | | | | | 6 | 0.72 | 5.5 | 0.02 |
| | −0.46 | −0.20 | −0.55 | | | | | | | | | 0.18 | | 6 | 0.72 | 5.6 | 0.01 |
| | −0.42 | | −0.65 | | | −0.22 | | | | | | | | 5 | 0.71 | 5.6 | 0.01 |
| **Presanella** | **−0.58** | **−0.27** | **−0.53** | | **0.28** | | | | | **−0.17** | **0.19** | | **−0.41** | **9** | **0.83** | **0.0** | **0.36** |
| | −0.59 | −0.25 | −0.58 | | 0.19 | | | | | −0.14 | | | −0.41 | 8 | 0.81 | 1.7 | 0.15 |
| | −0.57 | −0.27 | −0.51 | | 0.26 | | | | | | 0.16 | | −0.41 | 8 | 0.81 | 1.9 | 0.14 |
| | −0.58 | −0.26 | −0.55 | | 0.18 | | | | | | | | −0.41 | 7 | 0.80 | 2.3 | 0.11 |
| | −0.52 | −0.26 | −0.63 | | | | | | | −0.14 | | | −0.41 | 7 | 0.80 | 3.4 | 0.07 |
| | −0.51 | −0.26 | −0.60 | | | | | | | | | | −0.41 | 6 | 0.78 | 3.7 | 0.06 |
| | −0.67 | | −0.63 | | 0.19 | | | | | | | | −0.41 | 6 | 0.78 | 4.8 | 0.03 |
| | −0.58 | −0.26 | −0.55 | | 0.18 | | | | | | | | | 6 | 0.78 | 5.1 | 0.03 |
| | −0.63 | | −0.66 | | | | 0.16 | | | | | | −0.41 | 6 | 0.77 | 5.6 | 0.02 |
| | −0.62 | | −0.70 | | | | | | | −0.14 | | | −0.41 | 6 | 0.77 | 5.8 | 0.02 |
| | −0.61 | | −0.68 | | | | | | | | | | −0.41 | 5 | 0.76 | 5.9 | 0.02 |

**Table 1 Set of most parsimonious body mass models** *(Continued)*

| | | | | | | | | | | | | |
|---|---|---|---|---|---|---|---|---|---|---|---|---|
| **Brenta** | **−0.40** | | | **0.43** | **0.21** | | **0.13** | **1.06** | **7** | **0.69** | **0.0** | **0.24** |
| | −0.32 | | | 0.38 | 0.27 | −0.11 | | 1.06 | 7 | 0.69 | 0.1 | 0.23 |
| | −0.34 | | | 0.38 | 0.25 | | | 1.06 | 6 | 0.67 | 0.1 | 0.23 |
| | | −0.34 | | 0.34 | 0.22 | | | 1.06 | 6 | 0.66 | 1.1 | 0.14 |
| | | | −0.32 | 0.36 | 0.21 | | | 1.06 | 6 | 0.65 | 2.8 | 0.06 |
| | −0.44 | | | 0.41 | | | 0.20 | 1.06 | 6 | 0.65 | 3.3 | 0.05 |
| | | −0.37 | | 0.29 | | | | 1.06 | 5 | 0.62 | 4.6 | 0.02 |
| | −0.37 | | | 0.34 | | | | 1.06 | 5 | 0.61 | 5.4 | 0.02 |
| | | | −0.36 | 0.32 | | | | 1.06 | 5 | 0.60 | 5.9 | 0.01 |

Top model set containing models with a ΔAIC that is ≤6 and lower than all simpler nested versions [51]. The number of parameters in each model (K), $R^2$ values, ΔAICs and Akaike model weights are shown. The most parsimonious model for each site is displayed in bold. 'Sex' represents sex-specific model intercepts.

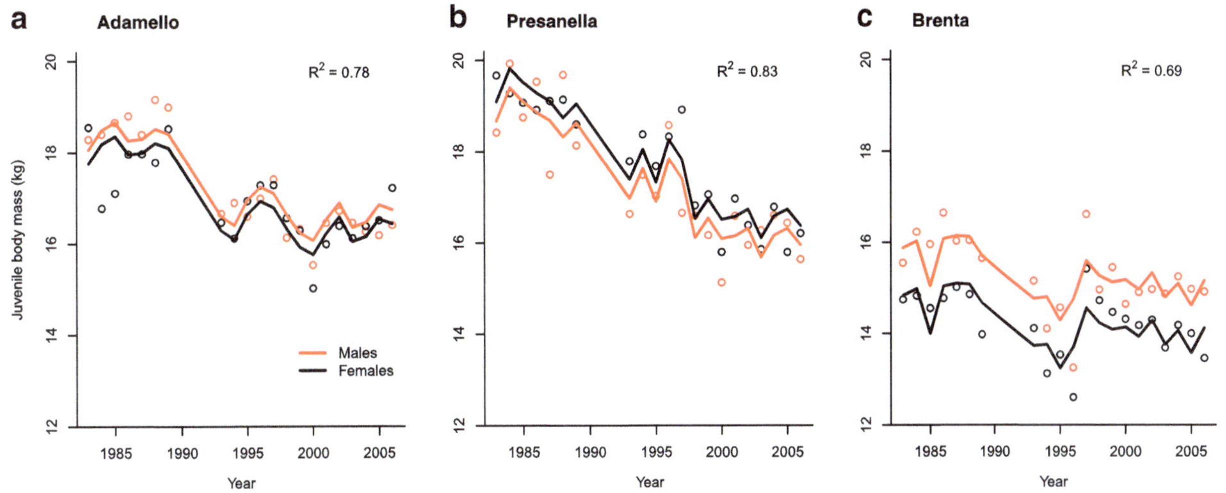


**Figure 4 Temporal variation in observed and fitted juvenile body mass.** Fitted juvenile body mass predictions of best models for males (red) and females (black) and observed body mass data in **a)** Adamello, **b)** Presanella and **c)** Brenta. Lines are predictions and points are observed data for years used in modelling (i.e. years for which there are no missing data for any predictor). $R^2$ shown.

In Adamello and Presanella, NDVI effects are generally weaker and inconsistent, although there is some support for a positive effect of long growing seasons in Presanella and, surprisingly, for a negative effect of INDVI in Adamello.

## Discussion

Pronounced long-term body mass declines were identified in all three populations. Models of juvenile body mass performed well in explaining these declines (Figure 4). However, the poor performances of residual models show that deviations from the general trends (i.e. years in which juveniles are particularly heavy or particularly light relative to the trend) cannot be explained as accurately. Mass declines appear to be linked most strongly to increases in growing season temperatures and population density; temperature and/or density have the strongest negative effects on body mass and deviations from yearly trend in all sites (Table 1, Figure 5). In comparison, there is no evidence of a strong negative effect of vegetation productivity or phenology on body mass, suggesting that body mass declines were not mediated by vegetation change as indexed by NDVI. Our results provide an interesting contrast to the majority of studies to date, which have implicated changes in food availability and quality as the most likely cause of body size declines [7].

Growing season temperature appears to be strongly linked to mass declines in all three populations (Table 1, Figures 5 and 6). However, despite significant temperature change (Figure 2a), there has been no consistent change in either vegetation productivity or phenology (Figure 3). Furthermore, there is no relationship between

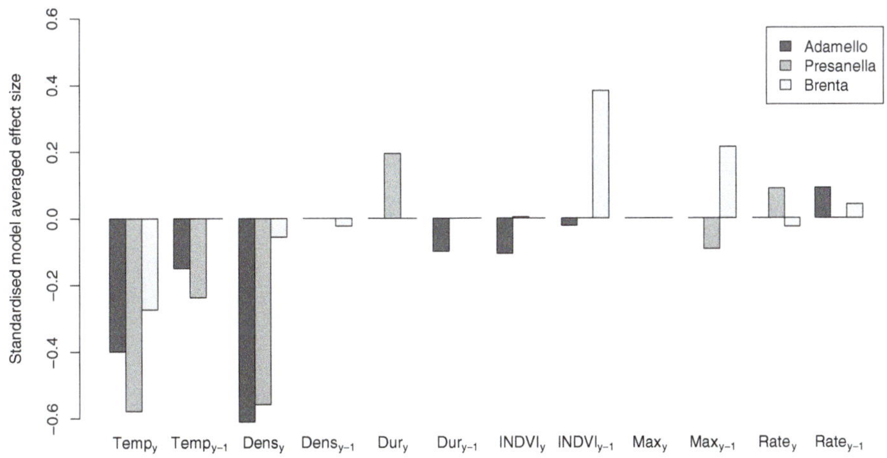


**Figure 5 Model averaged standardised predictor effect sizes.** Akaike model averaged standardised predictor coefficients calculated from all body mass models within the top model set for each site (see Table 1 for full model selection table).

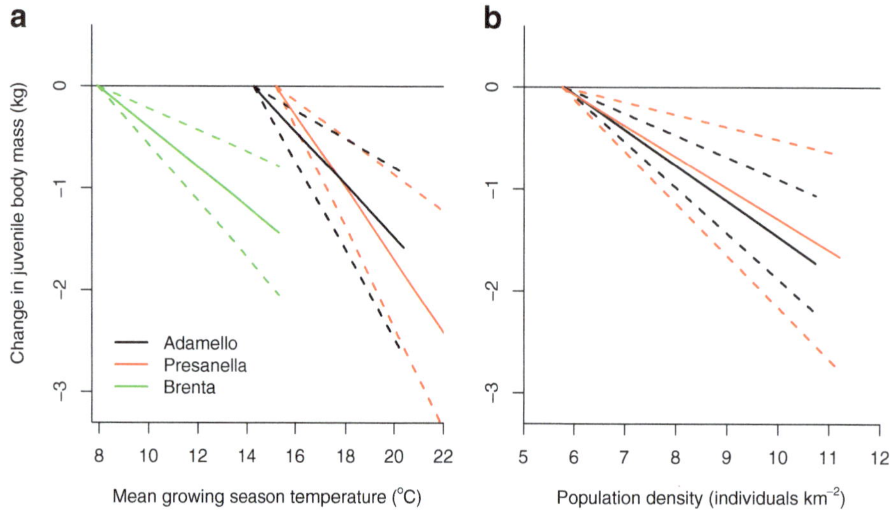


**Figure 6 Modelled effects of temperature and population density on juvenile body mass.** Modelled effects of **a)** mean growing season temperature and **b)** population density on change in juvenile body mass since 1983 in Adamello (black), Presanella (red) and Brenta (green). Solid lines are predictions of the most parsimonious body mass model for each site, with other predictors set to mean values. Dashed lines are 95% confidence intervals calculated from 1000 bootstrapped replicates [59].

temperature and any NDVI metric. As such, we find no evidence for our first putative driver of mass change: that climate change has affected body mass via effects on resource productivity or phenology. Instead, our results provide support for our second putative driver: that climate change could be directly affecting chamois behaviour or physiology, limiting their ability to acquire resources. In another alpine ungulate, the ibex (*Capra ibex*), temperatures above 15-20°C result in heat discomfort in males, reducing the time they can spend foraging [31]. In ungulates, behavioural changes, such as allocating less time to foraging, play an important role in thermoregulation because they can be more energetically efficient than autonomic thermoregulation [60,61]. Higher daily temperatures during spring and summer may have led to juvenile chamois spending more time resting and less time foraging than in the past, reducing their ability to store energy reserves and invest in growth. Indeed, chamois, like many ungulates, reduce their feeding activity during the hottest period of the day [62]. Furthermore, chamois spend less time foraging when it is hotter, independent of time of day [63]. Our results suggest that temperature affects the ability of juveniles to acquire resources themselves rather than via lagged effects on their mothers (Figure 5). It is possible that juvenile ungulates are more susceptible to higher temperatures due to having higher relative metabolic demands than adults [64,65]. Whilst we did not find any evidence for sex-specific environmental effects on body mass, that mass declines in juvenile males are consistently more pronounced than in females (Figure 1) suggests that, even in a relatively monomorphic ungulate species, a more energetically demanding growth phase could make males more susceptible to environmental stressors [20].

Population density also appears to be strongly linked to mass declines in juvenile chamois, providing support for our third putative driver: that changes in density affect per-capita resource availability (Table 1, Figures 5 and 6). The study populations have grown substantially following the implementation of stricter controls on hunting. As a result, increased intra-specific competition for resources may have led to reduced rates of per-capita food intake. Additionally, climate change may have affected chamois survival or fecundity, contributing to this population growth. Density dependence in body mass has been detected in a range of ungulate species (e.g. [27,36,66]) and even specifically in juvenile chamois [52]. At high latitudes and in alpine regions, this effect is most likely to manifest during winter, when snow cover reduces forage availability and increases forage patchiness, resulting in higher levels of agonistic interactions among individuals [67,68]. It is also possible that density dependence could be mediated by overgrazing; however, no changes in NDVI consistent with overgrazing, such as decreases in maximum NDVI or INDVI [69], were detected. Overgrazing can also encourage the colonisation of grazing-tolerant, less palatable species [70], which might not result in detectable changes in productivity. However, such changes would not alter our finding that increases in density appear to have influenced body mass via changes in per-capita resource availability. As with temperature, the effect of density in the year of shooting appears most important (Figure 5), suggesting that competition for forage following weaning, rather than a lagged influence of competition on mothers in the previous year, has a stronger influence on juvenile mass. Indeed, intra-specific competition for resources is most likely to result in the displacement of subordinate individuals, such as juveniles, from food patches [71-73]. Previously, the importance of lagged effects on juvenile condition has been stressed (e.g. [56,58]). However, our findings are consistent with recent work on juvenile chamois showing that environmental conditions during the second year of life, which influence the ability of juveniles to acquire resources directly, have an important effect on investment in growth and energy storage [37,52]. Only in Brenta is there no evidence for density dependence in body mass (Figures 5 and 6). Brenta is calcareous, harbouring a more nutrient-rich plant community [74], so forage availability could be less limiting here than in the other populations. A hunting regime maintaining this population below its carrying capacity, and limiting the effect of resource competition on mass, could also contribute to the less pronounced mass declines observed in Brenta (Figure 1).

There was no evidence for a strong role of vegetation productivity or phenology in the observed long-term mass declines (Table 1, Figure 5). Given that NDVI is only a proxy for vegetation productivity, it remains possible that other changes in vegetation quality and phenology have had more of an effect than suggested here. As with overgrazing, it is possible that warmer temperatures in spring and summer could lead to increases in the abundance of less palatable species [75], or reductions in the protein content of the species present [76], changes which NDVI might not detect. Although they do not appear to play a strong role in mass declines, there is some evidence that vegetation productivity and phenology influence variation in juvenile body mass. This is particularly the case in the calcareous area, Brenta, where there is strong evidence that productive growing seasons (in terms of INDVI and maximum NDVI) have positive effects on juvenile body mass (Table 1, Figure 5). This is consistent with findings from another calcareous Alpine area, where long growing seasons have been linked to higher juvenile chamois body mass [52]. In our study, it is intriguing that the relative importance of environmental factors should vary among neighbouring areas that differ predominantly only in their geological substrate. These differences may arise because calcareous vegetation, whilst being more nutrient-rich, varies more in quality in response to environmental variation than siliceous vegetation [74,77]. The variation in substrate in this study area has previously been shown to mediate differences in reproductive strategy, body mass [38] and horn length [37]; this study further highlights the importance of considering geological variation in studies of animal morphology, physiology and life history.

## Conclusions

We have detected strong links between recent environmental change and negative temporal body mass trends in juvenile chamois. Increases in both temperature, due to climate change, and population density, due to stricter controls on hunting, appear to be driving the mass declines. Since we found no evidence for an effect of changing resource productivity or phenology on body mass, the observed patterns may not be mediated by changes in vegetation. We propose that heightened thermoregulatory demands and intra-specific competition, to both of which juveniles are particularly susceptible, could be responsible. These findings add an interesting contrast to the large number of studies implicating climate-mediated changes in resource productivity or phenology as drivers of declines in animal body size. Interestingly, our findings highlight that changes in management could ameliorate negative impacts of climate change. For example, future increases in hunting offtake could reduce the intensity of resource competition, counteracting negative impacts of temperature warming on body mass. This study further highlights the importance of considering management when examining the influence of environmental change on species, which is an increasingly important focus of ecological research (e.g. [63,78,79]). Future reductions in body size are likely to have far-reaching effects on the functioning and productivity of ecosystems [7,8]; decreases in juvenile body condition could result in reduced survival of juveniles and depressed population growth rates. An increasing focus on the links between climate, management, body condition and demography will shed light on the influence of anthropogenic and climate change on populations.

## Additional files

**Additional file 1:** Annual variation in sample size of juvenile chamois.

**Additional file 2:** An illustration of the functional form describing intra-annual variation in NDVI.

**Additional file 3:** Variation in observed and fitted residuals of juvenile body mass temporal trends.

#### Competing interests
The authors declare that they have no competing interests.

#### Authors' contributions
TM, MA, SW and PS conceived the study. TM analysed the data and, with SW and PS, wrote the manuscript. MA and RC provided the data. All authors contributed to revisions and approved the final manuscript.

#### Acknowledgements
We are grateful to Trento Province, the Adamello Brenta Nature Park, and the "Associazione Cacciatori Trentini" (The regional hunting association), as well as to the hunting reserve of "Regole Spinale e Manez" for supplying data. We are indebted to M. Rocca, A. Brugnoli and all the game wardens who collected Alpine chamois measurements. TM was funded by a Natural Environment Research Council Doctoral Training Grant (NE/H524573/1) and supervised by SGW, PAS and MA. MA and RC were supported by the Italian

Ministry of Education, University and Research (PRIN 2010–2011, 20108 TZKHC). This study complies with all relevant national, regional and provincial Italian laws. The data used in this study are available on request from the corresponding author.

**Author details**

[1]School of Biological and Biomedical Sciences, Durham University, South Road, Durham DH1 3LE, UK. [2]Department of Science for Nature and Environmental Resources, University of Sassari, via Muroni 25, I-07100 Sassari, Sardinia, Italy.

**References**

1. Parmesan C, Yohe G: **A globally coherent fingerprint of climate change impacts across natural systems.** *Nature* 2003, **421**:37–42.
2. Menzel A, Sparks TH, Estrella N, Koch E, Aasa A, Ahas R, Alm-Kubler K, Bissolli P, Braslavska O, Briede A, Chmielewski FM, Crepinsek Z, Curnel Y, Dahl A, Defila C, Donnelly A, Filella Y, Jatcza K, Mage F, Mestre A, Nordli O, Penuelas J, Pirinen P, Remisova V, Scheifinger H, Striz M, Susnik A, Van Vliet AJH, Wielgolaski FE, Zach S: **European phenological response to climate change matches the warming pattern.** *Glob Chang Biol* 2006, **12**:1969–1976.
3. van de Pol M, Vindenes Y, Saether B-E, Engen S, Ens BJ, Oosterbeek K, Tinbergen JM: **Effects of climate change and variability on population dynamics in a long-lived shorebird.** *Ecology* 2010, **91**:1192–1204.
4. Musolin DL: **Insects in a warmer world: ecological, physiological and life-history responses of true bugs (*Heteroptera*) to climate change.** *Glob Chang Biol* 2007, **13**:1565–1585.
5. Ozgul A, Childs DZ, Oli MK, Armitage KB, Blumstein DT, Olson LE, Tuljapurkar S, Coulson T: **Coupled dynamics of body mass and population growth in response to environmental change.** *Nature* 2010, **466**:482–U485.
6. Ozgul A, Tuljapurkar S, Benton TG, Pemberton JM, Clutton-Brock TH, Coulson T: **The Dynamics of Phenotypic Change and the Shrinking Sheep of St. Kilda.** *Science* 2009, **325**:464–467.
7. Gardner JL, Peters A, Kearney MR, Joseph L, Heinsohn R: **Declining body size: a third universal response to warming?** *Trends Ecol Evol* 2011, **26**:285–291.
8. Sheridan JA, Bickford D: **Shrinking body size as an ecological response to climate change.** *Nat Clim Change* 2011, **1**:401–406.
9. Guthrie RD: **Rapid body size decline in Alaskan Pleistocene horses before extinction.** *Nature* 2003, **426**:169–171.
10. Bickford D, Howard SD, Ng DJJ, Sheridan JA: **Impacts of climate change on the amphibians and reptiles of Southeast Asia.** *Biodivers Conserv* 2010, **19**:1043–1062.
11. Bergmann C: **Ueber die Verhältnisse der Wärmeökonomie der Thiere zu ihrer Grösse (Concerning the relationship of heat conservation of animals to their size).** *Gottinger Studien* 1847, **3**:595–708.
12. Mayr E: **Geographical character gradients and climate adaptation.** *Evolution* 1956, **10**:105–108.
13. Ashton KG, Tracy MC, de Queiroz A: **Is Bergmann's rule valid for mammals?** *Am Nat* 2000, **156**:390–415.
14. Meiri S, Dayan T: **On the validity of Bergmann's rule.** *J Biogeogr* 2003, **30**:331–351.
15. Freckleton RP, Harvey PH, Pagel M: **Bergmann's rule and body size in mammals.** *Am Nat* 2003, **161**:821–825.
16. Teplitsky C, Mills JA, Alho JS, Yarrall JW, Merila J: **Bergmann's rule and climate change revisited: Disentangling environmental and genetic responses in a wild bird population.** *Proc Natl Acad Sci U S A* 2008, **105**:13492–13496.
17. Husby A, Hille SM, Visser ME: **Testing Mechanisms of Bergmann's Rule: Phenotypic Decline but No Genetic Change in Body Size in Three Passerine Bird Populations.** *Am Nat* 2011, **178**:202–213.
18. Coltman DW, O'Donoghue P, Jorgenson JT, Hogg JT, Strobeck C, Festa-Bianchet M: **Undesirable evolutionary consequences of trophy hunting.** *Nature* 2003, **426**:655–658.
19. Allendorf FW, England PR, Luikart G, Ritchie PA, Ryman N: **Genetic effects of harvest on wild animal populations.** *Trends Ecol Evol* 2008, **23**:327–337.
20. Clutton-Brock TH, Guinness FE, Albon SD: *Red deer. Behavior and ecology of two sexes.* USA & Edinburgh: University of Chicago & Edinburgh University Press; 1982.
21. Gaillard JM, Festa-Bianchet M, Delorme D, Jorgenson J: **Body mass and individual fitness in female ungulates: bigger is not always better.** *Proc R Soc London B Biol Sci* 2000, **267**:471–477.
22. Stewart KM, Bowyer RT, Dick BL, Johnson BK, Kie JG: **Density-dependent effects on physical condition and reproduction in North American elk: an experimental test.** *Oecologia* 2005, **143**:85–93.
23. Loison A, Langvatn R, Solberg EJ: **Body mass and winter mortality in red deer calves: disentangling sex and climate effects.** *Ecography* 1999, **22**:20–30.
24. Pettorelli N, Weladji RB, Holand O, Mysterud A, Breie H, Stenseth NC: **The relative role of winter and spring conditions: linking climate and landscape-scale plant phenology to alpine reindeer body mass.** *Biol Lett* 2005, **1**:24–26.
25. Pettorelli N, Pelletier F, von Hardenberg A, Festa-Bianchet M, Cote SD: **Early onset of vegetation growth vs. rapid green-up: Impacts on juvenile mountain ungulates.** *Ecology* 2007, **88**:381–390.
26. Toïgo C, Gaillard JM, Van Laere G, Hewison M, Morellet N: **How does environmental variation influence body mass, body size, and body condition? Roe deer as a case study.** *Ecography* 2006, **29**:301–308.
27. Mysterud A, Yoccoz NG, Stenseth NC, Langvatn R: **Effects of age, sex and density on body weight of Norwegian red deer: evidence of density-dependent senescence.** *Proc R Soc London B Biol Sci* 2001, **268**:911–919.
28. Yom-Tov Y, Yom-Tov S, MacDonald D, Yom-Tov E: **Population cycles and changes in body size of the lynx in Alaska.** *Oecologia* 2007, **152**:239–244.
29. Laternser M, Schneebeli M: **Long-term snow climate trends of the Swiss Alps (1931–99).** *Int J Climatol* 2003, **23**:733–750.
30. Parker KL, Robbins CT, Hanley TA: **Energy expenditures for locomotion by mule deer and elk.** *J Wildl Manage* 1984, **48**:474–488.
31. Aublet JF, Festa-Bianchet M, Bergero D, Bassano B: **Temperature constraints on foraging behaviour of male Alpine ibex (*Capra ibex*) in summer.** *Oecologia* 2009, **159**:237–247.
32. Luterbacher J, Dietrich D, Xoplaki E, Grosjean M, Wanner H: **European seasonal and annual temperature variability, trends, and extremes since 1500.** *Science* 2004, **303**:1499–1503.
33. Sinervo B, Mendez-de-la-Cruz F, Miles DB, Heulin B, Bastiaans E, Cruz MVS, Lara-Resendiz R, Martinez-Mendez N, Calderon-Espinosa ML, Meza-Lazaro RN, Gadsden H, Avila LJ, Morando M, De la Riva IJ, Sepulveda PV, Rocha CFD, Ibarguengoytia N, Puntriano CA, Massot M, Lepetz V, Oksanen TA, Chapple DG, Bauer AM, Branch WR, Clobert J, Sites JW: **Erosion of Lizard Diversity by Climate Change and Altered Thermal Niches.** *Science* 2010, **328**:894–899.
34. Kearney M, Shine R, Porter WP: **The potential for behavioral thermoregulation to buffer "cold-blooded" animals against climate warming.** *Proc Natl Acad Sci U S A* 2009, **106**:3835–3840.
35. Stuble KL, Pelini SL, Diamond SE, Fowler DA, Dunn RR, Sanders NJ: **Foraging by forest ants under experimental climatic warming: a test at two sites.** *Ecol Evol* 2013, **3**:482–491.
36. Kjellander P, Gaillard JM, Hewison AJM: **Density-dependent responses of fawn cohort body mass in two contrasting roe deer populations.** *Oecologia* 2006, **146**:521–530.
37. Chirichella R, Ciuti S, Grignolio S, Rocca M, Apollonio M: **The role of geological substrate for horn growth in ungulates: a case study on Alpine chamois.** *Evol Ecol* 2013, **27**:145–163.
38. Mason THE, Chirichella R, Richards SA, Stephens PA, Willis SG, Apollonio M: **Contrasting life histories in neighbouring populations of a large mammal.** *PLoS One* 2011, **6**:e28002.
39. Mason THE, Stephens PA, Willis SG, Chirichella R, Apollonio M, Richards SA: **Intraseasonal variation in reproductive effort: young males finish last.** *Am Nat* 2012, **180**:823–830.
40. Oftedal OT: **Body size and reproductive strategy as correlates of milk energy output in lactating mammals.** *Acta Zool Fenn* 1984, **171**:183–186.
41. Wu ZH, Huang NE, Long SR, Peng CK: **On the trend, detrending, and variability of nonlinear and nonstationary time series.** *Proc Natl Acad Sci U S A* 2007, **104**:14889–14894.
42. Jonzen N, Hedenstrom A, Hjort C, Lindstrom A, Lundberg P, Andersson A: **Climate patterns and the stochastic dynamics of migratory birds.** *Oikos* 2002, **97**:329–336.
43. Skogland T: **The effects of density-dependent resource limitations on the demography of wild reindeer.** *J Anim Ecol* 1985, **54**:359–374.
44. Fryxell J: **Food limitation and demography of a migratory antelope, the white-eared kob.** *Oecologia* 1987, **72**:83–91.

45. Sinclair AR, Dublin H, Borner M: **Population regulation of Serengeti Wildebeeest: a test of the food hypothesis.** *Oecologia* 1985, **65:**266–268.
46. Tucker CJ, Pinzon JE, Brown ME, Slayback DA, Pak EW, Mahoney R, Vermote EF, El Saleous N: **An extended AVHRR 8-km NDVI dataset compatible with MODIS and SPOT vegetation NDVI data.** *Int J Remote Sens* 2005, **26:**4485–4498.
47. Pettorelli N, Vik JO, Mysterud A, Gaillard JM, Tucker CJ, Stenseth NC: **Using the satellite-derived NDVI to assess ecological responses to environmental change.** *Trends Ecol Evol* 2005, **20:**503–510.
48. EEA: *Corine land cover 2006 (CLC2006) 100m - version 2/2010*; 2010.
49. Beck PS, Atzberger C, Høgda KA, Johansen B, Skidmore AK: **Improved monitoring of vegetation dynamics at very high latitudes: A new method using MODIS NDVI.** *Remote Sens Environ* 2006, **100:**321–334.
50. Burnham KP, Anderson DR: *Model selection and multimodel inference: a practical information-theoretic approach.* 2nd edition. New York: Springer; 2002.
51. Richards SA: **Dealing with overdispersed count data in applied ecology.** *J Appl Ecol* 2008, **45:**218–227.
52. Garel M, Gaillard JM, Jullien JM, Dubray D, Maillard D, Loison A: **Population abundance and early spring conditions determine variation in body mass of juvenile chamois.** *J Mammal* 2011, **92:**1112–1117.
53. Linderholm HW: **Growing season changes in the last century.** *Agr Forest Meteorol* 2006, **137:**1–14.
54. Crawley MJ: *Herbivory: the dynamics of animal-plant interactions.* Oxford: Blackwell; 1983.
55. R Development Core Team: *R: A language and environment for statistical computing.* Vienna, Austria: R Foundation for Statistical Computing; 2011.
56. Côté SD, Festa-Bianchet M: **Birthdate, mass and survival in mountain goat kids: effects of maternal characteristics and forage quality.** *Oecologia* 2001, **127:**230–238.
57. Weladji RB, Holand O: **Global climate change and reindeer: effects of winter weather on the autumn weight and growth of calves.** *Oecologia* 2003, **136:**317–323.
58. Bonenfant C, Gaillard JM, Klein F, Loison A: **Sex- and age-dependent effects of population density on life history traits of red deer *Cervus elaphus* in a temperate forest.** *Ecography* 2002, **25:**446–458.
59. Efron B, Tibshirani R: **Statistical Data Analysis in the Computer Age.** *Science* 1991, **253:**390–395.
60. Dussault C, Ouellet JP, Courtois R, Huot J, Breton L, Larochelle J: **Behavioural responses of moose to thermal conditions in the boreal forest.** *Ecoscience* 2004, **11:**321–328.
61. Maloney SK, Moss G, Cartmell T, Mitchell D: **Alteration in diel activity patterns as a thermoregulatory strategy in black wildebeest (*Connochaetes gnou*).** *J Comp Physiol A Neuroethol Sens Neural Behav Physiol* 2005, **191:**1055–1064.
62. Ruttimann S, Giacometti M, McElligott AG: **Effect of domestic sheep on chamois activity, distribution and abundance on sub-alpine pastures.** *Eur J Wildl Res* 2008, **54:**110–116.
63. Mason THE, Stephens PA, Apollonio M, Willis SG: **Predicting potential responses to future climate in an alpine ungulate: inter-specific interactions exceed climate effects.** *Glob Chang Biol* 2014, In press.
64. Hudson RJ: *Body size, energetics, and adaptive radiation.* Boca Raton, Florida: CRC Press, Inc.; 1985.
65. Solberg EJ, Jordhoy P, Strand O, Aanes R, Loison A, Saether BE, Linnell JDC: **Effects of density-dependence and climate on the dynamics of a Svalbard reindeer population.** *Ecography* 2001, **24:**441–451.
66. Forchhammer MC, Clutton-Brock TH, Lindstrom J, Albon SD: **Climate and population density induce long-term cohort variation in a northern ungulate.** *J Anim Ecol* 2001, **70:**721–729.
67. Weckerly FW: **Social bonding and aggression in female Roosevelt elk.** *Can J Zool* 1999, **77:**1379–1384.
68. Espmark Y: **Social behaviour of roe deer at winter feeding stations.** *Appl Anim Ethol* 1974, **1:**35–47.
69. Hilker T, Natsagdorj E, Waring RH, Lyapustin A, Wang Y: **Satellite observed widespread decline in Mongolian grasslands largely due to overgrazing.** *Glob Chang Biol* 2014, **20:**418–428.
70. Fernandez-Gimenez ME, Allen-Diaz B: **Testing a non-equilibrium model of rangeland vegetation dynamics in Mongolia.** *J Appl Ecol* 1999, **36:**871–885.
71. Veiberg V, Loe LE, Mysterud A, Langvatn R, Stenseth NC: **Social rank, feeding and winter weight loss in red deer: any evidence of interference competition?** *Oecologia* 2004, **138:**135–142.
72. Shi J, Dunbar R: **Feeding competition within a feral goat population on the Isle of Rum, NW Scotland.** *J Ethol* 2006, **24:**117–124.
73. Thouless C: **Feeding competition between grazing red deer hinds.** *Anim Behav* 1990, **40:**105–111.
74. Gensac P: **Plant and soil groups in the alpine grasslands of the Vanoise Massif, French Alps.** *Arctic Alpine Res* 1990, **22:**195–201.
75. Sage RF, Kubien DS: **Quo vadis C4? An ecophysiological perspective on global change and the future of C4 plants.** *Photosynth Res* 2003, **77:**209–225.
76. Barbehenn RV, Chen Z, Karowe DN, Spickard A: **C3 grasses have higher nutritional quality than C4 grasses under ambient and elevated atmospheric CO2.** *Glob Chang Biol* 2004, **10:**1565–1575.
77. Michalet R, Gandoy C, Joud D, Pages JP, Choler P: **Plant community composition and biomass on calcareous and siliceous substrates in the northern French Alps: Comparative effects of soil chemistry and water status.** *Arctic Antarctic Alpine Res* 2002, **34:**102–113.
78. Brodie J, Johnson H, Mitchell M, Zager P, Proffitt K, Hebblewhite M, Kauffman M, Johnson B, Bissonette J, Bishop C: **Relative influence of human harvest, carnivores, and weather on adult female elk survival across western North America.** *J Appl Ecol* 2013, **50:**295–305.
79. Juillet C, Choquet R, Gauthier G, Lefebvre J, Pradel R: **Carry-over effects of spring hunt and climate on recruitment to the natal colony in a migratory species.** *J Appl Ecol* 2012, **49:**1237–1246.

# Reproductive responses of birds to experimental food supplementation: a meta-analysis

Lise Ruffino[1*], Pälvi Salo[1], Elina Koivisto[1], Peter B Banks[2] and Erkki Korpimäki[1]

## Abstract

**Introduction:** Food availability is an important environmental cue for animals for deciding how much to invest in reproduction, and it ultimately affects population size. The importance of food limitation has been extensively studied in terrestrial vertebrate populations, especially in birds, by experimentally manipulating food supply. However, the factors explaining variation in reproductive decisions in response to food supplementation remain unclear. By performing meta-analyses, we aim to quantify the extent to which supplementary feeding affects several reproductive parameters in birds, and identify the key factors (life-history traits, behavioural factors, environmental factors, and experimental design) that can induce variation in laying date, clutch size and breeding success (i.e., number of fledglings produced) in response to food supplementation.

**Results:** Food supplementation produced variable but mostly positive effects across reproductive parameters in a total of 201 experiments from 82 independent studies. The outcomes of the food effect were modulated by environmental factors, e.g., laying dates advanced more towards low latitudes, and food supplementation appeared not to produce any obvious effect on bird reproduction when the background level of food abundance in the environment was high. Moreover, the increase in clutch size following food addition was more pronounced in birds that cache food, as compared to birds that do not. Supplementation timing was identified as a major cause of variation in breeding success responses. We also document the absence of a detectable food effect on clutch size and breeding success when the target species had poor access to the feed due to competitive interactions with other animals.

**Conclusions:** Our findings indicate that, from the pool of bird species and environments reviewed, extra food is allocated to immediate reproduction in most cases. Our results also support the view that bird species have evolved different life-history strategies to cope with environmental variability in food supply. However, we encourage more research at low latitudes to gain knowledge on how resource allocation in birds changes along a latitudinal gradient. Our results also emphasize the importance of developing experimental designs that minimise competition for the supplemented food and the risk of reproductive bottle-necks due to inappropriate supplementation timings.

**Keywords:** Effect size, Feeding experiment, Population regulation, Reproductive performance, Resource competition, Wildlife management

## Introduction

Resource availability is an important determinant of the demography and distribution of many species, and it also affects energy allocation strategies in individuals. The energy involved to obtain food and process it is tightly related to the physiology and behaviour of animals, and the proportion of total energy and nutrients that is allocated to reproduction (i.e., reproductive effort) causes variation in life histories across individuals [1]. The importance of food limitation in nature continues to puzzle ecologists, as it can vary in time and across species, breeding stages and environments. Food limitation is likely less important in systems where predation or critical resources other than food keeps populations below their carrying capacity (e.g., [2]). Conversely, food limitation can be inferred when a shortage of food or any critical nutrients results in lower reproductive performance and/or reduced survival [1,3,4], that ultimately leads to a decline in growth rate of the focal population. Food limitation can have short (e.g., reproductive effort)

* Correspondence: lisruffino@gmail.com
[1]Section of Ecology, Department of Biology, University of Turku, FI-20014 Turku, Finland
Full list of author information is available at the end of the article

and long-term (e.g., adult survival and future reproduction) life-history consequences. Variation in food availability in space and time is therefore an important cue for animals to adjust their reproductive decisions: it can provide information on where and when to breed, with whom to breed, and how much to invest in offspring (e.g., [1,5,6]).

Birds have proven to be useful models in exploring the ecological and biological factors influencing the level of reproductive investment. Some evidence for the effects of food supply on the reproduction of bird populations comes from correlations where natural spatial or temporal variations in food availability have been linked to observed changes in population growth. For instance, the breeding success of specialised avian predators is closely correlated with the marked population fluctuations of their main prey, i.e., small rodents (e.g., [7,8]). Moreover, heavy rainfalls in arid systems positively correlate with increases in bird fecundity, presumably via enhanced food supply (e.g., [9]). Other examples of resource pulse events (i.e., sudden large increases in food supply) that have triggered a reproductive response in birds include insect outbreaks [10] and seed mast events [11].

While such correlations provide apparent evidence about the extent of food limitation in bird reproduction, they do not show causality. In addition, they are, in a sense, special cases as resource pulse events and prey population cycles generate highly contrasted levels of resource abundance, as opposed to more subtle differences in food availability that may be more difficult to identify. On the other hand, supplementary feeding of birds, which involves provisioning with types of foods in quantities that would not be available naturally, has proven to be a useful alternative method to quantitatively assess the importance of food supply in the reproductive decisions of wild bird populations, by taking into account potential confounding factors. In cases where food abundance in nature is used by birds as a cue for food conditions, manipulating food supply can experimentally address how environmental variation (in food supply) shapes life history [1].

The effects of food supplementation on various components of bird reproduction have been experimentally studied in a variety of taxa and environments, and on various components of reproduction. However, the results have been somewhat equivocal: while most studies have reported advancements of laying dates, increases in chick growth rate and breeding success [12], others have found no clear fitness benefits, e.g. no obvious effect on clutch size [13,14], or have even described negative effects on clutch size and breeding success [15,16]. The reasons for inconsistent results are not well known but may relate to the life-history traits of the species, the environmental conditions or the experimental design. The level of intra- and inter-specific competition for food, which can depend on the efficiency of the food delivery and the capacity of the recipient species to defend the food, might also explain the variable responses of birds to supplemented food [17].

Reports on the effects of supplemental food on bird reproduction have primarily been qualitative or based on vote-counting [1,4,12,18,19], thus general predictions about the effects of food supply on different components of reproduction across bird taxa and environments have not been tested quantitatively (but see [20]). In addition, recent quantitative analyses on the relative effects of latitude, broodedness and migratory status on the advancement of laying date in response to food supplementation have revealed contrasting results, leading to some debate [20-22]. In this paper we present a meta-analysis that synthesises the outcomes of food supplementation experiments on several bird reproductive parameters across 48 species. We specifically aim to (i) quantify the extent to which supplementary feeding induces an advancement of egg-laying and an increase in other components of the bird breeding cycle, i.e. clutch size, egg size, hatching success, brood size, chick body mass and breeding success (i.e., number of fledglings produced), and (ii) identify the key factors (including life-history traits, behavioural factors, environmental factors, experimental design) that can induce variation in laying date, clutch size and breeding success in response to food supplementation. These results will provide key-elements for a better understanding of the impact of food limitation on wild bird populations, and they can have important implications for the conservation of endangered populations whose long-term persistence might require food provisioning by practitioners.

## Results

In total, we retained 82 independent studies (201 experiments) that tested the effect of food supplementation on bird reproduction. These experiments were carried out primarily on birds of prey (19 studies), corvids (11 studies) and other passerines (38 studies), with only 15 experiments on seabirds (7 studies), eight on wetland birds (5 studies) and two on other bird families (i.e., Alcedinidae, Phasianidae). Studies were biased towards northern latitudes (≥50°N; n = 44 studies), with only 10% (n = 8) of studies at latitudes below 30°N or 30°S. Forty one studies (50%) tested the effect of food supplementation over at least two years.

The effect of food supplementation was not consistent across all reproductive parameters. The 95% confidence intervals indicated that the mean effect size of food supplementation was significantly positive for four reproductive parameters: laying date, clutch size, chick body mass and breeding success (Figure 1). Conversely, food supplementation had no detectable effect on egg size, hatching success or brood size (Figure 1).

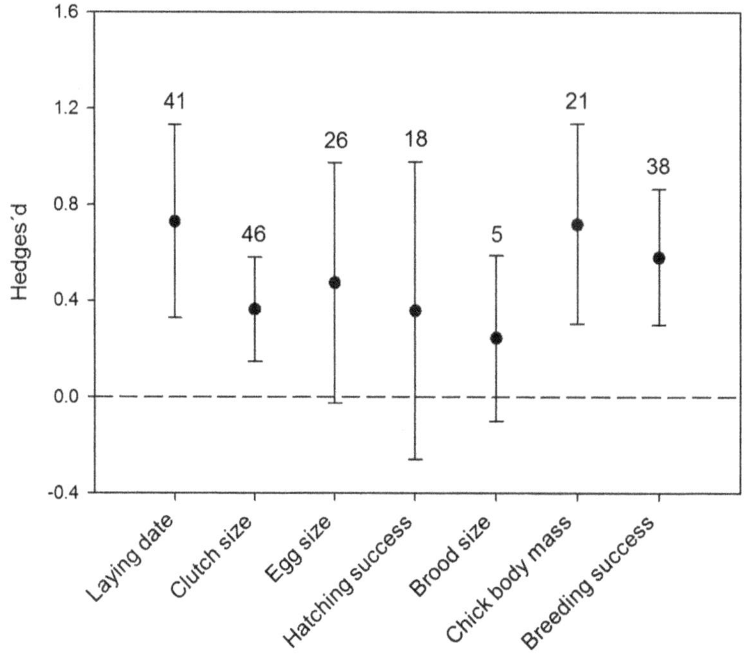


**Figure 1 Overall effects of food supplementation on reproductive parameters of birds.** Mean effect sizes (Hedges´*d*) and 95% confidence intervals are presented, with sample size (number of experiments) above bars.

The amplitude of laying date advancement was mainly explained by the latitude of the study area (Tables 1, 2), whereby egg-laying was most advanced at lower latitudes (Figure 2). By comparison, migratory status, elevation of the study area, broodedness and diet received no support for explaining variation in laying date (Table 2). Despite the moderate AICc weighting of accessibility to the feed, this variable did not significantly explained variation in laying date ($Q = 2.02$; $P = 0.36$; see also Additional file 1).

Accessibility to the feed and food caching behaviour received strong support as predictors of clutch size variation (Tables 1, 2). There was no detectable effect of food supplementation on clutch size when the level of food accessibility was deemed low or intermediate, or in birds that do not cache food (Figure 3). Conversely, birds that cache and had full access to the food increased their clutch size in response to food supplementation (Figure 3). Maximal clutch size, latitude, body mass and diet all received low support as predictors of clutch size variation (Table 2; see also Additional file 1).

Timing of food supplementation appeared as a strong predictor of the variation in breeding success, i.e. number of fledglings produced (Tables 1, 2). Studies that provided food only before and/or during egg-laying produced no detectable effect, whereas positive effect sizes were recorded for studies where food was either added from hatching to fledging or throughout the breeding season (Figure 4). Accessibility to the feed had a moderate AICc weight (0.47) in the series of models tested. Further inspection of the graphical results indicated that when the level of food accessibility was low or intermediate, food supplementation had on average no effect on breeding success, as opposed to when the level of accessibility to the food was high ($Q = 9.59$; $P = 0.02$; Figure 4). Diet, food caching or body mass did not explain variation in breeding success following food supplementation (Table 2; see also Additional file 1).

After controlling for the type of the reproductive parameter measured, food supplementation produced larger increases in reproductive responses when the background level of resources was lower than average, while supplementation had no detectable effect when the level of resources was higher than average ($Q = 35.6$; $d.f. = 3$; $P < 0.0001$). This effect was detectable in all three reproductive parameters (laying date, clutch size, breeding success; Figure 5).

## Discussion

We aimed to synthesise the outcomes of food supplementation experiments in several bird reproductive parameters by using quantitative meta-analyses. We found mostly positive effects of food supplementation experiments on bird reproduction, indicating that within the pool of bird species and environments reviewed, extra food is often allocated to immediate reproduction. The effects of supplementary food were still variable across reproductive parameters; this is consistent with the idea that varying energetic costs are associated with different

**Table 1 Performance of models explaining variation in laying date, clutch size and breeding success of birds**

| Model rank | Model | log-likelihood | AICc | ΔAICc | *wi* |
|---|---|---|---|---|---|
| Laying date | | | | | |
| 1 | rma.mv(d ~ Latitude + FoodAccess) | −37.02 | 88.84 | 0.00 | 0.13 |
| 2 | rma.mv(d ~ Latitude) | −40.44 | 90.06 | 1.22 | 0.07 |
| 3 | rma.mv(d ~ Latitude + FoodAccess + Elevation) | −36.10 | 90.20 | 1.36 | 0.07 |
| 4 | rma.mv(d ~ Latitude + FoodAccess + Migratory + Broodedness) | −36.13 | 90.26 | 1.42 | 0.07 |
| 5 | rma.mv(d ~ Latitude + FoodAccess + Migratory) | −36.13 | 90.26 | 1.42 | 0.07 |
| 6 | rma.mv(d ~ Latitude + Elevation) | −39.35 | 90.58 | 1.74 | 0.06 |
| 7 | rma.mv(d ~ Latitude + FoodAccess + Diet) | −36.48 | 90.97 | 2.13 | 0.05 |
| 8 | rma.mv(d ~ Latitude + Migratory) | −39.64 | 91.16 | 2.32 | 0.04 |
| 9 | rma.mv(d ~ Latitude + Diet) | −39.73 | 91.34 | 2.50 | 0.04 |
| 10 | rma.mv(d ~ Latitude + FoodAccess + Broodedness) | −36.70 | 91.39 | 2.55 | 0.04 |
| 30 | rma.mv(d ~ 1) | −57.93 | 122.53 | 7.08 | 0.00 |
| Clutch size | | | | | |
| 1 | rma.mv(d ~ FoodAccess + FoodCaching) | −26.77 | 67.94 | 0.00 | 0.13 |
| 2 | rma.mv(d ~ FoodAccess + FoodCaching + MaxClutchSize) | −25.72 | 68.84 | 0.90 | 0.08 |
| 3 | rma.mv(d ~ FoodCaching + MaxClutchSize) | −28.68 | 68.98 | 1.04 | 0.08 |
| 4 | rma.mv(d ~ FoodAccess) | −28.75 | 69.13 | 1.19 | 0.07 |
| 5 | rma.mv(d ~ FoodAccess + FoodCaching + Latitude) | −26.32 | 70.04 | 2.10 | 0.04 |
| 6 | rma.mv(d ~ FoodAccess + Latitude) | −27.97 | 70.35 | 2.41 | 0.04 |
| 7 | rma.mv(d ~ FoodAccess + FoodCaching + MaxClutchSize + Latitude) | −24.86 | 70.37 | 2.43 | 0.04 |
| 8 | rma.mv(d ~ FoodAccess + FoodCaching + Mass) | −26.58 | 70.56 | 2.62 | 0.03 |
| 9 | rma.mv(d ~ FoodCaching) | −30.78 | 70.58 | 2.64 | 0.03 |
| 10 | rma.mv(d ~ FoodAccess + MaxClutchSize) | −28.10 | 70.61 | 2.67 | 0.03 |
| 50 | rma.mv(d ~ 1) | −50.21 | 107.01 | 11.52 | 0.00 |
| Breeding success | | | | | |
| 1 | rma.mv(d ~ Timing + FoodAccess) | −39.21 | 93.53 | 0.00 | 0.13 |
| 2 | rma.mv(d ~ Timing) | −42.14 | 93.57 | 0.04 | 0.12 |
| 3 | rma.mv(d ~ Timing + Diet) | −40.81 | 93.69 | 0.17 | 0.12 |
| 4 | rma.mv(d ~ Timing + FoodAccess + Diet) | −37.86 | 94.19 | 0.66 | 0.09 |
| 5 | rma.mv(d ~ Timing + FoodCaching) | −41.40 | 94.86 | 1.33 | 0.06 |
| 6 | rma.mv(d ~ Timing + Mass) | −41.49 | 95.04 | 1.52 | 0.06 |
| 7 | rma.mv(d ~ Timing + FoodAccess + FoodCaching) | −38.30 | 95.09 | 1.56 | 0.06 |
| 8 | rma.mv(d ~ Timing + FoodCaching + Diet) | −40.04 | 95.19 | 1.67 | 0.05 |
| 9 | rma.mv(d ~ Timing + Diet + Mass) | −40.18 | 95.47 | 1.94 | 0.05 |
| 10 | rma.mv(d ~ Timing + FoodAccess + Mass) | −38.60 | 95.68 | 2.15 | 0.04 |
| 25 | rma.mv(d ~ 1) | −53.35 | 113.44 | 20.08 | 0.00 |

Only the 10 best models and the null models are shown.

stages of the breeding cycle, and it also shows that bird species have evolved different life-history strategies to cope with environmental variability in food supply. The average increase in bird breeding success (response ratio: $\bar{X}_{(EXP)}/\bar{X}_{(CONT)} = 1.33$ ) was of slightly higher magnitude to what Prevedello *et al.* [23] found for the reproductive response of small mammals following food addition (response ratio = 1.19), probably because of the use of different types of reproductive responses (i.e., number of breeding individuals used in Prevedello *et al.* [23] vs. number of offspring in this study). Overall, these positive responses support the idea that both bird and mammal populations can be constrained by food during the energy-demanding period of reproduction. Our analyses revealed

**Table 2 Variable weights from model selection analyses**

| Model type | Variables | Variable weight from AICc |
|---|---|---|
| Laying date | | |
| | Latitude | 0.92 |
| | Food accessibility | 0.54 |
| | Migratory status | 0.36 |
| | Elevation | 0.31 |
| | Broodedness | 0.28 |
| | Diet | 0.23 |
| Clutch size | | |
| | Food accessibility | 0.74 |
| | Food caching | 0.62 |
| | Maximal clutch size | 0.42 |
| | Latitude | 0.29 |
| | Bodymass | 0.22 |
| | Diet | 0.21 |
| Breeding success | | |
| | Timing | 0.92 |
| | Food accessibility | 0.47 |
| | Diet | 0.44 |
| | Food caching | 0.32 |
| | Body mass | 0.20 |

Variable weights were calculated as the sum of AICc weights of all candidate models including a given variable.

that the effects of food supplementation were modulated by (i) environmental factors, such as the background level of food abundance and factors that vary with latitude, (ii) behavioural traits in birds, such as food caching, and (iii) some components of the experimental design, such as the timing of food addition and the level of accessibility to the feed.

Our results showed an overall advancement of laying date in response to food addition across bird taxa, which is in accordance with the results of previous reviews [12,18,20,24]. Breeding earlier generally increases current and future reproductive outputs. For example, it can allow larger clutches [8,25], better chick survival [26,27], more re-nesting attempts [28,29] and more time for the parents to moult and prepare for the winter after breeding [26]. An advancement of egg-laying has also been documented in Europe in relation to increasing average spring temperatures [29-31] that lead to earlier plant growing seasons and forage availability.

We found some support for a latitudinal effect on the advancement of laying dates in response to food supplementation experiments. In accordance with the results of Schoech & Hahn [20], we documented a smaller advancement in egg-laying at higher latitudes. Most birds breeding in northern environments reproduce during a single favourable seasonal window when food is abundant, and they most likely show lower flexibility in life-history stages, such as initiation of egg-laying, than resident birds at lower latitudes. Hence, as suggested by Schoech & Hahn [20], species that rely on endogenous rhythms or environmental stimuli unrelated to food abundance, such as day length at high latitudes, may be limited in their ability to advance their reproduction as an adaptive response to untimely food abundance peaks. However, further research is urgently needed to corroborate these findings since the data currently available are biased against tropical environments. Despite Boutin's [18] recommendations, to our knowledge only two studies [32,33] have since tested the

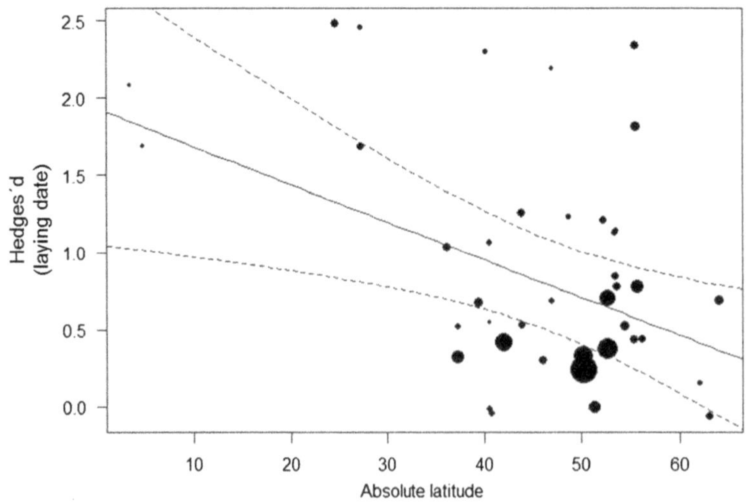


**Figure 2 Effect of latitude on advancement of laying date.** Effect of latitude of the study location on the degree of advancement of egg-laying date in response to food supplementation. Each dot represents one experiment and larger dots represent lower within-study variance, which has been used as a weighing factor in the analysis. Slope of the meta-regression = −0.024 (SE = 0.009). The regression line is bounded by 95% confidence intervals. Sample size is 41 experiments.

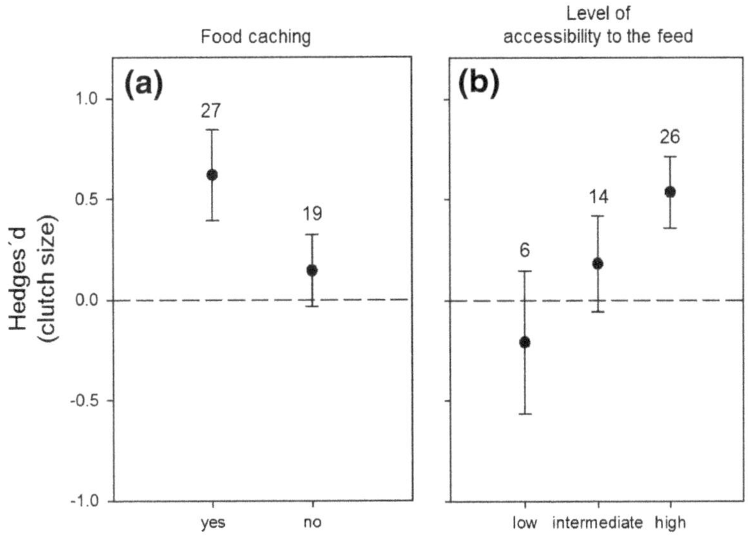


**Figure 3 Effects of food caching and food accessibility on variation in clutch size.** Effect of **(a)** food caching and **(b)** the level of accessibility to the feed on variation in clutch size in response to food supplementation in birds. Mean effect sizes (Hedges´*d*) and 95% confidence intervals are presented, with sample size (number of experiments) above bars.

effect of food addition on the timing of egg-laying in birds below 20°N or 20°S.

Our meta-analyses did not reveal any obvious differences in the effect of food on laying date between migratory and resident species, or between single- and multi-brooded species. There has been an extensive debate over the relative roles of latitude, migratory status and broodedness in modulating the advancement of egg-laying in birds. While the findings of Schoech & Hahn [20] (see also [22]) support the latitudinal gradient hypothesis, Dhondt [21] found no conclusive evidence that latitude influences the response, but identified broodedness as an important factor shaping the response of reproductive timing to food supplementation (see also [24]). These varying results and conclusions among studies probably arise both from the misassignment of some species as multi-brooded in Dhondt [21,22], and the statistical methods used to analyse the data, especially the failure to reduce pseudo-replication and incorporate within-study variances in the calculation of the food supplementation effect. In this paper, we dealt with the statistical

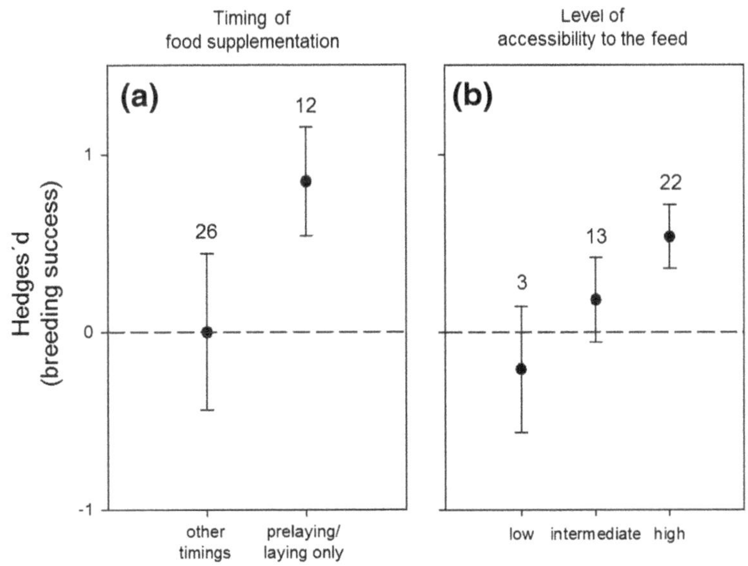


**Figure 4 Effects of timing and food accessibility on variation in breeding success.** Effect of **(a)** timing of food supplementation and **(b)** level of accessibility to the feed on breeding success variation in response to food supplementation in birds. Mean effect sizes (Hedges´*d*) and 95% confidence intervals are presented, with sample size (number of experiments) above bars.

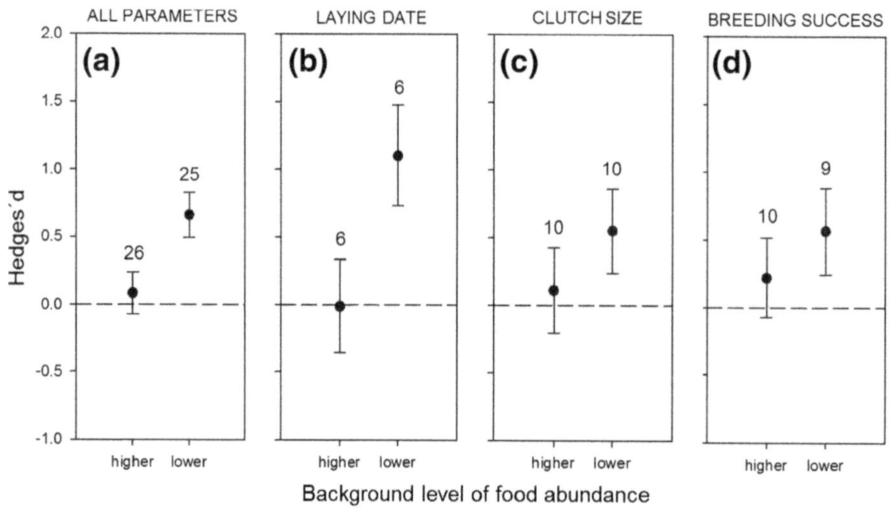


**Figure 5 Effect of the background level of resources on reproductive responses of birds.** Effect of the background level of resources in the environment (higher/lower than average) on variation in **(a)** three bird reproductive parameters, including **(b)** laying date, **(c)** clutch size and **(d)** breeding success in response to food supplementation experiments. Mean effect sizes (Hedges´*d*) and 95% confidence intervals are presented, with sample size (number of experiments) above bars.

issues by (i) controlling for shared phylogeny and study design, (ii) running multi-factor meta-analyses to account for the potential effects of other predictors, (iii) incorporating the within-study variation (SD) into the calculation of the effect sizes, and (iv) weighting the meta-analyses by the inverse of the within-study variance. Although we did not observe any apparent effect of broodedness, the magnitude of food supplementation effects on egg-laying advancement has been shown to differ between single- and multi-brooded species due to their contrasting breeding timing constrains and strategies [24,34]. As only eight experiments on multi-brooded species were available for our meta-analysis, additional experiments are needed to elucidate how broodedness modulates the reproductive response of birds to varying resource abundances in nature.

The effect of food supplementation on the number of eggs laid per clutch showed a moderate positive response. The observed increase in clutch size indicates either that the number of eggs a female can produce in a season is limited by nutrient or energy availability, or that food abundance during egg-laying is a fairly good predictor of resource availability later in the breeding season. In any case, clutch size is believed to be adaptive in many species and should correspond to a trade-off between the physiological constraints of producing more eggs and the energy demands during chick rearing [35]. Indeed, parental survival and future reproductive performance can be reduced due to the increased parental effort needed to rear more chicks to independence [35,36], as has been shown for great tits *Parus major* [37] and jackdaws *Corvus monedula* [38]. Our analyses revealed that food-caching birds, such as magpies, scrub-jays, shrikes, kestrels and some species of tits, increased clutch size in response to food supplementation, while non-caching species did not. Similarly, groups of acorn woodpeckers (*Melanerpes formicivorus*) with acorn stores lay significantly larger clutches than groups that lack stores during the breeding season [39]. Supplemented food can be gathered as soon as it is available and stored for periods when energy demand is higher. Food-caching may enable birds to predict the food conditions later in the season with more accuracy, and as a result stored food may initiate the production of larger clutches. However, it is still unclear whether larger food stores correlate with better overall productivity [40] and the results of our meta-analyses did not allow a conclusion on that matter (see Additional file 1).

The mechanisms driving clutch size variation among individuals, populations and species are complex, as they involve interactions between environmental, behavioural, physiological and genetic determinants. The absence of response of food supplementation in some species (e.g., small passerines, see Additional file 2) may be explained by the inability of the female to use food abundance during egg-laying as a breeding cue. Instead, the female may determine clutch size based on male quality (e.g., [41]), the presence of competitors (e.g., [42]), or by relying on habitat cues that predict environmental conditions and food availability later in the season [43,44]. Moreover, since the number of eggs produced per clutch is often tightly related to the date of egg-laying (i.e., smaller clutches are usually produced later in the season), the response of birds to food supplementation may also differ between early and late breeders in the same population. Unfortunately, we could not test this prediction since average laying dates were in most cases (93%) only provided for the entire breeding season. Finally, there is strong evidence that clutch size has a heritable component in wild populations of several bird species [45] and the use of "animal models" has proven to be valuable to address evolutionary-related questions (e.g., [46,47]).

Studies that delivered supplemental food merely during the pre-laying and/or egg-laying periods showed, on average, no detectable increase in breeding success (i.e., number of fledglings produced), whereas studies where food was added from hatching to fledging or throughout the breeding season documented net increases in breeding success. Supplementing food during pre-laying or egg-laying may actually create a bottle-neck for the birds since the advancement of egg-laying could create a mismatch between the nestling period and the maximal availability of nestling-suitable foods [16], which in turn increases the cost for the parents during chick rearing. However, as shown by von Brömssen & Jansson [26], the supplementation of food during pre-laying can also improve bird fitness in a different way, i.e. food-supplemented parents do not increase the absolute number of offspring produced but are able to produce individual offspring that are in better condition than young of non-supplemented parents.

Boutin [18] argued that food supplementation experiments should account for the possibility that the observed population responses, or the lack thereof, can merely be the result of competition and/or ineffective food delivery. Here, we documented the absence of a detectable effect of food supplementation on the clutch size and breeding success of birds when the target species did not have full access to the food provided, due to interactions at the feeding stations with con- or hetero-specifics. Manipulating food supply can attract a variety of species, both birds and mammals, from outside of the focal experimental area. When food is limiting, competition is likely to increase and can prevent subordinate individuals or species from utilising the feed. Based on our results, the risk of competition for the supplementary food tended to be low in those bird species that show strong territoriality or aggressive behaviour towards intruders, and in cavity-nesting species, to which food can be delivered inside nest boxes. We found that corvids were more likely to increase their breeding success in response to food supplementation than other bird groups (see Additional file 2), probably because corvids often exclude other birds from their territories and thereby from established feeders [48], but also because they can use novel food sources exhaustively due to their strong cognitive abilities and opportunist foraging behaviour.

Food supplementation increased bird reproductive responses when the background level of resources was lower than average, while supplementation had no detectable effect when food was unusually abundant in nature. The threshold hypothesis predicts that only individuals experiencing background levels of food availability below a saturation point will respond to supplementary feeding [18,25]. The relationship between food availability and reproductive traits may thus be non-linear and other factors, such as predation or competition for territories, become limiting when the background level of food is high [1,28,49]. The reproductive decisions of individuals in a given population may vary between years due to changing environmental conditions. This highlights the importance of measuring natural food abundance during any supplementation experiment as well as the value of conducting experiments over multiple years in varying environmental conditions.

### Conclusions and future prospects

Our results confirm that food supply plays a crucial role in avian ecology and evolution, through its effects on individual behaviours and life-history traits. They also imply that the extent of food limitation varies across the multiple stages of bird reproductive cycle, but also across bird taxa and environmental conditions. Latitude appears to modulate the effect of food supplementation on bird reproduction: birds living at low latitudes may be more plastic in advancing their timing of breeding, as compared to birds breeding at higher latitudes where factors other than food may constrain life-history parameters. On the other hand, it has been suggested (in line with the "slow pace of life in the tropics" syndrome [50]) that tropical resident birds should invest extra energy into self-maintenance rather than in immediate reproductive costs [51], and the rare manipulative experiments to address the question of resource allocation trade-offs in the tropics have produced contrasting results (e.g., [51,52]). Certainly, more food supplementation studies in tropical and equatorial environments are needed to be able to conclude on that matter. Comparative studies of closely-related bird species breeding in different environments will also help understanding how resource allocation changes along a latitudinal gradient.

Our findings further revealed that the results of food supplementation experiments could be affected by inadequate experimental designs. This is directly relevant to situations where food supplementation is applied as a conservation tool to boost the breeding performance of endangered birds. For example, if the objective of a supplementary feeding program is to enhance the overall productivity of a bird population, we do not recommend delivering the food during the pre-laying or egg-laying period only as it can create a bottle-neck for the parents later in the breeding season, if the abundance of resources in the environment does not match the energy needs of the parents for rearing more chicks. Instead, we encourage conservation managers to extend the food delivery until the hatching or chick-rearing period. Furthermore, because food supplementation can alter the interactions between individuals and species, it is crucial to ensure that the risk of competition for the supplemented feed is limited. In species where males tend to monopolise the feeders, this issue can be mitigated either by using large numbers of feeders (e.g., >3 per territory) so that females can have unrestricted access to supplemental food, or by delivering the feed to the females directly. Moreover, delivering the food over large treatment areas (with small perimeter-area ratios) instead of small clumped areas, can reduce the flow of intruders and hence the relative effect of immigration (e.g., [23]). Preventing non-target species or conspecifics from monopolising the supplemented food can be achieved by using feeders designed to supplement the feed to experimental animals only (e.g., automated feeders, [53]). Given that the feed is often delivered *ad libitum* over the entire experimental area, it can be difficult to identify which animals are utilising it and, hence, assess the effect of food on life-history parameters, such as reproduction, growth, condition, at the individual level. Recently, Robb *et al.* [54] have pointed out the potential for stable isotopes to trace the use of food supplements in animals and measure the variation in food use between recipient individuals. Alternatively, the use of radio frequency identification systems (RFID) can be useful in quantifying bird visits to feeders [55]. Further research is also needed to test how the spatial dispersion of supplementary food (e.g., patchy vs. scattered; [56]) affects intra- and inter-specific competition and, hence, the outcome of the food supplementation experiment. Supplementing endangered populations should, however, always be applied with caution as some recent studies have warned against potential adverse consequences of food supplementation on nestling health [57] and sex-ratio [58], breeding success [59], and the carry-over effects of winter food provision on the investment in egg production [60].

Another major issue related to the experimental design is the nutritional quality of the feed, which unfortunately could not be addressed in this review due to the lack of data provided by the studies included, but should be cautiously considered in future studies. Growing evidence suggests that the ecological impacts of food supplementation can depend on the specific nutritional profile of provisioned foods [48,60-64] and that the energetic content of the food (in joules or calories) does not necessarily constitute an indicator of food quality [61,63]. For example, the availability of essential amino acids [62,65] and vitamins or antioxidants [60] may constrain egg production in birds. The geometric framework for nutrition (see, e.g., [66]) explores how animals simultaneously regulate the intake of multiple nutrients, and has recently been applied in birds [67]. This modelling approach contributes further understanding of the relationships between the nutritional ratios of natural food and supplementary feeding, and reproduction in birds, and hence can assist with devising effective supplementary food to endangered species [67].

Despite the large body of scientific literature on the impacts of food supply in bird life-history traits and reproductive strategies, some groups of birds and types of environments remain severely under-studied. Large-scale supplementation experiments, testing the effect of food addition on several populations or species of the same groups of birds in variable climates or environmental conditions, would be highly desirable. Finally, despite the accumulation of food supplementation studies over the past thirty years, most of this research has focused on immediate reproductive investments during the current breeding season; very little is known about the carry-over effects of food on offspring and parental survival when supplementation is finished, and about how food limitation may constrain future reproduction (but see [68]).

## Materials and methods

### Literature search

We searched for studies that tested the effects of experimental food supplementation on reproduction of birds by conducting a literature search on the Web of Science and Google Scholar using combinations of the following keywords: "experiment*", "manipulat*", "food", "supplement*", "feeding", "addition", "bird", "reproduction" and "population". We also screened the reference lists of all retrieved articles and previous reviews (e.g., [12,18,20]) for other relevant publications. Searching ended in June 2013.

We restricted our search to food manipulations that provided sufficient data to calculate an effect size. When the means, error measurements or sample sizes were not provided in the publication, the authors were contacted. We discarded studies that reported the reproductive parameters of animals feeding in refuse dumps or of animals studied in habitats of different quality (with no food addition), as well as studies admitting that the supplemented food was not sufficient in either quality or quantity to adequately test for the effects of food supplementation. Out of 337 records identified, a total of 82 independent studies met our criteria (including 201 experiments). Data obtained directly from the authors of the reviewed publications amounted to 8% (16 experiments) of the whole dataset. Three of the authors (LR, EKoivisto, PS) extracted all data from the text, tables and figures (using DigitizeIt software) of the original publications. The process of record collection and study exclusion for the qualitative synthesis and quantitative meta-analyses is detailed in a PRISMA-type flow chart (Figure 6).

### Variables and effect sizes

The experiments were scored for variables that related to the behaviour and life-history traits of each study species

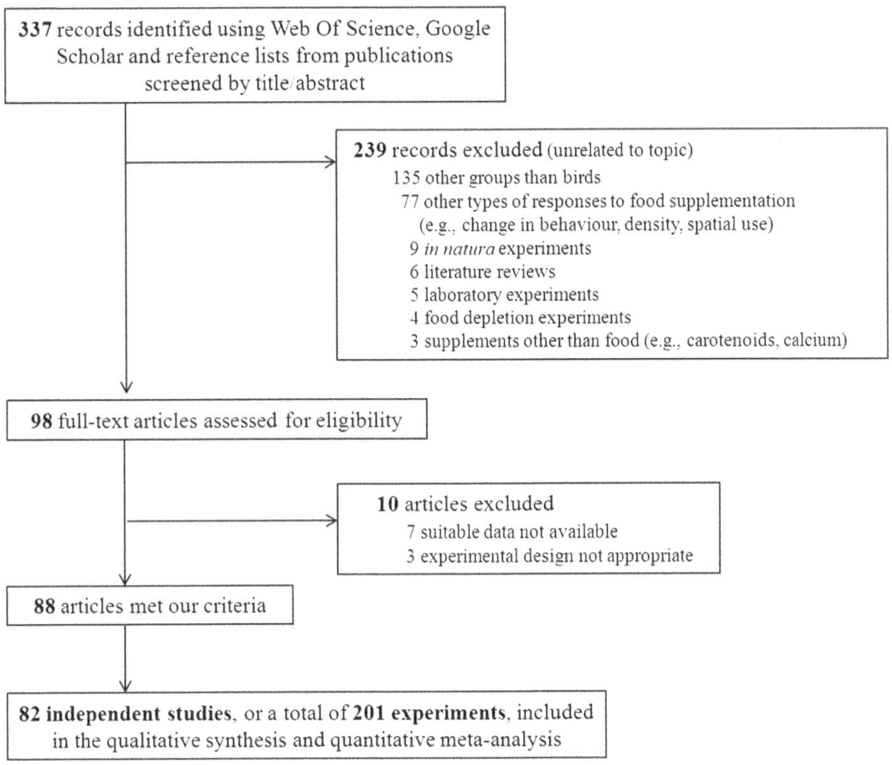


**Figure 6 Flow chart detailing the review and data selection process.** Flow chart detailing the process of record collection and study elimination for the qualitative synthesis and quantitative meta-analyses of food supplementation effects on bird reproduction. Several articles can be considered as a single independent study if they share the same experimental design and were conducted in the same study area.

by collating data from original publications or other literature (i.e., bird handbooks, internet; see Additional file 3 for more details). Each species was assigned to one of the six established bird types (corvids, passerines other than corvids, birds of prey, seabirds, wetland birds, others; see Additional file 3 for details on the categories). Behavioural variables included migratory status (migratory/year-round resident), food caching (caching/non-caching) and propensity to accept a new type of food (generalist diet/specialist diet), while life-history traits measured maximal clutch size, broodedness (one/more than one clutch per year) and body mass. All experiments were also scored for variables related to the experimental setup, such as timing of food supplementation (before and/or during egg-laying only/other timings, i.e., from hatching to fledging or throughout the breeding season), the level of accessibility to the feed for the target species (low/intermediate/high), environmental conditions, such as latitude and elevation (m a.s.l.) of the study site, and the relative background level of food resources (higher/lower than average) (see Additional file 3). The background level of resources was assigned for all studies that estimated the abundance of the most common food item or prey for the target species or reported clear differences in environmental quality (i.e., due to drastic changes in weather conditions) between study years. Food accessibility was considered low when food was substantially exploited by dominants, immigrants and/or non-target species. Accessibility was deemed high when supplemented food was delivered to the nest or nest box, or to the adults directly, and when direct observations indicated that most or all of the food was consumed by the target species. Accessibility was considered intermediate when food was only occasionally exploited by dominants, immigrants and/or non-target species, and not provided directly to the nest or nest box, or to the adults.

The reproductive responses considered in our analyses included laying date, clutch size, egg size, hatching success, brood size, chick body mass and breeding success. Hatching success was defined as the number of hatchlings divided by the total number of eggs laid. Several estimates of breeding success were used, i.e. the number of fledglings per nest, the number of fledglings divided by the total number of hatchlings or by the number of pairs that initiated reproduction or successfully fledged young, if the number of hatchlings was not reported.

Frequently, ecologists are interested in the actual difference in mean performance or the proportional change in performance in response to a treatment. This is why Hedges'*d* and the log response ratio (ln *R*) have been so commonly used as effect sizes in ecological meta-analyses. Since this study is interested in the factors causing variation in life-history traits of bird species, we need to incorporate a measure of variance in the calculation of the effect size, so that a change in any reproductive parameter can be interpreted as the fitness benefit to that species. For example, clutch size might be very constrained in one species and thus a 10% change in response to food supplementation (e.g., change from 1 ± 0.5 SD to 1.1 ± 0.5 SD) is to be valued more than a 10% change when clutch size is less constrained (e.g., change from 1 ± 1 SD to 1.1 ± 1 SD).

We used the unbiased standardized mean difference Hedges'*d* (hereafter *d*) as a measure of effect size in our meta-analyses. This effect size can be interpreted as the difference between the reproductive responses of supplemented birds versus non-supplemented birds, measured in units of standard deviations. Thus, large differences and low variability generate the largest effect sizes [69,70]. Positive *d* values indicate that food supplementation has a positive effect on the reproduction of a study species, while negative values indicate greater values in controls compared to the experimental group. By convention, a large effect is indicated by $|d| > 0.8$, a moderate effect by $|d| = 0.2\text{-}0.8$, and a small effect by $|d| = 0\text{-}0.2$ [71,72].

### Analyses

We used the metafor package [73] in R 3.0.2 to perform the meta-analyses and estimate a pooled effect size for each reproductive parameter and explanatory variable. In all meta-analyses, we controlled for repeated observations on the same species from the same study area by including Study ID as a random variable in the models. We also controlled for correlated structures due to shared phylogenetic history by specifying a phylogenetic correlation matrix of the species included in the meta-analysis and including Species as a random variable in our models (see Additional file 4 for the phylogenetic bird trees used and their characteristics). Variation in response magnitudes among bird types was confirmed by the heterogeneity statistics *Q* computed on the laying date, clutch size and breeding success datasets (Additional file 2). The following additional procedures also helped minimising the risk of pseudo-replication in the datasets: (i) each type of reproductive parameter was analysed separately, thereby avoiding the inclusion of several parameters taken from the same species or study area in the same dataset, (ii) in cases where the same reproductive parameters were measured on the same species over several years, an average effect size was calculated over all study years, (iii) in cases where a study tested different food qualities or supplementation timings on the same species, we selected the experiment with the highest effect size. In all meta-analyses, effect sizes were weighted by the inverse of their within-variance. Mean effect sizes showed significant effects of food addition when their 95% confidence intervals did not overlap zero. Model convergence was checked by inspecting the profile of the maximum restricted log-likelihood of the full models.

First, we estimated the overall effects of food supplementation on seven reproductive parameters, namely laying date (n = 41 experiments), clutch size (n = 46), egg size (n = 26), hatching success (n = 18), brood size (n = 5), chick body mass (n = 21) and breeding success (n = 38), by performing weighted random-effect meta-analyses without fixed explanatory variables (i.e., null random models). Second, we tested the effect of several explanatory variables on laying date, clutch size and breeding success (i.e., parameters for which $N \geq 30$), separately. For laying date, we tested the effect of migratory status, broodedness, diet, accessibility to the feed, latitude and elevation. Meta-analyses for clutch size tested the effect of caching behaviour, maximal clutch size, body mass, accessibility to the feed, diet and latitude. Explanatory variables for breeding success included caching behaviour, accessibility to the feed, timing of food addition, diet and body mass. For each analysis, we constructed a set of models that included all possible combinations of explanatory variables. We used conditional Akaike's Information Criterion, corrected for small sample size (AICc), to compare the goodness of fit among models. Models were ranked in relation to each other using ΔAICc values. AICc weights ($w_i$) were calculated to assess the relative likelihood of each model considered. Finally, we used multi-model inference to quantify the relative explanatory importance of variables over the set of candidate models. Variable weights were calculated as the sum of AICc weights of all the models including a given variable.

The background level of resources can affect the magnitude of the bird response to food supplementation (e.g., [25]; see also Results) but only 18 studies (out of 82) monitored or commented on the natural resource levels in their study area for the whole study period. For the 41 studies conducted over several years, we remained conservative in our approach and calculated an average effect size over the whole study period. Studies that were performed within one year and did not report the background level of resources (*n* = 38) were assumed to be conducted under average environmental conditions. Three studies reported experiments only at a persistently and distinctly lower [74] (poor quality habitat with low prey densities) or higher [75,76] (peak of main prey abundance in rodent specialised predators) background level of resources than average and were discarded to avoid biasing the analyses. Therefore, we made sure that the effect sizes used in all the above-mentioned meta-analyses matched averaged environmental conditions. The final sample size for the analyses described above was 195 experiments from 79 studies (Additional file 3).

Third, we explored the effect of varying background levels of food resources (lower/higher than average, as stated in the original publications) on the success of food supplementation in a separate meta-analysis. Reproductive parameter (laying date, clutch size, breeding success) was included as a fixed covariate in the meta-analysis since responses to food supplementation varied across reproductive parameters (see Results). This dataset consisted of 51 experiments from 18 studies (Additional file 3).

Finally, we examined the possibility of publication bias in our datasets by generating funnel and normal quantile plots (Additional file 5).

### Availability of supporting data

The data sets supporting the results of this article are included within the article (and its additional files).

## Additional files

**Additional file 1: Model selection procedure and mean effect sizes of the variables that did not show any support in explaining variation in the datasets.**

**Additional file 2: Results of the meta-analyses exploring the effect of food supplementation on laying date, clutch size and breeding success across several bird types.**

**Additional file 3: List of all the publications included in the meta-analyses testing the effect of food supplementation on several reproductive parameters.**

**Additional file 4: Phylogenetic trees used in the phylogenetic meta-analyses.**

**Additional file 5: Examination of publication bias.**

#### Competing interests

The authors declare that they have no competing interests.

#### Authors' contributions

EKor and LR designed the study. LR, PS and EKoi collected the data. LR performed the meta-analyses and wrote the manuscript. LR, PS, EKoi, PB and EKor contributed substantially to revisions. All authors read and approved the final manuscript.

#### Acknowledgements

The authors want to sincerely thank Julien Terraube, Quinn Fletcher and two anonymous reviewers for their constructive comments on a draft manuscript. Thanks to Patrice Bourgault, Martin Brinkhof, Alexandre Millon, Jan-Åke Nilsson, Markus Ritz, Stephan Schoech, Claire Spottiswoode and Liana Zanette for providing additional data that enabled calculating effect sizes. The study was financially supported by Norden Top-Level Research Initiative, "Effect Studies and Adaptation to Climate Change" (Nordic Center of Excellence TUNDRA) to LR and EKoi, by the Academy of Finland to PS (grant no. 138638) and to EKor (grant no. 250709) and by the Australian Academy of Science to PB (DP0881455).

#### Author details

[1]Section of Ecology, Department of Biology, University of Turku, FI-20014 Turku, Finland. [2]School of Biological Sciences, The University of Sydney, Sydney, Australia.

#### References

1. Martin TE: **Food as a limit on breeding birds: a life-history perspective.** *Annu Rev Ecol Syst* 1987, **18:**453–487.
2. Osenberg CW, Mittelbach GG: **The relative importance of resources limitation and predator limitation in food chains.** In *Food Webs: Integration of Patterns and Dynamics.* Edited by Polis GA, Winemiller KO. New York: Chapman and Hall; 1996:134–148.
3. Sinclair ARE: **The regulation of animal populations.** In *Ecological Concepts: The Contribution of Ecology to an Understanding of the Natural World.* Edited by Cherrett JM, Bradshaw AD. Oxford, UK: Blackwell Scientific Publications; 1989:197–241.
4. Newton I: *Population Limitation in Birds.* London: Academic Press; 1998:597.
5. Asdell SA: *Patterns of Mammalian Reproduction.* 2nd edition. Ithaca, New York: Cornell University Press; 1964.
6. Lack D: *Population Studies of Birds.* Oxford: Clarendon Press; 1966.
7. Korpimäki E, Norrdahl K: **Numerical and functional responses of kestrels, short-eared owls, and long-eared owls to vole densities.** *Ecology* 1991, **72:**814–826.
8. Korpimäki E, Wiehn J: **Clutch size of kestrels: seasonal decline and experimental evidence for food limitation under fluctuating food conditions.** *Oikos* 1998, **83:**259–272.
9. Illera JC, Díaz M: **Reproduction in an endemic bird of a semiarid island: a food-mediated process.** *J Avian Biol* 2006, **37:**447–456.
10. Hogstad O: **Numerical and functional responses of breeding passerine species to mass occurrence of geometrid caterpillars in a subalpine birch forest: a 30-year study.** *Ibis* 2005, **147:**77–91.
11. Curran LM, Leighton M: **Vertebrate responses to spatiotemporal variation in seed production of mast-fruiting Dipterocarpaceae.** *Ecol Monogr* 2000, **70:**101–128.
12. Robb GN, McDonald RA, Chamberlain DE, Bearhop S: **Food for thought: supplementary feeding as a driver of ecological change in avian populations.** *Front Ecol Environ* 2008, **6:**476–484.
13. Ewald PW, Rohwer S: **Effects of supplemental feeding on timing of breeding, clutch-size and polygyny in red-winged blackbirds *Agelaius phoeniceus*.** *J Anim Ecol* 1982, **51:**429–450.
14. Nilsson J-Å, Svensson E: **Energy constraints and ultimate decisions during egg-laying in the blue tit.** *Ecology* 1993, **74:**244–251.
15. Clamens A, Isenmann P: **Effect of supplemental food on the breeding of Blue and Great Tits in Mediterranean habitats.** *Ornis Scand* 1989, **20:**36–42.
16. Nilsson J-Å: **Energetic bottle-necks during breeding and the reproductive cost of being too early.** *J Anim Ecol* 1994, **63:**200–208.
17. Dhondt AA: *Interspecific Competition in Birds.* Oxford: Oxford University press; 2012.
18. Boutin S: **Food supplementation experiments with terrestrial vertebrates: patterns, problems, and the future.** *Can J Zool* 1990, **68:**203–220.
19. Meijer T, Drent R: **Re-examination of the capital and income dichotomy in breeding birds.** *Ibis* 1999, **141:**399–414.
20. Schoech SJ, Hahn TP: **Food supplementation and timing of reproduction: does the responsiveness to supplementary information vary with latitude?** *J Ornithol* 2007, **148:**625–632.
21. Dhondt AA: **Broodedness, not latitude, affects the response of reproductive timing of birds to food supplementation.** *J Ornithol* 2010, **151:**955–957.
22. Schoech SJ: **A reply to Dhondt: broodedness and latitude affect the response of reproductive timing of birds to food supplementation.** *J Ornithol* 2010, **151:**959–961.
23. Prevedello JA, Dickman CR, Vieira MV, Vieira EM: **Population responses of small mammals to food supply and predators: a global meta-analysis.** *J Anim Ecol* 2013, **82:**927–936.
24. Svensson E: **Avian reproductive timing: when should parents be prudent?** *Anim Behav* 1995, **49:**1569–1575.
25. Nager RG, Ruegger C, Van Noordwijkt AJ: **Nutrient or energy limitation on egg formation: a feeding in great tits experiment.** *J Anim Ecol* 1997, **66:**495–507.
26. von Brömssen A, Jansson C: **Effects of food addition to Willow Tit *Parus montanus* and Crested Tit *P. cristatus* at the time of breeding.** *Ornis Scand* 1980, **11:**173–178.
27. Barba EL, Gildelgado JA, Monros JS: **The costs of being late – consequences of delaying great tit *Parus major* first clutches.** *J Anim Ecol* 1995, **64:**642–651.
28. Arcese P, Smith JNM: **Effects of population density and supplemental food on reproduction in song sparrows.** *J Anim Ecol* 1988, **57:**119–136.
29. Dunn PO, Møller AP: **Changes in breeding phenology and population size of birds.** *J Anim Ecol* 2014, **83:**729–739.

30. Forchhammer MC, Post E, Stenseth NC: **Breeding phenology and climate….** *Nature* 1998, **391**:29–30.
31. Both C, Artemyev AV, Blaauw B, Cowie RJ, Dekhuijzen AJ, Eeva T, Enemar A, Gustafsson L, Ivankina EV, Järvinen A, Metcalfe NB, Nyholm NEI, Potti J, Ravussin P-A, Sanz JJ, Silverin B, Slater FM, Sokolov LV, Török J, Winkel W, Wright J, Zang H, Visser ME: **Large-scale geographical variation confirms that climate change causes birds to lay earlier.** *Proc Biol Sci* 2004, **271**:1657–1662.
32. Komdeur J: **Breeding of the Seychelles Magpie Robin *Copsychus sechellarum* and implications for its conservation.** *Ibis* 1996, **138**:485–498.
33. Scheuerlein A, Gwinner E: **Reduced nestling growth of East African Stonechats *Saxicola torquata axillaris* in the presence of a predator.** *Ibis* 2006, **148**:468–476.
34. Verboven N, Tinbergen JM, Verhulst S: **Food, reproductive success and multiple breeding in the great tit *Parus major*.** *Ardea* 2001, **89**:387–406.
35. Monaghan P, Nager RG: **Why don't birds lay more eggs?** *Trends Ecol Evol* 1997, **12**:270–274.
36. Santos ESA, Nakagawa S: **The costs of parental care: a meta-analysis of the trade-off between parental effort and survival in birds.** *J Evol Biol* 2012, **25**:1911–1917.
37. Visser ME, Lessells CM: **The costs of egg production and incubation in great tits (*Parus major*).** *Proc R Soc B* 2001, **268**:1271–1277.
38. Boonekamp JJ, Salomons M, Bouwhuis S, Dijkstra C, Verhulst S: **Reproductive effort accelerates actuarial senescence in wild birds: an experimental study.** *Ecol Lett* 2014, **17**:599–605.
39. Koenig WD: **Morphological and dietary correlates of clutch size in North American woodpeckers.** *Auk* 1987, **104**:757–765.
40. Vander Wall SB: *Food Hoarding in Animals.* London: The University of Chicago Press, Ltd.; 1990.
41. Soler JJ, De Neve L, Martínez JG, Soler M: **Nest size affects clutch size and the start of incubation in magpies: an experimental study.** *Behav Ecol* 2001, **12**:301–307.
42. Perrins CM: **Population fluctuations an clutch-size in the great tit, *Parus major* L.** *J Anim Ecol* 1965, **34**:601–647.
43. Perrins CM, Jones PJ: **The inheritance of clutch size in the Great Tit.** *Condor* 1974, **46**:225–229.
44. Preston KL, Rotenberry JT: **Independent effects of food and predator-mediated processes on annual fecundity in a songbird.** *Ecology* 2006, **87**:160–168.
45. Postma E, van Noordwijk AJ: **Genetic variation for clutch size in natural populations of birds from a reaction norm perspective.** *Ecology* 2005, **86**:2344–2357.
46. Charmantier A, Perrins P, McCleery RH, Sheldon BC: **Evolutionary response to selection on clutch size in a long-term study of the mute swan.** *Am Nat* 2006, **167**:453–465.
47. Garant D, Hadfield JD, Kruuk LEB, Sheldon BC: **Stability of genetic variance and covariance for reproductive characters in the face of climate change in a wild bird population.** *Mol Ecol* 2008, **17**:179–188.
48. Reynolds SJ, Schoech SJ, Bowman R: **Diet quality during pre-laying and nestling periods influences growth and survival of Florida scrub-jay (*Aphelocoma coerulescens*) chicks.** *J Zool* 2003, **261**:217–226.
49. Schultz ET: **The effect of energy reserves on breeding schedule: is there a saturation point?** *Funct Ecol* 1991, **5**:819–824.
50. Ricklefs RE, Wikelski M: **The physiology/life-history nexus.** *Trends Ecol Evol* 2002, **17**:462–468.
51. Class AM, Moore IT: **Effects of food supplementation on a tropical bird.** *Oecologia* 2013, **173**:355–362.
52. Scheuerlein A, Gwinner E: **Is food availability a circannual zeitgeber in tropical birds? A field experiment on stonechats in tropical Africa.** *J Biol Rhythms* 2002, **17**:171–180.
53. Isaac JL, Johnson CN, Grabau PJ, Krockenberger AK: **Automated feeders: new technology for food supplementation experiments with mammals.** *Wildl Res* 2004, **31**:437–441.
54. Robb GN, McDonald RA, Inger R, Reynolds TJE, Newton J, McGill RAR, Chamberlain DE, Harrison TJE, Bearhop S: **Using stable-isotope analysis as a technique for determining consumption of supplementary foods by individual birds.** *Condor* 2011, **113**:475–482.
55. Bonter DN, Bridge ES: **Applications of radio frequency identification (RFID) in ornithological research: a review.** *J F Ornithol* 2011, **82**:1–10.
56. Harris DB, Macdonald DW: **Interference competition between introduced black rats and endemic Galápagos rice rats.** *Ecology* 2007, **88**:2330–2344.
57. Blanco G, Lemus JA, García-Montijano M: **When conservation management becomes contraindicated: impact of food supplementation on health of endangered wildlife.** *Ecol Appl* 2011, **21**:2469–2477.
58. Clout MN, Elliott GP, Robertson BC: **Effects of supplementary feeding on the offspring sex ratio of kakapo: a dilemma for the conservation of a polygynous parrot.** *Biol Conserv* 2002, **107**:13–18.
59. Harrison TJE, Smith JA, Martin GR, Chamberlain DE, Bearhop S, Robb GN, Reynolds SJ: **Does food supplementation really enhance productivity of breeding birds?** *Oecologia* 2010, **164**:311–320.
60. Plummer KE, Bearhop S, Leech DI, Chamberlain DE, Blount JD: **Fat provisioning in winter impairs egg production during the following spring: a landscape-scale study of blue tits.** *J Anim Ecol* 2013, **82**:673–682.
61. Bolton M, Houston DC, Monaghan P: **Nutritional constraints on egg formation in the lesser black-backed gull: an experimental study.** *J Anim Ecol* 1992, **61**:521–532.
62. Ramsay SL, Houston DC: **The effect of dietary amino acid composition on egg production in blue tits.** *Proc R Soc B Biol Sci* 1998, **265**:1401–1405.
63. Houston D, Mcinnes K, Elliott G, Eason D, Moorhouse R, Cockrem J: **The use of a nutritional supplement to improve egg production in the endangered kakapo.** *Biol Conserv* 2007, **138**:248–255.
64. Williams TD: **Variation in reproductive effort in female zebra finches (*Taeniopygia guttata*) in relation to nutrient-specific dietary supplements during egg laying.** *Physiol Zool* 1996, **69**:1255–1275.
65. Jones PJ, Ward P: **Level of reserve protein as proximate factor controlling timing of breeding and clutch-size in red-billed quelea Quelea quelea.** *Ibis* 1976, **118**:547–574.
66. Simpson SJ, Raubenheimer D: **A multi-level analysis of feeding behaviour: the geometry of nutritional decisions.** *Philos Trans R Soc Lond B Biol Sci* 1993, **342**:381–402.
67. Raubenheimer D, Simpson SJ: **The challenge of supplementary feeding: can geometric analysis help save the kakapo?** *Notornis* 2006, **53**:100–111.
68. Brommer JE, Karell P, Pietiäinen H: **Supplementary fed Ural owls increase their reproductive output with a one year time lag.** *Oecologia* 2004, **139**:354–358.
69. Hedges LV, Olkin I: *Statistical Methods for Meta-Analysis.* Orlando, FL: Academic Press Inc.; 1985.
70. Borenstein M, Hedges LV, Higgins JPT, Rothstein HR: *Introduction to Meta-Analysis.* UK: John Wiley & Sons, Ltd; 2009:421.
71. Cohen J: *Statistical Power Analysis for the Behavioral Sciences.* N.Y., New York: Academic Press Inc.; 1969.
72. Gurevitch J, Morrow LL, Wallace A, Walsh JS, Url S: **A meta-analysis of competition if field experiments.** *Am Nat* 1992, **140**:539–572.
73. Viechtbauer W: **Meta-Analysis Package for R.** *J Stat Softw* 2010, **36**:1–47.
74. Newton I, Marquiss M: **Effect of additional food on laying dates and clutch sizes of Sparrowhawks.** *Ornis Scand* 1981, **12**:224–229.
75. Korpimäki E: **Breeding performance of Tengmalm's Owl *Aegolius funereus*: effects of supplementary feeding in a peak vole year.** *Ibis* 1989, **131**:51–56.
76. Hörnfeldt B, Hipkiss T, Fridolfsson A-K, Eklund U, Ellegren H: **Sex ratio and fledging success of supplementary-fed Tengmalm's owl broods.** *Mol Ecol* 2000, **9**:187–192.

# 10

# Unveiling the factors shaping the distribution of widely distributed alpine vertebrates, using multi-scale ecological niche modelling of the bat *Plecotus macrobullaris*

Antton Alberdi, Ostaizka Aizpurua, Joxerra Aihartza and Inazio Garin*

**Abstract**

Several alpine vertebrates share a distribution pattern that extends across the South-western Palearctic but is limited to the main mountain massifs. Although they are usually regarded as cold-adapted species, the range of many alpine vertebrates also includes relatively warm areas, suggesting that factors beyond climatic conditions may be driving their distribution. In this work we first recognize the species belonging to the mentioned biogeographic group and, based on the environmental niche analysis of *Plecotus macrobullaris*, we identify and characterize the environmental factors constraining their ranges. Distribution overlap analysis of 504 European vertebrates was done using the Sorensen Similarity Index, and we identified four birds and one mammal that share the distribution with *P. macrobullaris*. We generated 135 environmental niche models including different variable combinations and regularization values for *P. macrobullaris* at two different scales and resolutions. After selecting the best models, we observed that topographic variables outperformed climatic predictors, and the abruptness of the landscape showed better predictive ability than elevation. The best explanatory climatic variable was mean summer temperature, which showed that *P. macrobullaris* is able to cope with mean temperature ranges spanning up to 16°C. The models showed that the distribution of *P. macrobullaris* is mainly shaped by topographic factors that provide rock-abundant and open-space habitats rather than climatic determinants, and that the species is not a cold-adapted, but rather a cold-tolerant eurithermic organism. *P. macrobullaris* shares its distribution pattern as well as several ecological features with five other alpine vertebrates, suggesting that the conclusions obtained from this study might be extensible to them. We concluded that rock-dwelling and open-space foraging vertebrates with broad temperature tolerance are the best candidates to show wide alpine distribution in the Western Palearctic.

**Keywords:** Alpine distribution, Alpine long-eared bat, Biogeography, Chiroptera, Distribution, Modelling, Mountain long-eared bat, Zoogeography

## Introduction

Studying the causes shaping distribution patterns is one of the major tasks in biogeography [1]. Many organisms exhibit similar distributions that despite their current ranges could be the result of different historical, ecological and physiological processes. Identifying the common factors constraining their range is key to understanding these geographic analogies. The joint characteristic attribute of alpine species is the geographic restriction to high mountain environments, which were covered with ice-sheets during long periods in the Pleistocene [2]. The study of the biogeographic patterns of Eurasian alpine species has mainly targeted plants and invertebrates [3-6] that due to either low or no mobility [7,8] are largely conditioned by edaphic and climatic factors [9]. Most vertebrates, however, are unique in their higher mobility, and due to their distinct ecological and evolutionary traits, the biogeographic patterns of vertebrates are likely to differ from those of plants and invertebrates.

* Correspondence: inazio.garin@ehu.es
Department of Zoology and Animal Cell Biology, Faculty of Science and Technology, University of The Basque Country UPV/EHU, Sarriena auzoa s/n, 48940 Leioa, The Basque Country

The distribution patterns of high mountain vertebrates in the Western Palearctic vary considerably. Endemics restricted to one or few mountain systems are common mainly among vertebrates with low mobility [10,11]. Most volant species however show wider distributions. Many of them are ubiquitous and can be commonly found in other habitats in addition to alpine environments [12], and few arctic-alpine species show broad boreal distributions with isolated relict populations in alpine areas at lower latitudes [13]. Finally, there is a group of species sharing a broad geographic distribution restricted to Southern Palearctic mountain systems, hereafter referred to as palealpine species, whose distribution pattern has so far received little attention. The distribution of a single bat species, the alpine long-eared bat (*Plecotus macrobullaris*) is also known to fit this biogeographic pattern [14]. *P. macrobullaris* is a recently described species [15-17] known to forage in alpine meadows [18] and roost mainly in rock crevices and talus slopes [19]. Unlike other bat species in the Western Palearctic [20], this species has breeding populations in the subalpine and alpine belts of mountain systems in the South-western Palearctic [14].

The factors explaining why palealpine vertebrates like *P. macrobullaris* are restricted to mountain environments, yet absent from the rest of the region, have so far not been addressed. Although species inhabiting alpine environments are commonly regarded as cold-adapted species [21], this relationship is not always accurate. Whilst species with arctic-alpine distributions are physiologically restricted to cold conditions and remain active in the alpine belt during the winter [13], this is not necessarily the case for all alpine species [22]. In fact, the current distribution of many palealpine species also includes relatively warm areas in South-eastern Europe and the Middle East [23], suggesting that factors beyond climatic conditions likely dictate the distribution of these species.

Identifying the factors constraining geographic ranges requires using the correct resolution and extent [24] because the distribution of species is shaped by processes acting at different scales [25-27]. Climatic factors mainly drive the continental scale distribution due to the physiological limits of species [28]. Ecological features, instead, usually act more locally and require finer resolution [29,30]. Additionally, the effect of topographic factors such as slope or aspect also depends on scale [31]. Hence, a multiple resolution and extent approach allows the input of variables in the correct scale in which they act, providing more representative outputs of the environmental factors limiting the distribution of species [32].

In this study we test whether climatic variables or habitat and topographic factors are the main drivers of the distribution of palealpine vertebrates, so that the main factors constraining their distribution are identified. First, we identify the vertebrate species that exhibit palealpine distribution analysing range similarities. And secondly, in order to understand why palealpine species are closely linked to mountain environments, we apply our modelling approach at two spatial scales using *P. macrobullaris* as model species. At the broad-scale, across the Western Palearctic, we use coarse-grain presence records of the species for continental-level modelling, while at the fine-scale, around the Pyrenees, we use precise roosting records and finer spatial resolution to understand the factors shaping the local distribution.

## Results

The Sørensen Similarity Index (SSI) value between the distribution area we generated based on updated distribution information of *P. macrobullaris* [14] and the area available at IUCN Red List was 0.66. Thus, that was the error-factor we used for calculating the corrected SSI (cSSI) values of the other species, since we consider it more adjusted to the actual distribution similarity between species (Figure 1). The distribution overlap analysis showed that four birds and a single mammal share the geographic distribution with *P. macrobullaris*, namely the birds *Montifringilla nivalis*, *Pyrrhocorax graculus*, *Tichodroma muraria*, *Prunella collaris* and the vole *Chionomys nivalis* (Figure 1). All these species showed cSSI values above 0.6, while the average cSSI values for birds and mammals were 0.15±0.15 and 0.12±0.14 respectively. The average cSSI value for bats was 0.18±0.12.

### Broad-scale distribution modelling

Analysed evaluators showed that the model with the best predictive ability was the one including one topographic (ABR), five climatic (B4, B10, B12, B15 and B17) and one habitat-related (LAND) variables, generated with a regularization multiplayer of $\beta=1$ (Table 1, Additional file 1: Table S1). The predictive value of abruptness outperformed elevation in all the analysed cases. The Spearman correlation values ($\rho$) between elevation and mean summer temperature were −0.97 in the regional scale and −0.43 in the continental scale.

Areas with suitability values below 0.2886 (max SSS) were considered unsuitable, and 89.4% of the total known distribution records were located in suitable cells. The best model showed suitable areas in the main mountain ranges of the Southern Palearctic, as well as some areas of Northern Europe (Figure 2). All metrics indicated that abruptness was the variable with the best explanatory power (Table 2, Additional file 2: Figure S2). The second most important variable, based on all analysed metrics, was the mean temperature of the warmest quarter (B10), and the remaining climatic and habitat-related variables obtained lower values in all the evaluated metrics. The response curve of the best explanatory variable

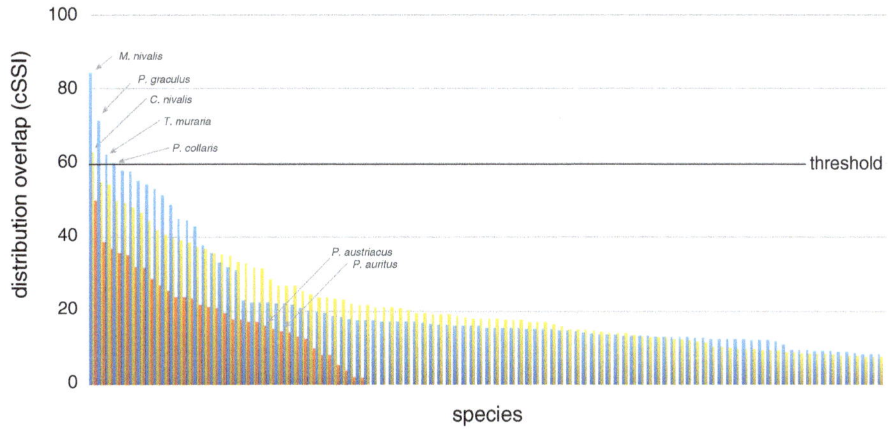


**Figure 1 Geographic overlap between the distribution of *P. macrobullaris* and the distributions of analysed European vertebrates.** Blue bars indicate birds, yellow bars indicate non-chiropteran mammals and red bars indicate bats. The names of the five species with the highest distribution resemblance, and the positions of *P. auritus* and *P. austriacus* are shown. Note that the species with cSSI values equal to 0 (distributions not-overlapped) are not shown and the list of species of birds and non-chiropteran mammals is limited to 100 species.

**Table 1 Composition, evaluation scores (AUCtest, AICc and MPA) and variable contribution ranks of the best models**

| | | | | | | |
|---|---|---|---|---|---|---|
| **Broad-scale modelling** | | | | | | |
| **Type** | **Model** | **β** | **$AUC_{test}$** | **AICc** | **MPA** | **Variable contribution rank** |
| Climatic | B8, B10, B12, B17 | 1 | 0.881 | 2688.1285 | 0.3383 | B10 (45.3) > B12 > B8 > B17 |
| | B4, B10, B12, B15, B17 | 1 | 0.880 | 2619.7250 | 0.2972 | B10 (43.7) > B12 > B17 > B15 > B4 |
| Topographic | ABR, ELEV | 1 | 0.873 | 2434.7099 | 0.365 | ABR (84.4) > ELEV |
| | ABR, ELEV | 2 | 0.866 | 2478.7364 | 0.343 | ABR (86.1) > ELEV |
| Climatic + Topographic | B4, B8, B10, B12, B15, B17, ELEV, ABR | 1 | 0.905 | 2542.8793 | 0.201 | ABR (71.7) > ELEV > B10 > B17 > B12 > B15 > B4 > B8 |
| | B4, B8, B10, B12, B15, B17, ABR | 2 | 0.911 | 2409.3266 | 0.237 | ABR (79.9) > B10 > B12 > B15 > B17 > B4 > B8 |
| Climatic + Habitat | B4, B8, B10, B12, B15, B17, LAND | 1 | 0.899 | 2426.4305 | 0.271 | B10 (32.7) > B12 > LAND > B8 > B17 > B15 > B4 |
| | B4, B10, B12, B15, B17, LAND | 1 | 0.889 | 2472.1983 | 0.237 | B10 (37.0) > B12 > LAND > B17 > B15 > B4 |
| Topographic + Habitat | ELEV, ABR, LAND | 1 | 0.873 | 2333.0673 | 0.321 | ABR (86.7) > ELEV > LAND |
| | ABR, LAND | 2 | 0.876 | 2379.0112 | 0.321 | ABR (96.6) > LAND |
| Climatic + Topographic + Habitat | B4, B8, B10, B12, B15, B17, ABR, LAND | 1 | 0.903 | 2334.3659 | 0.178 | ABR (70.6) > B10 > B17 > LAND > B15 > B12 > B8 > B4 |
| | **B4, B10, B12, B15, B17, ABR, LAND** | **1** | **0.912** | **2330.3513** | **0.208** | **ABR (75.1) > B10 > BIO12 > B17 > B15 > LAND > B4** |
| **Fine-scale modelling** | | | | | | |
| **Type** | **Model** | **β** | **$AUC_{test}$** | **AICc** | **MPA** | **Variable contribution rank** |
| Topographic | ELEV, SLOP, ORI | 1 | 0.9294 | 1087.2041 | 0.1381 | ELEV (56.3) > SLO > ORI |
| | ELEV, SLOP | 1 | 0.9298 | 1108.6065 | 0.1574 | ELEV (54.9) > SLO |
| Habitat | DIS-URBAN, DIS-FOREST, DIS-ROCK | 1 | 0.8968 | 1161.1387 | 0.2449 | dis-rock (74.3) > dis-urban > dis-forest |
| | DIS-URBAN, DIS-ROCK | 1 | 0.8943 | 1321.0146 | 0.2471 | dis-rock (89.3) > dis-urban |
| Topographic + Habitat | ELEV, SLOP, ORI, DIS-URBAN, DIS-FOREST, DIS-ROCK | 1 | 0.9429 | 1289.9253 | 0.0871 | dis-rock (32.0) > SLO > ELEV > dis-urban > dis-forest > ORI |
| | **ELEV, SLOP, DIS-URBAN, DIS-ROCK** | **1** | **0.9560** | **1199.8611** | **0.1026** | **dis-rock (34.4) > SLO > ELEV > dis-urban** |

Only the two models that obtained the best values in each variable-type combination are shown. The rest of the models can be found in Tables S1 and S2 in the Additional file 1. The contribution percentage of the best-ranked variable in each model is indicated in brackets. The selected models are marked in bold.

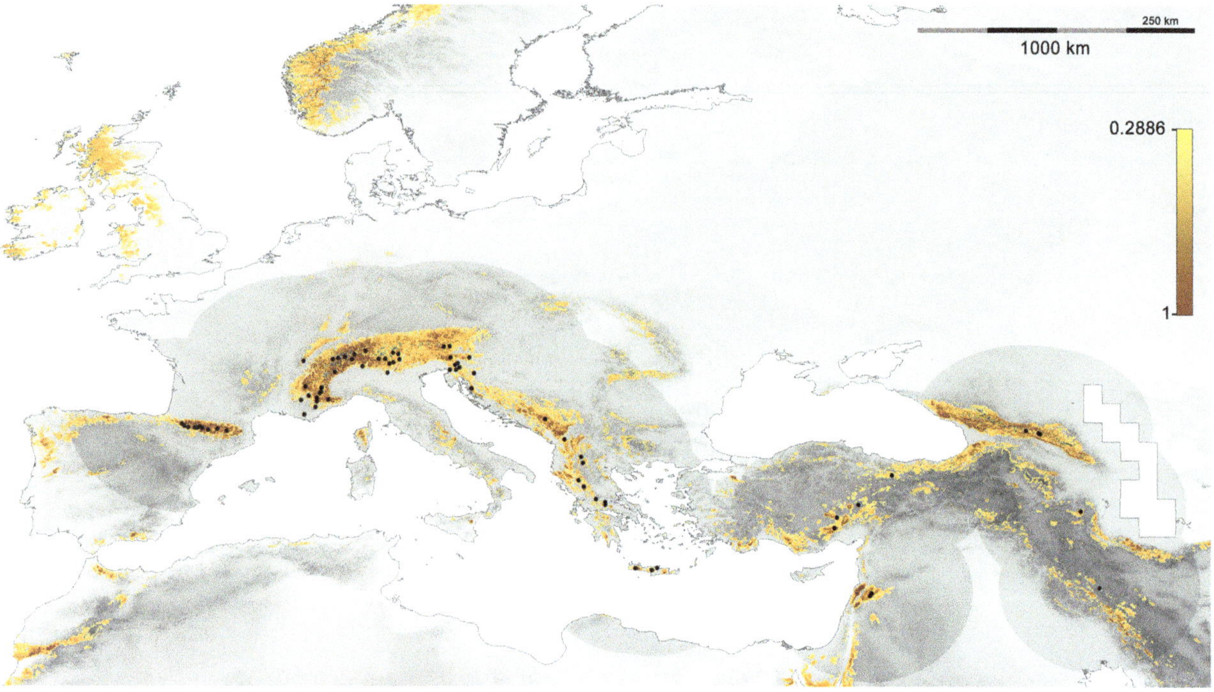


**Figure 2 Distribution model for *P. macrobullaris*, presented on a greyscale elevation map.** Only suitable areas (SV >0.2886) are shown in a colour gradient from light yellow (low suitability) to brown (high suitability). Presence location records used for modelling are represented by black dots and the darkened region is the area used for calibrating the model.

showed that suitability increased with abruptness, starting with zero probability of presence in flat regions (Figure 3A). The most suitable mean temperature of the warmest quarter was 14°C, but suitable values ranged from 6 to 22°C (Figure 3B).

### Fine-scale roosting habitat suitability modelling

The best model was built using two habitat (distance to rocky and urban areas) and two topographic variables (elevation and slope) (Table 1, Additional file 1: Table S2). The ENM showed good predictive ability (AUC =0.956).

**Table 2 Different metrics on the contribution of variables to the best models**

| Variable | Relative contribution | Permutation importance | Jackknife training gain with only variable | Jackknife training gain without this variable |
|---|---|---|---|---|
| Broad-scale modeling | | | | |
| ABR | 75.10 | 45.45 | 1.05 | 1.05 |
| B10 | 9.01 | 28.17 | 0.64 | 1.25 |
| B12 | 3.86 | 7.34 | 0.59 | 1.34 |
| B17 | 4.63 | 6.45 | 0.41 | 1.34 |
| B15 | 2.34 | 2.28 | 0.09 | 1.34 |
| LAND | 3.70 | 1.57 | 0.20 | 1.33 |
| B4 | 1.33 | 8.71 | 0.11 | 1.34 |
| Fine-scale modeling | | | | |
| DIS-ROCK | 34.36 | 42.86 | 0.97 | 1.71 |
| SLOPE | 31.68 | 14.36 | 1.16 | 1.53 |
| ELEV | 25.50 | 28.49 | 1.02 | 1.68 |
| DIS-URBAN | 8.44 | 14.27 | 0.30 | 1.73 |

All values are averages of the 50 replicates of the best models. The relative contribution is obtained from the increase of the regularized gain when each variable is added to the model. The permutation importance is obtained by randomly permuting the values of that variable among the training points and measuring the decrease in training AUC produced by the permutation. The Jackknife training gain with only variable is the training gain that the model achieves when using only that variable, and the Jackknife training gain without this variable is the training gain that the model achieves when using the rest of variables except that one. Consequently, in the first three metrics larger values indicate higher contribution of variables to the model, while in the last one, the lower values indicate greater importance of variables to the model.

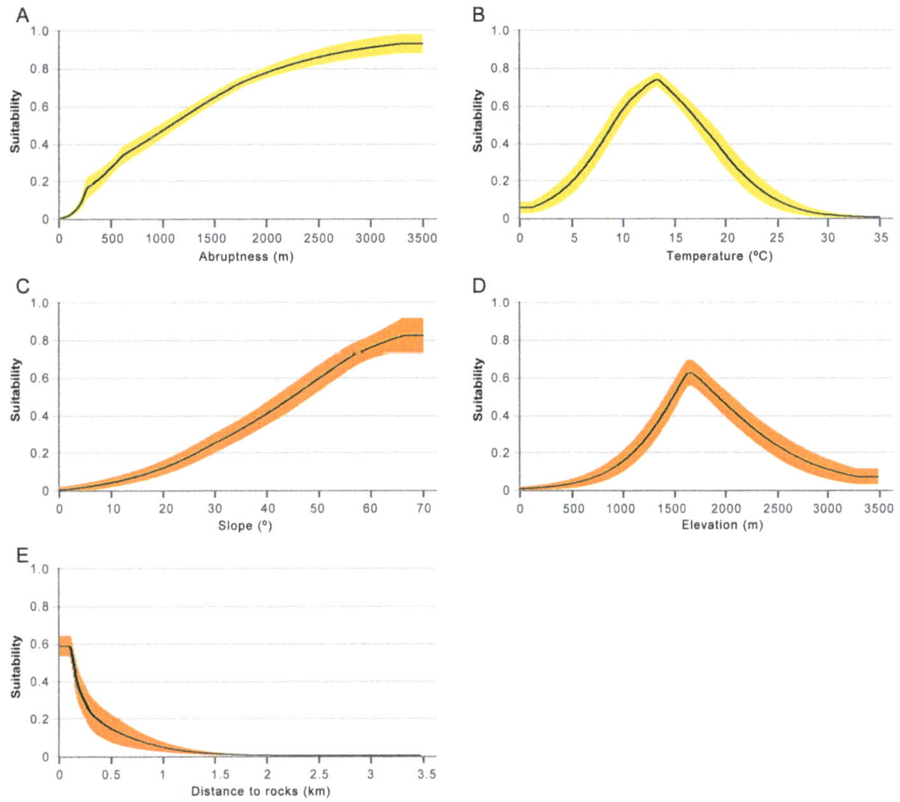


**Figure 3 Response curves of variables with the highest predictive ability. (A)** Abruptness and **(B)** mean temperature of the warmest quarter, as estimated by the broad-scale modelling; **(C)** slope, **(D)** elevation and **(E)** distance to rocks (DIS-ROCK), as estimated by the fine-scale modelling.

The three main factors shaping roosting habitat suitability for *P. macrobullaris* were the distance to rock areas (DIS-ROCK), slope and elevation. DIS-ROCK was the variable with the highest relative contribution to the best model, and Jackknife tests stressed the importance of slope and elevation (Table 2, Additional file 2: Figure S3). The closer the rocks and higher the slope the greater the probability of finding suitable roosting sites (Figure 3C and E). The response curve for elevation showed suitable areas for roosting in an elevation range from 1300 to 2400 m, with maximum values between 1500 and 2000 m (Figure 3D).

## Discussion

The distribution overlap analysis showed that *P. macrobullaris* shares its distribution pattern with five other vertebrates: the birds *Montifringilla nivalis*, *Pyrrhocorax graculus*, *Tichodroma muraria* and *Prunella collaris*, as well as the rodent *Chionomys nivalis*. Despite slight differences can be observed between the ranges of the mentioned species, they are all widely distributed but restricted to mountain environments (Figure 4). Therefore, the conclusions obtained from the ENM of *P. macrobullaris* are probably extensible to all of them. Additionally, as it is discussed below, all species exhibit similar ecological features. A handful of other birds, such as *Anthus spinoletta* (cSSI =0.58), *Pyrrhocorax pyrrhocorax* (0.57) or *Turdus torquatus* (0.53) were also linked to mountain environments, although their geographical distribution extends beyond the main mountain chains. The distribution range of the rest of analysed vertebrates showed lower resemblance with palealpine species (Figure 1).

The generated broad-scale model predicted that suitable areas for *P. macrobullaris* are found in almost all mountain ranges in the Western Palearctic, including some ranges where the species has not being hitherto reported. The species might be present in the Atlas Mts. in Morocco as well as in several mountain chains in the Iberian and Italian Peninsulas, as occurs with some other palealpine species (Figure 4). Two of the palealpine birds inhabit the Atlas while the Cantabrian Mts. and the Apennines host all the mentioned species. On the other hand, the presence of the species in the suitable mountain ranges of Scotland and Scandinavia seems implausible. Both ranges were covered by ice caps during Pleistocene glaciations [33], and the large unsuitable flatland separating them from the southern mountain

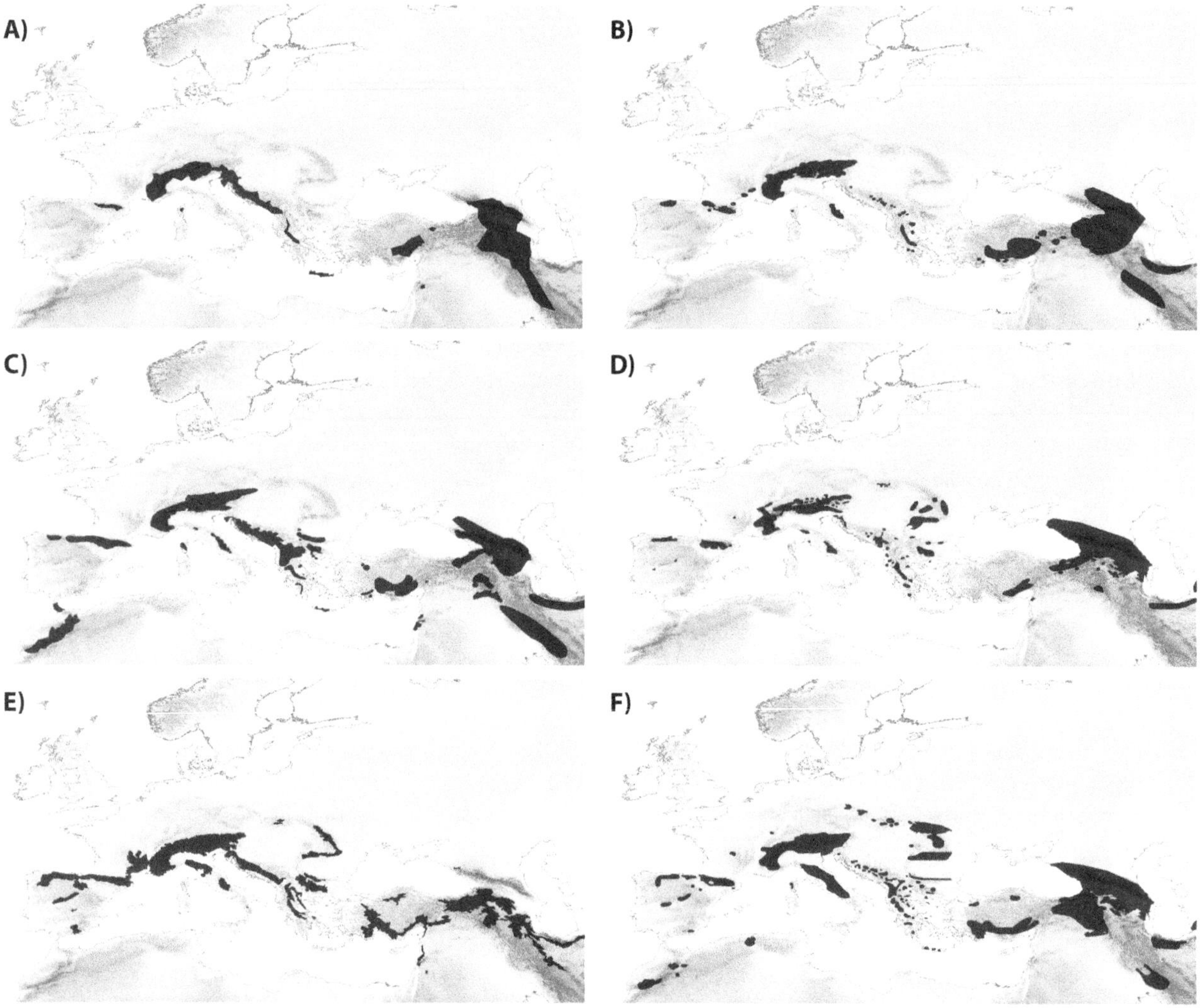


**Figure 4 Geographical distributions of six vertebrate species with palealpine distribution.** **(A)** Alpine long-eared bat *Plecotus macrobullaris*, **(B)** white-winged snowfinch *Montifringilla nivalis*, **(C)** alpine chough *Pyrrhocorax graculus*, **(D)** wallcreeper *Tichodroma muraria* (breeding areas), **(E)** snow vole *Chionomys nivalis* and **(F)** alpine accentor *Prunella collaris* (breeding areas). Map A was generated from data published by Alberdi et al. [14], and maps B-F were obtained from IUCN Red List of Threatened Species [69].

ranges could have prevented their postglacial colonisation by palealpine species. Additionally, harder climatic conditions in winter than in Southern Europe might be another reason that hampers colonisation. This view is supported by the distribution of other palealpine species, which are absent from both Scotland and Scandinavia despite the relatively high mobility of most of them (Figure 4).

Our results differ from the previously published ecological niche model of *P. macrobullaris* in Switzerland [34], which predicted a distribution restricted to lower elevations than what our model did for the Alps. Nevertheless, both research are hardly comparable as they only used localities of colonies detected in buildings at valley bottoms for generating their ENM, which might led to a biased model of the distribution of *P. macrobullaris*, since the species also uses natural roosting resources [19], and it has been also reported above the treeline in the Alps [14].

### The role of topographic variables

Our models identified the abruptness of the landscape as the main constraint acting on the broad-scale distribution of *P. macrobullaris*, over-performing elevation in all cases. Unlike other variables linked to the elevation gradient (abruptness and slope), the effect of elevation itself was not consistent across spatial scales. Even though elevation is broadly used as a predictor in species distribution modelling [35], it does not directly affect the distribution of species on its own, but acts as a surrogate for several other habitat and climatic factors [36]. Thus, elevation may be a good predictor in regional-scale modelling [37], but its predictive ability decreases when moving from the regional to the continental scale, because it loses its ecological relevance (in the case of summer mean temperature, correlation value with elevation drops from ρ = −0.97 in the Pyrenees to ρ = −0.43 across the whole Palearctic). The great variation in the suitable elevation range between massifs (Figure 5) supports this statement.

Abruptness likely acts as a surrogate of several ecological variables. One of them is high rock availability, an important resource for palealpine species. In fact, the fine-scale modelling disclosed the importance of the proximity of rocky environments and the steepness of the terrain, surfacing the significance of rock walls and their debris. The five vertebrates identified as palealpine species depend on rocks for sheltering [38-40]. Cliffs and talus slopes are also a key roosting resource for *P. macrobullaris* [19], even though it also uses buildings for sheltering [34,41,42]. The relative area of rock outcrops is extensive at high mountain environments, favouring roost availability for saxicolous species. This behaviour probably allowed palealpine species to expand their ranges throughout the narrow and steep mountain massifs in the Western Palearctic. For instance, most snowfinches (genera *Montifringilla*, *Pyrgilauda* and *Onychostruthus*) roost on the ground and are behaviourally adapted to living in steppes [38], but only the saxicolous *M. nivalis* has expanded its distribution towards the European mountain massifs.

Another characteristic shared by all palealpine species is their preference for foraging in open areas, mainly in habitats with short grassy vegetation and rocks [18,43-45]. Climatic factors hamper the development of wood patches in alpine areas, providing suitable grounds for open-space foragers. Steep environments at lower elevation can also provide similar open habitats aided by the outcrop of bedrock and the decrease in soil thickness.

Finally, the high elevation range of mountain areas allows species to perform short-distance altitudinal migrations. Boreal species cannot avoid adverse atmospheric conditions by performing short migrations, and instead need to be physiologically adapted to extreme cold temperatures. In contrast, palealpine species tend to commute to lower elevations when conditions become harsher [39,46,47], and therefore physiological adaptations to the extreme cold may not be a compulsory requirement for them. Despite altitudinal migration have been documented in bats [48], such a behaviour has not been studied in *P. macrobullaris*

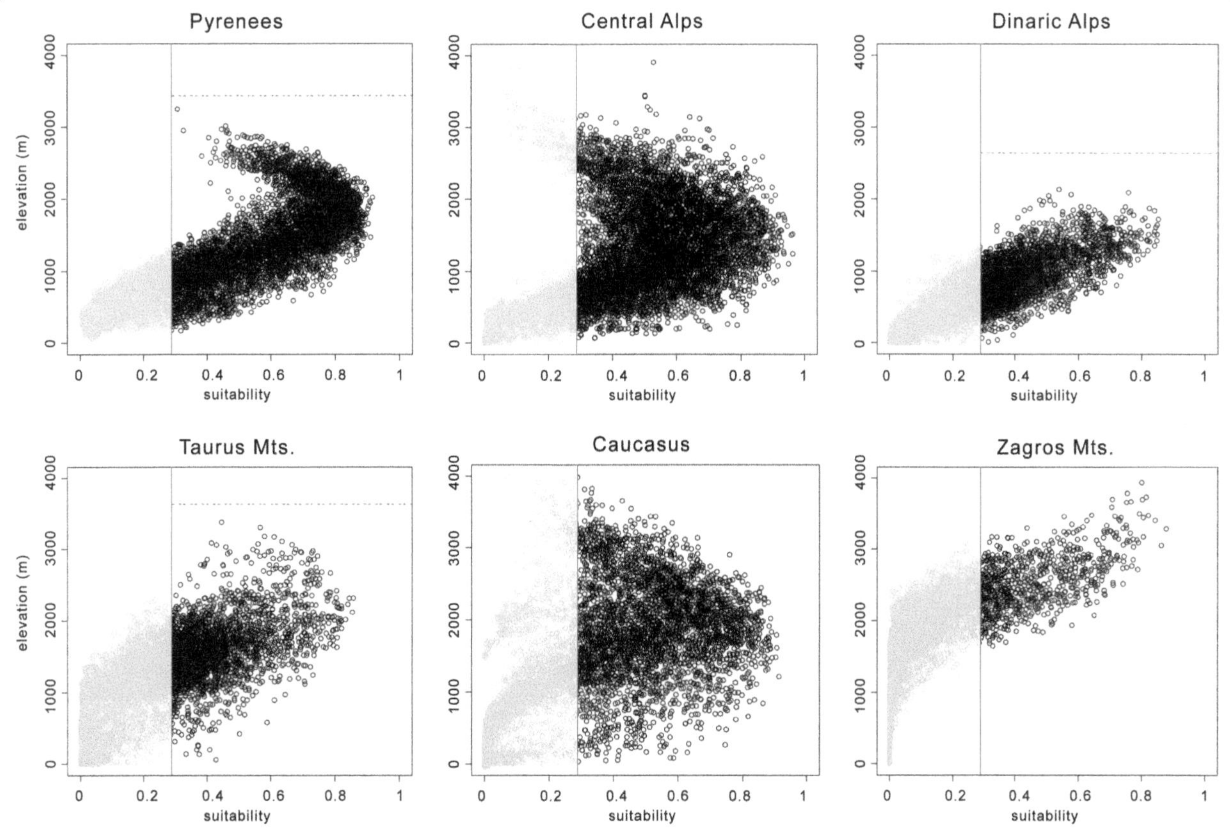


**Figure 5 Relationship between suitability and elevation in six mountain massifs.** Vertical red lines indicate the lower suitability boundary. Horizontal dotted lines indicate the maximum elevation of the mountain range.

so far. Nevertheless, it seems implausible bats could remain the whole year in alpine environments, and ecological features as well as comparative biogeography suggest that *P. macrobullaris* may actually perform seasonal altitudinal migrations [18].

### The role of climatic variables

Despite the lower contribution of climatic variables compared to topographic factors, the former also influenced the broad-scale model. Temperature mainly encompasses the altitudinal domains of the species' distribution. Different climatic patterns along the elevation gradients of each mountain range (with differences of up to 13°C in mean summer temperature at the same elevation; Additional file 2: Figure S4) produced contrasting suitability-elevation relationships between mountain chains across the predicted range of *P. macrobullaris* (Figure 5). For instance, suitable areas in the Zagros Mountains (Iran) were restricted to areas above 1100 m, whereas near-sea-level areas of the Alps and the Dinaric Alps were considered suitable. This fact explains the poor predictive ability of elevation in the continental scale models. The fine-scale model also reflected the limiting effect of temperature in the Pyrenees, since no areas above 2600 m were considered suitable despite containing abundant steep areas.

Broad-scale modelling showed that *P. macrobullaris*, and by extension palealpine species, are able to cope with mean summer temperatures that span from 6 to 22°C (Figure 3B), revealing their eurithermic nature. Even though palealpine vertebrates were previously considered as cold-adapted species [21], their distribution is not restricted to cold environments, suggesting they are cold-tolerant rather than cold-adapted. Climatic models have predicted that alpine species like *P. graculus* and *M. nivalis* might extend their distribution to high latitude areas due to climate change [21]. Even though those areas might become climatically suitable for alpine species, our results suggest that such a scenario is highly improvable. We show that topography and its correlating ecological factors are shaping mostly the distribution of such species, and thus, the large unsuitable flatland between boreal areas and Southern European mountain ranges would hamper putative colonisation events. The wide thermal tolerance likely facilitated the colonisation of most of the mountain massifs in the Central and Southern Palearctic. Conversely, the southern isolated populations of an arctic-alpine vertebrate such as the rock ptarmigan (*Lagopus muta*) are limited to mountain ranges that hold extensive areas with summer mean temperatures below 10°C, i.e. the Pyrenees and the Alps in Europe [49].

### The human factor

In addition to topographic and climatic factors, human activities probably affected the distribution of palealpine vertebrates. Similar to other palealpine species [43,50,51], *P. macrobullaris* is also an open-space forager [18]. The natural treeline in the Pyrenees is expected to be around 2000–2200 m; however, historical landscape management for pasture has lowered the upper forest limit to around 1500 m [52]. It is noteworthy that the most suitable elevation range predicted by both models for the Pyrenees coincides with this modified belt (1500 – 2000 m), which suggests that it may offer the best ecological conditions for *P. macrobullaris* and other palealpine vertebrates in terms of greater food abundance and diversity [53] and milder climatic conditions than at higher elevations [54]. Hence, future changes in landscape management due to pastoral abandonment will likely affect the foraging ecology of *P. macrobullaris* [18] and other alpine vertebrates [44].

### Biotic interactions

The importance of biotic interactions in the distribution of species is a largely recognised phenomenon [55]. Processes like competitive exclusion [56] or predator–prey relationships [57] can affect the distribution of species, in addition to climatic and habitat-related factors. Accordingly, Rutishauser et al. [34] reported that the mostly allopatric distribution of *P. macrobullaris* and *P. austriacus* in Switzerland might be the result of competitive exclusion. In fact, both species share ecological traits [18,58,59]. However, under this scenario sympatric populations would be the exception rather than the rule and the niche of *P. macrobullaris* would expand where it is released from interspecific competition. Neither one nor the other seems to occur. There are broad areas where both species are found in sympatry, such as the Pyrenees (own data), the French Prealps [60] and the Dinaric Alps [61], and the ecological niche of *P. macrobullaris* as well as its link to mountain environments is kept beyond the range of *P. austriacus*, e.g. the eastern part of the distribution of *P. macrobullaris*. Competitive exclusion might occur in some specific areas like Switzerland, or across the altitudinal gradient in precise contact zones [62]. We do not discard that competitive exclusion shows up where resources are limiting, but in our opinion there are no signs indicating a general effect on the ecological niche and continental-level distribution of *P. macrobullaris*.

### The environmental niche of palealpine species

Our results show that climatic factors are not the main drivers of the distribution of palealpine vertebrates. Generally, the broad-scale geographic distribution of species tend to be limited by climatic variables [28], while non-climatic factors usually shape their distribution at the regional or local level [63]. However, our models show that in the case of palealpine species topographic factors act at a broader scale. Unlike arctic-alpine organisms, palealpine species are able to thrive in relatively warm as well as cold temperatures, and this eurithermicity is responsible for the limited ability of climatic variables to predict their distribution at the broad-scale.

Although palealpine species share the same biogeographic pattern driven by topographic factors, they are expected to show differential responses to climatic conditions. Variations in behaviour as well as geographical and altitudinal distribution suggest that some species cope better with cold than others, e.g. the white-winged snowfinch compared to the wallcreeper [23]. Differences in thermal tolerance between groups of organisms are also expected to occur due to differences in their ecology, reproduction and physiology. Major differences between birds and bats probably stem from their distinct circadian rhythm, thermoregulation abilities and reproductive systems. Birds are diurnal, homeothermic and oviparous animals [64], while bats are highly heterothermic nocturnal organisms with viviparous reproduction [65]. Even though facultative heterothermy may offer an advantage over homeothermic organisms when exploiting alpine environments, this advantage is only applicable to non-reproductive bats. Breeding females cannot decrease their body temperature as much as non-breeding bats due to the metabolic demands of foetal development and lactation [66], so they may be hindered from exploiting high mountain habitats. Consequently, elevational segregation might occur in the coldest regions of their distribution, which would explain the high number of *P. macrobullaris* maternity colonies found below the treeline in the Alps [41,42]. Nevertheless, such a pattern was not so noticeable in the slightly milder range of the Pyrenees [19]. Finally, models generated in this study reflect the environmental niche during the favourable season. During winter, palealpine vertebrates behave differently depending on species and taxa: bats hibernate while birds migrate to lower elevations.

## Conclusions

The use of multi-scale modelling allowed us to identify features shaping the biogeographic pattern shared by alpine vertebrates with wide geographic distribution, which could pave the way for future studies and help fine-tune conservation measures for this group of animals. The wide temperature tolerance identified in this study suggests that rather than physiological limitations relating to warmer conditions, ecological factors, such as changes in treeline elevation and the incursion of lowland species, may be the main challenges palealpine species will have to cope with under future conditions resulting from climate change [67]. Palealpine species are found in high-mountain environments not because they are constrained by cold temperatures, but because these cold-tolerant species find in alpine environments their preferred foraging and sheltering habitats, characterized by abundant open-space and rocky areas.

## Methods

### Distribution similarity

We calculated the distribution area overlap of *P. macrobullaris* and 503 European terrestrial vertebrates using the Sørensen Similarity Index (SSI - [68]) in order to identify the species sharing the distribution pattern with *P. macrobullaris*. We generated a reference GIS layer based on updated information about the distribution of *P. macrobullaris* [14] and compared it with breading season distribution layers of 203 mammals available at IUCN Red List [69] and 298 birds obtained from BirdLife International [70]. We also compared the distribution area of *P. macrobullaris* we generated with the area available at IUCN Red List to assess the uncertainty of the data and create a corrected SSI (cSSI) that reflects better the distribution similarities. Visual inspection of the geographical distributions of the species with the highest cSSI values showed that cSSI = 0.60 marks a resemblance threshold between widely distributed species limited to mountain ranges and species with a range that extends beyond the main mountain chains.

### Environmental Niche Modelling (ENM)

The extent of the broad-scale models was set as the Western Palearctic region (20–70° N, –20–60° E), while fine-scale models were limited to the Pyrenees and the surrounding area (40–42.1° N, –1.1–2.2 °E). All predictive models were generated with the presence-only species distribution modelling software Maxent (Maximum Entropy Algorithm) version 3.3.3 [71,72]. When running the software, we set maximum iterations to 5000, used 10000 background points, and ran 50 replicates of each model, following a resampling method randomly selecting a subsample of 25% of location records for model validation in each round. The output of all models was set in *raw* format in order to analyse them using the software ENMTools. Correlation between variables was analysed using ENMTools v1.3 [73], and the variable with *a priori* the highest ecological meaning was selected when the Spearman's correlation coefficient (ρ) between two variables was greater than 0.65. The remaining spatial and statistical analyses were performed using ArcView GIS 3.2 and R 2.9.2 software (packages *raster* and *dismo*) available from CRAN [74].

### Broad-scale modelling

We used presence location records from all of the known distribution area of *P. macrobullaris* [14]. In order to minimise errors and ensure the best model performance, we took several measures following Merow et al. [75]. We applied 5 filters to all the available records with the following criteria: (1) *Molecular verification*: species identification was checked by molecular means to avoid misidentifications with morphologically similar congeners; (2) *Resolution*: maximum location uncertainty of 1 km; (3) *Pseudoreplication*: a minimum distance of 10 km between presence records was set based on the home-range information of the species [60,76]; (4) *Record source-type homogeneity*: similar amount of different record types (e.g. roost observations, net captures); and (5) *Geographical homogeneity*: similar amount of records in each geographic area. Sampling effort East of the Bosphorus has been considerably lower than in Europe, and the number of European records were four-fold the number of eastern locations after applying the filtering. In order to correct the sampling bias [77], a bias grid was introduced to the algorithm where the relative sampling probability West of the Bosphorus was considered four times higher than the eastern area. This helped outweigh the relative value of Anatolia, the Near East and the Caucasus compared to Europe. Spatial autocorrelation analyses were discarded because the disjunct distribution of alpine species is actually grouped in clusters limited to high mountain massifs.

We modelled the distribution of *P. macrobullaris* at a resolution of 30-arc-second (approximately 1 km) using 74 presence records. In order to avoid large unsuitable areas, such as the Siberian steppe or Sahara desert, that could yield artificially inflated model evaluations [78,79], we set a 500 km buffer around presence locations in the calibration area (Figure 2); i.e. the area from which background points used for generating the model were randomly selected. After filtering highly correlated predictors, we used three types of variables (6 climatic, 2 topographic and 1 habitat-related – Table 3) for generating 90 different models (each one including 50 replicates) using different variable combinations and regularization parameters. Climatic variables and elevation data were obtained from WorldClim database [80]. Land Cover data was obtained from GLCNMO [81]. Abruptness of landscape (ABR) was calculated from the elevation grid computing the maximum elevation difference within a 5 km buffer from each cell. Land cover categories were reclassified to 6 categories: forests, open natural areas, agricultural areas, urban areas, ice/snow and water.

### Fine-scale modelling

We used roost location data for modelling the fine-scale environmental niche of *P. macrobullaris* (Additional file 2: Figure S1). Roosts were identified using the *homing-in* radio-tracking method [19]. Only roosts located in natural structures were used, as they represent the original as well as the predominant roost type of *P. macrobullaris* [19]. In order to minimise pseudoreplication, one roost for each grid cell was used for the analysis, resulting in a total of 43 records. We used 6 variables (3 topographic and 3 habitat-related – Table 3) to generate 45 different models. Climatic variables were not used in the fine-scale modelling for being highly correlated with elevation. All variables were edited and homogenised to 100 m resolution. Elevation data was obtained from CGIAR-CSI [82], and the rest of topographic variables were derived from it. Habitat variables were obtained from Corine Land Cover 2006 [83] and categories were reclassified in ArcView GIS 3.2.

**Table 3 Characteristics of the variables used in the two modelling scales**

| **Broad-scale modelling** | | | | |
|---|---|---|---|---|
| **Predictor** | **Predictor description** | **Class** | **Resolution** | **Source** |
| **ABR** | Abruptness | Topographic | 30 arc-sec | Generated from elevation data |
| **ELEV** | Elevation | Topographic | 30 arc-sec | Wordclim SRTM |
| **B4** | Temperature seasonality | Climatic | 30 arc-sec | Worldclim Bioclim 4 |
| **B8** | Mean temperature of the wettest quarter | Climatic | 30 arc-sec | Worldclim Bioclim 8 |
| **B10** | Mean temperature of the warmest quarter | Climatic | 30 arc-sec | Worldclim Bioclim 10 |
| **B12** | Annual precipitation | Climatic | 30 arc-sec | Worldclim Bioclim 12 |
| **B15** | Precipitation seasonality | Climatic | 30 arc-sec | Worldclim Bioclim 15 |
| **B17** | Precipitation of the Driest Quarter | Climatic | 30 arc-sec | Worldclim Bioclim 17 |
| **LAND** | Land cover | Habitat | 30 arc-sec | GLCNMO |
| **Fine-scale modelling** | | | | |
| **Predictor** | **Description** | **Class** | **Resolution** | **Source** |
| **ELEV** | Elevation | Topographic | 100 m | SRTM 90 m DEM (CGIAR-CSI) |
| **SLO** | Slope | Topographic | 100 m | Generated from elevation data |
| **ORI** | Orientation | Topographic | 100 m | Generated from elevation data |
| **DIS-ROCK** | Distance to rock areas | Habitat | 100 m | Obtained from Corine LandCover 2006 |
| **DIS-FOREST** | Distance to forest areas | Habitat | 100 m | Obtained from Corine LandCover 2006 |
| **DIS-URBAN** | Distance to urban areas | Habitat | 100 m | Obtained from Corine LandCover 2006 |

Abruptness was used in the broad-scale modelling for reflecting the elevation range available for the bats in each geographic location, and it was based on the home-range information of *P. macrobullaris* [60,76]. Conversely, we used slope in the fine-scale modelling because we considered it might play a key role providing suitable rock roosts to *P. macrobullaris* [19].

### Model evaluation and selection

Models were evaluated using both threshold-independent and dependent means that deal with several aspects of model performance, including model accuracy, model complexity and prediction success [84-86]. Accuracy was evaluated using the Area Under the Curve (AUC) of the Receiver Operator Characteristics (ROC) [32,87-90]. Corrected Akaike's Information Criteria (AICc) [91] was used for evaluating model complexity in ENMTools. For that task a single model out of the 50 replicates was selected using a Principal Components Analysis (PCA), in order to identify the replicate with the closest values to the average. The PCA was performed using five matrices: regularised training gain, test gain, test AUC, 10 percentile training presence logistic threshold and 10 percentile training presence area. We also used the threshold-dependent minimal predicted area (MPA) evaluator [92]. Finally, we validated the best broad-scale model using all published distribution records (349) of *P. macrobullaris* [14].

### Model output and variable importance

The best models were repeated and plotted in a logistic format to provide the estimates of the probability of occurrence as predicted by the variables in a map [72,93]. We used the threshold selection method max SSS to discern suitable and unsuitable areas [94]. The relative importance of each variable was checked using heuristic (percent contribution), permutation (permutation importance) and jackknife approaches (Table 2) [95,96]. Response curves for the best explanatory variables were plotted in order to determine the response of suitability values to changes in specific predictors, and thus identify the most suitable conditions for the species (Figure 3).

## Additional files

**Additional file 1:** Two tables showing the composition, evaluation and variable contribution ranks of the 90 broad-scale models **(Table S1)** and 45 fine-scale models **(Table S2)**.

**Additional file 2: Figure S1.** map showing the roosting records used for the fine scale-modelling, **Figure S2.** results of the jackknife tests of variable importance of the best broad-scale model, **Figure S3.** results of the jackknife tests of variable importance of the best fine-scale model, and **Figure S4.** mean summer temperatures at different elevations and mountain ranges.

**Competing interests**

The authors declare that they have no competing interests.

**Authors' contributions**

AA, JA and IG designed the study. All authors participated in the fieldwork. AA performed the analyses. All authors drafted the manuscript. All authors have read and approved the final manuscript.

**Acknowledgements**

We wish to thank Orly Razgour for her insightful comments on the manuscript and for proofreading the text. The Basque Government (research projects IT385-07, IT301-10 and IT754-13, and predoctoral grants BFI-2010-190 and BFI-2009-252) and the University of the Basque Country UPV/EHU supported this work.

**References**

1. Lomolino MV, Riddle BR, Whittaker RJ, Brown JH: *Biogeography.* Sunderland: Sinauer Associates; 2010.
2. Hughes PD, Woodward JC, van Calsteren PC, Thomas LE, Adamson KR: **Pleistocene ice caps on the coastal mountains of the Adriatic Sea.** *Quat Sci Rev* 2010, **29:**3690–3708.
3. Schmitt T, Muster C, Schönswetter P: **Are Disjunct Alpine and Arctic-Alpine Animal and Plant Species in the Western Palearctic Really "Relics of a Cold Past?".** In *Relict Species Phylogeography and Conservation Biology.* Edited by Habel JC, Assmann T. Berlin: Springer; 2009:239–252.
4. Ehrich D, Gaudeul M, Assefa A, Koch MA, Mummenhoff K, Nemomissa S, Consortium I, Brochmann C: **Genetic consequences of Pleistocene range shifts: contrast between the Arctic, the Alps and the East African mountains.** *Mol Ecol* 2007, **16:**2542–2559.
5. Varga ZS, Schmitt T: **Types of oreal and oreotundral disjunctions in the western Palearctic.** *Biol J Linn Soc* 2008, **93:**415–430.
6. Ozenda P: *Die Vegetation Der Alpen Im Europäischen Gebirgsraum.* Stuttgart: Fischer; 1988.
7. Roland J, Keyghobadi N, Fownes S: **Alpine parnassius butterfly dispersal: effects of landscape and population size.** *Ecology* 2000, **81:**1642–1653.
8. Finn DS, Theobald DM, Black WC IV, Poff NL: **Spatial population genetic structure and limited dispersal in a Rocky Mountain alpine stream insect.** *Mol Ecol* 2006, **15:**3553–3566.
9. Dubuis A, Giovanettina S, Pellissier L, Pottier J, Vittoz P, Guisan A: **Improving the prediction of plant species distribution and community composition by adding edaphic to topo-climatic variables.** *J Veg Sci* 2012, **24:**593–606.
10. Carranza S, Amat F: **Taxonomy, biogeography and evolution of *Euproctus* (Amphibia: Salamandridae), with the resurrection of the genus *Calotriton* and the description of a new endemic species from the Iberian Peninsula.** *Zool J Linnean Soc* 2005, **145:**555–582.
11. Kryštufek B, BUŽAN EV: **Rarity and decline in palaeoendemic Martino's vole *Dinaromys bogdanovi*.** *Mammal Rev* 2008, **38:**267–284.
12. Meffert PJ, Marzluff JM, Dziock F: **Unintentional habitats: value of a city for the wheatear (*Oenanthe oenanthe*).** *Landscape Urban Plan* 2012, **108:**49–56.
13. Favaron M, Scherini GC, Preatoni D, Tosi G, Wauters LA: **Spacing behaviour and habitat use of rock ptarmigan (*Lagopus mutus*) at low density in the Italian Alps.** *J Ornithol* 2006, **147:**618–628.
14. Alberdi A, Garin I, Aizpurua O, Aihartza J: **Review on the geographic and elevational distribution of the mountain long-eared bat *Plecotus macrobullaris*, completed by utilising a specific mist-netting technique.** *Acta Chiropterol* 2013, **15:**451–461.
15. Kiefer A, Veith M: **A new species of long-eared bat from Europe (Chiroptera: Vespertilionidae).** *Myotis* 2002, **39:**5–16.
16. Spitzenberger F, Haring E, Tvrtkovic N: **Plecotus microdontus (Mammalia, Vespertilionidae), a new bat species from Austria.** *Natura Croatica* 2002, **11:**1–18.
17. Spitzenberger F, Strelkov P, Haring E: **Morphology and mitochondrial DNA sequences show that *Plecotus alpinus* Kiefer & Veith, 2002 and *Plecotus microdontus* Spitzenberger, 2002 are synonyms of *Plecotus macrobullaris* Kuzjakin, 1965.** *Natura Croatica* 2003, **12:**39–53.
18. Alberdi A, Garin I, Aizpurua O, Aihartza J: **The foraging ecology of the mountain long-eared Bat *Plecotus macrobullaris* revealed with DNA mini-barcodes.** *PLoS One* 2012, **7:**e35692.

19. Alberdi A, Aihartza J, Aizpurua O, Salsamendi E, Brigham M, Garin I: **Living above the treeline: roosting ecology of the alpine bat *Plecotus macrobullaris*.** *Eur J Wildl Res* 2014, doi:10.1007/s10344-014-0862-8.
20. Dietz C, Dietmar N, von Helversen O: *Bats of Britain, Europe and Northwest Africa.* London: A&C Black; 2009.
21. Smith SE, Gregory RD, Anderson BJ, Thomas CD: **The past, present and potential future distributions of cold-adapted bird species.** *Divers Distrib* 2013, **19:**352–362.
22. Bienkowski P, Marszalek U: **Metabolism and energy budget in the snow vole.** *Acta Theriol* 1974, **19:**55–67.
23. Arlott N: *Collins Field Guide: Birds of the Palearctic: Passerines.* London: Collins; 2007.
24. Boyce M: **Scale for resource selection functions.** *Divers Distrib* 2006, **12:**269–276.
25. Levin SA: **The problem of pattern and scale in ecology.** *Ecology* 1992, **73:**1943–1967.
26. Hortal J, Roura-Pascual N, Sanders NJ, Rahbek C: **Understanding (insect) species distributions across spatial scales.** *Ecography* 2010, **33:**51–53.
27. Lomba A, Pellissier L, Randin C, Vicente J, Moreira F, Honrado J, Guisan A: **Overcoming the rare species modelling paradox: a novel hierarchical framework applied to an Iberian endemic plant.** *Biol Cons* 2010, **143:**2647–2657.
28. Gaston KJ: *The Structure and Dynamics of Geographic Ranges.* Oxford: Oxford University Press on Demand; 2003.
29. Pearson RG, Dawson TP, Liu C: **Modelling species distributions in Britain: a hierarchical integration of climate and land-cover data.** *Ecography* 2004, **27:**285–298.
30. Graf RF, Bollmann K, Suter W, Bugmann H: **The importance of spatial scale in habitat models: capercaillie in the Swiss Alps.** *Landscape Ecol* 2005, **20:**703–717.
31. Piedallu C, Gégout J-C: **Efficient assessment of topographic solar radiation to improve plant distribution models.** *Agric For Meteorol* 2008, **148:**1696–1706.
32. Razgour O, Hanmer J, Jones G: **Using multi-scale modelling to predict habitat suitability for species of conservation concern: The grey long-eared bat as a case study.** *Biol Cons* 2011, **144:**2922–2930.
33. Mangerud J, Jakobsson M, Alexanderson H, Astakhov V, Clarke GKC, Henriksen M, Hjort C, Krinner G, Lunkka J-P, Möller P, Murray A, Nikoloskaya O, Saarnisto M, Svendsen JI: **Ice-dammed lakes and rerouting of the drainage of northern Eurasia during the last glaciation.** *Quat Sci Rev* 2004, **23:**1313–1332.
34. Rutishauser MD, Bontadina F, Braunisch V, Ashrafi S, Arlettaz R: **The challenge posed by newly discovered cryptic species: disentangling the environmental niches of long-eared bats.** *Divers Distrib* 2012, **18:**1107–1119.
35. Hof AR, Jansson R, Nilsson C: **The usefulness of elevation as a predictor variable in species distribution modelling.** *Ecol Model* 2012, **246:**86–90.
36. Hörsch B: **Modelling the spatial distribution of montane and subalpine forests in the central Alps using digital elevation models.** *Ecol Model* 2003, **168:**267–282.
37. Newbold T, Gilbert F, Zalat S, El-Gabbas A, Reader T: **Climate-based models of spatial patterns of species richness in Egypt's butterfly and mammal fauna.** *J Biogeogr* 2009, **36:**2085–2095.
38. Yan-Hua Q, Zuo-Hua Y, De Ritis SF: **Distribution patterns of snow finches (genus *Montifringilla*) in the Tibetan Plateau of China.** *Avocetta* 2002, **26:**11–18.
39. Saniga M: **Seasonal distribution, altitudinal and horizontal migration of the wallcreeper (*Tichodroma muraria*) in the Malá Fatra Mountais, Slovak Carpathians.** *Folia Zool* 1995, **44:**237–246.
40. Luque-Larena JJ, López P, Gosálbez J: **Microhabitat use by the snow vole *Chionomys nivalis* in alpine environments reflects rock-dwelling preferences.** *Can J Zool* 2002, **80:**36–41.
41. Mattei-Roesli M: **Situazione del genere *Plecotus* (Chiroptera) nel Cantone Ticino (Svizzera).** *Boll Soc Ticin Sci Nat* 2010, **98:**31–34.
42. Presetnik P, Koselj K, Zagmajster M: *Atlas Netopirjev (Chiroptera) Slovenije. Atlas of Bats (Chiroptera) of Slovenia.* Ljubiana: Miklavž na Dravskem polju: Center za kartografijo favne in flore; 2009.
43. Rolando A, Laiolo P, Formica M: **A comparative analysis of the foraging behaviour of the Chough *Pyrrhocorax pyrrhocorax* and the Alpine Chough *Pyrrhocorax graculus* coexisting in the Alps.** *Ibis* 1997, **139:**461–467.
44. Laiolo P, Dondero F, Ciliento E, Rolando A: **Consequences of pastoral abandonment for the structure and diversity of the alpine avifauna.** *J Appl Ecol* 2004, **41:**294–304.
45. Pérez-Aranda D: **Biología, ecología, genética y conservación del topillo nival ("*Chionomys nivalis*") en Peñalara y Sierra Nevada.** In *PhD Thesis.* Madrid: Complutense University, Department of Zoology and Physical Anthropology; 2009.
46. Henry PY: **Differential migration in the polygynandrous Alpine Accentor *Prunella collaris*.** *Bird Study* 2011, **58:**160–170.
47. Laiolo P, Rolando A, Carisio L: **Winter movements of the alpine chough: implications for management in the Alps.** *J Mount Ecol* 2001, **6:**21–30.
48. McGuire LP, Boyle WA: **Altitudinal migration in bats: evidence, patterns, and drivers.** *Biol Rev* 2013, **88:**767–786.
49. Bech N, Boissier J, Drovetski S, Novoa C: **Population genetic structure of rock ptarmigan in the "sky islands" of French Pyrenees: implications for conservation.** *Anim Conserv* 2009, **12:**138–146.
50. Janeau G, Aulagnier S: **Snow vole - *Chionomys nivalis*.** *Ibex JME* 1997, **4:**1–11.
51. Janiga M, Novotná M: **Feeding preferences and foraging behaviour in the Alpine Accentor *Prunella collaris*.** *Ornis Fennica* 2006, **83:**170–180.
52. Ninot JM, Carrillo E, Font X, Carreras J, Ferré A, Masalles RM, Soriano I, Vigo J: **Altitude zonation in the Pyrenees. A geobotanic interpretation.** *Phytocoenologia* 2007, **37:**371–398.
53. Beck J, Altermatt F, Hagmann R, Lang S: **Seasonality in the altitude–diversity pattern of Alpine moths.** *Basic Appl Ecol* 2010, **11:**714–722.
54. Barry RG: *Mountain Weather and Climate.* 3rd edition. New York: Cambridge University Press; 2008.
55. Araújo MB, Luoto M: **The importance of biotic interactions for modelling species distributions under climate change.** *Global Ecol Biogeogr* 2007, **16:**743–753.
56. Wisz MS, Pottier J, Kissling WD, Pellissier L, Lenoir J, Damgaard CF, Dormann CF, Forchhammer MC, Grytnes J-A, Guisan A, Heikkinen RK, Høye TT, Kühn I, Luoto M, Maiorano L, Nilsson M-C, Normand S, Öckinger E, Schmidt NM, Termansen M, Timmermann A, Wardle DA, Aastrup P, Svenning J-C: **The role of biotic interactions in shaping distributions and realised assemblages of species: implications for species distribution modelling.** *Biol Rev* 2012, **88:**15–30.
57. de Araújo CB, Marcondes-Machado LO, Costa GC: **The importance of biotic interactions in species distribution models: a test of the Eltonian noise hypothesis using parrots.** *J Biogeogr* 2013, **41:**513–523.
58. Dietrich S, Szameitat DP, Kiefer A, Schnitzler HU, Denzinger A: **Echolocation signals of the plecotine bat, *Plecotus macrobullaris* Kuzyakin, 1965.** *Acta Chiropterol* 2006, **8:**465–475.
59. Razgour O, Clare EL, Zeale M, Hanmer J, Schnell I, Rasmussen M, GIlbert T, Jones H: **High-throughput sequencing offers insight into mechanisms of resource partitioning in cryptic bat species.** *Ecol Evol* 2011, **1**(4):556–570.
60. Arthur L, Lemaire M: *Les Chauves-Souris De France, Belgique, Luxembourg Et Suisse.* Meze: Biotope Editions; 2009.
61. Tvrtkovic N, Pavlinic I, Haring E: **Four species of long-eared bats (*Plecotus*, Geoffroy, 1818; Mammalia, Vespertilionidae) in Croatia: field identification and distribution.** *Folia Zool* 2005, **54:**75–88.
62. Pavlinic I, Tvrtkovic N: **Altitudinal distribution of four *Plecotus* species (Mammalia, Vespertilionidae) occurring in Croatia.** *Croatian Natural History Museum* 2004, **13:**395–401.
63. Gaston KJ: *Rarity (Population and Community Biology Series).* 1st edition. Falmouth, UK: Springer; 1994.
64. Bicudo JE, Buttemer WA, Chappell MA, Pearson JT, Bech C: *Ecological and Environmental Physiology of Birds.* New York: Oxford University Press; 2010.
65. Kunz TH, Fenton MB: *Bat Ecology.* Chicago: University of Chicago Press; 2006.
66. Dzal YA, Brigham RM: **The tradeoff between torpor use and reproduction in little brown bats (*Myotis lucifugus*).** *J Comp Physiol B* 2012, **183:**279–288.
67. Reif J, Flousek J: **The role of species ecological traits in climatically driven altitudinal range shifts of central European birds.** *Oikos* 2012, **121:**1053–1060.
68. Sørensen T: **A method of establishing groups of equal amplitude in plant sociology based on similarity of species and its application to analyses of the vegetation on Danish commons.** *Biol Skr* 1948, **5:**1–34.
69. **IUCN Red List of Threatened Species.** [http://www.iucnredlist.org]
70. **BirdLife International.** [http://www.birdlife.org]
71. Phillips SJ, Anderson RP, Schapire RE: **Maximum entropy modeling of species geographic distributions.** *Ecol Model* 2006, **190:**231–259.

72. Phillips S, Dudík M: **Modeling of species distributions with Maxent: new extensions and a comprehensive evaluation.** *Ecography* 2008, **31**:161–175.
73. Warren D, Glor R, Turelli M: **ENMTools: a toolbox for comparative studies of environmental niche models.** *Ecography* 2010, **33**:607–611.
74. **The Comprehensive R Archive Network.** [http://cran.r-project.org/]
75. Merow C, Smith MJ, Silander JA Jr: **A practical guide to MaxEnt for modeling species' distributions: what it does, and why inputs and settings matter.** *Ecography* 2013, **36**:1058–1069.
76. Preatoni D, Spada M, Wauters L, Tosi G, Martinoli A: **Habitat use in the female Alpine long-eared bat (*Plecotus macrobullaris*): does breeding make the difference?** *Acta Chiropterol* 2011, **13**:355–364.
77. Kramer-Schadt S, Niedballa J, Pilgrim JD, Schröder B, Lindenborn J, Reinfelder V, Stillfried M, Heckmann I, Scharf AK, Augeri DM, Cheyne SM, Hearn AJ, Ross J, Macdonald DW, Mathai J, Eaton J, Marshall AJ, Semiadi G, Rustam R, Bernard H, Alfred R, Samejima H, Duckworth JW, Breitenmoser-Wuersten C, Belant JL, Hofer H, Wilting A: **The importance of correcting for sampling bias in MaxEnt species distribution models.** *Divers Distrib* 2013, **19**:1366–1379.
78. Lobo JM, Jiménez-Valverde A, Real R: **AUC: a misleading measure of the performance of predictive distributingon models.** *Global Ecol Biogeogr* 2008, **17**:145–151.
79. VanDerWal J, Shoo L, Graham CH, Williams S: **Selecting pseudo-absence data for presence-only distribution modeling: How far should you stray from what you know?** *Ecol Model* 2009, **220**:589–594.
80. Hijmans RJ, Graham CH: **The ability of climate envelope models to predict the effect of climate change on species distributions.** *Glob Chang Biol* 2006, **12**:1–10.
81. **Global Land Cover by National Mapping Organizations (GLCNMO).** [http://www.iscgm.org/]
82. **Consortium for spatial information (CGIAR-CSI).** [http://www.cgiar-csi.org]
83. **Corine Land Cover 2006.** [http://sia.eionet.europa.eu/CLC2006]
84. Murphy AH, Winkler RL: **A general framework for forecast verification.** *Mon Weather Rev* 1987, **115**:1330–1338.
85. Pearce J, Ferrier S: **Evaluating the predictive performance of habitat models developed using logistic regression.** *Ecol Model* 2000, **133**:225–245.
86. Warren D, Seifert S: **Ecological niche modeling in Maxent: the importance of model complexity and the performance of model selection criteria.** *Ecol Appl* 2011, **21**:335–342.
87. Rebelo H, Jones G: **Ground validation of presence-only modelling with rare species: a case study on barbastelles *Barbastella barbastellus* (Chiroptera: Vespertilionidae).** *J Appl Ecol* 2010, **47**:410–420.
88. Boubli J, de Lima M: **Modeling the geographical distribution and fundamental niches of *Cacajao* spp. And *Chiropotes israelita* in Northwestern Amazonia via a maximum entropy algorithm.** *Int J Primatol* 2009, **30**:217–228.
89. Wisz M, Hijmans R, Li J, Peterson A, Graham C, Guisan A: **Effects of sample size on the performance of species distribution models.** *Divers Distrib* 2008, **14**:763–773.
90. Giovanelli J, De SM, Haddad C, Alexandrino J: **Modeling a spatially restricted distribution in the Neotropics: How the size of calibration area affects the performance of five presence-only methods.** *Ecol Model* 2010, **221**:215–224.
91. Burnham KP, Anderson DR: *Model Selection and Multi-Model Inference.* 2nd edition. Berlin: Springer; 2002.
92. Engler R, Guisan A, Rechsteiner L: **An improved approach for predicting the distribution of rare and endangered species from occurrence and pseudo-absence data.** *J Appl Ecol* 2004, **41**:263–274.
93. Elith J, Phillips SJ, Hastie T, Dudík M, Chee YE, Yates CJ: **A statistical explanation of MaxEnt for ecologists.** *Divers Distrib* 2010, **17**:43–57.
94. Liu C, White M, Newell G: **Selecting thresholds for the prediction of species occurrence with presence-only data.** *J Biogeogr* 2013, **40**:778–789.
95. Yost AC, Petersen SL, Gregg M, Miller R: **Predictive modeling and mapping sage grouse (*Centrocercus urophasianus*) nesting habitat using Maximum Entropy and a long-term dataset from Southern Oregon.** *Ecol Inform* 2008, **3**:375–386.
96. Moratelli R, Andreazzi CS D, de Oliveira JA, Cordeiro J: **Current and potential distribution of *Myotis simus* (Chiroptera, Vespertilionidae).** *Mammalia* 2011, **75**:227–234.

# Roles of ecdysteroids for progression of reproductive cycle in the fresh water crustacean *Daphnia magna*

Eri Sumiya[1,2], Yukiko Ogino[1,2†], Hitoshi Miyakawa[2†], Chizue Hiruta[2], Kenji Toyota[1,2], Shinichi Miyagawa[1,2] and Taisen Iguchi[1,2*]

**Abstract**

**Introduction:** *Daphnia magna* exhibits a parthenogenetic reproductive cycle linked to a moulting cycle, but regulatory mechanisms of neither moulting nor reproductive cycle are understood in daphnids. Moulting is regulated by ecdysteroids in insects. A previous study showed that a titre of ecdysteroids changed during the reproductive cycle in *D. magna*; however, no clear correlation among titre, moulting and reproductive cycles has been proved in daphnids. To understand endocrine mechanisms underlying the coordinated reproductive cycle, we analysed the expression of genes coding for enzymes in ecdysteroids synthesis or inactivation pathways, and the effects of 20-hydroxyecdysone (20E) on moulting and ovulation in *D. magna*.

**Results:** We cloned orthologues of *neverland* (*nvd*) and *shade* (*shd*) in the ecdysteroids synthesis pathway, and *Cyp18a1* in the ecdysteroids inactivation pathway previously identified in insects. Gene expression of *Cyp18a1* changed conversely with the fluctuation in ecdysteroids titre during the intermoulting period. Tissue-specific expression analysis of *nvd* showed a prominent expression in the gut. Furthermore, treatment of adult female *D. magna* with 20E inhibited moulting and/or ovulation.

**Conclusions:** Our cloning and phylogenetic analyses showed that *nvd* and *shd* as well as *Cyp18a1* are evolutionary conserved in *D. magna*, suggesting that these genes appeared in arthropods before the radiation of insects. The gene expression analysis during the reproductive cycle indicated that Cyp18a1 possibly regulates the decline of ecdysteroid titre before moulting and ovulation. Furthermore, the expression of *nvd* in the gut suggested that ecdysone might be synthesised in the gut. Exogenous 20E-treatment resulted in the failure of not only moulting, but also ovulation, suggesting that a low level of ecdysteroids before moulting is required for moulting and ovulation in *D. magna*.

**Keywords:** *Daphnia magna*, Ecdysteroids, Moulting, Reproductive cycle, 20-hydroxyecdysone

## Introduction

Crustacea and Insecta comprising monophyletic group, the Pancrustacea, are the most diverse animal group in aquatic habitats [1]. Reproductive diversity as an adaptation to environmental conditions may have facilitated the radiation of crustaceans [2]. Reproduction of decapod crustaceans is regulated by complex endocrine mechanisms involving ecdysteroids, neuropeptides, biogenic amines, opioids, juvenile hormones, and vertebrate-type steroids [3]. Daphnids show characteristic features in reproduction. They reproduce parthenogenetically every three days under favourable conditions, and the reproductive cycle occurs synchronously with the moulting cycle. After moulting, mother daphnids ovulate new eggs from the ovary into the brood pouch (ovulation). At the end of each reproductive cycle, the mother releases the brood from the pouch to the surrounding water. After this release, the mother moults again and repeats the cycle. These events occur in a very systematic order with intervals that span approximately three days depending on the condition.

* Correspondence: taisen@nibb.ac.jp
†Equal contributors
[1]Faculty of Life Science, Graduate University for Advanced Studies (SOKENDAI), 5-1 Higashiyama Myodaiji, Okazaki, Aichi 444-8787, Japan
[2]Okazaki Institute for Integrative Bioscience, National Institute for Basic Biology, National Institutes of Natural Sciences, and Department of Basic Biology, 5-1 Higashiyama Myodaiji, Okazaki, Aichi 444-8787, Japan

Consequently, a closely related system might regulate the moulting and reproductive cycle. A similar association of moulting with a reproductive cycle has also been reported in other crustaceans such as amphipods, isopods, shrimps, prawns and a sand crab [4,5]. However, regulation of the reproductive cycle remains largely unknown in most crustaceans.

Ecdysteroids (ecdysone and 20-hydroxyecdysone; 20E) are steroid hormones that regulate moulting and metamorphosis in insects. These have been extensively studied in insects, and are commonly called "moulting hormones" as their major role implies. Recent studies in insects revealed that ecdysteroids have a variety of functions besides moulting in the control of reproduction and embryogenesis [6]. In *Drosophila melanogaster*, ecdysteroids are required for differentiation of germ line stem cells [7,8] and prepupal development [9,10]. A pioneering study reported that a titre of ecdysteroids increased at the premoult stage and decreased back to the basal level before moulting in *Daphnia magna* [11], indicating that ecdysteroids are predominant moulting hormones. However, in crustaceans, reports on ecdysteroidal regulation in both moulting and reproduction are scarce [4]. The synchronised moulting and reproductive cycle in daphnids implies a possibility that ecdysteroids may regulate not only moulting, but also the reproductive cycle.

In insects, biosynthesis of ecdysteroids starts with dietary cholesterol. Rieske oxygenase DAF-36 (neverland, nvd) converts it to 7-dehydrocholesterol [12]. After several steps mediated by enzymes belonging to the cytochrome P450 family (CYP450), it is converted to ecdysone. Then, Cyp314A1 (shade, shd) finally converts ecdysone into 20E, which is the biologically active form binding to an ecdysone receptor (EcR) [13,14]. The complex of 20E and EcR recruits ultraspiracle (USP) in insects and daphnids or retinoid X receptor (RXR) in decapods, and acts as a heterodimer [15]. EcR and USP have been cloned in *D. magna* [16]. Three EcR paralogues with five distinct N-terminus sequences and one USP were identified [16]. These genes in the ecdysteroids biosynthesis pathway except for *nvd* belong to CYP450 encoding hydroxylases. *D. pulex* possesses five CYP450 genes: *spook, phantom, disembodied, shadow* and *shd* [17]. In decapod crustaceans, *Cyp306A1* (*phantom*) in kuruma prawn (*Marsupenaeus japonicus*) [18], and four CYP450s (*spook, phantom, disembodied* and *shadow*) and *nvd* in cray fish (*Pontastacus leptodactylus*) [19] have been identified to date. However, neither gene expression nor functional analyses have been explored in microcrustaceans such as copepoda, phyllopoda and cladocera [15].

The synthesis and secretion of ecdysone take place in a specialised organ in insects and decapods. Insect larvae synthesise ecdysone in the prothoracic gland although it is destined to degenerate in the adult of most species [20]. The synthetic organ in decapods is the Y-organ. However, the organ that synthesises and secretes ecdysone has not been identified in microcrustaceans, including daphnids. Some reports have speculated that these species may synthesise ecdysone in the epidermis [15,21,22] since ecdysone production in the epidermis was observed *in vitro* in the tick *Ornithodoros parkeri* [23]. Identification of the endocrine organ is essential to understand the endocrine system of daphnids.

As an entry point to understand the roles of ecdysteroids in the reproductive cycle in *D. magna*, we cloned genes of three key enzymes for 20E synthesis or inactivation: *nvd, shd* and *Cyp18a1*. Spacio-temporal expression analyses of these genes during the reproductive cycle indicated that ecdysone is possibly synthesised in the gut, and that the decline of the 20E titre before moulting [11] might be regulated by Cyp18a1. In addition, exogenous 20E treatment during the reproductive cycle resulted in a failure of moulting and ovulation, suggesting that a low level of ecdysteroids before moulting is required for successful moulting and ovulation in *D. magna*. The present study provides insight into the endocrine system that regulates the highly coordinated reproductive system in crustaceans.

## Results

### Molecular cloning and phylogenetic analysis of genes in 20E synthesis or inactivation pathway in *D. magna*

We cloned full-length of *nvd* encoding 444 amino acids (a.a.), *shd* encoding 526 a.a. and *Cyp18a1* encoding 551 a.a. from *D. magna*. Furthermore, we identified the partial cDNA sequence of a distinct *nvd*. Aligning amino acids of these genes with *D. pulex* and two insect species, *D. melanogaster* and *Bombyx mori*, revealed that both *D. magna* nvd contain a characteristic domain, Rieske [2Fe-2S], and a non-heme Fe (II) motif (Figure 1A) [12]. The obtained shd and Cyp18a1 belonging to the P450 family have a P/G-rich domain and five conserved P450 motifs: Helix-C, Helix-I, Helix-K, PERF and Heme-binding (Figure 1B) [24].

Phylogenetic analysis was conducted by comparing the amino acid sequences of the obtained nvd, shd and Cyp18a1 with other available arthropod sequences. In each obtained tree, *D. manga* and *D. pulex* clustered together with other arthropods, forming a monophyletic group. This suggests that the orthologous genes of nvd, shd and Cyp18a1 were identified in *D. magna*. Two of the *D. magna* nvd were assembled in the *Daphnia* nvd cluster. These nvd paralogoues were therefore designated nvd1 (full length, AB839171) and nvd2 (partial cDNA sequence, AB839172) (Figure 2A). We further found a duplication of nvd1, nvd1a (sequence ID in *D. pulex* genome database, NCBI_GNO_67514) and nvd1b (NCBI_GNO_462084).

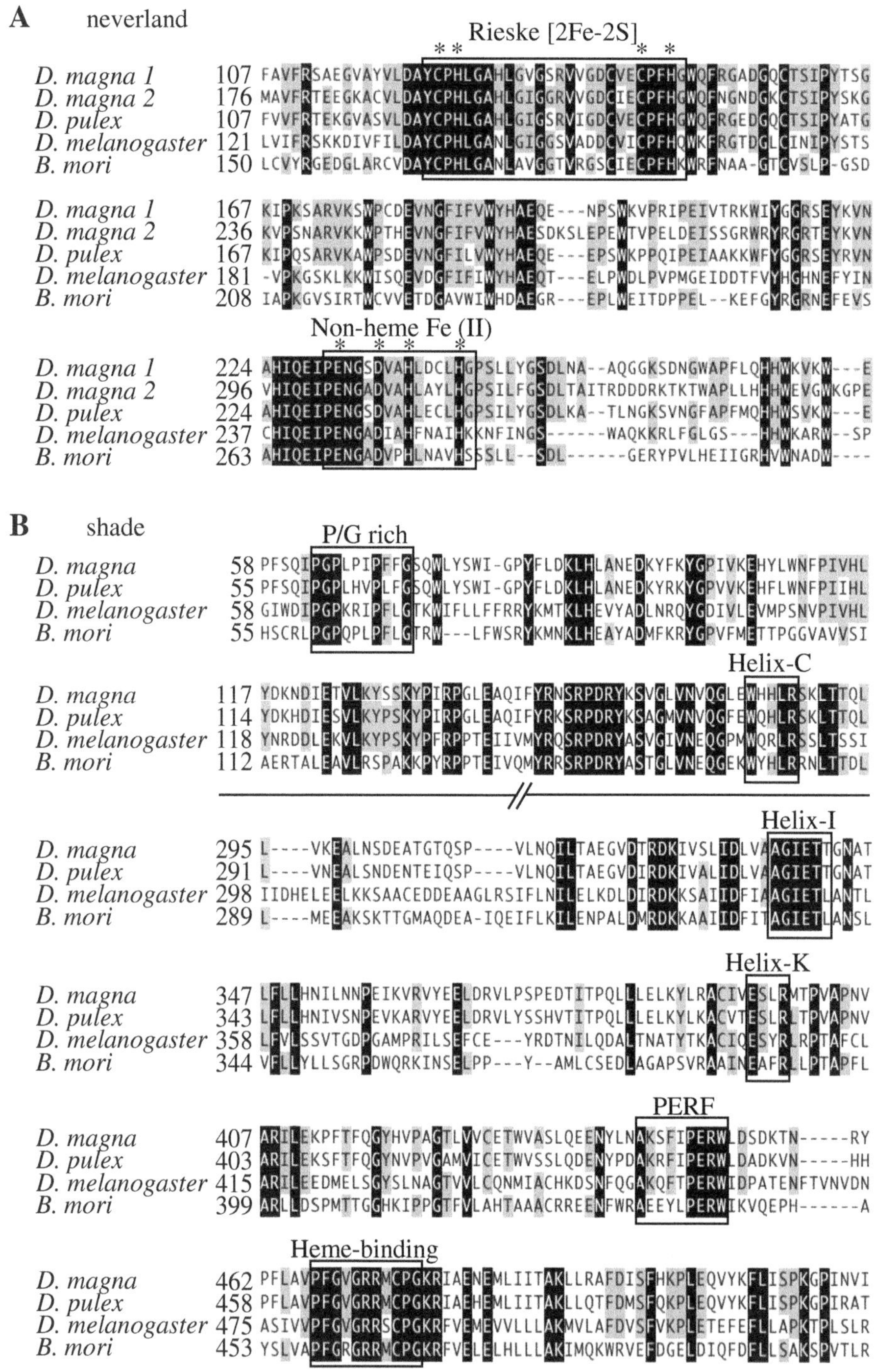


**Figure 1 Alignments of predicted amino acid sequences of (A) neverland and (B) shade of *Daphnia magna* with other arthropod species, *Daphnia pulex*, *Drosophila melanogaster* and *Bombyx mori*.** *D. magna 1* and *D. magna 2* indicate the two neverland paralogues, nvd 1 and nvd 2, respectively. The conserved Rieske [2Fe-2S] domain and the non-heme Fe (II) motif in nvd and conserved domain and motifs of the P450 family in shd are shown in boxes. Asterisks show the evolutionally conserved amino acids in nvd. Entries in a black background indicate identities.

In the case of shd, no paralogous genes were identified in the *D. magna* database. In agreement with the arthropod phylogenetic trend, tick shd is the most ancestral followed by daphnids shd, and insects' shd being the most descendant (Figure 2B). As seen in nvd and shd, the obtained *D. magna* Cyp18a1 pairs with *D. pulex* Cyp18a1 (Figure 2C) showing a close relationship.

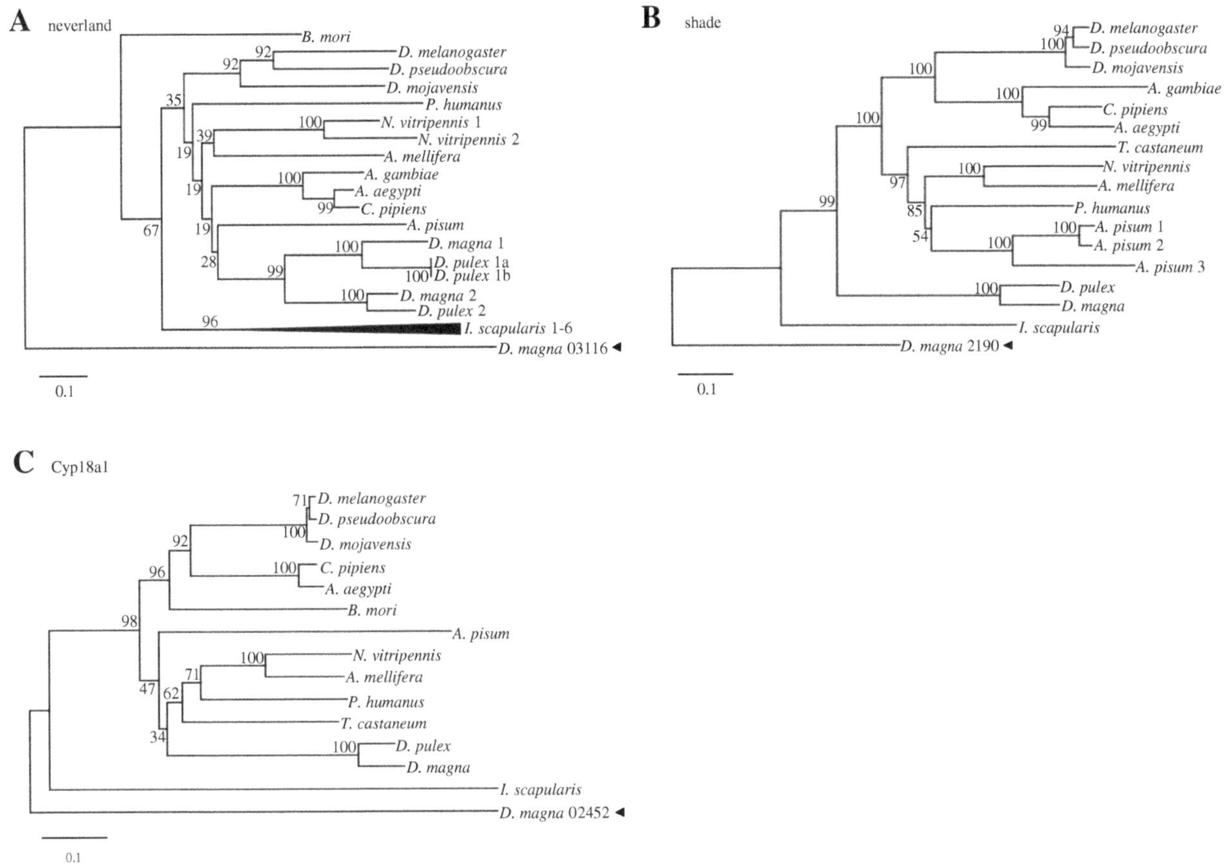


**Figure 2 Neibour joining trees of (A) neverland, (B) shade and (C) Cyp18a1.** These trees were constructed with Poisson model and the amino acid sequences excluding gap positions. Bootstrap values (%) show analyses of 1000 trials. Arrowhead represents an outgroup. Scale bar indicates branch length of 0.1 changes per residue. *D. magna* 03116, 2190 and 02452 are sequences in *D. magna* database showing high homology of amino acids with respective genes, and were used as outgroups.

### Time-course and tissue-specific expression analyses of *neverland, shade, Cyp18a1, EcR* and *USP* during intermoulting period

Adult *D. magna* has a moulting cycle of three days under current conditions, which takes 80 h on average between two consecutive moultings. Investigation of the expression profiles of the genes in the ecdysteroids synthesis or inactivation pathway during the intermoulting period was performed by quantitative real-time PCR. *nvd1, nvd2* and *shd* were highly expressed in the early intermoulting period at 0–10 hours after moulting (ham) (Figure 3A-C). The lowest expression level of *nvd1* was reached at 20 ham, and it increased at 70 ham. In contrast, the expression level of *nvd2* did not change significantly after the abrupt drop at 20 ham. The expression level of *shd* gradually decreased until 20 ham. Thereafter, it stayed at a rather constant level. *Cyp18a1* was also highly expressed at 0 ham, and its expression declined to the lowest level by 30 ham (Figure 3D). At the end of the moulting cycle, *Cyp18a1* expression showed a steep rise at 80 ham. The expression level of *EcR* was relatively high until 10 ham and after 50 ham (Figure 3E). The expression level of *USP* was highest at 0 ham but decreased to the lowest level by 10 ham after which it no longer shifted (Figure 3F).

A synthetic and target organs/tissues for ecdysteroids are unknown in daphnids. Therefore, tissue-specific expression of genes in the ecdysteroids synthesis or inactivation pathway was analysed in the major organs such as the ovary, gut and remaining tissues (i.e., tissues other than the ovary and gut) by quantitative real-time PCR. *nvd1* was strongly expressed in the gut (Figure 4A). *nvd2* expression was also observed primarily in the gut, but there was no significant difference in expression levels among tissues (Figure 4B). *shd* was predominantly expressed in the gut, and its expression was also detected in the ovary and remaining tissues (Figure 4C). Alternatively, the expression level of *Cyp18a1* was relatively low in the gut (Figure 4D). The expression level of *Cyp18a1* was the highest in the remaining tissues, but not significantly higher than in the ovary or gut. *EcR* and *USP* expression were detected in all tissues examined,

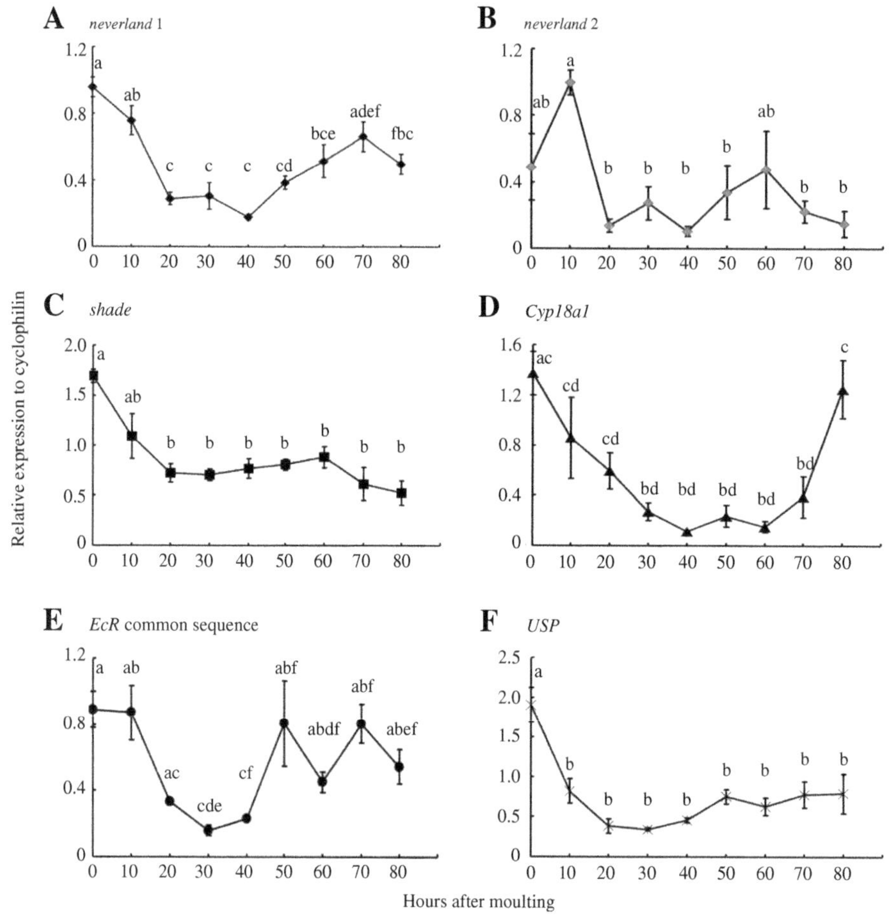


**Figure 3 Expression profiles relative to cyclophilin. (A)** *neverland* 1, **(B)** *neverland* 2, **(C)** *shade*, **(D)** *Cyp18a1*, **(E)** *EcR* common sequence and **(F)** *USP* in the intermoulting period of adult female *D. magna*. 0 hour after moulting (ham) refers to the time immediately after moulting and 80 ham represents the time before the next moulting. Bars represent standard errors. Different letters indicate significant differences among time points. $P < 0.05$ (ANOVA followed by the Tukey-Kramer test).

and the expression levels were highest in the remaining tissues (Figure 4E,F).

### Exogenous 20E treatment on adult *D. magna* in the mid-reproductive cycle

To understand the importance of fluctuation of the ecdysteroids titre on moulting and ovulation, adult female *D. magna* were treated with exogenous 20E in the middle of the reproductive cycle, which is a day before moulting and ovulation occur. After this treatment, unusual phenotypes were observed at a significantly higher rate ($P < 0.001$): normal moulting without ovulation or neither moulting nor ovulation (Figure 5). In the 20E-treated group, only 34% of 128 individuals underwent normal moulting followed by ovulation whereas 99% of 85 control individuals moulted and ovulated normally. The most frequent phenotype appeared in 53% of the 20E-treated individuals. Those individuals neither moulted nor ovulated. In addition, 13% of the 20E-treated individuals moulted normally but failed to ovulate. The toxic influence seemed minimal as none of the individuals died within 48 hours after the treatment.

The ovary and carapace were then analysed histologically. The whole ovary was full of yolk granules in control individuals (Figure 6A,B). The meiotic phase of the eggs was also examined. In the ovarian eggs of *D. pulex*, the nuclear envelope degenerates and becomes invisible at the point of moulting [25]. This was confirmed in *D. magna*. The nuclear envelope was invisible in control individuals before moulting (Figure 6C). After moulting, a meiotic apparatus was prominently visible in the ovarian egg (Figure 6D) and ovulated egg (Figure 6E). In comparison, yolk granules were also present in the ovary of 20E-treated individuals (Figure 6F,G). However, the nuclear envelope did not degenerate in the 20E-treated individuals that failed to moult (Figure 6H). Interestingly, the meiotic apparatus was observed in the ovarian egg of those 20E-treated individuals that failed to ovulate after normal moulting (Figure 6I). As a result, treatment with 20E caused *D. magna* to be unable to moult and/or

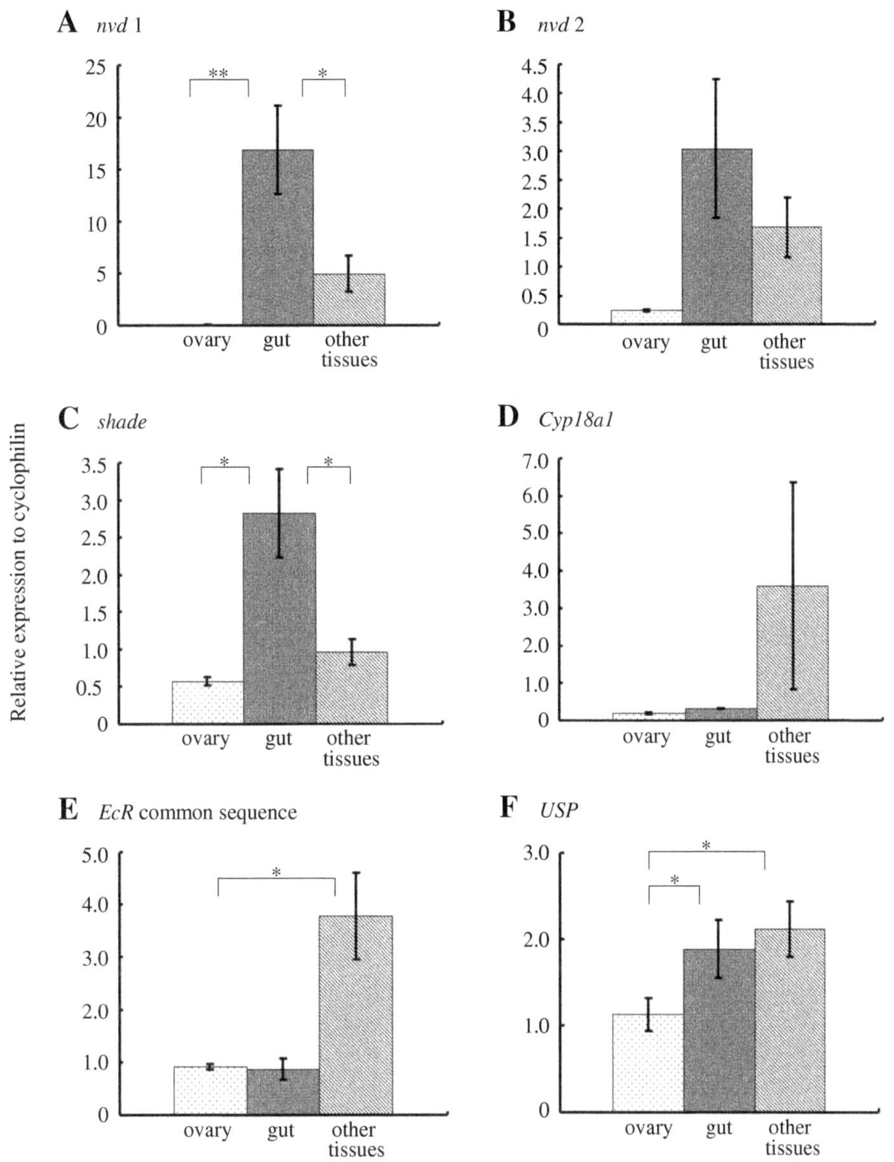


**Figure 4 Relative expression to cyclophilin.** **(A)** *nvd* 1, **(B)** *nvd* 2, **(C)** *shd*, **(D)** *Cyp18a1*, **(E)** *EcR* common sequence and **(F)** *USP* in *D. magna* adult female tissues. The ovary, gut and remaining tissues were excised at 70 hours after moulting in the intermoulting period. Bars represent standard errors. $^{*}P < 0.05$ $^{**}P < 0.01$ (ANOVA followed by the Tukey-Kramer test).

ovulate, and it ceased meiotic division at the time when the nuclear envelope breaks or meiotic metaphase. A new carapace, which was present in the control before moulting (Figure 6J,K), was also formed in the 20E-treated individuals (Figure 6L,M).

## Discussion

### Genes in ecdysteroids synthesis or inactivation pathway are evolutionary conserved in *D. magna*

The moulting hormones, ecdysone and 20E, have long been known to play critical roles in moulting and development of insects and crustaceans [26]. The final product of the ecdysteroids synthesis pathway, 20E, actively interacts with EcR/USP heterodimeric receptor in *D. magna* [16]. This indicates that the binding of 20E to EcR in arthropod species can activate the ecdysteroids signalling pathway. Although there is physiological evidence indicating that crustaceans synthesise 20E from cholesterol, the molecular basis of 20E synthesis in crustaceans is not fully understood. Recently, orthologues of the insect Halloween genes encoding P450 enzymes for ecdysteroids biosynthesis such as spook, disembodied, phantom, shadow and shade were obtained by searching the *D. pulex* database [17]. We newly identified genes covering the ecdysteroids synthesis or inactivation pathway, i.e. not only these P450 genes but also *nvd* and *Cyp18a1* in the *D. magna* EST

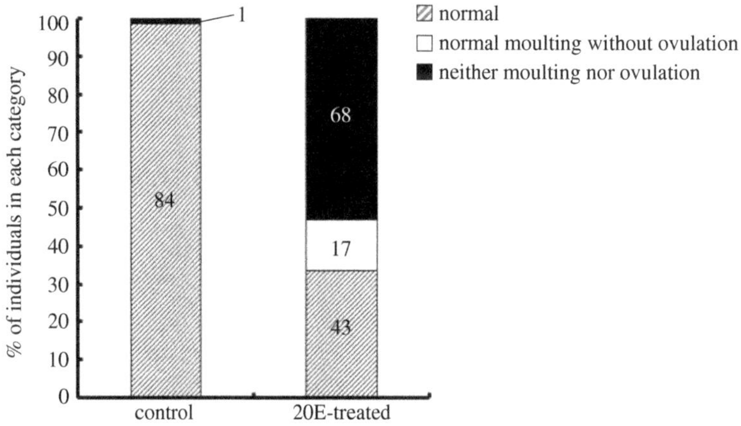


**Figure 5 Percentage of adult female *D. magna* categorised into different phenotypic classes after treatment with exogenous 20-hydroxyecdysone (20E) (*N* = 128) or ethanol as the control (*N* = 85).** Values in bars represent the number of *D. magna* used.

database. The present study focused on nvd, which catalyses the first step in the biosynthesis of ecdysteroids [12], shade mediating the hydroxylation of ecdysone to 20E [13], and Cyp18a1, a key enzyme for 20E inactivation [14] in insects. Our cloning and phylogenetic analyses showed that nvd and shd as well as Cyp18a1 genes are evolutionary conserved in *D. magna*. This suggests that *D. magna* possesses ecdysteroids synthesis and inactivation pathways similar to those conserved in insects. This result is in agreement with recent phylogenetic analysis suggesting that these genes appeared in arthropods before the radiation of insects. Interestingly, we identified two different paralogues of the *nvd* gene in *D. magna*. Our phylogenetic analysis indicates that gene duplication gave rise to these two different nvd paralogues, nvd1 and nvd2 in daphnids. This duplication is thought to have occurred after the divergence of daphnids and insects but before the split of *D. magna* and *D. pulex*. We also identified the two nvd1 paralogues, nvd1a and nvd1b, in *D. pulex*. In the *D. pulex* linage, a number of gene duplications occurred [27]. These duplication events are thought to have resulted in a complex lifecycle of daphnids.

### Cyp18a1 might regulate the decline of ecdysteroids titre before moulting

In crustaceans including *D. magna*, shrimp and crab species, ecdysteroids titre shows a single peak during the intermoulting period [4,28,29]. In contrast, ecdysteroids titre in insect larvae fluctuates in more complex patterns during embryogenesis and larval development. These rises and falls of ecdysteroids at specific times are mandatory for normal moulting and metamorphosis in insects [30]. Cyp18a1 inactivates 20E through 26-hydroxylation in insects to lower the ecdysteroid titre [14]. Gene expression analysis during the intermoulting period in *D. magna* showed that the relative expression of *Cyp18a1* changed conversely with the fluctuation in ecdysteroids titre (Figure 3D). In other words, the ecdysteroids titre is high in the mid-moulting cycle when *Cyp18a1* expression is low, but is low at the end of the moulting cycle when *Cyp18a1* expression is high. This result suggests that Cyp18a1 possibly regulates the decline of the ecdysteroids titre before moulting in *D. magna*.

Relative expression analysis of *shd* and *Cyp18a1* showed different patterns (Figure 3C,D). Each P450 gene in the ecdysteroids synthesis pathway exhibits a different expression pattern in insect larvae, and it varies among species [24,31-34]. Therefore, having a unique expression pattern for each of the genes appears to be a common phenomenon. Such variable expression patterns reflect which gene is responsible for a rate-limiting step or is a target of prothoracicotropic hormone (PTTH), which induces ecdysone synthesis in insects [33].

### Ecdysone might be synthesised in the gut, judged by the expression of *neverland*

In insect larvae, expression of *nvd* is exclusively detected in the ecdysone synthetic organ [12]. On the other hand, *shd* has been observed in a range of 20E-target tissues/organs [13,24,31]. In adult *D. melanogaster*, *nvd* is expressed in the ovary where ecdysone is produced [12]. A gene encoding phantom, which participates in ecdysone synthesis, in kuruma prawn is expressed in the Y-organ [18]. *D. magna nvd1* was primarily expressed in the gut of the adult, suggesting that ecdysone is possibly synthesised in the gut. The expression of *shd* was detected in the ovary as well as in the gut at 70 ham when oogenesis had been promoted in the ovary for upcoming ovulation, and the eggs had formed inside. Such spatial expression of *shd* indicates that ecdysone targets the ovary besides the gut and remaining tissues. Ecdysteroids might function in ovarian egg maturation in *D. magna*.

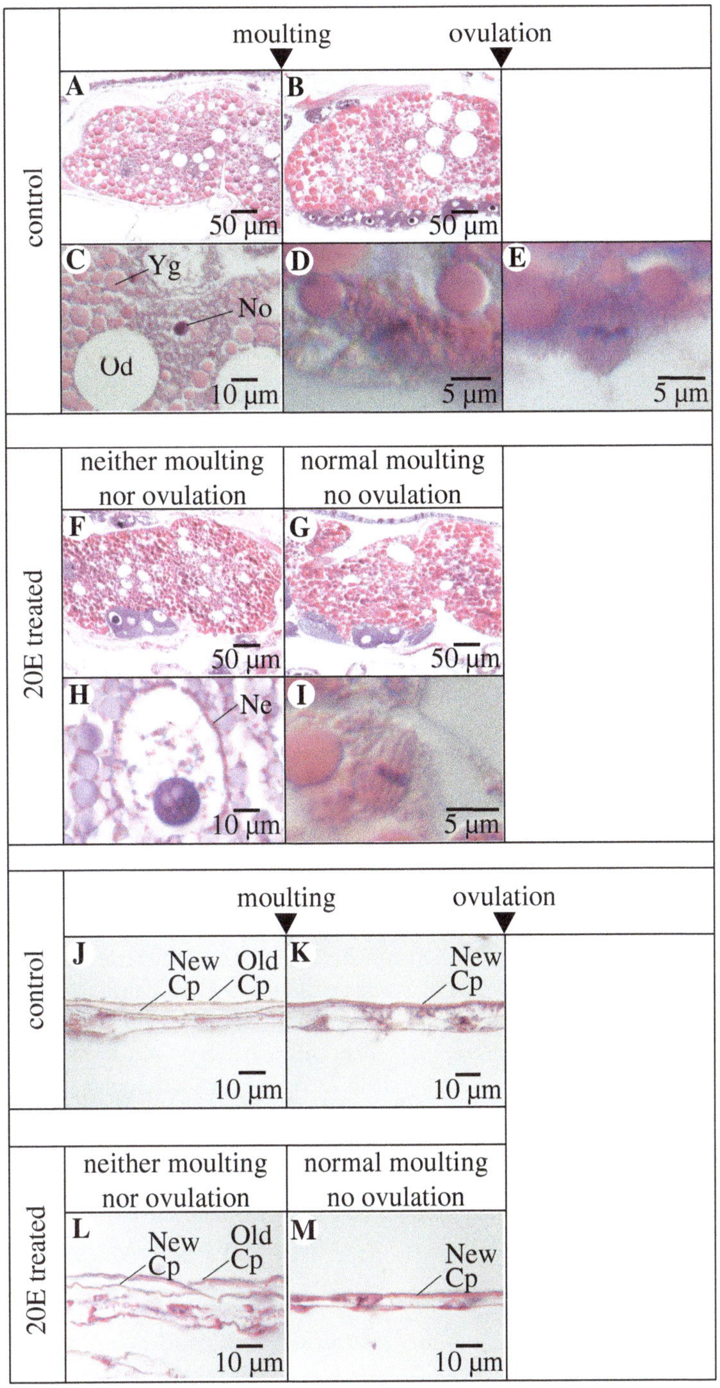


**Figure 6 Histological analysis of the ovary and carapace. (A,B)** Ovarian eggs in a control individual. **(C)** The nuclear envelope was invisible in controls before moulting. **(D,E)** A meiotic apparatus found in ovarian and ovulated eggs of controls, respectively, and **(F,G)** ovarian eggs of a 20E-treated individual. **(H)** The nuclear envelope had not degenerated in the ovarian egg in the 20E-treated individual that neither moulted nor ovulated. **(I)** A division apparatus appeared and chromosomes aligned at the metaphase plate in the 20E-treated individual that moulted normally but did not ovulate. This meiotic apparatus at metaphase was also seen in the control eggs released into the brood pouch immediately after ovulation. However, in the 20E-treated individuals, meiosis stopped at metaphase and did not proceed. **(J,K)** Carapace formation in controls and **(L,M)** 20E-treated individuals. **(J)** A carapace layer with a newly formed carapace underneath the old one formed before moulting in the control. **(K)** A new carapace was observed immediately after moulting in the control. **(L)** A new carapace was also formed in the 20E-treated individual that neither moulted nor ovulated, and **(M)** the 20E-treated individual that moulted normally but did not ovulate. Yg: yolk granule, No: nucleolus, Od: oil droplet, Ne: nuclear envelope, Cp: carapace. Tissue sections of exogenous 20E-treated and the control adult female *D. magna* were stained with hematoxylin and eosin.

**The decline of ecdysteroids level before moulting is required for ensuing moulting and ovulation**

Ecdysteroids stimulate vitellogenesis in most decapod species [35]. On the other hand, 20E inhibits vitellogenesis in *D. magna* [36]. In support of this, endogenous ecdysteroids titre is low when moulting and ovulation are about to occur in *D. magna* [11]. Ovulation occurs immediately after moulting when ecdysteroids titre has decreased. As such a correlation between ecdysteroids titre, moulting and reproductive events exists, we hypothesised that the fluctuation of ecdysteroids titre plays a key role in the endocrine systems regulating the reproductive and moulting cycles. Therefore, we analysed the effect of 20E-treatment on moulting and ovulation in *D. magna*. Treatment with 20E from the middle of the moulting cycle inhibited moulting and/or ovulation. This indicates that a decrease of ecdysteroids at the end of the reproductive cycle is required for moulting and ovulation. Indeed, *Cyp18a1* was highly expressed at the end of the moulting cycle. Histological analyses revealed that yolk granules existed in the ovary of both the control and 20E-treated *D. magna*. This implies that suppressed ovulation might not have been caused by inhibition of vitellogenesis by a high 20E titre. In the present study, those individuals that did not moult after treatment with 20E produced a new carapace underneath the old one. Therefore, 20E treatment did not affect the process of cuticle deposition.

Recently, a meiosis-like process of parthenogenetic oogenesis was documented in *D. pulex* [25]. In the present study, the nuclei-germinal vesicles broke down (GVBD: germinal vesicle breakdown), chromosomes condensed, and the nuclei were invisible before moulting in the ovarian eggs of control *D. magna*, as observed in *D. pulex*. In the ovary of 20E-treated individuals showing neither moulting nor ovulation, the meiotic process had stopped before GVBD. Although the ovary contained as many yolk granules as the control ovary, 20E inhibited the entry of the ovarian eggs into GVBD, which marks the onset of meiosis.

A meiotic apparatus at metaphase was observed soon after moulting and soon after ovulation in the control eggs in *D. magna* (Figure 6D,E). Such a meiotic apparatus at metaphase was also observed in the eggs of 20E-treated individuals that moulted normally but failed to ovulate (Figure 6I). In the 20E-treated individuals, however, meiosis stopped at metaphase and did not proceed further even after enough time had passed for meiosis to progress (Figure 6I). Such a terminated meiotic process might correspond to a failure in meiotic reinitiation. The first reinitiation appears as GVBD and the second as anaphase. Furthermore, meiotic reinitiation is correlated with the ecdysteroids titre in prawn (*Palaemon serratus*) and locust (*Locusta migratoria*) [37]. The present results suggest that the 20E titre might also be responsible for meiotic reinitiation in *D. magna*. Ovulation never occurs unless moulting precedes it. The time interval between moulting and ovulation is precisely 13 minutes in *D. pulex* at 18°C [25]. Therefore, moulting is expected to act as an ovulation trigger. There must be coordinated regulation between moulting and ovulation. Our results are compatible with this idea, but moulting alone is not enough to initiate ovulation. Some regulatory mechanisms might be present to promote ovulation after moulting. Alternatively, some inhibitory mechanisms might also be present for ovulation of abnormal eggs.

In *Drosophila*, neural signalling with octopamine, which causes oviduct-muscle contraction, controls ovulation [38,39]. In decapod crustaceans, octopamine in the lobster *Homarus americanus* also stimulates muscle contraction of ovarian walls [3]. Neural signalling besides a direct endocrine pathway could possibly drive the regulation of ovulation in *D. magna*. That is because ovarian eggs likely require some mechanical force, such as muscle contraction, to be released into the brood pouch. This suggests that 20E might affect ovulation through muscle contraction.

## Conclusions

In conclusion, we identified genes in the ecdysteroids synthesis or inactivation pathway in *D. magna*. Decline of the ecdysteroids level at the end of the reproductive cycle is necessary for ensuing moulting and ovulation. Our gene expression analyses suggest that fluctuation of the ecdysteroids titre during the reproductive cycle might be regulated by the expression of *Cyp18a1*. Judging by the predominant expression of *nvd*1 in the gut, we also propose that ecdysone might be synthesised in the *D. magna* gut although further studies are required to identify the specific site of ecdysteroidogenesis in the gut. Gene identification and expression analyses in ecdysteroids synthesis or inactivation pathways provide a concrete basis for understanding the contribution of ecdysteroids signalling in the highly coordinated reproductive system of crustaceans.

## Materials and methods

### Animals

*Daphnia magna* (NIES strain) obtained from the National Institute for Environmental Studies (NIES; Tsukuba, Japan) were maintained in M4 medium [40] at 20 ± 1°C under a 14-hour:10-hour light:dark photoperiod. *Chlorella vulgaris* was added daily into each culture. For experiments, each animal was incubated under the same conditions mentioned above but in 50 ml of M4 medium individually. The culture medium was substituted every three days. Under these conditions, *D. magna* produces female offspring every three days.

**Table 1 Primer sequences used for gene cloning**

| Genes | Forward | Reverse |
|---|---|---|
| *neverland* 1 | ATGGATGTCTTTTTCGTTTTCAAATCC | GGTGTAACCGTCTCGAAG |
| *neverland* 2 | GGACGCGTATTGCCCTCATTTGGGA | TAATCCCGGCCCGATTTGTTTGGCT |
| *Cyp18a1* | ATGATCGTGTGGAGTTTGGGA | TTTGGCGTTCTGGATCTCCAG |
| **3' RACE** | | |
| *neverland* 1 | GCCTCCATCATGAACTACGCCTATTCGG | |
| *Cyp18a1* | GCGACGGCCGTGAGAAAATGGAG | |
| | GCCGTTCACATGAACCCTGCCTTGC | |

### Cloning and phylogenetic analysis

Total RNA extracted from adult female *D. magna* using the RNAqueous Kit (Life Technologies, Carlsbad, CA, USA) was reverse transcribed with Super Script III (Life Technologies). To obtain gene fragments of *nvd* and *Cyp18a1*, oligonucleotide primers were designed based on the sequences of the *D. magna* EST database [41] (Table 1). PCR was performed with Platinum Taq DNA Polymerase High Fidelity (Life Technologies). The amplified products were cloned into pGEM-T easy (Promega, Tokyo, Japan), and sequenced using the BigDye Terminator Cycle Sequencing Kit (Life Technologies). Based on the sequence information of the obtained PCR products, the 5' and 3' ends of *nvd* 1 and *Cyp18a1* sequences were amplified using a rapid amplification cDNA ends (RACE) kit (Gene Racer Kit, Life Technologies) with Platinum Taq High Fidelity (Life Technologies). The obtained two *nvd* and *Cyp18a1* sequences were assigned the following accession numbers [DDBJ: *nvd1*: AB839171, *nvd2*: AB839172 and *Cyp18a1*: AB839173]. Sequence of *shd* of *D. magna* was available in the NCBI database (AB257771.1). Sequences of *EcR* and *USP* were defined in our previous study (AB274819-24) [16].

The amino acid sequences of *nvd*, *shd* and *Cyp18a1* in other arthropod species were identified in euGenes/Arthropod genomes (Indiana University, http://arthropods.eugenes.org/arthropods/). The amino acid sequences were aligned with MEGA5 software using MUSCLE method with the default options [42]. Neighbour joining trees were constructed from these alignments using Poisson model with bootstrap analyses of 1000 replicates along with complete deletion options.

### Quantitative PCR

Whole body adult *D. magna* samples were prepared at nine time points with 10-hour intervals during an 80-hour intermoulting period. For the tissue-specific gene expression analysis, adult *D. magna* at 70 ham was dissected, and ovary, gut and the remaining tissues were separately obtained. Three replicates were prepared for each time point for each experiment. Total RNA was extracted using the RNAqueous-Micro Kit (Life Technologies). cDNA was synthesised from 0.5 μg of total RNA using Super Script III (Life Technologies) with random primers. PCR was conducted with SYBR-Green PCR Master Mix (Life Technologies) using the 7000 real-time PCR system (Life Technologies). The primer sequences used for the quantitative PCR analyses are shown in Table 2. To detect *EcR*, a common sequence region of three *EcR* genes was utilized to design the primers. Among four housekeeping genes (*cyclophilin*, *Ltd4ds*, *LBT* and *L32*), *D. magna cyclophilin* was selected as a reference gene calculated with geNorm [43] and NormFinder [44]. Statistical analyses of gene expression at each time point were performed by ANOVA followed by the Tukey-Kramer post-hoc test using Excel 2004 (Microsoft Corp., Redmond, WA) with an add-in software Statcel 2 [45].

**Table 2 Primer sequences used for quantitative real-time PCR**

| Genes | Forward | Reverse |
|---|---|---|
| *cyclophilin* | GACTTTCCACCAGTGCCATT | AACTTTCCATCGCATCATCC |
| *neverland* 1 | AGCACAAGGCGGGAAGAGT | GCTTCCCATTTCACCTTCCA |
| *neverland* 2 | CGTCGGTGACTGCATCGA | TGCCGTCGTTCCCATTG |
| *shade* | GACTGCTGAAGGCGTTGACA | CGGCTGCCACTAGGTCGATA |
| *Cyp18a1* | TACCCGATCGTCGGTTACCT | GAGCGCCGTCAGCTCTTC |
| *EcR* common region | GAGGCGCTGCAGGCTTAC | GAGTTTGGCAAACTCCGTCATC |
| *USP* | GTTGGAGTCAAGGATGGTATCGT | AGCCGAGTTCCGGTGGAT |

### Treatment with exogenous ecdysteroid

A stock solution of 10 mM 20E (Sigma, St. Luis, MO, USA) dissolved in ethanol was kept at -20°C until use. This stock solution was directly added to each 50 ml of M4 medium containing one adult female *D. magna* at two days after moulting ($N = 128$). The final concentration was 0.5 μM 20E. Ethanol (0.0001%) served as the control ($N = 85$). The number of individuals was counted for three phenotypes in the following order of severity: normal moulting and ovulation (no effect), normal moulting following no ovulation (moderate effect), or neither moulting nor ovulation (severe effect). Differences in the incidence between the control and 20E-treated animals were analysed with the Mann–Whitney $U$ test using Excel 2004 (Microsoft Corp.) with Statcel 2 [45]. For tissue sectioning, the 20E-treated and control animals were fixed in Bouin's fluid overnight. The samples dehydrated in ethanol and xylene were embedded in paraffin. Serial sections of 5 μm thickness were subjected to haematoxylin and eosin staining. The sections were studied under a microscope (Axioplan2, ZEISS, Oberkochen, Germany) and the images were processed with callSens Standard, version 1.3 (Olympus, Tokyo, Japan). Six individuals of the 20E-treated animals were observed for each of the three phenotypes. Three control animals were examined at three stages: before moulting (sampled after the release of a brood), soon after moulting and before ovulation (sampled within 15 minutes after moulting) and soon after ovulation (sampled when ovulation finished).

### Abbreviations

Cyp450: CyctochromeP450; EcR: Ecdysone receptor; GVBD: Germinal vesicle breakdown; Ham: Hours after moulting; Nvd: Neverland; shd: Shade; USP: Ultra spiracle; 20E: 20-hydroxyecdysone.

### Competing interests

The authors declared that they have no competing interests.

### Authors' contributions

ES performed experiments, analysed data, and wrote the manuscript. YO and TI conceived the study. HM participated in the experimental design and phylogenetic analysis. CH, KT, SM participated in the interpretation of data. YO, HM, CH and TI also helped with the manuscript. All authors read and approved the final manuscript.

### Acknowledgements

*D. magna* and *D. pulex* sequence data were provided by The Center for Genomics and Bioinformatics at Indiana University which is supported in part by the METACyt Initiative of Indiana University, funded in part through a major grant from Lilly Endowment, Inc., and distributed via wFleaBase in collaboration with Dr. Donald Gilbert and the *Daphnia* Genomics Consortium https://wiki.cgb.indiana.edu/display/DGC/Home. This work was partly supported by a Japan Society for the Promotion of Science (JSPS) Grant-in-Aid to ES (No. 26·04101), and by grants from the Ministry of Education, Culture, Sports, Science and Technology (TI), the Ministry of the Environment of Japan (TI), a grant from the National Institute for Basic Biology (TI), and the Research Council of Norway (project 221455) -.Adverse Outcome Pathways for Endocrine Disruption in *Daphnia magna*, a conceptual approach for mechanistically-based risk assessment (TI).

### References

1. Kenny NJ, Quah S, Holland PW, Tobe SS, Hui JH: **How are comparative genomics and the study of microRNAs changing our views on arthropod endocrinology and adaptations to the environment?** *Gen Comp Endocrinol* 2013, **188**:16–22.
2. Bliss DE: **Ecological aspects of reproduction.** In *The Biology of Crustacea.* Edited by Vernberg FJ, Vernberg WB. New York: Academic Press, Inc; 1983:179–191.
3. Swetha CH, Sainath SB, Reddy PR, Reddy PS: **Reproductive endocrinology of female crustaceans: perspective and prospective.** *J Marine Sci Res Dev* 2011, **S3**:2.
4. Gunamalai V, Kirubagaran R, Subramoniam T: **Hormonal coordination of molting and female reproduction by ecdysteroids in the mole crab *Emerita asiatica* (Milne Edwards).** *Gen Comp Endocrinol* 2004, **138**:128–138.
5. Shyama SK: **Studies on moulting and reproduction in the prawn *Macrobrachium idella* (hellar).** *Mahasagar-Bull Natl Inst Oceanog* 1987, **20**:15–21.
6. Nijhout HF: *Insect Hormones.* Princeton University Press: Princeton; 1994.
7. Ables ET, Drummond-Barbosa D: **The steroid hormone ecdysone functions with intrinsic chromatin remodeling factors to control female germline stem cells in *Drosophila*.** *Cell Stem Cell* 2010, **7**:581–592.
8. König A, Yatsenko AS, Weiss M, Shcherbata HR: **Ecdysteroids affect *Drosophila* ovarian stem cell niche formation and early germline differentiation.** *EMBO J* 2011, **30**:1549–1562.
9. Agawa Y, Sarhan M, Kageyama Y, Akagi K, Takai M, Hashiyama K, Wada T, Handa H, Iwamatsu A, Hirose S, Ueda H: ***Drosophila* Blimp-1 is a transient transcriptional repressor that controls timing of the ecdysone-induced developmental pathway.** *Mol Cell Biol* 2007, **27**:8739–8747.
10. Akagi K, Ueda H: **Regulatory mechanisms of ecdysone-inducible *Blimp-1* encoding a transcriptional repressor that is important for the prepupal development in *Drosophila*.** *Dev Growth Deffir* 2011, **53**:697–703.
11. Martin-Creuzburg D, Westerlund SA, Hoffmann KH: **Ecdysteroid levels in *Daphnia magna* during a molt cycle: Determination by radioimmunoassay (RIA) and liquid chromatography–mass spectrometry (LC-MS).** *Gen Comp Endocrinol* 2007, **151**:66–71.
12. Yoshiyama T, Namiki T, Mita K, Kataoka H, Niwa R: **Neverland is an evolutionally conserved Rieske-domain protein that is essential for ecdysone synthesis and insect growth.** *Development* 2006, **133**:2565–2575.
13. Petryk A, Warren JT, Marques G, Jarco MP, Gilbert LI, Kahler J, Parvy J, Li Y, Dauphin-Villemant C, O'Connor MB: **Shade is the *Drosophila* P450 enzyme that mediates the hydroxylation of ecdysone to the steroid insect molting hormone 20-hydroxyecdysone.** *Proc Natl Acad Sci U S A* 2003, **100**:3773–13778.
14. Guittarda E, Blais C, Maria A, Parvy J, Pasricha S, Lumb C, Lafont R, Daborn PJ, Dauphin-Villemant C: **CYP18A1, a key enzyme of *Drosophila* steroid hormone inactivation, is essential for metamorphosis.** *Dev Biol* 2011, **349**:35–45.
15. Hopkins PM: **Crustacean ecdysteroids and their receptors.** In *Ecdysone: structures and functions.* Edited by Smagghe G. Netherlands: Springer; 2009:73–97.
16. Kato Y, Kobayashi K, Oda S, Tatarazako N, Watanabe H, Iguchi T: **Cloning and characterization of the ecdysone receptor and ultraspiracle protein from the water flea *Daphnia magna*.** *J Endocrinol* 2007, **193**:83–194.
17. Rewitz KF, Gilbert LI: ***Daphnia* Halloween genes that encode cytochrome P450s mediating the synthesis of the arthropod molting hormone: Evolutionary implications.** *BMC Evol Biol* 2008, **8**:60.
18. Asazuma H, Nagata S, Nagasawa H: **Inhibitory effect of molt-inhibiting hormone on phantom expression in the Y-organ of the kuruma prawn, *Marsupenaeus japonicus*.** *Arch Insect Biochem Physiol* 2009, **70**:220–233.
19. Tom M, Manfrin C, Giulianini PG, Pallavicini A: **Crustacean oxi-reductases protein sequences derived from a functional genomic project potentially involved in ecdysteroid hormones metabolism-A starting point for function examination.** *Gen Comp Endocrinol* 2013, **194**:71–80.
20. Marchal E, Badisco L, Verlinden H, Vandersmissen T, Soest SV, Wielendaele PV, Broeck JV: **Role of the Halloween genes, *Spook* and *Phantom* in ecdysteroidogenesis in the desert locust, *Schistocerca gregaria*.** *J Insect Physiol* 2011, **57**:1240–1248.
21. Lachaise F, Le Roux A, Hubert M, Lafont R: **The molting gland of crustaceans: localization, activity, and endocrine control (a review).** *J Crustacean Biol* 1993, **13**:198–234.
22. Lafont R, Mathieu M: **Steroids in aquatic invertebrates.** *Ecotoxicology* 2007, **16**:109–130.

23. Zhu XX, Oliver JH, Dotson EM: **Epidermis as the source of ecdysone in an argasid tick.** *Proc Natl Acad Sci U S A* 1991, **88:**3744–3747.
24. Iga M, Smagghe G: **Identification and expression profile of Halloween genes involved in ecdysteroid biosynthesis in *Spodoptera littoralis*.** *Peptides* 2010, **31:**456–467.
25. Hiruta C, Nishida C, Tochinai S: **Abortive meiosis in the oogenesis of parthenogenetic *Daphnia pulex*.** *Chromosome Res* 2010, **18:**833–840.
26. Subramonian T: **Crustacean ecdysteroids in reproduction and embryogenesis.** *Comp Biochem Physiol C* 2000, **125:**135–156.
27. Colbourne JK, Pfrender ME, Gilbert D, Thomas WK, Tucker A, Oakley TH, Tokishita S, Aerts A, Arnold GJ, Basu MK, Bauer DJ, Cáceres CE, Carmel L, Casola C, Choi J, Detter JC, Dong Q, Dusheyko S, Eads BD, Fröhlich T, Geiler-Samerotte KA, Gerlach D, Hatcher P, Jogdeo S, Krijgsveld J, Kriventseva EV, Kültz D, Laforsch C, Lindquist E, Lopez J, *et al*: **The ecoresponsive genome of *Daphnia pulex*.** *Science* 2011, **331:**555–561.
28. Soumoff C, Skinner DM: **Ecdysteroid titers during the molt cycle of the blue crab resemble those of other Crustacea.** *Biol Bull* 1983, **165:**321–329.
29. Huberman: **A shrimp endocrinology. A review.** *Aquaculture* 2000, **191:**191–208.
30. Gilbert LI: *Insect Endocrinology.* 1st edition. London: Academic; 2012.
31. Rewitz KF, Rybczynski R, Warren JT, Gilbert LI: **Identification, characterization and developmental expression of Halloween genes encoding P450 enzymes mediating ecdysone biosynthesis in the tobacco hornworm, *Manduca sexta*.** *Insect Biochem Mol Biol* 2006, **36:**188–199.
32. Ono H, Rewitz KF, Shinoda T, Itoyama K, Petryk A, Rybczynski R, Jarcho M, Warren JT, Marqués G, Shimell MJ, Gilbert LI, O'Connor MB: ***Spook* and *Spookier* code for stage-specific components of the ecdysone biosynthetic pathway in Diptera.** *Dev Biol* 2006, **298:**555–570.
33. Yamanaka N, Honda N, Osato N, Niwa R, Mizoguchi A, Kataoka H: **Differential regulation of ecdysteroidogenic P450 gene expression in the silk moth, *Bombyx mori*.** *Biosci Biotechnol Biochem* 2007, **71:**2808–2814.
34. Niwa R, Namiki T, Ito K, Shimada-Niwa Y, Kiuchi M, Kawaoka S, Kayukawa T, Banno Y, Fujimoto Y, Shigenobu S, Kobayashi S, Shimada T, Katsuma S, Shinoda T: ***Non-moulting glossy/shroud* encodes a short-chain dehydrogenase/reductase that functions in the 'Black Box' of the ecdysteroid biosynthesis pathway.** *Development* 2010, **137:**1991–1999.
35. Nagaraju GPC: **Reproductive regulators in decapod crustaceans: an overview.** *J Exp Biol* 2010, **214:**3–16.
36. Hannas BR, Wang YH, Thomson S, Kwon G, Li H, LeBlanc JA: **Regulation and dysregulation of vitellogenin mRNA accumulation in daphnids (*Daphnia magna*).** *Aquat Toxicol* 2011, **101:**351–357.
37. Lanot R, Clédon P: **Ecdysteroids and meiotic reinitiation in *Palaemon serratus* (Crustacean Decapoda natantia) and in *Locusta migratoria* (Insecta orthoptera). A comparative study.** *Invert Reprod Dev* 1989, **16:**169–175.
38. Lee HG, Seong CS, Kim YC, Davis RL, Han KA: **Octopamine receptor OAMB is required for ovulation in *Drosophila melanogaster*.** *Dev Biol* 2003, **264:**179–190.
39. Rodríguez-Valentín R, López-González I, Jorquera R, Labarca P, Zurita M, Reynaud E: **Oviduct contraction in *Drosophila* is modulated by a neural network that is both, octopaminergic and glutamatergic.** *J Cell Physiol* 2006, **209:**183–198.
40. Elendt BP, Bias WR: **Trace nutrient deficiency in *Daphnia magna* cultured in standard medium for toxicity testing, effects of the optimization of culture conditions on life history parameters of *Daphnia magna*.** *Water Res* 1990, **24:**1157–1167.
41. Kato Y, Kobayashi K, Watanabe H, Iguchi T: **Environmental sex determination in the branchiopod crustacean *Daphnia magna*: deep conservation of a *Doublesex* gene in the sex-determining pathway.** *PLoS Genet* 2011, **7:**e1001345.
42. Tamura K, Peterson D, Peterson N, Stecher G, Nei M, Kumar S: **MEGA5: Molecular evolutionary genetics analysis using maximum likelihood, evolutionary distance, and maximum parsimony methods.** *Mol Biol Evol* 2011, **28:**2731–2739.
43. Vandesompele J, de Preter K, Pattyn F, Poppe B, van Roy N, de Paepe A, Speleman F: **Accurate normalization of real-time quantitative RT-PCR data by geometric averaging of multiple internal control genes.** *Genome Biol* 2002, **3:**research0034.
44. Andersen CL, Jensen JL, Ørntoft TF: **Normalization of real-time quantitative reverse transcription-PCR data: a model-based variance estimation approach to identify genes suited for normalization, applied to bladder and colon cancer data sets.** *Cancer Res* 2004, **64:**5245–5250.
45. Yanai H: *Statcel, the Useful Add-in Software Forms on Excel.* 2nd edition. Tokyo: OMS; 2004.

# Repeated stressors in adulthood increase the rate of biological ageing

Michaela Hau[1,2*], Mark F Haussmann[4], Timothy J Greives[1,3], Christa Matlack[4], David Costantini[5,6], Michael Quetting[1,2], James S Adelman[1,7], Ana Catarina Miranda[2,8,9] and Jesko Partecke[2,8]

## Abstract

**Background:** Individuals of the same age can differ substantially in the degree to which they have accumulated tissue damage, akin to bodily wear and tear, from past experiences. This accumulated tissue damage reflects the individual's biological age and may better predict physiological and behavioural performance than the individual's chronological age. However, at present it remains unclear how to reliably assess biological age in individual wild vertebrates.

**Methods:** We exposed hand-raised adult Eurasian blackbirds (*Turdus merula*) to a combination of repeated immune and disturbance stressors for over one year to determine the effects of chronic stress on potential biomarkers of biological ageing including telomere shortening, oxidative stress load, and glucocorticoid hormones. We also assessed general measures of individual condition including body mass and locomotor activity.

**Results:** By the end of the experiment, stress-exposed birds showed greater decreases in telomere lengths. Stress-exposed birds also maintained higher circulating levels of oxidative damage compared with control birds. Other potential biomarkers such as concentrations of antioxidants and glucocorticoid hormone traits showed greater resilience and did not differ significantly between treatment groups.

**Conclusions:** The current data demonstrate that repeated exposure to experimental stressors affects the rate of biological ageing in adult Eurasian blackbirds. Both telomeres and oxidative damage were affected by repeated stress exposure and thus can serve as blood-derived biomarkers of biological ageing.

**Keywords:** Biomarker, Repeated stressors, Eurasian blackbird, Oxidative stress, Glucocorticoid, Telomere

## Introduction

Within populations of wild animals, individuals of the same age class can show vast variation in physiological and behavioural performance, as seen in self-maintenance processes like immune function and DNA repair or the degree of reproductive investment [1]. Understanding the causes and consequences of individual variation in behaviour and physiology has long been a subject of intensive research [2], and it has also permeated into the demographic and medical fields [3,4]. Variation among individuals can arise from differences in genetic quality and conditions during development, but also from both past and current health status and stressful experiences that may have caused cellular damage akin to bodily wear and tear. The extent of cellular damage that an individual has accumulated can affect physiological and behavioural traits through effects on physical condition, resource allocation and mortality risk [4-13]. However, although scientific and popular interest in the genetic, physiological and evolutionary factors underlying an individual's biological age has intensified [14-20], it is still unclear how it can be reliably assessed in wild vertebrates [20-24].

The mechanisms that underlie ageing (or senescence) are currently under lively debate [16,18,23,25,26]. One predominant concept is the oxidative stress theory of ageing, which assumes that cells continuously generate endogenous oxygen radicals which, especially when produced at elevated rates, damage essential molecules such

* Correspondence: mhau@orn.mpg.de
[1]Max Planck Institute for Ornithology, Evolutionary Physiology Group, Am Obstberg 1, 78315 Radolfzell, Germany
[2]Department of Biology, University of Konstanz, Universitätsstraße 10, 78464 Konstanz, Germany
Full list of author information is available at the end of the article

as DNA, lipids and proteins [10,14,27-29]. Another popular concept focuses on the attrition of telomeres [30-34]. Telomeres are repetitive sequences of DNA at the ends of eukaryote chromosomes that shorten with each replication event and when reaching a critical length, induce a permanent arrest in the cell cycle [35,36]. The oxidative stress theory of ageing and the telomere attrition theory are mechanistically linked since telomere shortening can be exacerbated by free radical attack [37]. A third group of hypotheses, the allostatic load/reactive scope theories suggest that the adaptive function of the catecholamine and glucocorticoid hormone systems to cope with disturbances can turn pathophysiological when becoming chronically over-activated in stressed individuals [38-41].

While these theories are not mutually exclusive and their proposed mechanisms likely interact in complex ways [21,22], they suggest potential blood-derived biomarkers that could serve as indicators of individual health and cumulative damage in wild populations: measures of oxidative stress load, telomere dynamics, and circulating concentrations of glucocorticoid hormones. While progress is being made in understanding the links between these markers and biological ageing, most studies have been limited by concentrating on single variables or certain physiological systems, or by collecting data cross-sectionally on a population level. Furthermore, to our knowledge most work on how stressful experiences in adulthood affect biomarkers of health and ageing has been conducted in humans, using correlative rather than experimental approaches [42]. In non-human vertebrates experimental studies are increasingly being carried out, but so far have mostly focused on the effects of stressful experiences early in life (e.g. [43-46]).

To improve our understanding of the factors that allow the determination of an individual's accumulated tissue damage and thus its biological age, we conducted an experimental study under common garden conditions on adult Eurasian blackbirds (*Turdus merula*). To generate differences in wear-and-tear among adult individuals with similar ontogenetic experiences (hand-raised birds kept under standardised conditions from early age onward), we exposed one group of birds to repeated immune and disturbance challenges, while a control group experienced identical conditions but was not subject to these stressors (Figure 1). Because we did not know the extent of accumulated tissue damage for each individual before we began our study, and since organisms in nature are exposed to a variety of stressors, we chose to apply a combination of the two stressors. In addition, based on the likely connections among the oxidative stress, telomere, and allostatic load theories [31], a particular strength of our approach was to explore all three of these potential mediators of biological ageing in concert (Figure 1).

Based on the oxidative stress, telomere, and allostatic load theories and the known relationships among their peripheral markers, we made the following predictions for the outcome of our experiment. First, we expected changes in general indicators of individual condition over time, such as a decrease in body mass [47] and an increase in locomotor activity [48]. These two measures are commonly used to provide an easy assessment of short-term changes in individual condition. Second, we expected that individuals from our stress-exposed group would have more telomere shortening than control birds by the end of the experiment. We chose telomere length as a likely endpoint, because studies in several vertebrate species including humans have linked both absolute telomere length as well as telomere shortening rates to individual life span and disease risk [5,7-9,11,12,49-51]. The impact of stressors on telomeres is thought to be mediated by pathways involving both glucocorticoid hormones and oxidative stress [13,26,28,31,46,49,52-55]. Glucocorticoid hormones may also increase oxidative stress and thereby further contribute to telomere shortening rates [46]. Therefore, since both circulating baseline and stress-induced glucocorticoid concentrations have been used traditionally to assess an individual's responses to stressors [38,56,57], we thirdly predicted that individuals from our stress-exposed group would show increased concentrations of baseline corticosterone (the main glucocorticoid in birds), muted peak corticosterone secretion after disturbance, and a delayed negative feedback (shutdown of corticosterone secretion once a stressor has abated). Such traits are typically considered hallmarks of chronically stressed individuals (e.g., [38,56,57], but see

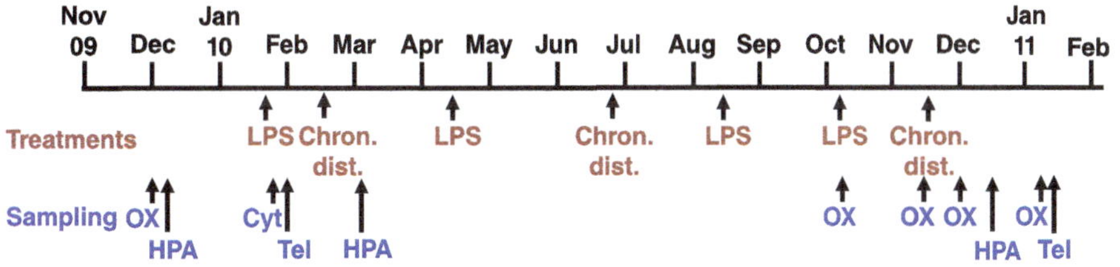


**Figure 1 Timeline of experiment.** Treatment types: LPS: LPS injections, Chron. dist.: chronic disturbance. Biomarker sampling: Tel: telomeres, OX: oxidative stress, HPA: glucocorticoid hormones, CYT: cytokines.

[47]). Fourth, we expected stress-exposed individuals to display a greater oxidative stress load as oxidative stress can increase following demanding experiences such as high physical activity [7,27,58], immune challenges [59-61], egg production [46,62], or embryonic exposure to stress hormones [46]. Oxidative stress is thought to affect telomere dynamics on the one hand by directly targeting telomeric DNA, which is particularly vulnerable to oxidative damage [37,63], and on the other hand by damaging telomerase, the enzyme that can rebuild telomeres [24,31,37]. We further predicted that stress-exposed birds will up-regulate antioxidant defences compared to control birds. An up-regulation of antioxidant defences may also only occur initially; as long-term stress exposure may eventually overwhelm self-maintenance systems.

## Results

As a long-term consequence of the treatment, telomere dynamics of birds from the two groups diverged: while birds from both treatment groups decreased genome-wide telomere lengths, stress-exposed birds showed a stronger decrease in telomere lengths from the beginning to the end of the experiment than control birds (Figure 2a; all statistical results including notes on transformations are presented in Table 1). All birds decreased circulating oxidative damage over time (Figure 2b), but stress-exposed birds showed a less robust decrease than control birds, thus maintaining comparatively higher levels of oxidative damage. Plasma non-enzymatic antioxidants decreased (Figure 2c) and glutathione peroxidase concentrations in red blood cells increased over time (Figure 2d), but there were no differences between groups, respectively. All birds tended to increase body condition over time, but there were no group differences (Figure 2e). However, locomotor activity in the two groups was differentially affected by treatment: control birds decreased while stress-exposed birds maintained or slightly increased their activity rates (Figure 2f). We did not detect changes over time or differences between groups in any measure of the endocrine stress axis (Figure 2g-k).

## Discussion

More rapid decreases in telomere lengths generally indicate a more rapid rate of cellular ageing [30-34]. We found that chronic stress exposure in hand-raised Eurasian blackbirds resulted in greater telomere loss compared to control birds, providing experimental evidence that one outcome of stress exposure in adult blackbirds is an increase in cumulative cellular damage. While we were not able to follow the survival of our experimental birds, it is possible that our stress-exposed blackbirds could have suffered a greater mortality risk. Indeed, previous studies on vertebrate species including humans have suggested that individuals with shorter telomeres or an increased rate of telomere shortening have shorter life spans [5,7-9,11,12,49-51]. Importantly, while a growing number of studies suggest that early life stress can hasten telomere loss [43-46], our study is one of the first experimental tests indicating that stress in adulthood results in more rapid telomere loss and an increase in cellular ageing.

Over the course of the current experiment, stress-exposed birds maintained higher levels of plasma oxidative damage than control birds (Figure 2; Table 1). This corroborates previous findings in which relationships between immune challenges, psychological stress and oxidative damage have been demonstrated [10,28,59-61]. We can currently only speculate as to why there was a general decrease in plasma levels of oxidative compounds over time in both groups. It is possible that this decrease is the result of prolonged captivity (in control birds it was paralleled by a decrease in locomotor activity). Nevertheless, stress-exposed birds showed a significantly smaller decrease, thus ending up with a higher level of circulating oxidative compounds than control birds at the end of the experiment.

We determined circulating concentrations of oxidative damage and non-enzymatic antioxidants from plasma while glutathione peroxidase and telomere measures were obtained from red blood cells. By determining biomarkers from blood samples we aimed at quantifying systemic effects of the stressors rather than effects on specific tissues. Furthermore, this technique enabled us to repeatedly sample individuals to assess longitudinal effects of stressful experiences. Plasma and red blood cells are both components of the same type of connective tissue and preliminary data show a significant covariation of measures derived from blood components with oxidative processes in other tissues [64]. Also, a covariation among oxidative markers of red blood cells and of plasma similar to those used in the present study was demonstrated in another songbird species, although the strength of covariation can change with the intensity of a challenge [58]. Furthermore, telomere lengths in blood cells and other tissues are correlated in humans [65] and birds [66]. Given that both oxidative stress and telomere measures correlate among tissues, we feel justified in using markers in blood (cells and plasma) as proxies for what is happening in the whole organism.

Contrary to our predictions, specific effects of treatment could not be detected in other potential biomarkers like antioxidants, and all four corticosterone traits that we assessed (Figure 2; Table 1). There are a few possible explanations for why our predictions regarding these markers were not supported. One possibility is that the benign conditions in captivity might have mitigated the effects of the stress exposure on these biomarkers. An alternative explanation could involve the limited sample sizes, which could have prevented us from

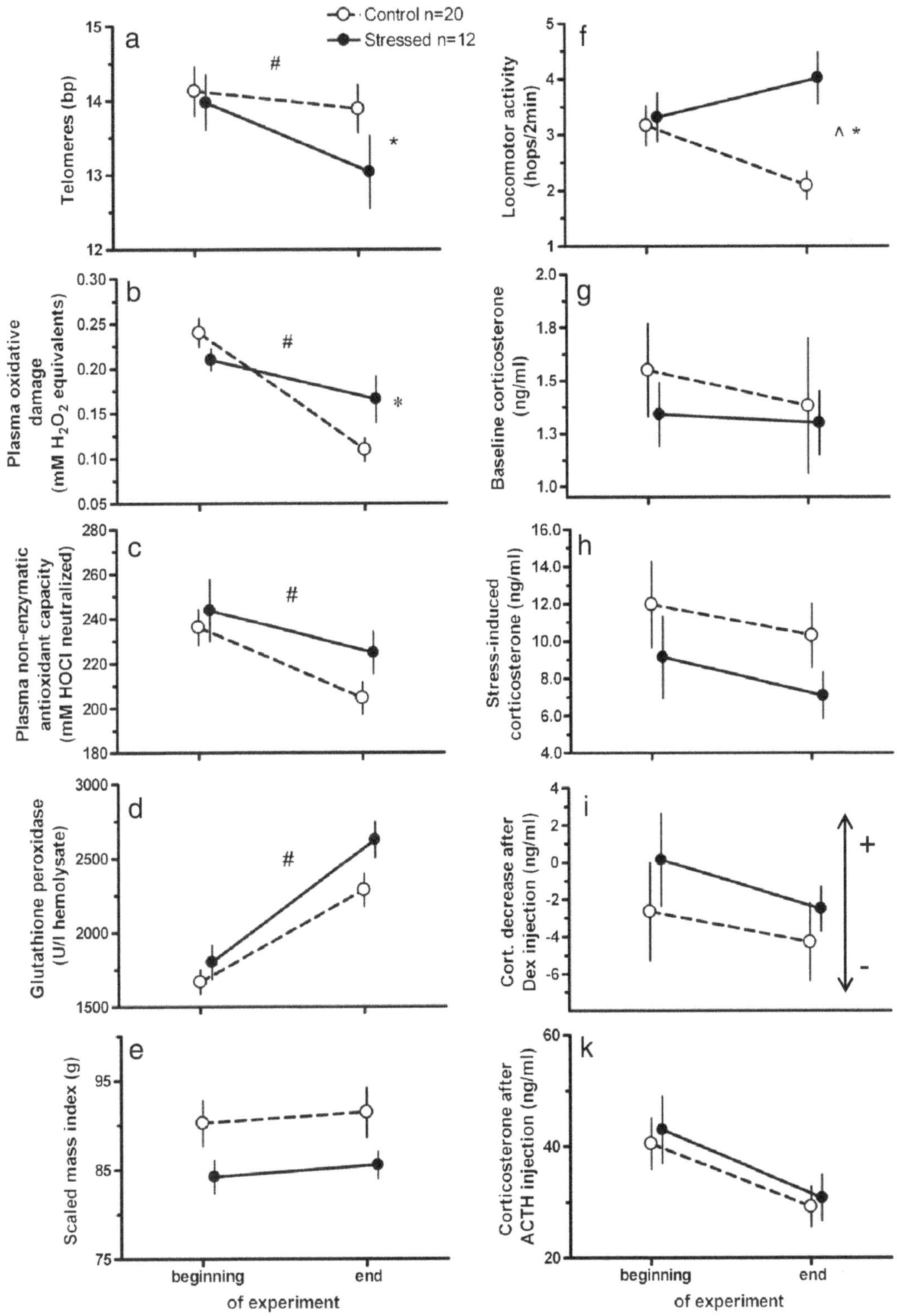


**Figure 2 Long-term effects of treatment on potential biomarkers. (a)** Telomere length, **(b)** plasma oxidative damage, **(c)** plasma non-enzymatic antioxidant capacity, **(d)** red blood cell glutathione peroxidase, **(e)** Scaled Mass Index, **(f)** locomotor activity, and plasma concentrations of **(g)** baseline corticosterone, **(h)** stress-induced corticosterone, **(i)** difference in corticosterone between stress-induced levels and concentrations after injection of dexamethasone (please note that positive values indicate strong (+) negative feedback whereas negative values show weak (–) negative feedback , see also arrow), **(k)** corticosterone after injection of ACTH. Open circles: control group (n = 20), filled circles: stress-exposed group (n = 12). All data are mean ± 1SEM. Results as derived from LMMs (see text): * interaction time*treament $p < 0.05$, ^ treatment $p < 0.05$, # time $p < 0.05$.

detecting significant group differences. Indeed, a visual examination of Figure 2 suggests that both plasma non-enzymatic antioxidant and red blood cell glutathione peroxidase concentrations tend to be higher in stress-exposed compared to control birds, in line with our prediction of an up-regulation of antioxidant processes in individuals that show higher concentrations of damaging compounds in their plasma. And finally, it is also possible that some biomarkers indeed were not affected much by the stress exposure, either because they do not represent appropriate biomarkers for cumulative tissue damage or because the organism is resilient and can cope with the treatment for these specific traits.

Our analyses of the short-term effects of the two treatment types (Additional file 1: Table S1, Additional file 2: Table S2) indicated that both treatments (immune and disturbance challenges) caused significant group divergences in body mass, in plasma oxidative damage and in cytokines (measured only after immune challenge), but not in antioxidants or endocrine traits (only measured after the chronic disturbance treatment). Interestingly, plasma oxidative damage showed opposite short-term responses in stress-exposed birds, increasing following LPS injection (Additional file 1: Table S1) but decreasing following chronic disturbance (Additional file 2: Table S2). Due to logistical constraints we ended up collecting samples for plasma oxidative damage following the first injection of LPS in January 2010, at the beginning of the experiment (see Figure 1), while plasma oxidative damage following chronic disturbance were assessed during the last chronic disturbance period in November 2010 (Additional file 2: Table S2, Figure 1). While an increase in plasma oxidative damage following LPS injection is expected after an activation of the immune system [59-61], a decrease following a disturbance stressor is not. However, this decrease in plasma oxidative damage in stress-exposed birds is actually the result of elevated levels at the beginning of the chronic disturbance period and a subsequent decrease to the levels that control birds displayed throughout (Additional file 2: Table S2). Hence, part of this pattern might be explained by stress-exposed birds having increased levels of oxidative stress at this late stage of the experiment before the chronic disturbance started. Why these levels then decreased during the 10 days of disturbance stressor is currently unclear, but may be related to increased clearance from the circulation since stress-exposed birds tended to have higher levels of plasma non-enzymatic antioxidants and red blood cell glutathione peroxidase than control birds (Additional file 2: Table S2).

Over the course of the year-long experiment, we also observed changes in general indicators of individual condition. Body mass increased over time, but there were no discernible differences between the two groups. However, stress-exposed birds displayed greater locomotor activity than control birds by the time the experiment was ended (Figure 2; Table 1). Further experiments are needed to determine whether increased activity rates are the end result of exposure to stressors or whether they also directly contributed to increased oxidative damage in the stress-exposed group as previously shown in another song bird species ([58]).

Among the strengths of the present study are the experimental approach but also the repeated assessment of individuals. Our results confirm that repeated sampling of focal individuals is important for interpreting studies on cumulative damage and cellular senescence. In the present study, many of the potential biomarkers assessed showed significant changes over time, even in control individuals (Figure 2). We consider it likely that they changed as part of the natural ageing process since we aimed at sampling the birds at the same time of year at the

**Table 1 Effects of treatment on potential biomarkers**

| | Effect of treatment | Effect of time | Interaction treatment*time |
|---|---|---|---|
| Genome-wide telomere length (bp) | $p > 0.28$ | **$F_{(1,32)} = 28.81$, $p < 0.0005$** | **$F_{(1,32)} = 11.47$, $p = 0.002$** |
| Plasma oxidative damage (mM $H_2O_2$ equivalents) | $p > 0.2$ | **$F_{(1,30.99)} = 42.46$, $p < 0.0005$** | **$F_{(1, 30.99)} = 6.84$, $p = 0.014$** |
| Plasma non-enzymatic antioxidants (mM HOCl neutralized) | $p > 0.1$ | **$F_{(1,29.75)} = 8.49$, $p = 0.007$** | $p > 0.3$ |
| Glutathione peroxidase (U/l hemolysate) | $p > 0.1$ | **$F_{(1,30.71)} = 64.54$, $p < 0.0005$** | $p > 0.26$ |
| Scaled mass index | $p > 0.12$ | $F_{(1,32)} = 3.65$, $p = 0.065$ | $p > 0.7$ |
| Locomotor activity (hops/min.) | **$F_{(1,32)} = 5.18$, $p = 0.03$** | $p > 0.1$ | **$F_{(1,32)} = 21.84$, $p < 0.0005$** |
| Baseline corticosterone (ng/ml) | $p > 0.7$ | $p > 0.3$ | $p > 0.5$ |
| Stress-induced corticosterone (ng/ml) | $p > 0.3$ | $p > 0.4$ | $p > 0.8$ |
| Difference between stress-induced corticosterone and concentrations after dexamethasone injection (ng/ml) | $p > 0.1$ | $p > 0.6$ | $p > 0.7$ |
| Corticosterone after ACTH injection (ng/ml) | $p > 0.5$ | $p > 0.09$ | $p > 0.6$ |

Effects of stress-exposure (repeated LPS injections and 10-day chronic disturbance protocols combined) on potential biomarkers (all log-transformed) assessed at the beginning and the end of the experiment, as determined from LMM statistics (for details see text). Significant effects are highlighted in bold. Data are shown in Figure 2.

beginning and the end of the experiment to control for any seasonal variation. However, the repeated measures design allowed us to disentangle effects of time from effects that were specific to treatment.

Analyses of intra-class correlation coefficients (ICCs; [67]) of potential biomarkers can provide a measure of their within-individual repeatability and thus their reliability in indicating an individual's health. Keeping in mind the limitations of our data set (limited sample sizes and number of repeated measurements), ICC analyses suggested that telomere lengths, scaled mass index and stress-induced corticosterone concentrations were highly repeatable traits of individuals (Additional file 3: Table S3). Plasma oxidative damage and glutathione peroxidase levels also showed significant ICCs, although with lower coefficients. Hence, since the variables telomere length and oxidative damage were the ones that showed significant treatment effects as well as significant repeatability, this suggest that these measures, in particular telomere length, are relevant biomarkers for an individual's health and senescence status.

## Conclusions

Effects of treatment on telomere dynamics and oxidative damage were observed in adult, known-age individuals held under relatively benign conditions in the laboratory including mild temperatures, absence of predation, and food *ad libitum*. Consequences of exposure to stressors may be more pronounced in free-living individuals that are exposed to an array of more severe challenges and a greater limitation of resources than the captive individuals in our study. Other long-term stressors present in nature, for example social challenges, predation pressure, infections, parasites, as well as anthropogenic disturbances, noise, or pollutants may similarly affect telomere dynamics and oxidative damage, possibly reducing the fitness of individuals in the wild. Experimental stress-exposure resulted in telomere shortening and increased concentrations of plasma oxidative damage, which together with the significant repeatability in these measures, suggest that these traits can serve as biomarkers for health and senescence in individuals of a passerine bird, and possibly other taxa as well. Future research will be important for unravelling the causal connections among the different biomarkers.

## Methods

### Study species

The Eurasian blackbirds used for this experiment were 2.5 years old at the start of the experiment (life span in the wild has been estimated at 1.8-2.8 years, but upon reaching adulthood, Eurasian blackbirds live can live up to 17 years [68]). The birds (sample size: 49) were originally collected in 2007 at an age of 5–11 days from 9 urban and 9 rural nest (see Additional file 4 for details) and hand-raised under identical conditions (see [69]). Here we only analysed data from individuals for which we were able to obtain data for all biomarkers before and after the experiment (n = 32, 20 control [11 male, 9 female], 12 stressed [8 male, 4 female] birds). This conservative approach allowed us to keep the sample size and identity of birds identical across different types of analyses. However, analyses of the entire data set (including individuals for which biomarkers could not be obtained for all time points) gave similar results. Because of the limited sample size and the concomitant low power in complex statistical tests we also refrained from analysing differences in responses to treatment of urban and rural birds (but see [70]).

Birds were kept in individual recording cages and were exposed to a simulated local natural photoperiod in Radolfzell, Germany (see Additional file 4 for details). Before the experiment started we measured tarsus length of all birds as a measure of structural size to the nearest 0.1 mm with a dial calliper and body mass with a digital balance to the nearest 0.1 g. Body mass was assessed repeatedly during the experiment, when we caught the birds for blood sampling. When establishing experimental groups we made sure that sex and origin were balanced among control and stress-exposed groups. The stress-exposed group was exposed to one of two alternating stress treatments about every 6 weeks (for timeline and sequence of treatments see Figure 1): a.) an acute immune challenge (see below; four times over the course of the experiment), and b.) a chronic stress disturbance regime (see below; a total of three times; Figure 1).

### Stress treatments

We injected a dose of 2.0 μg LPS (Sigma L2880) diluted in phosphate-buffered saline (PBS) per gram body mass into the breast muscle (see Additional file 4 for details). Control birds were injected with PBS only (but only for the first two periods when cytokine concentrations after injection were compared between treatment groups). The concentration of LPS used successfully induces an acute phase response including fever and sickness behaviour in other songbird species [71,72]. For verification of LPS injections on cytokine concentrations in this study see Additional file 1: Table S1.

Our chronic stress disturbance regime represented a milder version of existing chronic stress protocols for small passerines (i.e. fewer disturbances per day and fewer days with disturbances than in [73,74]). Every day, for 10 consecutive days, we applied each of the following four treatments to birds in the stressed group, in a random order and at random times, but always during daylight hours: 30 min of chasing (waving a catching net with a yellow plastic bag attached in front of and over the top of the cage for a conspicuous and noisy disturbance), 30 min of crowding (adding 2–3 other birds to a single cage), 60 min of restraint (putting an individual into a cloth bag) and 60 min of loud radio playing in the room. We used this protocol in trying to avoid habituation to a single stressor.

**Biomarkers**

In total, stress-exposed birds were subject to 4 immune challenges and 3 chronic disturbance stress periods during the year-long experiment. All control and stress-exposed birds were sampled for biomarkers at the same time periods (see Figure 1). Blood samples were collected from the wing vein in heparinised capillary tubes and stored on ice until centrifugation. The plasma fraction was frozen for later hormone and oxidative stress analyses, the red blood cells either stored in cryoprotectant buffer for telomere analyses or frozen and stored at −80°C for glutathione peroxidase analyses (see Additional file 4 for details and sampling times). To determine the long-term effects of treatment, we measured the oxidative status of plasma and red blood cells at the beginning (Dec 2009) and at the end of the experiment (Jan 2011, Figure 1). To verify the efficacy of our treatments and to confirm that our treatments were capable of inducing changes in oxidative status, blood samples were collected in October 2010 to assess the short-term effects of LPS injection on oxidative status, and in November 2010 samples were collected to assess the short-term effects of the chronic stress protocol.

Plasma oxidative damage was determined using the d-ROMs test (Diacron International, Grosseto, Italy). This assay mostly measures oxidative damage compounds generated early in the oxidative cascade (i.e., hydroperoxides); these compounds are precursors of several end-products of lipid peroxidation, such as malondialdehyde, hydroxynonenal and isoprostanes. The reaction of a dilution series of cumene hydroperoxide with the d-ROM reagents was highly linear (range: 0 to 4.5 μM, $R^2$ = 0.9996; physiological values in vertebrates). The OXY-Adsorbent test (Diacron International) was used to quantify the ability of plasma non-enzymatic antioxidant compounds (vitamins, carotenoids, uric acid, thiol proteins) to react in vitro with HOCl (oxidant of pathologic relevance in biological systems). The concentration of glutathione peroxidase in red blood cells was quantified using the Ransel assay (Randox Laboratories, Crumlin, UK). For further details on all procedures including assay quality see Additional file 4.

Total locomotor activity of individual birds was recorded continuously and binned in two-minute-intervals using passive infrared motion detectors (see Additional file 4 for more details). For each individual bird, we calculated mean activity over five complete days during the same periods (20-24th Jan) at the beginning (2010) and the end of the experiment (2011). Such averaging of activity rates was done to limit any unexpected differences in circadian variations in locomotor activity among individuals and between groups. Comparing the activity levels between the two groups at the same time of year allowed us to distinguish whether differences in individual activity were due to treatment versus age effects while controlling for possible seasonal changes in activity.

Telomeres were measured with the Telomere Restriction Fragment (TRF) assay and the procedure was carried out on whole blood as in previous studies [46], see Additional file 4 for more details, including assay quality). Leukocytes account for less than 0.2% of the blood volume in birds, and so TRF measurements largely reflect erythrocyte telomeres.

Our protocol for the assessment of the functioning of the endocrine stress axis was modified from existing protocols [73,75,76]. We conducted the stress axis assessment on a subset of individuals from both groups each day, sampling as many individuals as was feasible within three minutes of entering the room. Remaining individuals were sampled on consecutive days in an identical manner, with sample collection of the entire study population being completed within a total of 3–4 days for each sampling period (always changing the sequence of sampled individuals between each sampling period). Samples (100 μl blood) for baseline corticosterone concentrations (BaseCort) were always taken within 3 min of entering the experimental room. Stress-induced samples (100 μl blood, StressCort) were taken after 30 min of restraint in individual cloth bags. Birds then were injected intramuscularly with a dose of 200 μg/kg dexamethasone (DEX; stock solution 4 mg/ml, Bela-Pharm, Germany) dissolved in 0.9% saline (Braun), to induce negative feedback and a concomitant reduction in plasma Cort concentrations. Blood samples were taken 90 minutes following injection with DEX (DEXCort, calculated as [StressCort - corticosterone concentrations after DEX injection] to determine the strength of negative feedback; please note that due to this calculation positive values indicate strong negative feedback while negative values indicate that corticosterone concentrations after DEX injections were actually higher than after StressCort). Following this sample, we injected the individuals intramuscularly with a dose of 100 IU/kg of adrenocorticotropic hormone (ACTH; Sigma-Aldrich) to assess the maximal capacity of the adrenal gland to secrete corticosterone. A final blood sample was taken 30 min following ACTH injection (ACTHCort). Following this final blood sample, all birds were returned to their home cages. We determined plasma hormone concentrations using radioimmunoassays (see [77], following a double ether extraction of all samples (see Additional file 4 for more details including assay quality).

### Statistics

Body condition was calculated using the Scaled Mass Index [78]. Data were tested for normal distribution and homogeneity of variances using a combination of Kolmogorov-Smirnov and Levene tests as well as visual observations of histograms of the models' residuals. Data were then transformed (as indicated in the Results section) appropriately before statistical analysis.

Short-term effects of the two treatment types on potential biomarkers were analysed separately (for detailed statistical methods and results see Additional file 4) and for reasons of divergent sample sizes and sampling times could not be included with analyses of long-term effects.

To analyse the long-term effects of the treatment, we compared repeated measurements of individuals from our two groups taken at the beginning with the ones taken at the end of the experiment. By analysing samples taken 1 year apart from each other we could eliminate possible seasonal variations) in telomeres, body mass, oxidative status, and stress hormones. We used Linear Mixed Models (LMMs, using random slopes and random intercepts if those improved the model fit) as described above, always including treatment, time, the interaction of treatment with 'time' as fixed factors and individual ID, as well as ID nested within nest ID as random variables. The variable 'sex' and the interactions of 'sex*time', 'sex*group' and 'sex*time*group' were first included but then removed from the models, as they did not explain a significant proportion of the variation.

All statistics were done in SPSS v 21.0 (Chicago). Data are publically available at the Dryad data repository (https://datadryad.org).

## Additional files

**Additional file 1: Table S1.** Efficacy of the immune challenge.

**Additional file 2: Table S2.** Efficacy of the 10-day chronic disturbance protocol on stress hormones and oxidative stress load.

**Additional file 3: Table S3.** Intraclass correlation coefficients (ICCs, [95% confidence intervals]) for all biomarkers.

**Additional file 4: General methodology.**

### Competing interests

The authors declare that they have no competing interests.

### Authors' contributions

MH, TJG and JP designed the study, MH, TJG, MQ, ACM and JP collected samples, MFH, ACM, DC, MQ, JSA analysed samples and data, MH performed statistics, MH wrote the manuscript with contributions from TJG, MFH, DC, JSA, ACM and JP. All authors read and approved the final manuscript.

### Acknowledgements

For crucial assistance during experiments we thank J. Lodde, E. Fricke, H. Schmid, G. Beltrami, M. Romero, D. Santos, D. Dominoni, K.H. Siebenrock, M. Echeverry, K. Mortega, N. Perfito, G. Bentley, A. Schmidt, F. Wikelski, M. van Toor. Excellent bird care was provided by T. Vogler, K. Greeven, E. Jäckle, A. Falk, P. Sigwart, M. Schoenwald, R. Dunz. Many thanks for helpful discussions to M. Romero, W. Goymann and the BEEP lab group. We appreciate the constructive comments of several anonymous reviewers on previous versions of the manuscript. Funding was provided by the Max Planck Society to M.H. and by the Volkswagen Foundation (Initiative 'Evolutionary Biology') to J.P.

### Author details

[1]Max Planck Institute for Ornithology, Evolutionary Physiology Group, Am Obstberg 1, 78315 Radolfzell, Germany. [2]Department of Biology, University of Konstanz, Universitätsstraße 10, 78464 Konstanz, Germany. [3]Department of Biological Sciences, North Dakota State University, 1340 Bolley Drive, Fargo, ND 58202, USA. [4]Department of Biology, Bucknell University, Lewisburg, PA 17837, USA. [5]Department of Biology, University of Antwerp, Universiteitsplein 1, 2610 Wilrijk, Belgium. [6]University of Glasgow, Institute of Biodiversity, Animal Health & Comparative Medicine, Glasgow QG12 8Q, UK. [7]Department of Biological Sciences, 4092B Derring Hall, Virginia Tech, Blacksburg, VA 24061-0406, USA. [8]Max Planck Institute for Ornithology, Department of Migration and Immuno-ecology, Am Obstberg 1, 78315 Radolfzell, Germany. [9]Departamento de Biologia, Centro de Ciências Biológicas e da Saúde, Universidade Federal do Maranhão, Campus do Bacanga, São Luís, Maranhão, Brazil.

### References

1. Stearns SC. The Evolution of Life Histories. Oxford: Oxford University Press; 1992.
2. Bennett AF. Interindividual variability: an underutilized resource. In: Feder ME, Bennett AF, Burggren WW, Huey RB, editors. New Directions in Ecological Physiology. Cambridge, UK: Cambridge University Press; 1987. p. 147–69.
3. Vaupel JW, Carey JR, Christensen K, Johnson TE, Yashin AI, Holm NV, et al. Biodemographic trajectories of longevity. Science. 1998;280:855–60.
4. Seeman TE, McEwen BS, Rowe JW, Singer BH. Allostatic load as a marker of cumulative biological risk: MacArthur studies of successful aging. Proc Natl Acad Sci U S A. 2001;98:4770–5.
5. Haussmann MF, Winkler DW, Vleck CM. Longer telomeres associated with higher survival in birds. Biol Lett. 2005;1:212–4.
6. Gruenewald TL, Seeman TE, Ryff CD, Karlamangla AS, Singer BH. Combinations of biomarkers predictive of later life mortality. Proc Natl Acad Sci U S A. 2006;103:14158–63.
7. Ilmonen P, Kotrschal A, Penn DJ. Telomere Attrition Due to Infection. Plos One. 2008;3:e2143.
8. Bize P, Criscuolo F, Metcalfe NB, Nasir L, Monaghan P. Telomere dynamics rather than age predict life expectancy in the wild. Proceed Royal Soc B-Biol Sci. 2009;276:1679–83.
9. Salomons HM, Mulder GA, van de Zande L, Haussmann MF, Linskens MHK, Verhulst S. Telomere shortening and survival in free-living corvids. Proceed Royal Soc B-Biol Sci. 2009;276:3157–65.
10. Metcalfe NB, Alonso-Alvarez C. Oxidative stress as a life-history constraint: the role of reactive oxygen species in shaping phenotypes from conception to death. Funct Ecol. 2010;24:984–96.
11. Heidinger BJ, Blount JD, Boner W, Griffiths K, Metcalfe NB, Monaghan P. Telomere length in early life predicts lifespan. Proc Natl Acad Sci U S A. 2012;109:1743–8.
12. Bauch C, Becker PH, Verhulst S. Telomere length reflects phenotypic quality and costs of reproduction in a long-lived seabird. Proceed Royal Soc B-Biol Sci. 2013; 280 DOI: 10.1098/rspb.2012.2540
13. Epel ES. Psychological and metabolic stress: A recipe for accelerated cellular aging? Hormones-Int J Endocrinol Metab. 2009;8:7–22.
14. Finkel T, Holbrook NJ. Oxidants, oxidative stress and the biology of ageing. Nature. 2000;408:239–47.
15. Finch CE, Crimmins EM. Inflammatory exposure and historical changes in human life-spans. Science. 2004;305:1736–9.
16. Monaghan P, Charmantier A, Nussey DH, Ricklefs RE. The evolutionary ecology of senescence. Funct Ecol. 2008;22:371–8.
17. Bouwhuis S, Choquet R, Sheldon BC, Verhulst S. The forms and fitness cost of senescence: Age-specific recapture, survival, reproduction, and reproductive value in a wild bird population. Am Nat. 2012;179:E15–27.
18. Gems D, Partridge L: Genetics of Longevity in Model Organisms: Debates and Paradigm Shifts. In Annual Review of Physiology, Volume 75. Edited by Julius D. Palo Alto: Annual Reviews; 2013: 621–644

19. Nussey DH, Froy H, Lemaitre JF, Gaillard JM, Austad SN. Senescence in natural populations of animals: Widespread evidence and its implications for bio-gerontology. Ageing Res Rev. 2013;12:214–25.
20. Butler RN, Sprott R, Warner H, Bland J, Feuers R, Forster M, et al. Biomarkers of aging: From primitive organisms to humans. J Gerontol Series Biol Sci Med Sci. 2004;59:560–7.
21. Vleck CM, Haussmann MF, Vleck D. Avian senescence: underlying mechanisms. J Ornithol. 2007;148:S611–24.
22. Holmes D, Martin K. A bird's-eye view of aging: What's in it for ornithologists? Auk. 2009;126:1–23.
23. Austad SN. Methusaleh's Zoo: How nature provides us with clues for extending human health span. J Comp Pathol. 2010;141:S10–21.
24. Sanders JL, Newman AB: Telomere Length in Epidemiology. A Biomarker of Aging, Age-Related Disease, Both, or Neither? Epidemiol Reviews 2013; 10.1093/epirev/mxs008
25. Bonsall MB. Longevity and ageing: appraising the evolutionary consequences of growing old. Phil Trans Royal Soci B Biol Sci. 2006;361:119–35.
26. Ricklefs RE. The evolution of senescence from a comparative perspective. Funct Ecol. 2008;22:379–92.
27. Costantini D. Oxidative stress in ecology and evolution: lessons from avian studies. Ecol Lett. 2008;11:1238–51.
28. Schiavone S, Jaquet V, Trabace L, Krause KH. Severe life stress and oxidative stress in the brain: from animal models to human pathology. Antioxidants Redox Signal. 2013;18:1475–90.
29. Selman C, Blount JD, Nussey DH, Speakman JR. Oxidative damage, ageing, and life-history evolution: where now? Trends Ecol Evol. 2012;27:570–7.
30. Monaghan P, Haussmann MF. Do telomere dynamics link lifestyle and lifespan? Trends Ecol Evol. 2006;21:47–53.
31. Haussmann MF, Marchetto NM. Telomeres: Linking stress and survival, ecology and evolution. Curr Zool. 2010;56:714–27.
32. Monaghan P. Telomeres and life histories: the long and the short of it. In Year in Evolutionary Biology. Volume 1206. Edited by Schlichting CD, Mousseau TA; 2010: 130–142.
33. Boonekamp JJ, Simons MJP, Hemerik L, Verhulst S. Telomere length behaves as biomarker of somatic redundancy rather than biological age. Aging Cell. 2013;12:330–2.
34. Vleck C, Vleck D, Palacios M. Evolutionary ecology of senescence: a case study using tree swallows, *Tachycineta bicolor*. J Ornithol. 2011;152:203–11.
35. Capper R, Britt-Compton B, Tankimanova M, Rowson J, Letsolo B, Man S, et al. The nature of telomere fusion and a definition of the critical telomere length in human cells. Genes Dev. 2007;21:2495–508.
36. Hornsby PJ. Replicative senescence of human and mouse cells in culture: significance for aging research. Mech Ageing Dev. 2003;124:853–5.
37. von Zglinicki T. Oxidative stress shortens telomeres. Trends Biochem Sci. 2002;27:339–44.
38. McEwen BS, Wingfield JC. The concept of allostasis in biology and biomedicine. Horm Behav. 2003;43:2–15.
39. Romero LM, Dickens MJ, Cyr NE. The reactive scope model – A new model integrating homeostasis, allostasis, and stress. Horm Behav. 2009;55:375–89.
40. Juster RP, McEwen BS, Lupien SJ. Allostatic load biomarkers of chronic stress and impact on health and cognition. Neurosci Biobehav Rev. 2010;35:2–16.
41. Koolhaas JM, Bartolomucci A, Buwalda B, de Boer SF, Flugge G, Korte SM, et al. Stress revisited: A critical evaluation of the stress concept. Neurosci Biobehav Rev. 2011;35:1291–301.
42. Monaghan P. Organismal stress, telomeres and life histories. J Exp Biol. 2014;217:57–66.
43. Nettle D, Monaghan P, Gillespie R, Brilot B, Bedford T, Bateson M. An experimental demonstration that early-life competitive disadvantage accelerates telomere loss. Proceed Royal Soc B-Biol Sci. 2015;282:20141610.
44. Herborn KA, Heidinger BJ, Boner W, Noguera JC, Adam A, Daunt F, et al. Stress exposure in early post-natal life reduces telomere length: an experimental demonstration in a long-lived seabird. Proceed Royal Soc B-Biol Sci. 2014;281:20133151.
45. Boonekamp JJ, Mulder GA, Salomons HM, Dijkstra C, Verhulst S. Nestling telomere shortening, but not telomere length, reflects developmental stress and predicts survival in wild birds. Proceed Royal Soc B-Biol Sci. 2014;281:20133287.
46. Haussmann MF, Longenecker AS, Marchetto NM, Juliano SA, Bowden RM. Embryonic exposure to corticosterone modifies the juvenile stress response, oxidative stress and telomere length. Proceed Royal Soci B-Biol Sci. 2012;279:1447–56.
47. Dickens MJ, Romero LM. A consensus endocrine profile for chronically stressed wild animals does not exist. Gen Comp Endocrinol. 2013;191:177–89.
48. Adams NJ, Farnworth MJ, Rickett J, Parker KA, Cockrem JF. Behavioural and corticosterone responses to capture and confinement of wild blackbirds *(Turdus merula)*. Appl Anim Behav Sci. 2011;134:246–55.
49. Epel ES, Blackburn EH, Lin J, Dhabhar FS, Adler NE, Morrow JD, et al. Accelerated telomere shortening in response to life stress. Proc Natl Acad Sci U S A. 2004;101:17312–5.
50. Bauch C, Becker PH, Verhulst S. Within the genome, long telomeres are more informative than short telomeres with respect to fitness components in a long-lived seabird. Mol Ecol. 2014;23:300–10.
51. Barrett ELB, Burke TA, Hammers M, Komdeur J, Richardson DS. Telomere length and dynamics predict mortality in a wild longitudinal study. Mol Ecol. 2013;22:249–59.
52. Cattan V, Mercier N, Gardner JP, Regnault V, Labat C, Maki-Jouppila J, et al. Chronic oxidative stress induces a tissue-specific reduction in telomere length in CAST/Ei mice. Free Radic Biol Med. 2008;44:1592–8.
53. Costantini D, Fanfani A, Dell'Omo G. Effects of corticosteroids on oxidative damage and circulating carotenoids in captive adult kestrels *(Falco tinnunculus)*. J Comparative Physiol B-Biochem Syst Environ Physiol. 2008;178:829–35.
54. Costantini D, Marasco V, Moller AP. A meta-analysis of glucocorticoids as modulators of oxidative stress in vertebrates. J Comparative Physiol B-Biochem Syst Environ Physiol. 2011;181:447–56.
55. Geiger S, Le Vaillant M, Lebard T, Reichert S, Stier A, Le Maho Y, et al. Catching-up but telomere loss: half-opening the black box of growth and ageing trade-off in wild king penguin chicks. Mol Ecol. 2012;21:1500–10.
56. Sapolsky RM, Romero LM, Munck AU. How do glucocorticoids influence stress responses? Integrating permissive, suppressive, stimulatory, and preparative actions. Endocr Rev. 2000;21:55–89.
57. Romero LM. Physiological stress in ecology: lessons from biomedical research. Trends Ecol Evol. 2004;19:249–55.
58. Costantini D, Monaghan P, Metcalfe NB. Loss of integration is associated with reduced resistance to oxidative stress. J Exp Biol. 2013;216:2213–20.
59. Sirak AA, Beavis AJ, Robertson FM. Enhanced hydroperoxide production by peripheral blood leucoyctes following exposure of murine epidermis to 12-O-tetradecanoylphorbol-13-acetate. Carcinogenesis. 1991;12:91–5.
60. Costantini D, Møller AP. Does immune response cause oxidative stress in birds? A meta-analysis. Comparative Biochem Physiol Part Mol Int Physiol. 2009;153:339–44.
61. van de Crommenacker J, Horrocks NPC, Versteegh MA, Komdeur J, Tieleman BI, Matson KD. Effects of immune supplementation and immune challenge on oxidative status and physiology in a model bird: implications for ecologists. J Exp Biol. 2010;213:3527–35.
62. Travers M, Clinchy M, Zanette L, Boonstra R, Williams TD. Indirect predator effects on clutch size and the cost of egg production. Ecol Lett. 2010;13:980–8.
63. Jemielity S, Kimura M, Parker KM, Parker JD, Cao XJ, Aviv A, et al. Short telomeres in short-lived males: what are the molecular and evolutionary causes? Aging Cell. 2007;6:225–33.
64. Veskoukis AS, Nikolaidis MG, Kyparos A, Kouretas D. Blood reflects tissue oxidative stress depending on biomarker and tissue studied. Free Radic Biol Med. 2009;47:1371–4.
65. Daniali L, Benetos A, Susser E, Kark JD, Labat C, Kimura M, et al. Telomeres shorten at equivalent rates in somatic tissues of adults. Nat Commun. 2013;4:6.
66. Reichert S, Criscuolo F, Verinaud E, Zahn S, Massemin S. Telomere length correlations among somatic tissues in adult zebra finches. Plos One. 2013;8:7.
67. Nakagawa S, Schielzeth H. Repeatability for Gaussian and non-Gaussian data: a practical guide for biologists. Biol Rev. 2010;85:935–56.
68. Stephan B. Die Amsel. Hohenwarsleben: Westarp Wissenschaften; 1999.
69. Miranda AC, Schielzeth H, Sonntag T, Partecke J. Urbanization and its effects on personality traits: a result of microevolution or phenotypic plasticity? Glob Chang Biol. 2013;19:2634–44.
70. Costantini D, Greives TJ, Hau M, Partecke J. Does urban life change blood oxidative status in birds? J Exp Biol. 2014;217:2994–2997.

71. Adelman JS, Bentley GE, Wingfield JC, Martin LB, Hau M. Population differences in fever and sickness behaviors in a wild passerine: a role for cytokines. J Exp Biol. 2010;213:4099–109.
72. Owen-Ashley NT, Hasselquist D, Raberg L, Wingfield JC. Latitudinal variation of immune defense and sickness behavior in the white-crowned sparrow (*Zonotrichia leucophrys*). Brain Behav Immun. 2008;22:614–25.
73. Rich EL, Romero LM. Exposure to chronic stress downregulates corticosterone responses to acute stressors. Am J Physiol Regul Integr Comp Physiol. 2005;288:R1628–36.
74. Cyr NE, Earle K, Tam C, Romero LM. The effect of chronic psychological stress on corticosterone, plasma metabolites, and immune responsiveness in European starlings. Gen Comp Endocrinol. 2007;154:59–66.
75. Dickens MJ, Delehanty DJ, Romero LM. Stress and translocation: alterations in the stress physiology of translocated birds. Proceed Royal Soc B-Biol Sci. 2009;276:2051–6.
76. Dickens MJ, Earle KA, Romero LM. Initial transference of wild birds to captivity alters stress physiology. Gen Comp Endocrinol. 2009;160:76–83.
77. Ouyang JQ, Sharp PJ, Dawson A, Quetting M, Hau M. Hormone levels predict individual differences in reproductive success in a passerine bird. Proceed Royal Soc B-Biol Sci. 2011;278:2537–45.
78. Peig J, Green AJ. New perspectives for estimating body condition from mass/length data: the scaled mass index as an alternative method. Oikos. 2009;118:1883–91.

# Developmental dynamics of myogenesis in the shipworm *Lyrodus pedicellatus* (Mollusca: Bivalvia)

Andrea Wurzinger-Mayer[1], J Reuben Shipway[2], Alen Kristof[1], Thomas Schwaha[1], Simon M Cragg[2] and Andreas Wanninger[1*]

**Abstract**

**Background:** The shipworm *Lyrodus pedicellatus* is a wood-boring bivalve with an unusual vermiform body. Although its larvae are brooded, they retain the general appearance of a typical bivalve veliger-type larva. Here, we describe myogenesis of *L. pedicellatus* revealed by filamentous actin labelling and discuss the data in a comparative framework in order to test for homologous structures that might be part of the bivalve (larval) muscular ground pattern.

**Results:** Five major muscle systems were identified: a velum retractor, foot retractor, larval retractor, a distinct mantle musculature and an adductor system. For a short period of larval life, an additional ventral larval retractor is present. Early in development, a velum muscle ring and an oral velum musculature emerge. In late stages the lateral and dorsal mantle musculature, paired finger-shaped muscles, an accessory adductor and a pedal plexus are formed. Similar to other bivalve larvae, *L. pedicellatus* exhibits three velum retractor muscles, but in contrast to other species, one of them disappears in early stages of *L. pedicellatus*. The remaining two velum retractors are considerably remodelled during late larval development and are most likely incorporated into the elaborate mantle musculature of the adult.

**Conclusions:** To our knowledge, this is the first account of any larval retractor system that might contribute to the adult bodyplan of a (conchiferan) mollusk. A comparative analysis shows that a pedal plexus, adductors, a larval velum ring, velum retractors and a ventral larval retractor are commonly found among bivalve larvae, and thus most likely belong to the ground pattern of the bivalve larval musculature.

**Keywords:** Teredinid, Mollusk, Evodevo, Evolution, Ontogeny, Larva, Veliger, Lophotrochozoa

## Background

With 15.000 extant species, Bivalvia represents the second largest class-level taxon within the Mollusca. Bivalve species are aquatic, laterally compressed lophotrochozoans with two shell valves that are hinged dorsally and interconnected by a ligament, which acts antagonistically against the adductor muscles. The dimyarian arrangement with one anterior and one posterior adductor muscle is characteristic of many bivalves [1]. During development, most marine bivalves undergo a lecithotrophic trochophore stage that is followed by a planktotrophic (sometimes lecithotrophic) veliger larva. Exceptions include the putative basal protobranchs with a pericalymma and certain freshwater bivalves with a parasitic glochidium larva [2,3].

Representatives of the Teredinidae and some Pholadidae (Heterodonta) burrow into and feed on wood and thereby as adults greatly deviate from the typical bivalve anatomy and mode of life. As such, the teredinid *Lyrodus pedicellatus* has an elongated, worm-like body extending well beyond the shell, which acts as a highly specialized drilling tool. The shell surrounds a short, well-developed foot that serves as an attachment organ to the wood and helps facilitate boring [4,5]. The two adductors, typical for heterodonts, are anisomyarian and the smaller anterior adductor is divided into two distinct subportions [1,6]. The valves are interconnected by the adductor muscles and not by the hinge as is the case in other bivalves. The gills not only serve as respiratory and filtering organs, but also harbour bacterial symbionts and house the larvae in brood pouches. While the adult *L. pedicellatus* has a highly derived body, its later-stage larvae bear significant gross morphological resemblance to trochophore and veliger larvae of most other

* Correspondence: andreas.wanninger@univie.ac.at
[1]Department of Integrative Zoology, Faculty of Life Sciences, University of Vienna, Althanstrasse 14, 1090 Vienna, Austria
Full list of author information is available at the end of the article

bivalve representatives, despite the fact that *L. pedicellatus* larvae are brooded to the advanced pediveliger or even metamorphically competent stage and have no direct access to the water column (Additional file 1: Figure S1). Thereby, an oval-shaped and densely ciliated retractable velum characterizes the veliger larva of *L. pedicellatus* [7]. During metamorphosis the velum is resorbed [8]. The two convex lobes of the D-shaped (veliger stage) or roundish (pediveliger-stage) larval shell are hinged and cover the soft body of the larva.

While myogenesis has been described for a number of mollusks (e.g., [9-19]), surprisingly few recent investigations have dealt with bivalve muscle development. Other aspects of bivalve larval anatomy and ontogeny have been addressed by classical light (e.g. [20-23]) as well as electron and confocal laser scanning microscopy (e.g. [24-27]). Herein, we describe myogenesis from early stage larvae to early juveniles in the shipworm *Lyrodus pedicellatus.* The results are compared with the scarce available data on other bivalve taxa (see Figure 1) in order to assess potential homologies and to shed light on the ground pattern of the bivalve (larval) muscular bodyplan.

## Methods

### Animal fixation and F-actin staining

Adult *Lyrodus pedicellatus* were obtained from an infested pier composed of Greenheart wood (*Chlorocardium rodiei*) at Portsmouth, UK. Adults were harvested in June and October 2012 and were dissected in order to yield the larvae. Subsequent cultures were reared in aquaria at the Institute of Marine Sciences, University of Portsmouth. Seawater was taken directly from Langstone Harbour, maintained at a temperature between 15-18℃ and a salinity of 36 PSU and kept aerated throughout. Tanks were regularly provided with small panels of wood (pine, measuring 2.5 cm ?2.5 cm ?2.5 cm) for larval settlement.

Larvae were relaxed in 7.5% $MgCl_2$, fixed in 4% paraformaldehyde in 0.1M phosphate buffer (PB) and stored in 0.5M PB +0.1% $NaN_3$ at 4℃. For decalcification, the larvae were incubated in 0.5M ethylene glycol tetraacetic acid (EGTA) overnight at room temperature. Specimens were then rinsed three times (10 min each) in 0.1M PB. Afterwards, the animals were permeabilized in PBT (0.1M PB +0.2% Triton X-100) overnight at room temperature. F-actin was labelled by fluorescence-coupled phalloidin (Alexa Fluor 488; Molecular Probes, Eugene, OR, USA) in a 1:40 dilution of PBT for 6 to 10 hours in the dark at room temperature. Cell nuclei were stained with DAPI (4′, 6-diamidino-2-phenylindole) (Sigma, St. Louis, MO, USA), followed by three washes in 0.1M PB (10 min each). Stained specimens were mounted on glass slides using Fluoromount G (Southern Biotech, Birmingham, AL, USA).

For details on preparation of the scanning electron micrographs used in Supplemental Figure S1, please see Additional file 2.

### Analysis and digital image processing

A Leica DM 1000 epifluorescence microscope (Leica, Wetzlar, Germany) was used for gross morphological analysis of 318 animals. 64 of them were selected for detailed examinations using a Leica TCS SP5 II confocal laser scanning microscope (CLSM) (Leica, Wetzlar, Germany). Optical sections between 0.2 and 0.5 μm of the wholemounts were generated and digitally merged into maximum

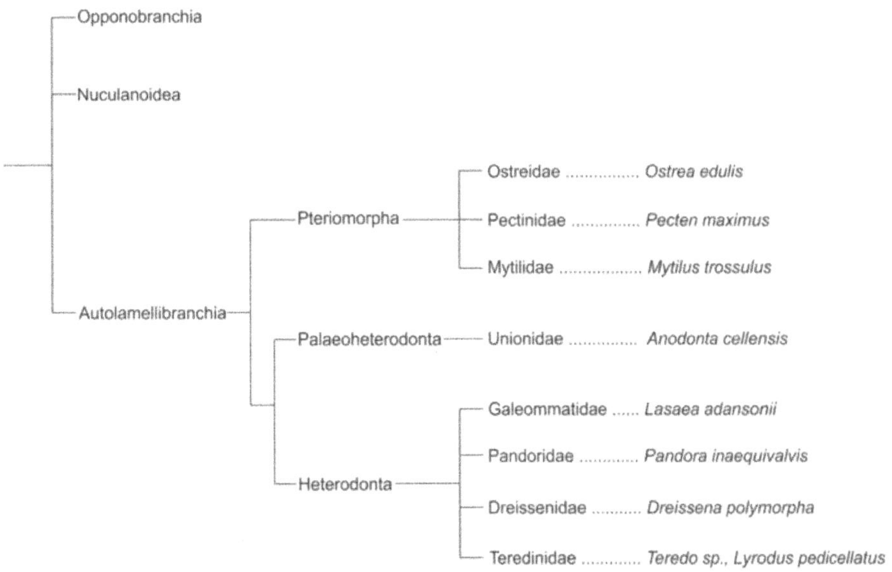


**Figure 1 Suggested relationships of selected bivalve taxa after Giribet [1], with species for which significant data on myogenesis are available specifically mentioned.**

projections. In order to depict the 3-dimensionality of the specimens and to accentuate the regions of interests, the confocal stacks were analyzed with the 3D-reconstruction software Imaris 7.3 (Bitplane, Zřich, Switzerland). Adjustment and optimization of contrast and brightness as well as the conversion into black and white images was performed with Photoshop CS 6 (Adobe Systems, San Jose, CA, USA). Line drawings were generated with Corel Draw X 6 (Corel Corporation, Ottawa, Ontario, Canada).

### Staging of larvae

Staging of specimens based on a timescale referring to oocyte fertilization was not possible due to internal development and brooding in *Lyrodus pedicellatus*. Therefore, size measurements were the prime means of dividing the specimens into individual developmental stages. Besides the developing complexity of the musculature, the formation and loss of morphological features such as the velum and the shell, as well as the degree of calcification, colour and shape of the latter are used herein for defining individual larval stages.

### Orientation of larvae

There exists some confusion concerning axis determination and orientation in bivalve larvae, which largely results from the fact that the morphological dorso-ventral designation used in adult bivalves (whereby ventral marks the axis running from mouth to anus and dorsal the opposite region, often the site of the hinge) has been directly transferred to the orientation in the (pedi-) veliger stages. In postmetamorphic and adult bivalves, the mouth (and foot) usually comes to lie close to the anterior pole, often in the vicinity of an anterior adductor system. Accordingly, the mouth occupies a ventro-anterior position in many postmetamorphic bivalves. In bivalve larvae, the mouth is situated immediately behind the velum, which is shed during metamorphosis. Thus, the apical region, terminating with the apical tuft, is often referred to as ventral in bivalve larvae (e.g., [23]; see also [28]). This, however, is in stark contrast to axis determination used in almost all other lophotrochozoan larvae, where the apical tuft marks anterior and the opposite pole posterior. In order to conform to this terminology widely used in comparative lophotrochozoan larval anatomy, we use the latter designation herein (apical tuft = anterior, opposite of apical tuft = posterior, position of the foot = ventral), although we are aware that this may not correspond to the condition exhibited in adult (postmetamorphic) bivalves.

## Results

The youngest developmental stage investigated was characterized by the lack of a velum or a shell, was less than 60 μm in length and did not show any muscle-specific signal. In the earliest veliger stage investigated (Figure 2A), ellipsoid shells are formed. They are interconnected in the posterior region of the larva and envelope not more than one third of the body (Figure 3A). Underneath the cap-like velum a delicate velum muscle ring is present (Figure 3A, B). In addition, several well-developed muscle bundles have formed at this stage. The anterior adductor muscle spans between the left and the right valve in the dorsal part of the shell and is divided into two bundles (Figures 2A and 3B). A pair of dorsal velum retractors attaches to the area of the developing shell, extends into the velum and runs along the inner surface of the anterior adductor (Figures 2A and 3A, B). Clearly separated from the dorsal velum retractor, a pair of ventral velum retractors appears and extends toward the central portion of the velum (Figures 2A and 3A). Ventrally, clearly distinct from the ventral velum retractor, a ventral larval retractor, which consists of two bundles, develops and projects toward the anterior part of the velum (Figures 2A and 3A).

In early veliger larvae, the ventral larval retractor disappears (Figures 2B and 3C, D). The dorsal velum retractors elongate and few fibres project into the velum, where the velum ring musculature is now fully developed and thicker than in the previous stage (Figures 2B and 3C, D). The ventral velum retractors now consist of two distinct, parallel bundles, which extend far into the velum (Figures 2B and 3C). Delicate muscle fibres, which originate at the base of the left and right dorsal branches of the ventral velum retractors and at the lower third of both ventral branches, represent the anlage of the foot retractors, which project toward the midline (Figures 2B and 3C, D). The anterior adductor expands between the shell valves, which now reach the velum and are rounded in shape (Figure 3C, D).

In slightly older veligers, the shell has a distinct concave indentation in the posterior region near the attachment area of the velum retractors (Figure 3E). In this stage, the two bundles of the ventral velum retractors intercross (Figures 2C and 3E). The left and right muscle bundles, which later will develop into the foot retractors, fuse in the median plane (Figures 2C and 3F). The anterior adductor increases in size and the velum ring musculature still underlies the edge of the velum (Figures 2C and 3E, F).

In mid-veligers, the oral part of the velum ring disintegrates, while its remaining fibres connect to the oral velum musculature, which consists of a muscle grid that appears to interconnect postoral soft tissues (Figures 2D and 3G). The base of the foot retractors splits and forms a second ventral attachment zone, close to the ventral velum retractor (Figure 3G, H). Along the median plane the foot retractor elongates and adjacent to it, muscle fibres appear and connect to the oral velum musculature (Figures 2D and 3G). The velum retractors increase in size and fan out into the velum. The two portions of the ventral velum retractors fuse but maintain different

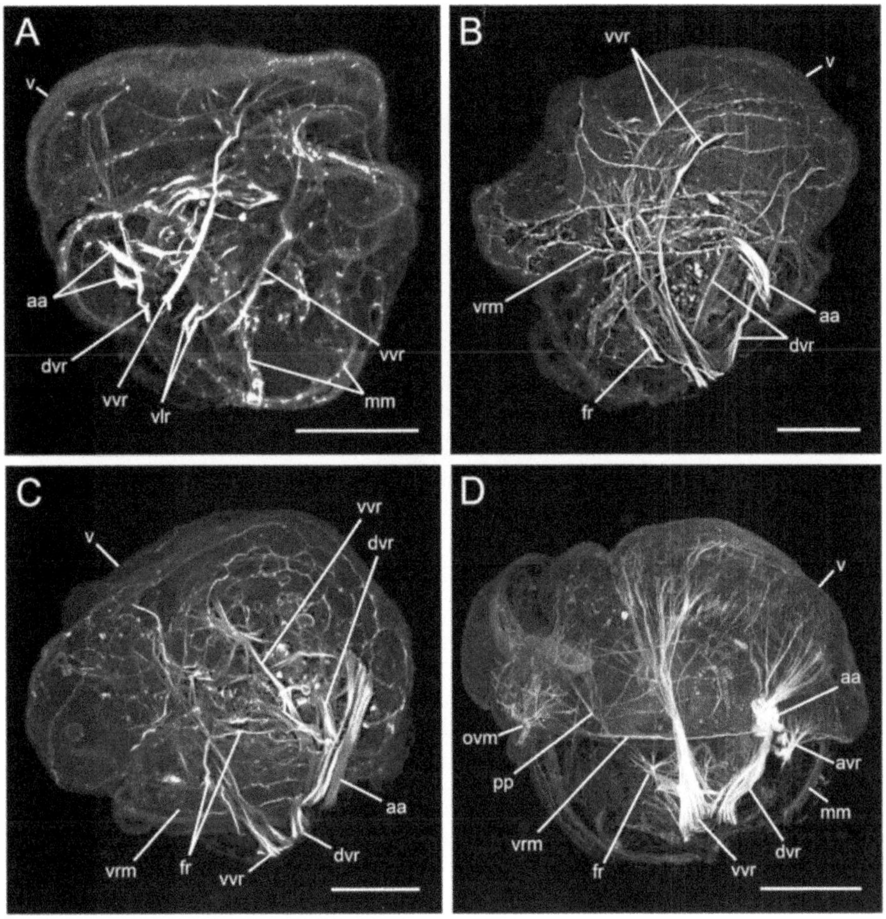


**Figure 2 Confocal micrographs showing myogenesis in four subsequent early larval phases of *Lyrodus pedicellatus*.** Scale bars represent 20 μm, except in **D**, in which it equals 50 μm. All aspects are in lateral view, **C** is slightly tilted in anterodorsal direction. Anterior is upwards in all aspects. Ventral is to the left, except in **A**, where it is to the right. **(A)** Earliest veliger stage investigated, showing two muscle bundles of the paired ventral larval retractors (vlr) and a pair of ventral (vvr) and dorsal velum retractors (dvr), which extend into the velum (v). The anterior adductor muscle (aa), divided into two bundles, is visible in the dorsal position of the larva and the mantle musculature (mm) surrounds the mantle margin. **(B)** Early veliger with the velum ring musculature (vrm) and first muscle fibres of the paired foot retractor (fr). Note that the pair of ventral velum retractors now each consists of two parallel muscle bundles. **(C)** Marginally more advanced veliger than the specimen shown in B with fibres of the left and right foot retractor fusing in the median plane. **(D)** Mid-veliger with accessory velum retractor (avr) that is attached to the outer surface of the anterior adductor and spreads into the velum. The foot retractor elongates along the median plane. At the ventral side of the velum, a muscle grid (oral velum musculature; ovm) that has formed contact to the velum ring musculature and to the first fibers of the pedal plexus (pp) develops in the postoral region. Note the distinct separation of the pedal plexus and the foot retractors.

shapes (Figures 2D and 3G). An unpaired accessory velum retractor appears in close proximity to the well-defined anterior adductor (Figures 2D and 3G, H). The shell becomes more round and grows anteriorly (Figure 3G, H).

In the late veliger stage, the dorsal and ventral pair of the velum retractors are both pronounced and numerous distal branches project into the velum. The unpaired accessory velum retractor is distinct and the anterior adductor is well-developed (Figures 4A and 5A, B). The posterior adductor forms ventrally from the insertion area of the ventral velum retractor and attaches to both valves (Figures 4A and 5B). The muscle fibres that follow the trajectory of the foot retractors become thicker and form three strong branches along the median plane. These are the anlage of the pedal plexus, which at this stage has no contact to the foot retractor (Figures 4A and 5A, B). Distally, at its ventral side, the pedal plexus is still connected to the muscle grid in the velum that now reaches its largest dimension. When expanded, it overhangs two-thirds of the shell (Figure 5A).

In the mid-pediveliger, the velum shrinks in the oral region such that the remaining parts of the oral velum musculature become denser and lie immediately in front of the tips of the ventral velum retractor (Figures 4B and 5C). The velum ring musculature starts to disintegrate from the ventral side (Figures 4B and 5C). Both the dorsal and the ventral velum retractor are remarkably well pronounced at this stage (Additional file 1: Figure S1B). Their main trunks split

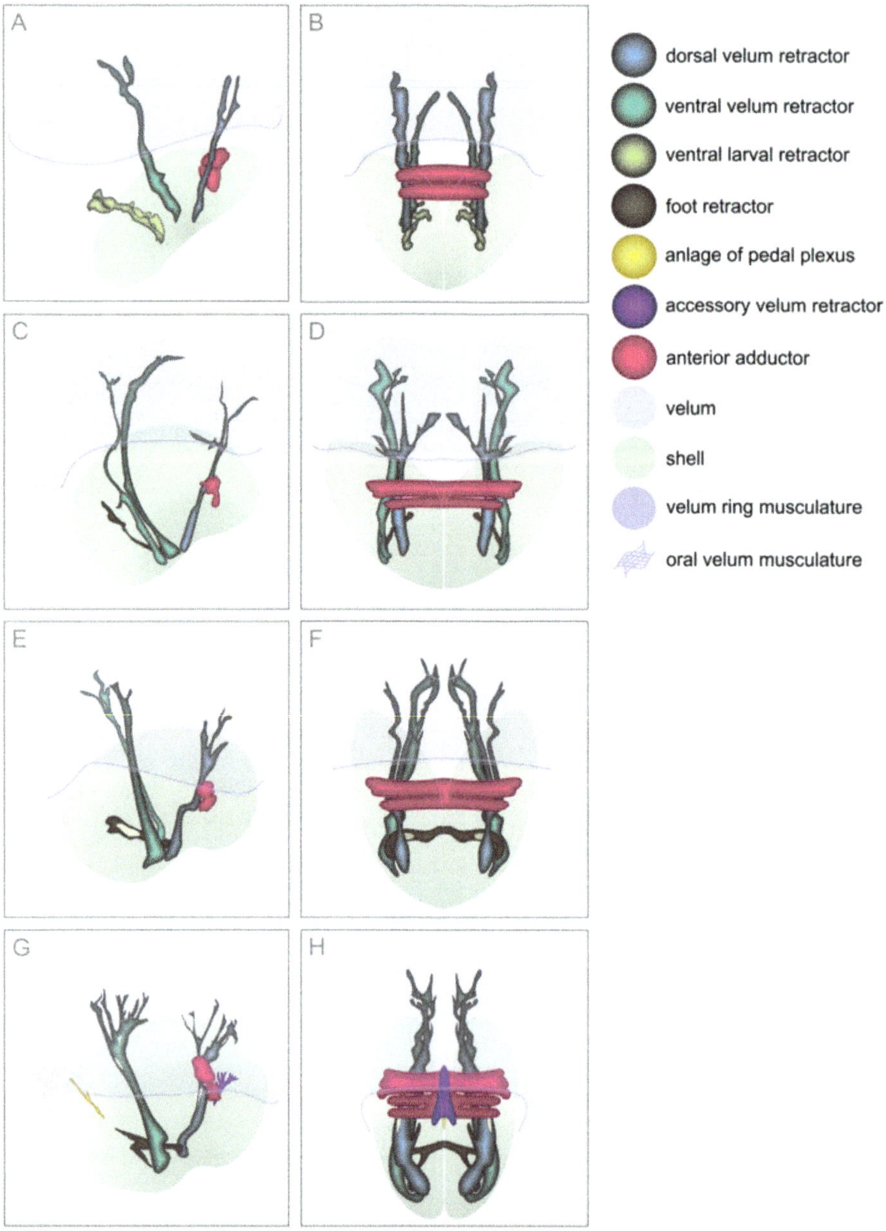


**Figure 3 Graphic representation based on 3D-reconstructions of larval myogenesis in *Lyrodus pedicellatus*.** Anterior is upwards in all aspects. **A**, **C**, **E**, **G** is in a lateral view and ventral is to the left. **B**, **D**, **F**, **H** is in dorsal view. **(A)** Earliest veliger stage investigated with oval-shaped valves and velum with a delicate velum muscle ring. Ventral and dorsal velum retractors extend into the velum. Ventral larval retractors project ventrally. The anterior adductor muscle has contact to the dorsal velum retractor. **(B)** Earliest veliger stage investigated showing the two portions of the anterior adductor muscle which spans between the left and right valve. Paired velum retractors project into the velum. **(C)** Early veliger larva with the anlage of the foot retractors that originate at the paired ventral velum retractors, which now consist of two distinct bundles. Note that the ventral larval retractor has already disintegrated in this early stage. **(D)** Early veliger larva with the developing foot retractors which project towards the midline. **(E)** Slightly further developed veliger showing two intercrossing bundles of the ventral velum retractors. **(F)** Veliger with the left and right foot retractor that meet in the midline and fuse. **(G)** Developing median muscle fibres that form the anlage of the pedal plexus in the mid-veliger larva. An oral velum musculature appears. On both sides, the two branches of the ventral velum retractor fuse. Note that the accessory velum retractor is attached to the outer surface of the anterior adductor muscle. **(H)** Mid-veliger with unpaired accessory velum retractor.

into several delicate bundles, which project toward the median plane from the left and the right. They enter the velum, where the branches fan out (Figures 4B and 5C, D). The dorsal velum retractor maintains close contact to the anterior adductor, to which a small muscle band is attached anteriorly (Figures 4B and 5C, D). The band runs along the edge of the left and right shell and elongates as a U-shaped mantle musculature toward the ventral side

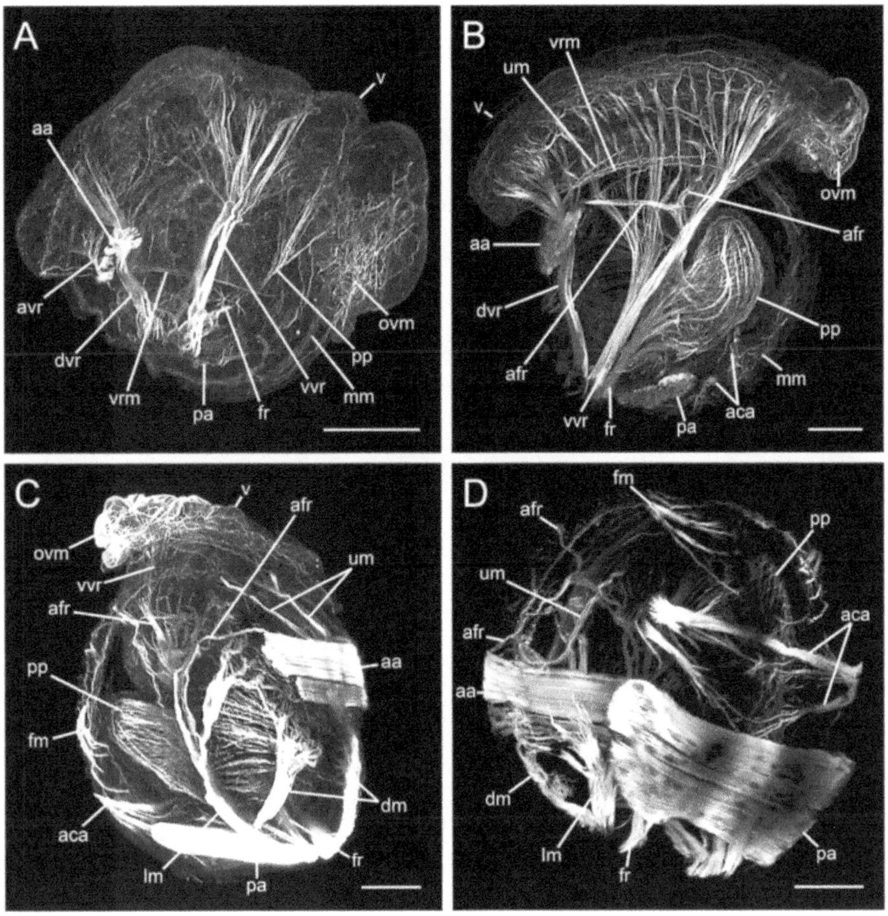


**Figure 4 Myogenesis from late veliger to the early juvenile in Lyrodus pedicellatus (confocal micrographs).** Scale bars: 50 μm. **A**, **B**: Lateral view. **C**: Dorso-lateral view. **D**: Ventro-lateral view. Anterior up and ventral right, except for **C**, where ventral is left. **(A)** Late veliger with paired dorsal (dvr) and ventral velum retractors (vvr). An accessory velum retractor (avr) is attached to the anterior adductor (aa). The pedal plexus (pp) contacts the oral velum musculature (ovm). The pedal plexus projects toward the foot retractor (fr). The posterior adductor (pa) has developed ventrally of the ventral velum retractor. Compact fibres indicate the margin of the mantle musculature (mm). **(B)** Mid-pediveliger. The velum (v) becomes smaller. The velum ring musculature (vrm) loses its solid appearance. The accessory velum retractor (avr) disappears, the dorsal (dvr) and ventral velum retractor (vvr) are predominant. An accessory adductor (aca) has developed. Pedal plexus and foot retractor have fused and form the foot musculature. The accessory foot retractors (afr) divide halfway into two branches. A diffuse mantle musculature (mm) and a U-shaped mantle musculature (um) originate at the anterior edge of the anterior adductor. **(C)** Late pediveliger showing disintegration of larval structures. The velum (v) shrinks, the oral velum musculature (ovm) is pushed together, the velum ring musculature disappears. The U-shaped mantle musculature (um), connected to the anterior adductor (aa), elongates. The velum retractors have largely disintegrated, except for one bundle of the ventral velum retractor (vvr). Lateral (lm) and dorsal mantle musculature (dm) lie close to the shell. Foot retractors (fr), accessory foot retractors (afr) and pedal plexus (pp) form the foot musculature. Posterior adductor (pa) and accessory adductor (aca) are well-developed. A finger-shaped mantle musculature (fm) is formed. **(D)** Early juvenile with velum lost. U-shaped mantle musculature (um) fuses with finger-shaped mantle musculature (fm).

(Figures 4B and 5D). The accessory velum retractor disappears. The posterior adductor, which lies close to the inner surface of the shell, grows and surpasses the size of the anterior adductor (Figures 4B and 5C, D). Anterior to the posterior adductor a solid two-pronged accessory adductor reaches between the valves (Figures 4B and 5C, D). The pedal plexus fuses with the foot retractor and a complex muscular grid develops (Figures 4B and 5C, Additional file 1: Figure S1B). The two insertion sites of the foot retractor are fused at their base (Figure 5C). At this point, the foot retractor is clearly separated from the ventral velum retractor (Figure 5D, Additional file 1: Figure S1B). An accessory foot retractor develops (Additional file 1: Figure S1B) on the left and right side of the lateral area of the pedal plexus, runs parallel to the inner surface of the shell and furcates halfway into two branches on either side. One end projects toward the anterior edge of the anterior adductor, the other attaches to the shell (Figures 4B

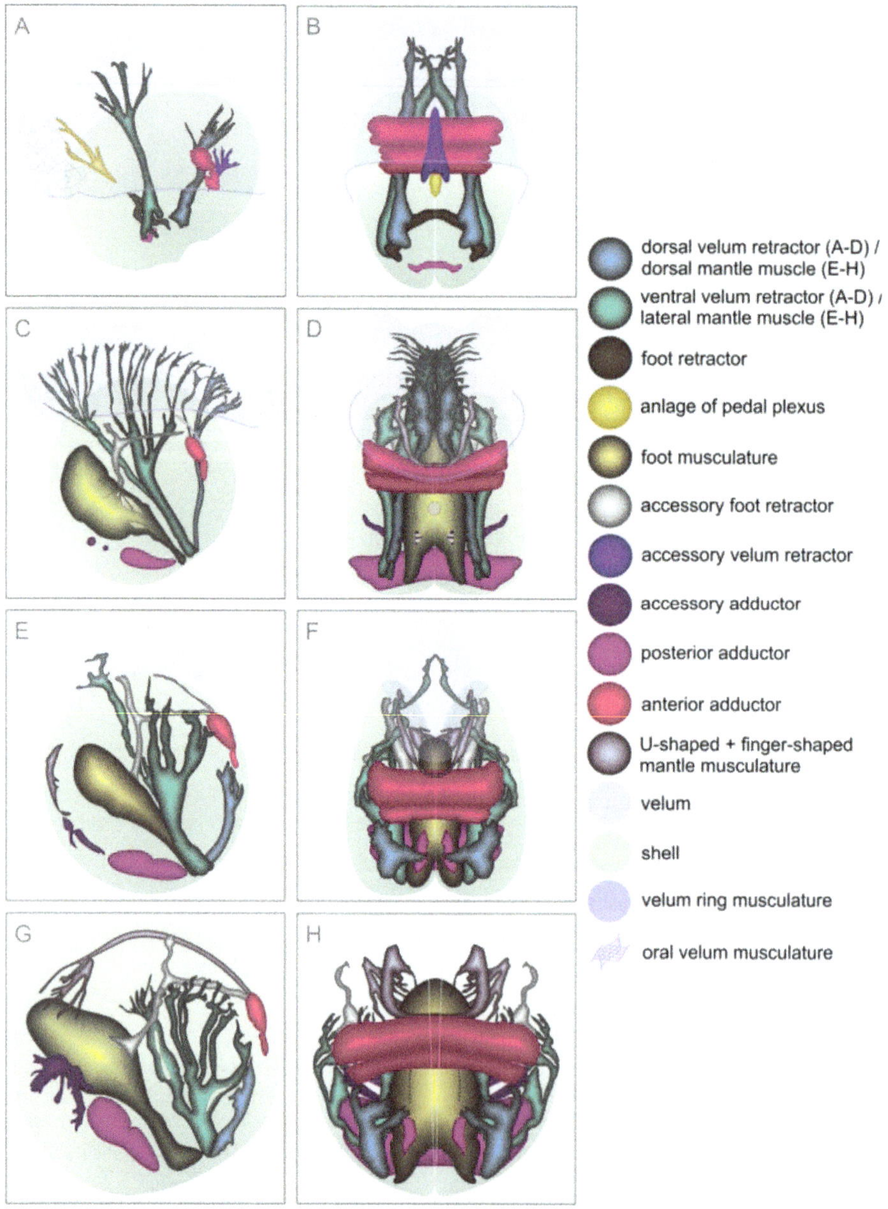


**Figure 5 Graphic representation based on 3D-reconstructions of larval myogenesis in *Lyrodus pedicellatus*.** **A**, **C**, **E**, **G** are in lateral view, anterior is upwards, ventral to the left. **B**, **D**, **F**, **H** are in dorsal view, anterior is upwards. **(A)** Late veliger with round shells and a fully developed velum. The foot consists of two separated portions, the foot retractor and the anlage of the pedal plexus. **(B)** Late veliger with developing posterior adductor muscle which extends between the left and the right valve. **(C)** Mid-pediveliger with shrinking velum. The foot retractors and the pedal plexus have fused and together with a two-pronged accessory foot retractor form the foot musculature. The paired velum retractors are fully developed. An accessory adductor muscle appears. **(D)** Mid-pediveliger with a solid anterior and posterior adductor muscle. The accessory adductor muscle spans between the valves. Note the developing U-shaped mantle musculature. The accessory velum retractor has disappeared. **(E)** Late pediveliger with a largely atrophied velum. The velum ring musculature has disintegrated. For the most part, the velum retractors are resorbed distally. Paired finger-shaped mantle musculature develops. **(F)** Late pediveliger with remaining basal parts of the velum retractors which transform into the lateral and dorsal mantle musculature. This paired mantle musculature gains volume and has shifted more laterally compared to previous stages. **(G)** Early juvenile with shed velum. Ventrally, the U-shaped mantle musculature forms contact to the finger-shaped mantle musculature. **(H)** Early juvenile with lateral and dorsal mantle musculature at the periphery, parallel to the inner surface of the spherical shell.

and 5C). The convex indentation in the posterior area of the shell flattens and the outline of the valves is almost round (Figure 5C).

In late pediveliger stages, the velum shrinks from the aboral side, degenerates further and the velum ring musculature disappears (Figures 4C and 5E). The dorsal velum retractors no longer have contact to the velum and distally become entirely resorbed (Figures 4C and 5E, F). Conversely, the basal part now appears more solid and becomes incorporated into the mantle as dorsal mantle musculature (Figures 4C and 5E, F). For the most part, the ventral velum retractor is no longer connected to the velum (Figures 4C and 5E). Only a single branch reaches the central part of the velum (Figures 4C and 5E, F). Similar to the dorsal velum retractor, the stem of the ventral velum retractor becomes thicker and develops into the well-pronounced lateral mantle musculature (Figures 4C and 5E, F). Due to its atrophying distal parts, the dorsal velum retractor loses the contact to the anterior adductor. Especially the anterior bundle of the anterior adductor increases in diameter (Figures 4C and 5E, F). The posterior adductor is massive and the accessory adductor is well-developed (Figures 4C and 5E). The accessory foot retractor connects to the anterolateral edge of the anterior adductor (Figures 4C and 5E). Ventral to the tip of the foot, a pair of finger-shaped mantle musculature develops near the inner rim of the left and right shell and the U-shaped mantle musculature extends along the shell margin toward the ventral side (Figures 4C and 5E, F).

In the last developmental stage analyzed, the velum is shed and the branches of the velum retractors are lost (Figures 4D and 5G, H). The shell is round and the animal now appears spherical (Figures 4D and 5G, H). The anterior and the posterior adductors are massive, the accessory adductor is pronounced and the foot is predominant with compact retractors (Figures 4D and 5G, H). The U-shaped mantle musculature, which extends from the anterior surface of the anterior adductor, forms contact with the paired ventral finger-shaped mantle musculature (Figures 4D and 5G). A solid, lateral mantle musculature on both sides of the inner shell surface originates dorsal of the foot and fuses with the base of the dorsal mantle musculature (Figures 4D and 5G). After a short distance it divides into a solid dorsal and ventral branch, whereby the latter branch again subdivides into three parts, which again split and terminate in the region of the accessory foot retractor (Figures 4D and 5G, H). All branches of the lateral and dorsal mantle musculature lie laterally at the periphery (Figures 4D and 5H), except for the remnant of the single branch of the ventral velum retractor, which maintains the straight position next to the median plane (Figure 5H).

The mantle musculature originates alongside the mantle margin. Diffuse mantle muscle fibres appear all over the mantle and in later stages the fibres exhibit a more or less regular pattern in some areas (Figures 2D and 4A, B).

Taken together, the data presented herein show that early-stage *Lyrodus pedicellatus* veliger larvae, with a length of less than 60 μm, have already developed the anlagen of all major larval muscles (Figure 6A, Table 1). The ventral larval retractor disintegrates in slightly older larvae (Figure 6B, Table 1). The foot retractors, which originate at the base of the larval retractors, start to develop in the early veliger larva (Figure 6B, Table 1). The formation of the foot is successively elaborated (Figure 6B-H, Table 1). The velum retractors continuously expand with some portions being subsequently resorbed while others are transformed into mantle musculature (Figure 6, Table 1).

## Discussion

The analysis of myogenesis provided herein is based on an unprecedented sequence of bivalve larval stages from early veliger to early juveniles. Five major muscle systems, which consist of altogether 17 distinct subunits, were identified and their genesis including their disintegration or transformation are documented and in the following compared to other bivalve and molluscan representatives.

### Comparative analysis of myogenesis in Bivalvia

#### *Adductor systems*

In *Lyrodus pedicellatus*, the adult dimyarian arrangement with an anterior and a posterior adductor muscle is already established during larval development. Interestingly, the pteriomorphan *Ostrea edulis* has a monomyarian arrangement with only one adductor muscle as an adult, but during the larval phase, two adductor muscles form, of which the anterior one degenerates during metamorphosis [23]. In the glochidium larva of the freshwater mussel *Anodonta cellensis* (Palaeoheterodonta), the larval adductor completely disintegrates during the parasitic phase and the adult adductors form *de novo* [22] (Table 2). In *L. pedicellatus*, the anterior adductor is formed early in larval development, and at that stage is the only muscle capable of closing the shell (Table 1). It is divided into two distinct parts, similar to the pteriomorph *Pecten maximus* and the heterodont *Dreissena polymorpha* [21,24,29]. The anterior adductor in *L. pedicellatus* remains smaller than the posterior adductor. While the posterior adductor is a single muscle bundle in scallop larvae, it splits into two parts in *O. edulis*, and is divided into several bundles in the heterodont *D. polymorpha* ([21,23,29]). In *L. pedicellatus* the posterior adductor muscle appears to be divided into multiple units. Our study provides the first account of an accessory adductor for any larval bivalve and may constitute an apomorphy of either *L. pedicellatus*, the genus *Lyrodus*, or even a higher taxonomic unit (Table 2). Although the accessory adductor develops late in the larval phase (Table 1), it is nevertheless well-developed and of complex structure by the end of this stage.

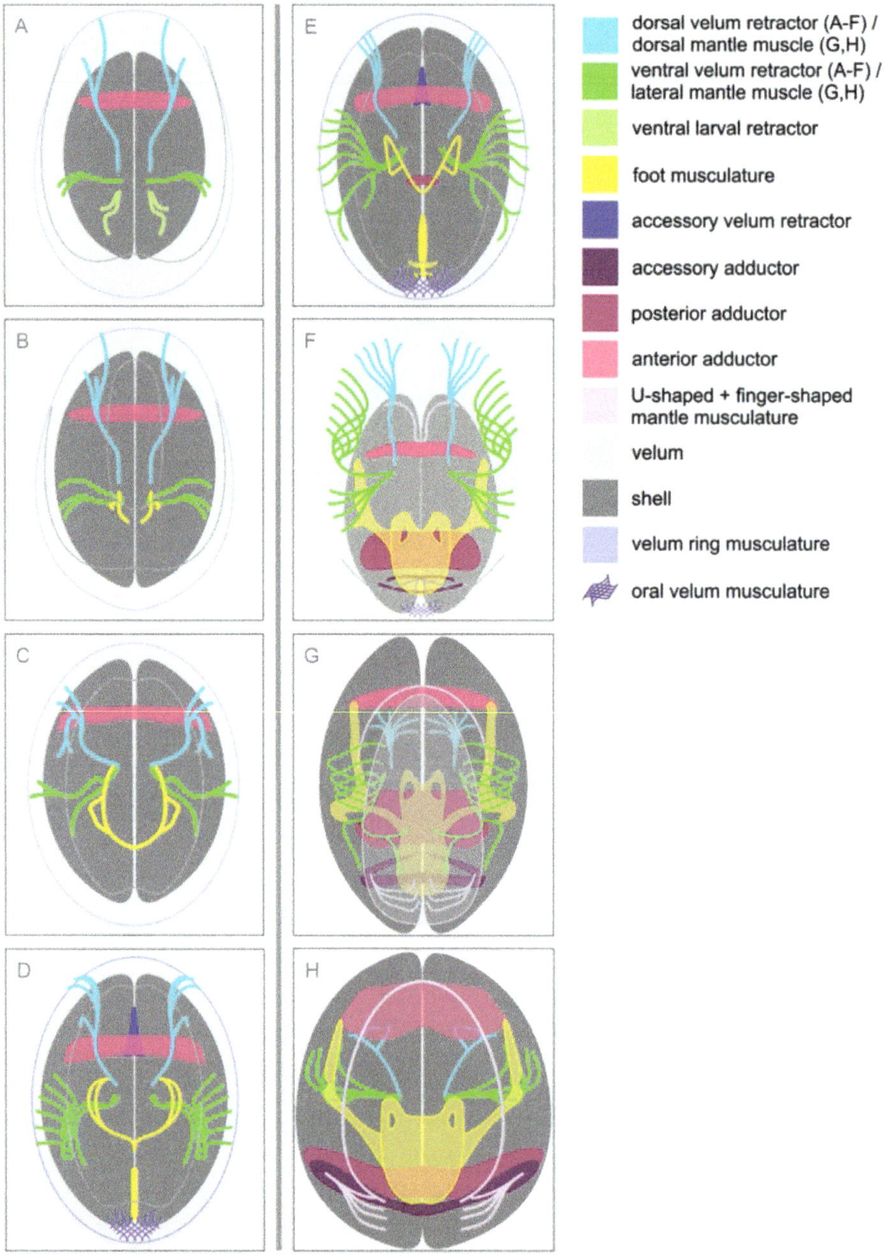


**Figure 6 Schematic representations of myogenesis from the early veliger to the early juvenile in *Lyrodus pedicellatus*.** Anterior view, apical faces the viewer and dorsal is upwards in all aspects. Folds of the velum are displayed by grey lines. **(A)** Very early veliger with paired dorsal and ventral velum retractors. Note that the paired ventral larval retractor only exists in this early stage. The anterior adductor muscle spans between the valves. **(B)** Early veliger showing paired anlagen of the foot retractor as a first component of the foot musculature. **(C)** Marginally older veliger with fused left and right foot retractors in the middle. **(D)** Mid-veliger with median muscle fibres in the longitudinal projection of the foot retractor. The unpaired accessory velum retractor is attached to the anterior adductor muscle. An oral velum musculature appears. **(E)** Late veliger larva. The median muscle fibres of the foot increase in size, manifesting as pedal plexus which is not yet in contact with the foot retractors. The posterior adductor muscle develops. **(F)** Mid-pediveliger with accessory adductor muscle that reaches between the valves. The pedal plexus unites with the foot retractors and together they form the foot. The paired accessory foot retractors have formed on both sides. The U-shaped mantle musculature forms and extends ventrally. The accessory velum retractor disappears. **(G)** Late pediveliger with reduced velum. Note the finger-shaped mantle musculature. The accessory foot retractor forms contact to the anterior adductor muscle on both sides. **(H)** Early juvenile. Note that the velum has been shed. Partially atrophied velum retractors have transformed into lateral and dorsal mantle muscles. Ventrally, the U-shaped mantle musculature connects to the finger-shaped mantle musculature.

**Table 1 Larval myogenesis in *Lyrodus pedicellatus***

| Muscles | | Stage 1 | Stage 2 | Stage 3 | Stage 4 | Stage 5 | Stage 6 | Stage 7 | Stage 8 |
|---|---|---|---|---|---|---|---|---|---|
| Ventral larval retractor | | + | - | - | - | - | - | - | - |
| Velum ring musculature | | + | + | + | + | + | + / - | - | - |
| Dorsal velum retractor | | + | + | + | + | + | + | - | - |
| Ventral velum retractor | | + | + | + | + | + | + | + / - | - |
| Anterior adductor | | + | + | + | + | + | + | + | + |
| Foot retractor | anlage + → median fusion ++ → prolongation +++ → fusion with pedal plexus and accessory foot retractor ++++ | - | + | ++ | +++ | +++ | ++++ | ++++ | ++++ |
| Pedal plexus | anlage + → expansion ++ → fusion with foot retractor and accessory foot retractor +++ | - | - | - | + | ++ | +++ | +++ | +++ |
| Oral velum musculature | | - | - | - | + | + | + / - | - | - |
| Accessory velum retractor | | - | - | - | + | + | - | - | - |
| Posterior adductor | | - | - | - | - | + | + | + | + |
| Accessory foot retractor | | - | - | - | - | - | + | + | + |
| Accessory adductor | | - | - | - | - | - | + | + | + |
| U-shaped mantle musculature | | - | - | - | - | - | + | + | + |
| Finger-shaped mantle musculature | | - | - | - | - | - | - | + | + |
| Dorsal mantle musculature | | - | - | - | - | - | - | + | + |
| Lateral mantle musculature | | - | - | - | - | - | - | - / + | + |

character: (+) present; (-) absent; (+/-) decreasing; (-/+) increasing.
stage **1** = earliest veliger investigated; stage **2** = early veliger; stage **3** = mid-veliger 1; stage **4** = mid-veliger 2; stage **5** = late veliger; stage **6** = mid-pediveliger; stage **7** = late pediveliger; stage **8** = early juvenile.

***Foot retractor system and pedal plexus***

The massive foot retractors in the larva of *Lyrodus pedicellatus* attach in a posterior area near the hinge and, due to their similar relative position in the larval body, are considered homologous to the posterior retractors of the pteriomorphans *Ostrea edulis* and *Mytilus edulis*, the heterodonts *Lasaea adansonii* and *Pandora inaequivalvis* and the palaeoheterodont *Anodonta cellensis* (the foot retractors in *O. edulis* have previously been described as cruciform retractor because their medial bundles cross the median body plane) ([22,23,25,30-32]; Table 2). Likewise for positional reasons, the paired accessory foot retractor in *L. pedicellatus* most likely corresponds to the anterior retractor described for *O. edulis, M. edulis, L. adansonii, P. inaequivalvis* and *A. cellensis* [22,25,30-32] (Table 2), although most studies do not provide enough details to unequivocally settle this issue. In *Pecten maximus* unidentified fibres, which emerge from either end of the anterior bundle of the anterior adductor, extend along the bodywall lining [24] and, according to their position, they are likely homologous to the accessory foot retractor of *L. pedicellatus*. In *Teredo*, another common shipworm,

**Table 2 Muscle systems in larvae of 9 bivalve species**

| *Species* family | Foot musculature: no. of pairs, except protractor (unpaired) | | | Mantle musculature: no. of pairs | | | |
|---|---|---|---|---|---|---|---|
| | Foot retractor | Access. foot retractor | Protractor | U-shaped | Finger-shaped | Lateral | Dorsal |
| *Ostrea edulis* Ostreidae | 1 | 1 | ? | ? | ? | ? | ? |
| *Pecten maximus* Pectinidae | ? | 1 | ? | ? | ? | ? | ? |
| *Mytilus trossulus* Mytilidae | ? | ? | ? | 1 | ? | ? | ? |
| *Anodonta cellensis* Unionidae | 1 | 1 | ? | ? | ? | ? | ? |
| *Lasaea adansonii* Galeommatidae | 1 | 1 | 1 | 1 | ? | ? | ? |
| *Pandora inaequivalvis* Pandoridae | 1 | 1 | ? | ? | ? | ? | ? |
| *Dreissena polymorpha* Dreissenidae | 1 | ? | ? | ? | ? | ? | ? |
| *Teredo sp.* Teredinidae | ? | 1 | ? | ? | ? | ? | ? |
| *Lyrodus pedicellatus* Teredinidae | 1 | 1 | - | 1 | 1 | 1 | 1 |

| *Species* family | Adductor unpaired | | | Velum retractor no. of pairs, except accessory retractor (unpaired) | | | | Oral velum musculature unpaired | Velum ring musculature | Ventral larval retractor no. of pairs |
|---|---|---|---|---|---|---|---|---|---|---|
| | Anterior | Posterior | Other | Dorsal | Ventral | Accessory | Total number described | | | |
| *Ostrea edulis* Ostreidae | 1 | 1 | ? | 1 | 1 | ? | 3 | ? | ? | ? |
| *Pecten maximus* Pectinidae | 1 | 1 | ? | ? | ? | ? | 4 | ? | - | 3 |
| *Mytilus trossulus* Mytilidae | 1 | 1 | ? | ? | ? | ? | 3 | ? | 1 | 3 |
| *Anodonta cellensis* Unionidae | 1 | 1 | 1 (larval) | - | - | - | - | ? | - | - |
| *Lasaea adansonii* Galeommatidae | 1 | 1 | ? | ? | ? | ? | 3 | ? | ? | ? |
| *Pandora inaequivalvis* Pandoridae | 1 | 1 | ? | 1 | 1 | ? | 3 | ? | ? | ? |
| *Dreissena polymorpha* Dreissenidae | 1 | 1 | ? | 1 | 1 | - | 2 | ? | ? | 1 |
| *Teredo sp.* Teredinidae | ? | ? | ? | 1 | 1 | - | 2 | ? | ? | - |
| *Lyrodus pedicellatus* Teredinidae | 1 | 1 | 1 (accessory) | 1 | 1 | 1 | 3 | + | + | 1 |

character: (+) present; (-) absent; (?) unknown.

a coniform hump extends and shows the same growth direction as the accessory foot retractor of *L. pedicellatus* [20] (Table 2). In *Dreissena polymorpha*, the foot retractors have been described as the last muscles to form in the larva [21]. This is in sharp contrast to *L. pedicellatus,* where the first muscle fibres of the foot retractors branch off the velum retractors in an early larval stage (Table 1). Also in scallop larvae, foot retractors probably originate from velum retractors [29].

A single protractor, present in *Lasaea adansonii* [25], was not found in *Lyrodus pedicellatus* (Table 2). In general, foot development in *L. pedicellatus* starts early in larval life and is successively refined, with the foot musculature reaching its sophisticated architecture at the end of the larval phase (Table 1).

#### *Velum ring musculature*

A muscle ring underneath the developing velum is already present in the earliest *Lyrodus pedicellatus* veliger stage investigated and disintegrates in the late pediveliger stage (Table 1). This muscle ring has no contact to the velum retractors, whereas in an unidentified autobranch bivalve the velum retractors have been shown to insert at the velum muscle ring [33]. The putative homologous prototroch muscle ring in *Mytilus trossulus* has been described as being transformed into larval retractors during the veliger stage, but this surprising finding calls for more detailed assessment ([26]; Table 2).

#### *Larval retractor systems*

The bivalve larval muscle systems that so far have been described in most detail are the retractors and protractors. The results obtained are as diverse as the species examined. Whereas retractors and protractors are striated in *Mytilus trossulus*, there is no evidence for a striation in *Lyrodus pedicellatus*, although transmission electron microscopy would be needed to unequivocally settle this issue (cf. [26] and data herein). Similar to *Teredo* [20], *Lyrodus pedicellatus* has two pairs of velum retractors from the onset of myogenesis, whereas the unpaired accessory velum retractor, only found in *L. pedicellatus,* is added later in development and disappears shortly afterwards (Tables 1 and 2). *Pecten maximus* and probably *Tridacna maxima*, *Tridacna squamosa* and *Tridacna crocea* have four pairs, while *Lasaea adansonii* has three larval retractors, despite its almost direct development and the lack of a velum [24,25,34,35]. Three pairs of velum retractors are also found in *Mytilus trossulus, Ostrea edulis, Pandora inaequivalvis* and *Dreissena polymorpha* [21,23,26,30]. The ventral pair in *D. polymorpha*, adjacent to the anus, is presumably homologous to one of the three posterior retractor pairs in *P. maximus*, which likewise attach near the anus [24]. Due to their position, the three posterior protractors in *M. trossulus* [26] may be homologous to both above-mentioned muscles and are therefore considered as ventral larval retractors herein (Table 2). While the ventral retractor in *D. polymorpha* is maintained until the late veliger stage and the protractors in *M. trossulus* and the posterior retractors in *P. maximus* are present until the pediveliger stages, the ventral larval retractor of *L. pedicellatus,* which grows in the same direction, only exists in a very early veliger stage ([21,24,26]; Table 1). In contrast to 3-4 pairs of velum retractors and diverse larval retractors in pteriomorphans, the heterodont *L. pedicellatus* eventually displays two pairs of velum retractors and one pair of short-lived ventral larval retractors. While the distal parts of the ventral and dorsal velum retractors become successively atrophied and eventually degenerate, the basal portion doubles in volume and develops into the future lateral and dorsal mantle musculature, thereby maintaining their distinct insertion sites at the shell.

#### *Mantle musculature*

As in *Mytilus trossulus* and *Lasaea adansonii*, the mantle edge in *Lyrodus pedicellatus* is held in position partially by a U-shaped mantle musculature ([25,26]; Table 2). In addition, lateral and dorsal mantle muscle bundles are present in *L. pedicellatus.* The finger-shaped mantle musculature on the ventral side has not been described for any other bivalve so far and may constitute a teredinid apomorphy, which may be constrained by the highly derived overall bodyplan of a wood-boring shipworm (Table 2). However, since detailed comparative studies using modern methods are still largely lacking for these enigmatic bivalves, this cannot be clarified at present.

### Comparison of the prototroch muscle ring and larval retractors among mollusks

*Lyrodus pedicellatus* larvae display a muscle ring that underlies the prototroch and, in later stages, the velum. Such a muscle is also known from larval Gastropoda [11,12,17,18], Polyplacophora [13], Neomeniomorpha [36] and Chaetodermomorpha [16], but not Scaphopoda [14] (the monoplacophoran condition remains unknown). This velum ring is usually connected to the main larval retractor in gastropods [17,18]. In addition, in the vetigastropod *Haliotis rufescens,* the caenogastropod *Polinices lewisii,* the patellogastropod *Patella caerulea,* the opisthobranch *Aplysia californica* and the nudibranch *Aeolidiella stephanieae*, the larval retractors do form connections to adjacent regions such as the foot musculature, in addition to their main insertion sites within the velar lobes [10-12,17,18,37]. The velum retractors of *L. pedicellatus* neither have contact with the velum ring musculature, nor do they insert at other muscle systems, but instead fan out into the velar tissue.

Crucial differences are also obvious with respect to the timing of disintegration of the larval retractors in the various conchiferan taxa. In the gastropods *Haliotis kamtschatkana,*

*Patella caerulea*, *Aeolidiella stephanieae* and *Nerita melanotragus*, the larval retractors are resorbed during metamorphosis [9,11,18,19]. However, in the caenogastropod *Polonices lewisii*, resorption of the larval retractor muscle already begins soon after hatching [10]. The adult muscles arise independently from the larval shell muscles in all bivalves and gastropods investigated so far. Although the future fate of the dorsal and ventral mantle musculature is unknown for *Lyrodus pedicellatus*, their remodelling into shell-inserted mantle retractors during late larval life may indicate that they are retained in the adult shipworm. If true, the former assumption that the conchiferan larval shell musculature is generally independent from that of the adult [38], would have to be reconsidered at least for bivalves. Overall, *L. pedicellatus* is a prime example that larval shell muscles may be remodelled and subsequently integrated into the postmetamorphic bodyplan, thus raising the question as to what extent this may also be the case in other conchiferan mollusks.

## Conclusion

The early degeneration of the velum, the ventral larval retractor and the smaller number of velum retractors reflect, compared to other planktotrophic bivalve clades, a trend toward an abbreviated larval phase in *Lyrodus pedicellatus*. In contrast to gastropods, at least certain portions of the larval retractor muscles of *Lyrodus pedicellatus* most likely do contribute to the postmetamorphic muscular bodyplan. The foot retractor originates in very close proximity to the larval velum retractor and the lateral and dorsal mantle musculature constitutes a remodelled derivative of the larval velum retractors in early juveniles. To what extent, if at all, these muscles are incorporated into the definite adult, worm-like body of *Lyrodus pedicellatus* may only be assessed by analysis of advanced juveniles, which were not available for the present study. This is also true for the fate of other larval muscles including the accessory adductor and the finger-shaped mantle musculature, which so far are only known from *L. pedicellatus*.

Although only few data on bivalve myogenesis are currently available, certain muscular subsets appear to be conserved among the larvae of the various bivalve subtaxa and thus were most likely part of the larval bivalve groundplan. These include a dense pedal plexus, distinct sets of adductors as well as various velum retractors. The larval velum ring was found in *Lyrodus* and *Mytilus* and its presence in most other molluskan taxa confirms homology of this system among bivalves and the entire Mollusca. The presence of distinct prototroch/velum retractors in Bivalvia and Gastropoda (but not Scaphopoda) and their absence in Aculifera suggests co-evolution of these systems with embryonic and/or larval shell(s), either in these two clades alone or at the base of Conchifera (with secondary loss in scaphopods and cephalopods).

## Additional file

**Additional file 1: Figure S1.** Scanning electron micrographs of late-stage *Lyrodus pedicellatus* larvae.

**Additional file 2: Description of the methods used to generate the scanning electron micrograph shown in Supplemental Figure 1.**

### Competing interests

The authors declare that they have no competing interests.

### Authors contributions

AWM performed research, analyzed data, drafted the manuscript and generated all figures. AK contributed to the graphic representations and aided in data interpretation. TS contributed to acquisition of confocal data and to specimen preparation. RS provided the study material and provided input on *Lyrodus pedicellatus* gross larval anatomy. SC provided supplemental SEM images and contributed to data interpretation. AW designed the study, supervised research, contributed to data interpretation and contributed to writing of the manuscript. All authors read and approved the final version of the manuscript.

### Acknowledgements

We thank the members of the Wanninger lab for methodological input. AK is the recipient of a Lise Meitner fellowship from the Austrian science foundation (FWF, M1523-B19).

### Author details

[1]Department of Integrative Zoology, Faculty of Life Sciences, University of Vienna, Althanstrasse 14, 1090 Vienna, Austria. [2]Institute of Marine Sciences, University of Portsmouth, Ferry Road, Portsmouth P04 9LY, UK.

### References

1. Giribet G: **Bivalvia.** In *Phylogeny and Evolution of the Mollusca*, Volume 6. Edited by Ponder W, Lindberg D. Berkeley and L.A., California, USA: University of California Press; 2008:105 141.
2. Salvini-Plawen L: **Was ist eine Trochophora? Eine Analyse der Larventypen mariner Protostomier.** *Zool Jb Anat* 1980, **103**:389 423.
3. Hoggarth M: **Description of some of the glochidia of the Unionida (Mollusca: Bivalvia).** *Malacologia* 1999, **41**:1 118.
4. Moll F: **Die Bohrmuschel (Genus *Teredo* Linn?).** *Naturwissenschaftliche Zeitschrift fr Forst- und Landwirtschaft* 1914, **11**(12):505 564.
5. Isham LB, Tierney JQ: **Some aspects of the larval development and metamorphosis of *Teredo (Lyrodus) pedicellata* de Quatrefages.** *Bull Mar Sci Gulf and Caribbean* 1953, **2**:574 589.
6. Morse M, Zardus JD: **Bivalvia.** In *Microscopic Anatomy of Invertebrates*, Volume 6A. Edited by Harrison FW, Kohn AJ. N.Y: Wiley- Liss; 1997:1 118.
7. Cragg S: **The Phylogenetic Significance of Some Anatomical Features of Bivalve Veliger Larvae.** In *Origin and Evolutionary Radiation of the Mollusca*. Edited by Taylor J. Oxford: Oxford University Press. The Malacological Society of London; 1996:371 380.
8. Sigerfoos C: **Natural history organisation, and development of the Teredinidaei, or ship-worms.** *Bull Bur Fish* 1907, **27**:191 231.
9. Page L: **Larval shell muscles in the abalone *Haliotis kamtschatkana*.** *Biol Bull* 1997, **193**:30 46.
10. Page L: **Sequential development programmes for retractor muscles of a caenogastropod: reappraisal of evolutionary homologues.** *Proc R Soc Lond B* 1998, **265**:2243 2250.
11. Wanninger A, Ruthensteiner B, Lobenwein S, Salvenmoser W, Dictus W, Haszprunar G: **Development of the musculature in the limpet *Patella* (Mollusca, Patellogastropoda).** *Dev Genes Evol* 1999, **209**:226 238.
12. Wanninger A, Ruthensteiner B, Dictus W, Haszprunar G: **The development of the musculature in the limpet *Patella* with implications on its role in the process of ontogenetic torsion.** *Invertebr Reprod Dev* 1999, **36**(1 3):211 215.
13. Wanninger A, Haszprunar G: **Chiton myogenesis: perspectives for the development and evolution of larval and adult muscle systems in molluscs.** *J Morphol* 2002, **251**:103 113.

14. Wanninger A, Haszprunar G: **Muscle development in *Antalis entalis* (Mollusca, Scaphopoda) and its significance for scaphopod relationship.** *J Morphol* 2002, **254:**53 64.
15. Page L: **Ontogenetic torsion in two basal gastropods occurs without shell attachements for larval retractor muscles.** *Evol Dev* 2002, **4**(3):212 222.
16. Nielsen C, Haszprunar G, Wanninger A: **Early development of the aplacophoran mollusc *Chaetoderma*.** *Acta Zoologica (Stockholm)* 2007, **88:**231 247.
17. Wollesen T, Wanninger A, Klussmann-Kolb A: **Myogenesis in *Aplysia californica* (Cooper, 1863) (Mollusca, Gastropoda, Opisthobranchia) with special focus on muscular remodeling during metamorphosis.** *J Morphol* 2008, **269:**776 789.
18. Kristof A, Klussmann-Kolb A: **Neuromuscular development of *Aeolidiella stephanieae* Vald?z, 2005 (Mollusca, Gastropoda, Nudibranchia).** *Frontiers in Zoology* 2010, **7:**5.
19. Page L, Ferguson S: **The other gastropod larvae: Larval morphogenesis in a marine neritomorph.** *J Morphol* 2013, **274:**412 428.
20. Hatschek B: **?ber Entwicklungsgeschichte von *Teredo*.** *Arb Zool Inst Wien* 1880, **3:**1 47.
21. Meisenheimer J: **Entwicklungsgeschichte von *Dreissensia polymorpha* Pall.** *Zeitschrift fr wissenschaftliche Zoologie* 1901, **69:**1 137.
22. Hebers K: **Entwicklungsgeschichte von *Anodonta cellensis* Schr?t.** *Zeitschrift fr wissenschaftliche Zoologie* 1913, **108:**1 174.
23. Erdmann W: **?ber die Entwicklung und die Anatomie der ansatzreifen Larve von *Ostrea edulis*.** *Helgol. Wiss. Meeresunters.* 1935, **19:**1 25.
24. Cragg S: **The adductor and retractor muscles of the veliger of *Pecten maximus* (L.) (Bivalvia).** *J Moll Stud* 1985, **51:**276 283.
25. Altn?der A, Haszprunar G: **Larval morphology of the brooding clam *Lasaea adansonii* (Gmelin, 1791) (Bivalvia, Heterodonta, Galeommatoidea).** *J Morphol* 2008, **269:**762 774.
26. Dyachuk V, Odintsova N: **Development of the larval muscle system in the mussel *Mytilus trossulus* (Mollusca, Bivalvia).** *Develop Growth Differ* 2009, **51:**69 79.
27. Dyachuk V: **Extracellular matrix is required for muscle differentiation in primary cell cultures of larval *Mytilus trossulus* (Mollusca: Bivalvia).** *Cytotechnology* 2013, **65:**725 735.
28. Zardus JD, Martel A: **Phylum Mollusca: Bivalvia.** In *Atlas of Marine Invertebrate Larvae.* Edited by Young CM, Sewell M, Rice M. Berkeley and L. A., California, USA: University of California Press; 2001:427 445.
29. Cragg S, Crisp D: **The Biology of Scallop Larvae.** In *Biology, Ecology and Aquaculture of Scallops.* Edited by Shumway S. Amsterdam: Elsevier; 1991:75 132.
30. Allen J: **The development of *Pandora inaequivalvis* (Linn?).** *J Embryol Exp Morph* 1961, **9**(Part 2):252 268.
31. Bayne B: **Some Morphological Changes That Occur at the Metamorphosis of the Larvae of *Mytilus edulis*.** In *Fourth European Marine Biology Symposium.* Edited by Crisp DJ. London: Cambridge University Press; 1971:259 280.
32. Cranfield H: **A study of the morphology, ultrastructure and histochemistry of the foot of the pediveliger of *Ostrea edulis*.** *Marine Biology* 1973, **22:**187 202.
33. Wanninger A, Koop D, Moshel-Lynch S, Degnan B: **Molluscan Evolutionary Development.** In *Phylogeny and Evolution of the Mollusca,* Volume 16. Edited by Ponder W, Lindberg D. Berkeley and L.A., California, USA: University of California Press; 2008:427 445.
34. La Barbera M: **Larval and post-larval development of the giant clams *Tridacna maxima* and *Tridacna squamosa* (Bivalvia: Tridacnidae).** *Malacologia* 1975, **15**(1):69 79.
35. Jameson SC: **Early life history of the giant clams *Tridacna crocea* Lamarck, *Tridacna maxima* (R?ding), and *Hippopus hippopus* (Linnaeus).** *Pacific Science* 1976, **30**(3):219 233.
36. Scherholz M, Redl E, Wollesen T, Todt C, Wanninger A: **Aplacophoran mollusks evolved from ancestors with polyplacophoran-like features.** *Curr Biol* 2013, **23:**2130 2134.
37. Degnan B, Degnan S, Morse D: **Muscle-specific regulation of tropomyosin gene expression and myofibrillogenesis differs among muscle systems examined at metamorphosis of the gastropod *Haliotis rufescens*.** *Dev Genes Evol* 1997, **206:**464 471.
38. Haszprunar G, Wanninger A: **Molluscan muscle systems in development and evolution.** *J Zool Syst Evol Res* 2000, **38:**157 163.

# 14

# Evolutionary history of Chaetognatha inferred from molecular and morphological data: a case study for body plan simplification

Samah Gasmi[1,2], Gabriel Nève[1], Nicolas Pech[1], Saïda Tekaya[2], André Gilles[1†] and Yvan Perez[1*†]

**Abstract**

**Background:** Chaetognatha are a phylum of marine carnivorous animals which includes more than 130 extant species. The internal systematics of this group have been intensively debated since it was discovered in the 18th century. While they can be traced back to the earlier Cambrian, they are an extraordinarily homogeneous phylum at the morphological level - a fascinating characteristic that puzzled many a scientist who has tried to clarify their taxonomy. Recent studies which have attempted to reconstruct a phylogeny using molecular data have relied on single gene analyses and a somewhat restricted taxon sampling. Here, we present the first large scale phylogenetic study of Chaetognatha based on a combined analysis of nearly the complete ribosomal RNA (rRNA) genes. We use this analysis to infer the evolution of some morphological characters. This work includes 36 extant species, mainly obtained from *Tara Oceans* Expedition 2009/2012, that represent 16 genera and 6 of the 9 extant families.

**Results:** Cladistic and phenetic analysis of morphological characters, geometric morphometrics and molecular small subunit (SSU rRNA) and large subunit (LSU rRNA) ribosomal genes phylogenies provided new insights into the relationships and the evolutionary history of Chaetognatha. We propose the following clade structure for the phylum: (((Sagittidae, Krohnittidae), Spadellidae), (Eukrohniidae, Heterokrohniidae)), with the Pterosagittidae included in the Sagittidae. The clade (Sagittidae, Krohnittidae) constitutes the monophyletic order of Aphragmophora. Molecular analyses showed that the Phragmophora are paraphyletic. The Ctenodontina/Flabellodontina and Syngonata/Chorismogonata hypotheses are invalidated on the basis of both morphological and molecular data. This new phylogeny also includes resurrected and modified genera within Sagittidae.

**Conclusions:** The distribution of some morphological characters traditionally used in systematics and for species diagnosis suggests that the diversity in Chaetognatha was produced through a process of mosaic evolution. Moreover, chaetognaths have mostly evolved by simplification of their body plan and their history shows numerous convergent events of losses and reversions. The main morphological novelty observed is the acquisition of a second pair of lateral fins in Sagittidae, which represents an adaptation to the holoplanktonic niche.

**Keywords:** Chaetognatha, Phylogenetics, Systematics, Procrustes surimposition, Homoplasy, Body plan simplification

## Background

Chaetognaths are small predators of major importance in the marine ecosystem [1,2]. They are abundant in every sea worldwide and can be traced back to the Cambrian radiation [3]. Most of them are planktonic but a few are benthic. Chaetognaths are particularly renowned for their peculiar morphological and developmental features. These characters, as well as the affinities of the group within the metazoans, have been extensively debated by zoologists since the discovery of the phylum in the 18th century [4]. So far the most recent phylogenetic analyses have also proved problematic for inferring their sister-group relationships within metazoans, which makes their positioning one of the most difficult issues in animal phylogeny [5]. Numerous alternative phylogenetic hypotheses have been proposed over a long history of debate (for review [5,6]). However, a recent hypothesis has emerged, based

* Correspondence: yvan.perez@imbe.fr
†Equal contributors
[1]Aix-Marseille Université, CNRS, IRD, Avignon Université, IMBE UMR 7263, 13331 Marseille cedex 3, France
Full list of author information is available at the end of the article

on morphological [7,8] and phylogenomic analyses [9-13], where chaetognaths have been considered an early diverging member of Protostomia. The circumoral brain and the intraepithelial ventral cords have been recognized to be two of the key apomorphies of Protostomia [14,15]. The nervous system in Chaetognatha is characterized by such a typical arrangement. However, even though the Chaetognatha partly share the Protostomia ground pattern, Perez et al. [5] concluded that "their derived genome and morphology do not include any convincing synapomorphy that would suggest a sister-group relationship to another metazoan taxon".

As their relationships within metazoans, their internal systematics is still very much debated [16-18]. Here, we recall the main hypotheses previously proposed based on morphological and, more recently, molecular data. According to Ritter-Zahony [19] and Hyman [20], Chaetognatha was traditionally divided into six genera representing four families: *Sagitta* (Sagittidae), *Pterosagitta* (Pterosagittidae), *Spadella*, *Eukrohnia* and *Heterokrohnia* (Eukrohniidae) and *Krohnitta* (Krohnittidae). Later, Tokioka [21] reevaluated the relationships between families by creating two new orders (Figure 1A): the plesiomorphic Phragmophora (presence of a transverse musculature, namely the phragms, and various kinds of glandular structures on the body surface) composed of Spadellidae and Eukrohniidae; and the derived Aphragmophora (absence of phragms and few glandular structures). Tokioka [21] suggested creating two Aphragmophora suborders according to the shape of teeth and hooks and the number of teeth rows. The suborder Flabellodontina only contains the family Krohnittidae, while the Pterosagittidae and Sagittidae belonged to the Ctenodontina. In a following work, Tokioka [22] suggested the paraphyly of Aphragmophora (Figure 1B), with the Ctenodontina were thought to be closer to the Phragmophora than to the Flabellodontina. Inspired by a previous suggestion of Alvariño [23], Tokioka [21] considered that some of the structural differences between *Sagitta* species were of significant systematic importance. This author divided *Sagitta* into nine new genera and gathered them into the Sagittidae. After the discovery of several new deep benthoplanktonic chaetognaths, Casanova [24] slightly modified Tokioka's hypothesis (Figure 1C). In his version, the members of the Phragmophora order were split into two new orders. First, the Biphragmophora (comprising the new Heterokrohniidae family, with transverse muscles in both trunk and tail) was included into the subclass Syngonata (with ducts between the genital glands). Second, the Monophragmophora (Spadellidae and Eukrohniidae, with transverse muscles in trunk only) was associated with the Aphragmophora into the subclass Chorismogonata (without such ducts). Using multivariate analyses based on body shape, Dallot and Ibanez [25] suggested the existence of three groups (*Sagitta*, *Eukrohnia*, and *Spadella*/*Bathyspadella*) and a close relationship between the

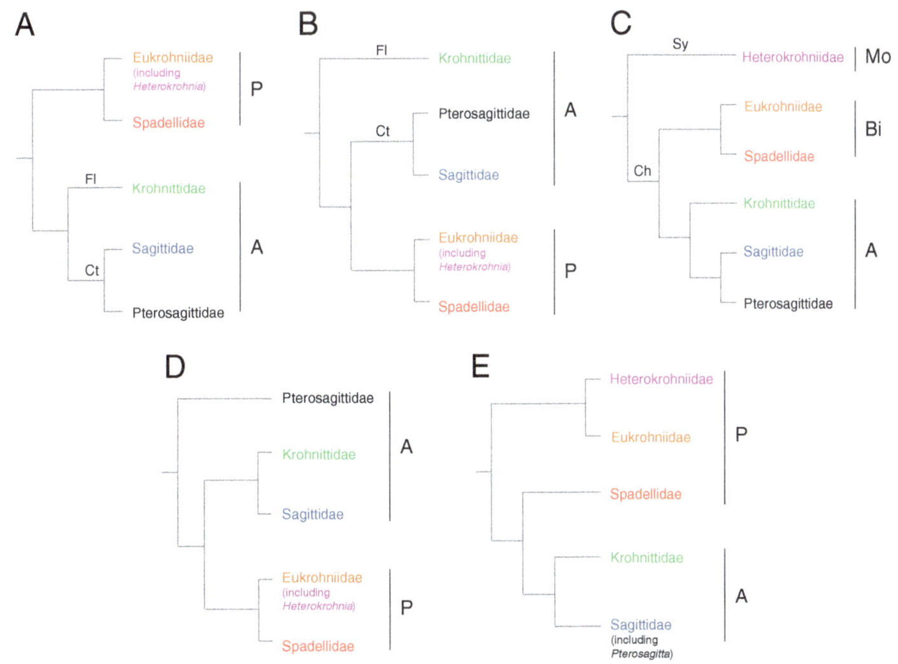


**Figure 1 Five main phylogenetic hypotheses of chaetognaths relationships.** Hypothesis based on morphological data: A, Tokioka [21], **B**, Tokioka [22], **C**, Casanova [24], **D**, Salvini-Plawen [26]. **E**, data presented here. A, Aphragmophora; Bi, Biphragmophora; Ch, Chorismogonata; Ct, Ctenodontina; Fl, Flabellodontina; Mo, Monophragmophora; P, Phragmophora; Sy, Syngonata.

planktonic species *Pterosagitta draco* and the benthic Spadellidae. They also questioned the inclusion of *Sagitta lyra* within the genus *Sagitta*. In another study, Salvini-Plawen [26] proposed the littoral-neritic *Pterosagitta draco* as the most plesiomorphic species and contradicted the ancestrality of phragms (Figure 1D). He also omitted the Aphragmophora suborders Ctenodontina and Flabellodontina of Tokioka [21] and the Syngonata/Chorismogonata hypothesis of Casanova [24]. Finally, Bieri [21,22] proposed the most recent revision of the chaetognaths classification. Following Alvariño [23] and Tokioka [21,22], he suggested new genera within Sagittidae. Several morphological criteria were taken into account: position and shape of the corona ciliata; position and shape of lateral fins and seminal vesicles; presence/absence and shape of the intestinal diverticula; trunk/tail length ratio; rayless-zones in the lateral fins; body aspect. This author also disregarded the Syngonata/Chorismogonata hypothesis.

Thus, based on a consensus between Tokioka, Casanova and Bieri's hypotheses, the extant Chaetognatha are represented by three orders (Biphragmophora, Monophragmophora, Aphragmophora) and nine families (Heterokrohniidae, Eukrohniidae, Pterokrohniidae, Spadellidae, Krohnittellidae, Krohnittidae, Pterosagittidae, Sagittidae, Bathybelosidae).

The first molecular study of chaetognaths systematics was conducted with a short portion of the large subunit ribosomal RNA 28S (LSU rRNA) gene [27]. These authors concluded that the LSU rRNA gene is duplicated in Chaetognatha, the division into Aphragmophora and Phragmophora is supported and several genera of the Sagittidae family described by Tokioka [21] and Bieri [16] are recovered. Papillon et al. [18] carried out a more extensive molecular study based on 26 sequences of the small subunit ribosomal RNA 18S (SSU rRNA) isolated from members of six extant families; they concluded that (1) similarly to LSU rRNA, a duplication of SSU rRNA gene occurred, suggesting that the whole ribosomal cluster is duplicated, (2) Tokioka's suborders Ctenodontina and Flabellodontina are not validated, (3) Casanova's hypotheses Syngonata/Chorismogonata and Monophragmophora/Biphragmophora are rejected, (4) the families Krohnittidae and Pterosagittidae are not supported, (5) three monophyletic groups are identified: Sagittidae/Krohnittidae, Spadellidae/Pterosagittidae and Eukrohniidae/Heterokrohniidae, (6) the order Aphragmophora without *Pterosagitta draco* is monophyletic. Since then, no molecular study has been made to further explore the systematics of this phylum. Finally, a recent barcoding analysis was highly successful at discriminating between the described species [28]. It notably revealed little geographical structure and showed that *Eukrohnia bathypelagica* and *Eukrohnia hamata* are probably young sister-species.

Thus, even after one century of heavy debates, it has not been possible to establish a stable and reliable hypothesis on the evolutionary history of Chaetognatha. In the present work, we have conducted an extensive molecular analysis based on LSU and SSU rRNA duplicated genes. We combined, for the first time, the molecular results with a morphological classification and geometric morphometrics. In the light of our results, we present a revised phylogeny and discuss the morphology-based character systems that have traditionally been used to classify this enigmatic phylum.

## Results

### Alignments and erroneous sequences

First, we identified erroneous sequences by constructing test trees from LSU and SSU rRNA genes and by then identifying which sequences from public databases came out in suspicious positions when compared with new sequences obtain in the present study. This approach revealed three chaetognath sequences that are erroneous due to contaminant or bad species diagnosis. These sequences that must be excluded from any future phylogenetic analyses were:

- The SSU rRNA of *Krohnitta pacifica* (class I DQ351879 and class II DQ351891) from [18] which most likely is a Sagittidae contaminant close to *Parasagitta setosa* when compared with three new sequences belonging to *Krohnitta subtilis*.
- The SSU rRNA of *Sagitta sp* (class I AY922316) from [29] which belongs to the Eukrohniidae family, close to *Eukrohnia hamata* and *Eukrohnia bathypelagica*.
- We also characterised new sequences for three different specimens of *Pterosagitta draco* that were used instead of the erroneous SSU rRNA sequences (class I DQ351885 and class II DQ351898) from [18].

After excluding these contaminants, we set up five alignments available upon request (dataset 1, LSU rRNA paralogous genes; dataset 2, SSU rRNA paralogous genes; dataset 3, SSU rRNA paralogy class I; dataset 4, SSU rRNA paralogy class II; dataset 5, concatenated alignment of SSU and LSU rRNA genes).

### Molecular phylogenetic analyses

Whatever the dataset considered, the Bayesian and approximate likelihood ratio (aLRT) trees obtained were almost identical to the maximum likelihood (ML) bootstrapped tree. Thus their statistical values were reported on the ML topology. However, in the case of the concatenated analysis, because only the aLRT and the Bayesian reconstructions showed a fully congruent topology, statistical values obtained from the three methods were reported on the Bayesian topology. The validity of the Krohnittidae and Heterokrohniidae families will not be discussed because both are only represented here by a single species.

*Large subunit rRNA*. The alignment of 58 sequences of LSU rRNA (49 specimens for the class I and 9 for the class II) was 410 base pairs (bp) long (dataset 1 = LSU rRNA Class I and II). The best model of evolution estimated with MODELTEST was the K2P + Γ model (with a lnL of −2457.20). Without the nine class II sequences, the same model was selected (with a lnL of −1483.09). Using the midpoint rooting, the two classes of LSU rRNA yielded two monophyletic groups (Figure 2). The rate of substitution for the unconstrained tree was equal to $8.16 \times 10^{-2}$ ([$8.02 \times 10^{-2}$ – $8.30 \times 10^{-2}$] 95% confidence interval). Only the topology based on class I sequences is discussed. Parametric (approximate likelihood ratio values, aLRTv and posterior probabilities, pp) and non-parametric branch tests (bootstrap values, bv) showed a recurrent lack of resolution for the deep nodes of trees based on LSU rRNA genes. To assess whether a consistent phylogenetic signal could be revealed by reducing the impact of noise, the same phylogenetic analyses were carried out, but this time after treatment of the dataset 1 with the program Aliscore. The new LSU rRNA alignment (Dataset 1bis) displayed 155 (37.8%) 'noisy' sites that were filtered out and did not impact on the selected model of evolution (K2P + Γ with a lnL of −1735.6). Most of these potential 'noisy' sites were located in the loop regions of the LSU rRNA gene class II due to *Xenokrohnia sorbei* LSU rRNA gene class I alignment. After removing these positions, the accuracy of the trees did not increase (data not shown). Thus, the lack of resolution for deep branching nodes is more likely linked to a lack of phylogenetic information than to a 'noisy' signal due to mismatches in sequence alignment (*i.e.*, soft polytomy). Because of this, relationships between orders and families were very difficult to assess. LSU rRNA sequences contained information only relevant to the discussion of the relationships between species of a given genus and, to a lesser extent, between genera of the same family. However, the trees constructed under ML and Bayesian criteria produced very similar topologies if considering only the well-resolved nodes. First, Aphragmophora (Krohnittidae + Sagittidae + Pterosagittidae) and Phragmophora (Heterokrohniidae + Eukrohniidae + Spadellidae) were not monophyletic. A close relationship between the Eukrohniidae and *X. sorbei*, the only representative of the Heterokrohniidae, was well supported in all methods (78/0.90/0.94, bv/aLRTv/pp hereinafter). At the family level, only the Spadellidae received high support whatever the method considered (99/1/1). The monophyly of Eukrohniidae was supported only in the Bayesian analysis (39/0.60/0.98). The monophyly of the Sagittidae, which was the most represented family (25 species/11 genera), cannot be confirmed because of very low support value in ML analysis (bv = 20) or the lack of node in aLRT and Bayesian analyses. It is however possible to characterize two main lineages within this family. Within Sagittidae, a well-supported assemblage consisted of *Ferosagitta*, *Aidanosagitta* and *Flaccisagitta* (83/0.95/1). A surprising result was the inclusion of *Pterosagitta draco*, the only representative of the Pterosagittidae, within Sagittidae, close to *Sagitta*, *Parasagitta* genera and *Mesosagitta minima*, but with weak support (33/0.77/0.75). Several genera validated in the Bieri's [16] nomenclature appeared monophyletic with marginal or high support values. These are *Ferosagitta* (two of the seven known species are represented, 68/0.74/0.97), *Solidosagitta* (two of the three known species are represented, 87/0.83/0.96), *Serratosagitta* (three of the five known species are represented, 97/0.88/1), *Pseudosagitta* (three of the four known species are represented, 99/0.98/1). Yet *Flaccisagitta* (two of the three known species are represented) received low support values from the ML analysis and marginal support values from Bayesian analysis (38/0.84 respectively) whereas this node was not recovered in the aLRT tree. The remaining genera are either monospecific (*Caecosagitta*, *Pterosagitta*) or did not appear monophyletic but were grouped in tight assemblages. It is especially the case of large and heterogeneous genera such as *Aidanosagitta* and *Parasagitta*. The *Mesosagitta* genus, represented here by *M. minima* and *M. decipiens*, was invalidated.

*Small subunit rRNA*. Three data sets of SSU rRNA sequences were analysed. First, phylogenetic reconstructions were conducted with dataset 2 comprising all available SSU rRNA sequences from both paralogy classes (dataset 2 = SSU rRNA Class I and II, 138 sequences from 33 species: 80 class I and 58 class II sequences). One of the recurrent problems when analysing the molecular data in Chaetognatha is the lack of relevant outgroups among bilaterians. So, this first analysis offered to root one paralogy class on the other, highlighting the branching order of five represented families with good statistical values and revealing the following sequence from the last common ancestor to the most derived family (Additional file 1): Eukrohniidae + Heterokrohniidae, Spadellidae, Krohnittidae and Sagittidae, the latter including Pterosagittidae. Similarly to the LSU rRNA analysis, Phragmophora (Heterokrohniidae + Eukrohniidae + Spadellidae) were paraphyletic while Aphragmophora (Pterosagittidae + Sagittidae + Krohnittidae) received high support in most methods applied (class I 80/1/0.98, class II 19/1/0.85).

Datasets 3 (SSU rRNA Class I, 1679 bp long) and 4 (SSU rRNA Class II, 1119 bp long) consisted of two independent alignments for each paralogy. Here, we only present the analyses based on these two datasets (Figures 3 and 4 respectively) because separate phylogenetic reconstructions based on one paralogous gene gave more robust and accurate results than the full data set comprising both SSU rRNA paralogy classes (dataset 2). The selected model was the K2P + Γ with a lnL of −7733.81

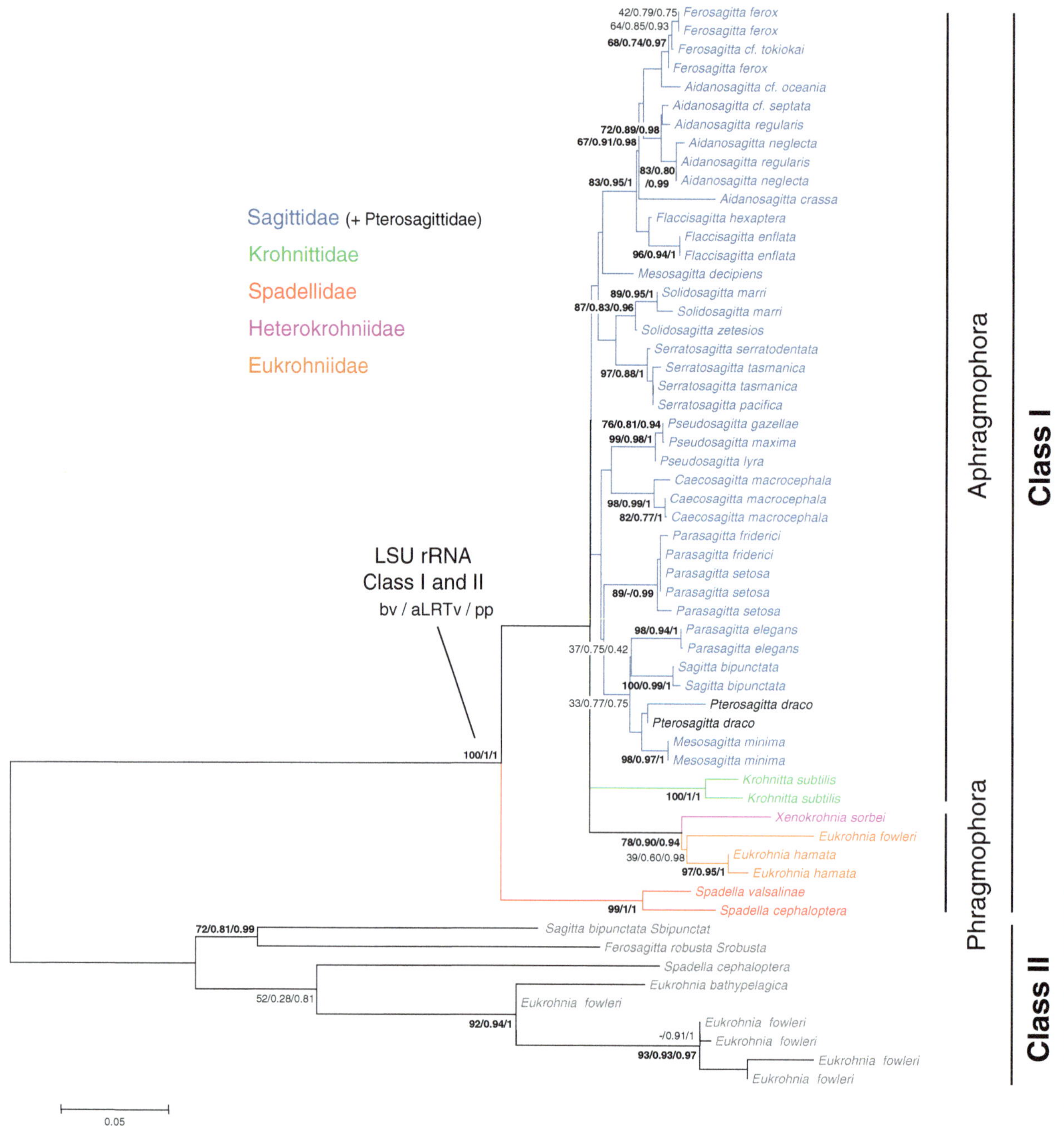


**Figure 2 Maximum likelihood tree calculated from LSU rRNA by the Mega program with the K2P + Γ model.** The alignment of 58 sequences of LSU rRNA was 410 base pairs long. The lnL value of this optimal tree is –2457.20. The two classes of LSU rRNA yielded trees that rooted each other. Support values obtained using different reconstruction approaches are indicated at nodes in the following order: maximum Likelihood bootstrap probabilities (bv), approximative likelihood ratio test (aLRTv), and Bayesian posterior probabilities (pp). Support values are displayed when bv/aLRTv ≥75 or pp ≥0.85. Node absence in a given method is indicated by -.

for the class I and the T92 + Γ with a lnL of –4073.31 for the class II. The substitution rates were equal to $1.14 \times 10^{-2}$ ([$4.19 \times 10^{-4}$ – $2.30 \times 10^{-2}$] 95% confidence interval) for the SSU rRNA class I and $1.03 \times 10^{-2}$ ([$4.83 \times 10^{-4}$ – $2.14 \times 10^{-2}$] 95% confidence interval) for the SSU rRNA class II. When looking at the overall alignments based on both SSU rRNA paralogy classes, they included 579 variable sites (34.48%). This number decreased to 446 in the SSU rRNA class I (26.56%) and to 252 in the SSU rRNA class II (22.52%). However, the number of highly supported nodes within the ML phylogenies (bv >75) increases from 44 when analysing both SSU rRNA classes together to 56 (32 in class I and 24 in class II) when analysing each class separately. In fact,

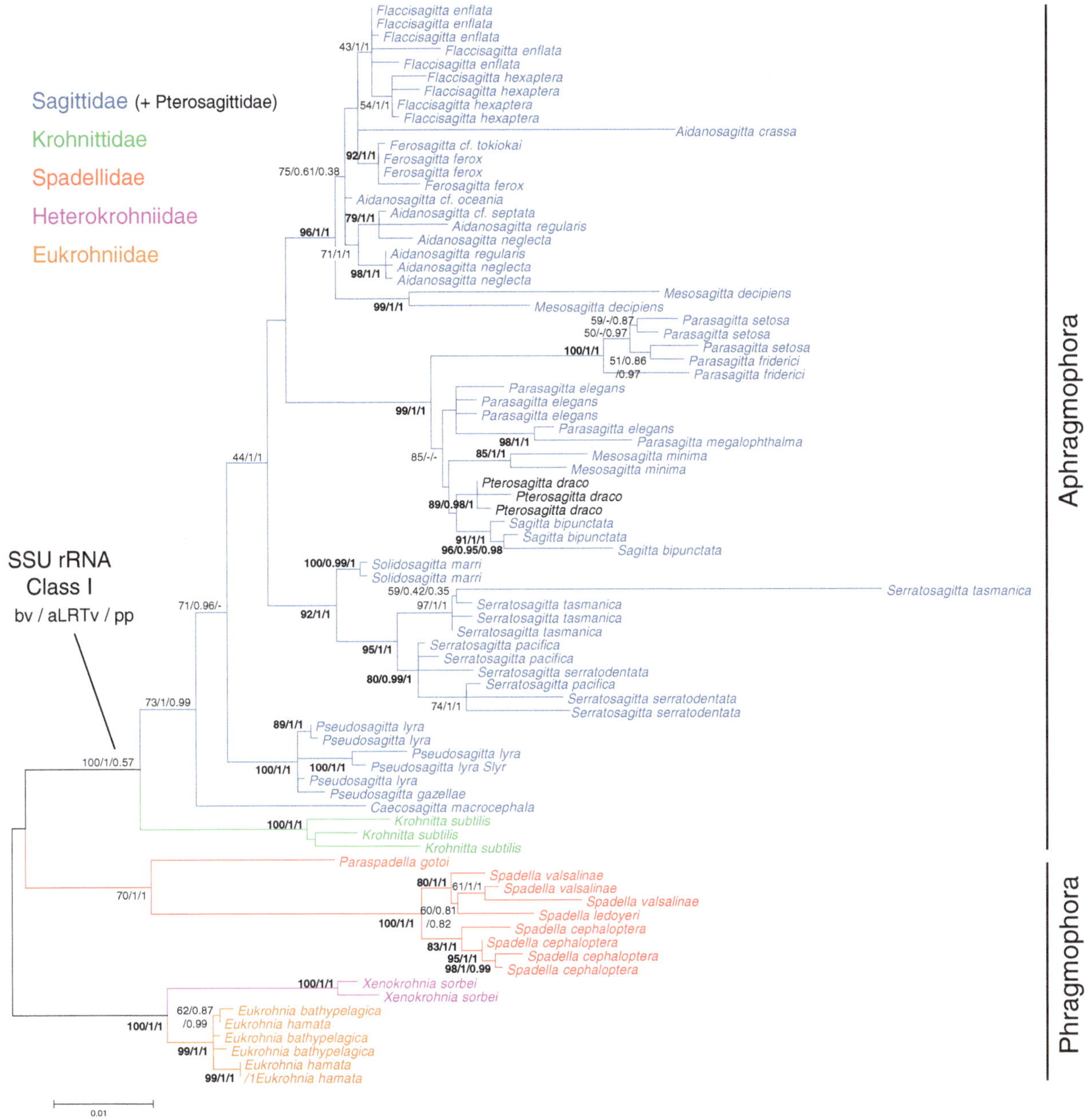


**Figure 3 Maximum likelihood tree calculated from SSU rRNA Class I by the Mega program with the K2P + Γ model.** The alignment of 80 sequences of LSU rRNA was 1679 base pairs long. The lnL value of this optimal tree was −7733.81. Support values obtained using different reconstruction approaches are indicated at nodes in the following order: maximum Likelihood bootstrap probabilities (bv), approximative likelihood ratio test (aLRTv) and Bayesian posterior probabilities (pp). Support values are displayed when bv/aLRTv ≥75 or pp ≥0.85. Node absence in a given method is indicated by -. Tree was rooted on the monophyletic assemblage consisting of Eukrohniidae + Heterokrohniidae. This rooting was also observed in most analyses when using the midpoint rooting function.

the evolutionary rates of the two paralogous genes are very close to each other. However, while transitions and tranversions are each responsible for 50% of the substitutions in class I, transitions are involved for 68% in class II. This explains the selection of two different evolutionary models by MODELTEST. Therefore, the difference in distance estimation between the classes I and II is more likely due to the lower frequency of transversion events in class II than to base composition. According to the topology obtained in the analyses of both classes of SSU rRNA genes (see Additional file 1), separate analyses of each paralogy classes were rooted on the monophyletic assemblage consisting of Eukrohniidae + Heterokrohniidae. This rooting was also observed

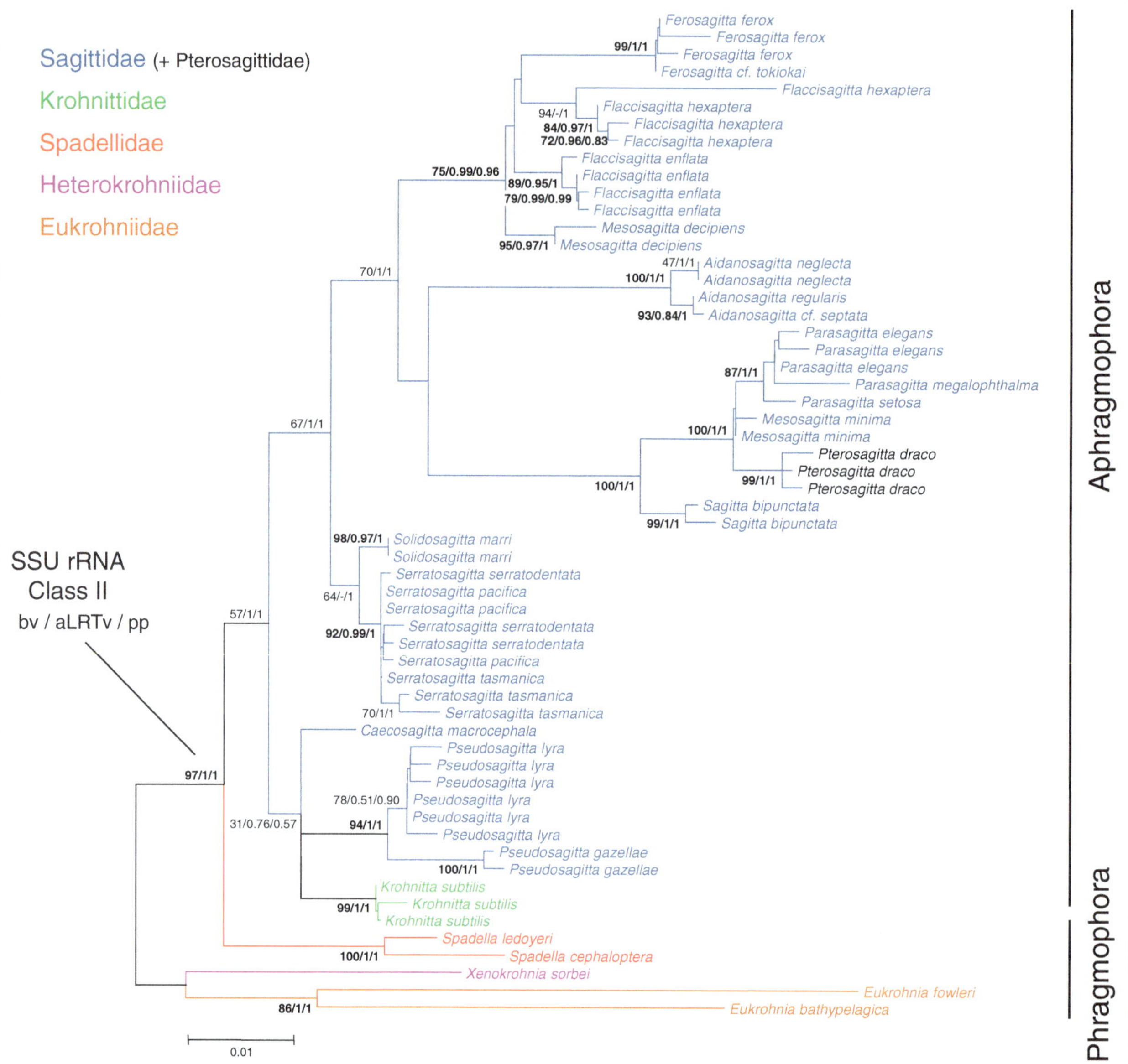


**Figure 4 Maximum likelihood tree calculated from SSU rRNA Class II by the Mega program with the T92 + Γ model.** The alignment of 58 sequences of LSU rRNA was 1119 base pairs long. The lnL value of this optimal tree was −4073.31. Support values obtained using different reconstruction approaches are indicated at nodes in the following order: maximum Likelihood bootstrap probabilities (bv), approximative likelihood ratio test (aLRTv) and Bayesian posterior probabilities (pp). Support values are displayed when bv/aLRTv ≥75 or pp ≥0.85. Node absence in a given method is indicated by -. Tree was rooted on the monophyletic assemblage consisting of Eukrohniidae + Heterokrohniidae. This rooting was also observed in most analyses when using the midpoint rooting function.

in most analyses when using the midpoint rooting function. The overall topologies obtained with SSU rRNA sequences belonging to class I and II (Figures 3 and 4 respectively) were more robust than with the LSU rRNA sequences. The monophyletic group consisting of *X. sorbei* and all species belonging to Eukrohniidae was unambiguously supported in all methods applied (class I 100/1/1, class II 97/1/1 hereinafter). As with the LSU analysis, Aphragmophora (Krohnittidae + Sagittidae + Pterosagittidae) received good support (100/1/0.57, 57/1/1). All the families (Eukrohniidae 99/1/1, 86/1/1; Spadellidae 70/1/1, 100/1/1; Sagittidae + Pterosagittidae 73/1/0.99, class I only) were highly supported by SSU trees. In all methods used, the branching order of these clades was extremely well conserved between both paralogy classes, with one main difference being the position of Krohnittidae. This monogeneric family comprising three species (but only represented here by *Krohnitta subtilis*) was found either as the sister-group to Sagittidae in the class I trees with high support (100/1/0.57) or within Sagittidae (57/1/1) in class II trees in a mongrel assemblage comprising *Caecosagitta* and *Pseudosagitta* genera but with low support (31/0.76/0.57). Several traditional genera were not monophyletic, for instance *Mesosagitta* (*M. minima* and *M. decipiens*) which was invalidated by our results. *Aidanosagitta* was paraphyletic in class I analysis and monophyletic with high support in class II analysis (100/1/1) while the opposite was obtained for *Flaccisagitta*, the monophyly of which received marginal bv and high aLRTv and pp values in class I analysis (43/1/1). *Parasagitta* was paraphyletic in the class I analysis and monophyletic in the class II analysis with high support (87/1/1). Three main subclusters were identified within Sagittidae. *Aidanosagitta*, *Flaccisagitta*, two *Ferosagitta* species (*F. tokiokai* and *F. ferox*) and *Mesosagitta decipiens* formed a first clade in the class I trees with high support (96/1/1). In class II trees, only *Flaccisagitta, Ferosagitta and M. decipiens* were clustered together with high support (75/0.99/0.96). *Parasagitta, P. draco, Sagitta bipunctata* and *M. minima* constituted a second subcluster with high support in all trees (99/1/1, 100/1/1). This second subcluster displayed a close relationship with the first one in the class II analysis (70/1/1). A third and last Sagittidae subcluster consisting of *Solidosagitta* and *Serratosagitta* genera was also found to be monophyletic but was highly supported only in class I analyses (92/1/1, 64/-/1). Finally, the class I analysis confirmed the monophyly of *Pseudosagitta* (*P. lyra* and *P. gazellae*) with the three main Sagittidae subclusters only in the ML and aLRT trees (71/0.96/-) and the most basal genus among all Sagittidae was the monospecific genus *Caecosagitta* (73/1/0.99). The trees resolution drastically decreased at the infrageneric level and phylogenies based on SSU rRNA data did not recover the monophyly of several species, for instance in *Serratosagitta pacifica* and *S. serratodentata* (class I and II), *M. minima* (class II), *Aidanosagitta regularis and A. neglecta* (class I), *Parasagitta elegans* (class I and II), *Parasagitta setosa and Parasagitta friderici* (Class I), *P. lyra* (class I), *F. ferox* (class I), *Eukrohnia hamata* and *E. Bathypelagica* (class I).

*Concatenated analysis.* As it is done for large coding genes that contain considerable phylogenetic signal, rRNA genes were concatenated to increase the accuracy of the phylogenetic reconstructions (Dataset 5 = SSU rRNA Class I, SSU rRNA Class II and LSU rRNA Class I concatenated). The total length of the concatenation based on the two paralogous SSU rRNA class I and class II (1670 and 1087 pb respectively) and LSU rRNA class I (366 pb) was 3123 bp. We conserved as many taxa as possible in the concatenated approach, *e.g.*, all species showing at least two genes and two specimens. *Eukrohnia bathypelagica* and *Paraspadella gotoi* displayed SSU rRNA sequences only for class I and *Eukrohnia fowleri* only for class II. The model of evolution was estimated on the full concatenated data set without any partitioning of the data. The selected model was the GTR + Γ + I with a lnL of −15473.70 with the dataset comprising 55 sequences and lnL of −13443.12 with the dataset comprising 53 sequences.

Trees resulting from concatenation (Figure 5) showed a very similar topology to the one obtained from the SSU rRNA genes and, to a certain extent, from the LSU rRNA genes, with a better accuracy not only at the family rank but also at the sub family and genus ranks. ML, aLRT and Bayesian analyses rooted on midpoint were consistent with the topology obtained in the rooted analysis including both SSU rRNA classes of paralogy and yielded the monophyly of the Aphragmophora (82/1/1) and paraphyly of the Phragmophora. Eukrohniidae, Spadellidae and Sagittidae families were monophyletic with high supports in most methods employed (−/0.99/0.90, 83/1/1 and 69/1/1 respectively). Heterokrohniidae was sister group to Eukrohniidae (80/1/1) and Krohnittidae was sister group to Sagittidae (82/1/1). Topologies based on the larger matrix were also well resolved for relationships within families and were similar to the ones obtained from single gene rRNA analyses. Specifically, within Eukrohniidae, *Eukrohnia fowleri* appeared basal with aLRTv and Bayesian posteriors (−/0.99/0.90) and *E. hamata* and *E. bathypelagica* showed close relationships (43/0.97/1). Moreover, these latter two species were paraphyletic to each other. *Paraspadella* was basal within Spadellidae (83/1/1). Within *Spadella* (100/1/1), *S. valsalinae* was found to be sister to *S. ledoyeri* (89/0.97/1). Finally, within the Sagittidae, the clade *Caecosagitta* + *Pseudosagitta* was supported with high Bayesian posteriors only (73/66/1) and appeared sister to all other Sagittidae species. As in the single gene analyses, three main lineages of Sagittidae were recovered with high supports. These were: *Solidosagitta* + *Serratosagitta* (93/0.99/1), *Parasagitta* (including *M. minima*) + *Sagitta* + *P. draco* (99/1/1) and *Flaccisagitta* + *Aidanosagitta* + *Ferosagitta* + *M. decipiens* (86/1/1). As with the separate analysis of SSU rRNA genes, the resolution of the relationships within Sagittidae decreased at the infrageneric level and the monophyly of several species was not supported. This was the case for the following species: *S. tasmanica and S. pacifica, M. decipiens, A. neglecta, P. elegans, P. setosa and P. friderici, E. hamata and E. bathypelagica.*

### Cladistic and phenetic analysis by morphology

Because it is impossible to choose an appropriate outgroup to root a cladistic analysis of chaetognaths morphological features, we performed a first analysis using the MinF option that minimizes the f-value [30]. Among the 23 characters, 21 were cladistically informative. The Heuristic MP search found 12,435 trees 65 steps long. The consistency index (CI, excluding uninformative characters) was equal to 0.5846 with a retention index (RI) of 0.8043, a homoplasy index (HI) of 0.4154 and a g1 of −0.4702.

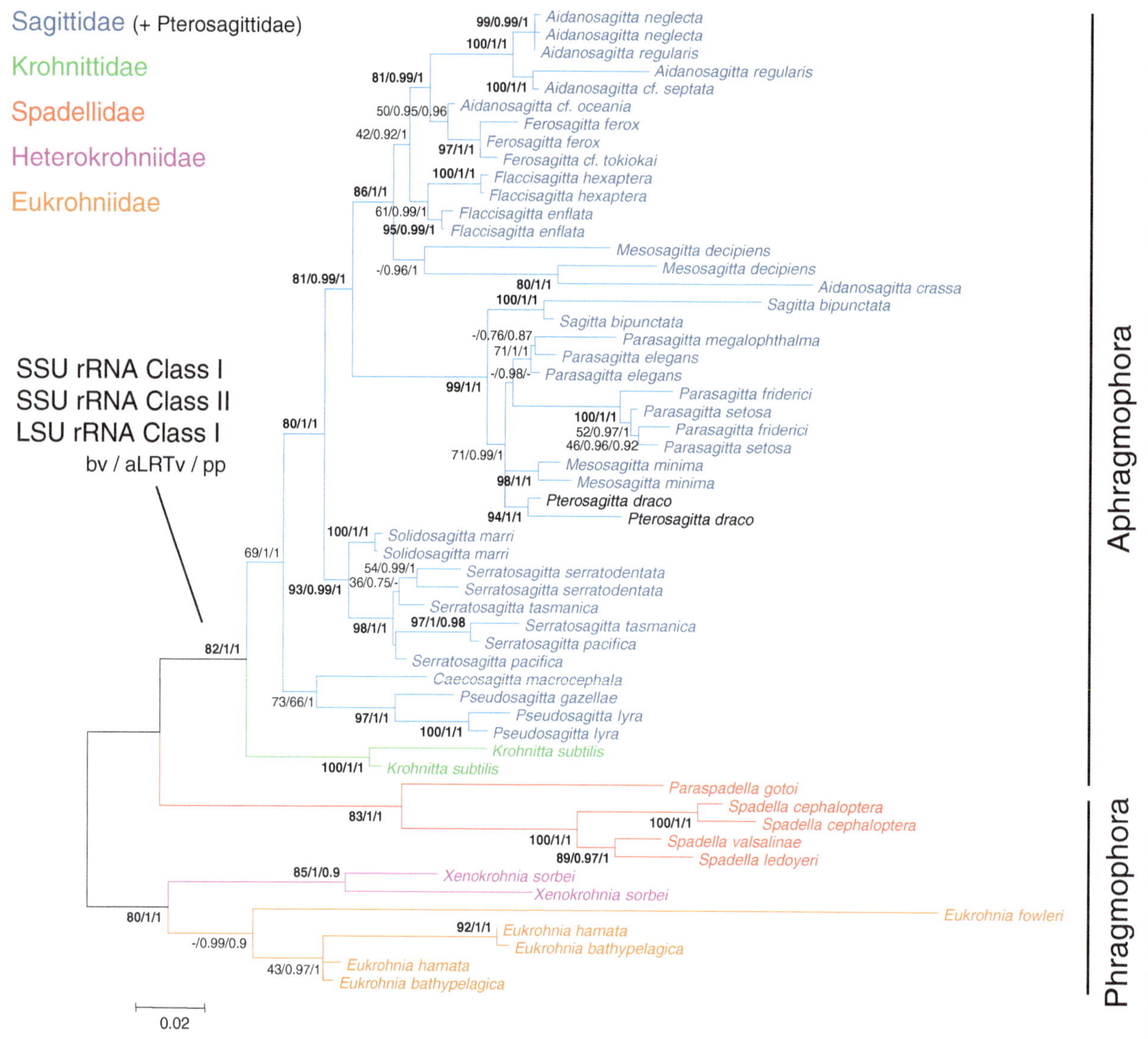


**Figure 5 Bayesian majority-rule consensus tree obtained from a supermatrix including both paralogy classes of SSU rRNA and LSU rRNA class I genes under the GTR + Γ + I model.** The alignment of 55 concatenated rRNA sequences was 3123 base pairs long. Support values obtained using different reconstruction approaches are indicated at nodes in the following order: maximum Likelihood bootstrap probabilities (bv), approximative likelihood ratio test (aLRTv) and Bayesian posterior probabilities (pp). Support values are displayed when bv/aLRTv ≥75 or pp ≥0.85. Node absence in a given method is indicated by -. Tree was rooted using the midpoint rooting function.

The CI excluding uninformative characters was equal to 0.5714. When we removed species with identical morphological coding (9 taxa, 26.47% of the data set) 20 characters were informative. We found 4,115 trees showing a CI of 0.603, a RI of 0.779, a HI of 0.397, and a g1 of −0.706. As identical morphological sequences had no impact on tree topology, we analysed the complete data set. The consensus topology rooted on midpoint yielded Phragmophora and Aphragmophora monophyletic (data not shown). On the basis of this topology, we conducted two constraint analyses rooted on Phragmophora for character-state optimization under ACCTRAN and DELTRAN settings (Figure 6A). In the following, we will only discuss the synapomorphies with a consistency index equal or higher than 0.6. Within Phragmophora, the relationships were well resolved with the monophyly of the Eukrohniidae (100, character #20 one row of posterior teeth) and the monophyly of *Xenokrohnia sorbei* + Spadellidae (100, character #7 heterosarcomeric secondary muscle). Within the latter, *Spadella* was paraphyletic with *Paraspadella gotoi* as the most derived genus of Spadellidae. The optimization of the character #8 changed when ACCTRAN setting was selected and appeared as a synapomorphy of the grouping *X. sorbei* + Spadellidae while it was a synapomorphy of the Spadellidae only when DELTRAN was selected. This slight difference between ACCTRAN and DELTRAN

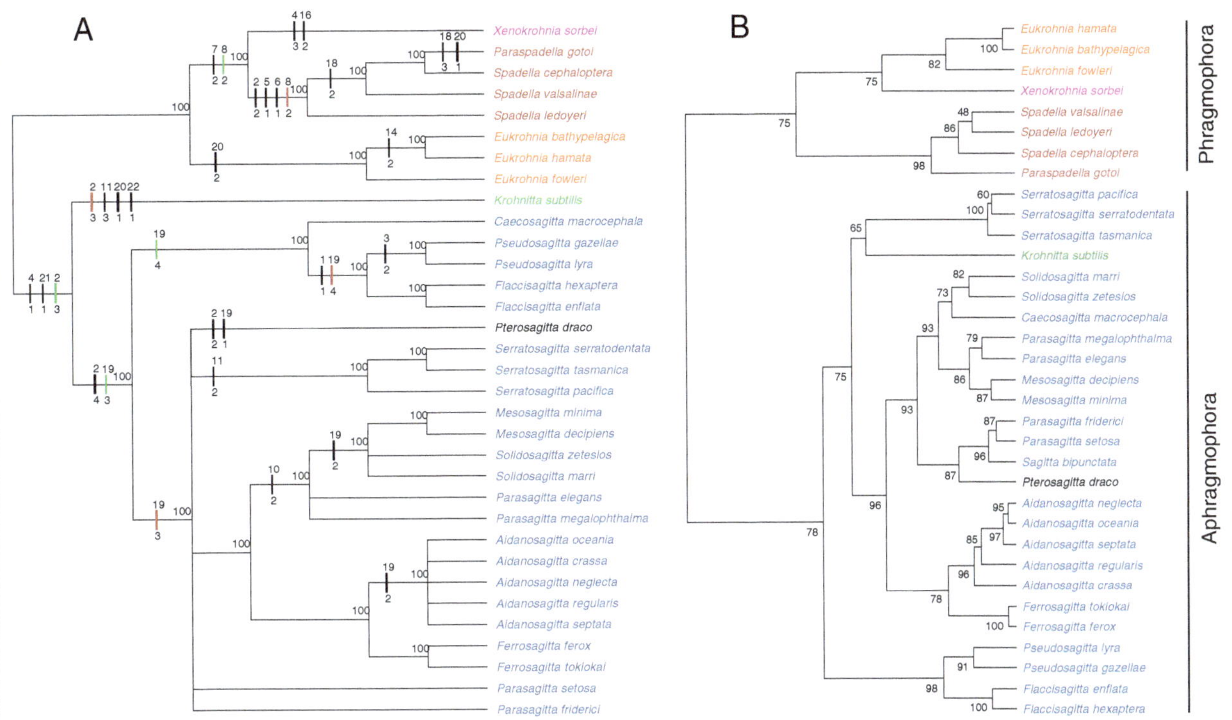


**Figure 6 Cladistic and phenetic analyses of morphological data. A**: Cladistical analysis of morphological data based on 23 qualitative characters. Majority rule consensus tree of 340 equally parsimonious tree (consistency index = 0.633; retention index = 0.837) obtained with the exclusion of *Sagitta bipunctata*. Majority rule consensus values are shown above internal branches. Only synapomorphies presenting a consistency index = 0.6 are shown on the branches. Black bars represent synapomorphies with the corresponding morphological character number above and the character state change below. Red and Green bars correspond to alternative hypotheses according to DELTRAN or ACCTRAN optimisation respectively. Characters coding is presented in Additional file 2. **B**: Phenetic analysis of morphological data based on 23 qualitative and 9 quantitative characters. Approximately unbiased test values are shown above internal branches. Characters coding is presented in Additional file 2.

optimization is likely due to the lack of data for character #8 in *X. sorbei* (see Additional file 2). Four synapomorphies were found with DELTRAN setting for the Spadellidae (100, characters #2 one short pair of lateral fins, the anterior end at the level of the caudal septum, #5 phragms with supercontraction, #6 longitudinal muscle with B fibres only, #8 organisation of RFamide-like neurons type B). We identified one species, *Sagitta bipunctata*, responsible for a decrease of the branching resolution within Aphragmophora (data not shown). When the analysis was conducted with *S. bipunctata*, *Krohnitta subtilis* and *Caecosagitta macrocephala* split first and constituted two distinct lineages. The next cladogenesis event received moderate support (70) and yielded a group consisting of *Pterosagitta draco* and the remaining Sagittidae were left in a broad polytomy. Nevertheless, three groupings received good or moderate support for instance *Flaccisagitta* with *Pseudosagitta* (100 characters #1 flaccid body), *Aidanosagitta* with *Ferosagitta* (95) and *Mesosagitta* with *Solidosagitta* (76). We found that characters #9 (intestinal diverticula) and # 13 (seminal vesicles close to tail fin and well separated from lateral ones) were primarily responsible for the HI increase. When removing *S. bipunctata* from the analysis, the number of resulting trees decreased to 340 (62 steps) showing a CI of 0.6129, a RI of 0.8222, a HI of 0.3871 and a g1 of −0.5039. We found three synapomorphies for Aphragmophora (100) under the ACCTRAN optimization (characters #2 one pair of lateral fins with the anterior end and the posterior end at equal distance from the caudal septum, #4 absence of phragms and #21 glandular structures on the body surface scarcely developed). Nevertheless, this state of character #2 appeared to be an autapomorphy of *K. subtilis* with DELTRAN settings. Sagittidae including *P. draco* were monophyletic with high support (100). This clade exhibited two synapomorphies under the ACCTRAN optimization (characters #2 two pairs of lateral fins, #19 corona ciliata type C) while one of which (characters #19) was synapomorphic to *C. macrocephala*, *Flaccisagitta* and *Pseudosagitta* under DELTRAN settings. The sister-group relationship between *Flaccisagitta and Pseudosagitta* was confirmed (100, characters #1 flaccid body) with *C. macrocephala* basal to this grouping (100). The next cladogenesis event grouped *P. draco* with the remaining Sagittidae (100), with character #19 as synapomorphy with DELTRAN settings. These differences between ACCTRAN and DELTRAN optimisations were a consequence of a lack of description of the corona ciliata in *C. macrocephala* (see Additional file 2). Once again the overall relationships between Sagittidae genera were not fully resolved, except for the sister-group relationships between *Ferosagitta* + *Aidanosagitta* (100, no synapomorphy found) and *Parasagitta* + *Solidosagitta* + *Mesosagitta* (100, character #10 vacuolated intestinal cells). *Serratosagitta* (100, character #11 serrated hooks) as well as *Aidanosagitta* were monophyletic (100) but the latter was defined by a convergence with *Mesosagitta* and *Solidosagitta* (character #19 corona ciliata type B). *Parasagitta* and *Solidosagitta* were not monophyletic. Finally, *P. draco* was characterized by two homoplasic states in respect to Spadellidae due to the reversion of the characters #2 (one short pair of lateral fins, the anterior end at the level of the caudal septum) and #19 (corona ciliata type A).

We conducted an alternative phenetic approach to include 9 quantitative characters and to estimate the overall similarities between taxa. The phenetic analysis of morphological variations was rooted on midpoint (Figure 6B). As in the cladistic analysis, the phenetic clustering resulted in a good estimation of the phylogenetic relationships at the order rank with the monophyly of Phragmophora (approximately unbiased test value = 75) and Aphragmophora (78) and for assigning a species to the correct family. The phenetic approach was more congruent to molecular trees than the cladistics. It yielded the sister-group relationships between *X. sorbei* and Eukrohniidae (75), congruent relationships within Spadellidae with basal *P. gotoi*, *Spadella* monophyletic and sister group relationships between *S. ledoyeri* and *S. valsalinae*. *P. draco* was also found nested in a group comprising *S. bipunctata* and two *Parasagitta* species (87). Apart from the case of Heterokrohniidae and Krohnittidae, each having a single representative in the analysis, the remaining families, Spadellidae and Eukrohniidae, were found to be natural groups with high support (98 and 82 respectively). Sagittidae were not recovered not only because of the inclusion of *P. draco* but also that of *Krohnitta subtilis* as sister to *Serratosagitta* species. However, the low support for this grouping (63) raises questions over this mongrel assemblage. All Sagittidae genera except *Parasagitta* were monophyletic and most nodes picturing their relationships received good support. Similar to the molecular and morphological cladistic analyses, the morphological phenogram yielded a close relationship between *Ferosagitta* and *Aidanosagitta*. However, the three main Sagittidae lineages highlighted in the molecular trees were not recovered.

### Morphometrics

A phenogram was constructed on the basis of body shape similarities, using the Riemannian shape distance $\rho$, computed on all species pairs (see Additional file 3 for comparison between different families). The purpose was not to give more insights about chaetognaths relationships but rather to test whether the two pairs of lateral fins observed in Sagittidae come from the division of a single fin or from the neoformation of the anterior pair. The phenogram of body shape similarities was more congruent with the molecular data when it was constructed using the anterior end of the posterior lateral fin as homologous to the anterior end of the unique lateral fin of species having only one pair (PH1; agglomerative coefficient = 0.84; $p = 0.0098^{***}$, Additional file 4A) than when it had been constructed with the anterior end of the anterior lateral fin as homologous to the anterior end of the unique lateral fin of species having only one pair (PH2; agglomerative coefficient = 0.89; $p = 0.4537$, Additional file 4B).

## Discussion

### A reassessment of chaetognaths relationships

The species studied in the present report belong to the Aphragmophora (Sagittidae, Krohnittidae and Pterosagittidae) and Phragmophora which have been divided in Biphragmophora (Heterokrohniidae) and Monophragmophora (Spadellidae and Eukrohniidae) [24]. We were therefore able to discuss the evolutionary history of six of the nine traditional families of the phylum (according to the views of Casanova [24] and Bieri [16]). The monophyly of the Krohnittidae cannot be debated because only one of the three known species, *Krohnitta subtilis*, is included in the present analysis. The same applies to Heterokrohniidae represented here only by *Xenokrohnia sorbei*. The Eukrohniidae and Spadellidae families have been confirmed, but not the Sagittidae and Pterosagittidae (a family comprising only one species, *Pterosagitta draco*). Indeed, the Sagittidae *sensu stricto* is a paraphyletic assemblage from which *P. draco* derives.

We observed poorly resolved basal nodes in the LSU rRNA trees and to a lesser extent in the SSU rRNA trees as well as a lack of relationships accuracy within Sagittidae. However, when molecular analyses were based on the concatenation of the two paralogous SSU rRNA class I and class II and LSU rRNA class I the resolution for deep branching nodes and the accuracy of relationships within Sagittidae were improved (see Figure 5). This indicates that when the amount of molecular data increases the phylogenetic signal does too. Thus, the low resolution of the single rRNA gene tree reconstructions is not due to a recent acceleration of diversification within Chaetognatha, as previously proposed [18,27], but rather to the short length of the aligned DNA segments.

Morphology and molecules produced some mixed phylogenetic results. This incongruence can be resolved by allowing several convergent losses and/or reversions of morphological characters. The lack of a character cannot be rigorously coded in cladistic due to the lack of primary homology hypothesis especially in a clade with few fossils and inappropriate outgroup. One consequence is that the amount of convergent losses can be underestimated. In such a situation, a phenetic approach should give a better estimation of the evolutionary history. In Chaetognatha, quantitative and qualitative morphological variations based on the degree of overall similarities have been more congruent with molecular topologies than cladistic did, especially concerning the relationships within Spadellidae and the position of *X. sorbei* as sister to Eukrohniidae. The position of *K. subtilis*, the only Krohnittidae relative, is also unstable according to the method used: either at the basis of all Sagittidae (cladistics on morphological data and molecular data) or sister group to *Pseudosagitta* within Sagittidae (phenetics on morphological data). Such an unstable position could be explained by an independent evolution of Krohnittidae during a long period from an early stage of Aphragmophora [21]. Finally, most inconsistencies observed between molecular and morphological approaches concern the Sagittidae relationships, which shows that there is no linear relationship between the degree of morphological divergence and the time of divergence within this family.

**Phragmophora – Aphragmophora split and the Ctenodontina/Flabellodontina hypothesis**

The clade Sagittidae + *P. draco* as sister group to Krohnittidae is highly supported by our molecular and morphological cladistic analyses and revives the Aphragmophora, a clade invalidated by Papillon et al. [18]. However the conclusion these authors made was mainly based on the positioning of *P. draco* within Spadellidae, which was in agreement with morphometry and body appearance [25]. However, the present molecular results demonstrate that the previous *P. draco* sequence was likely to be a contaminant. Moreover, our morphological analyses do not support the Dallot and Ibanez [25] conclusion, highlighting the convergence of some morphological traits between Spadellidae and *P. draco*. It is interesting to note that on the basis of posterior lateral fins restricted to the tail observed in *Demisagitta demipenna* (firstly described as *Aidanosagitta demipenna* [31]), Bieri [17] pointed out a possible relationships between *P. draco* and some species belonging to Sagittidae. He noted that if *D. demipenna* were to lose the anterior fin and develop a pair of large floating bristles, the species would be included into the *Pterosagitta* genus. However, according to the first description [31] and a subsequent revision [21], even the position of the posterior lateral fins of *D. demipenna* is unique in Sagittidae and similar to the one-fin species *P. draco*, other characters conform to those of *Aidanosagitta*. Finally, and more importantly, the inclusion of *P. draco* within Sagittidae is corroborated by a recent report on the organisation of the chaetognath nervous system [32]. Indeed, this study showed that the RFamidergic pattern of *P. draco* is similar to that of several Sagittidae species when compared to several Spadellidae species.

Traditionally, authors who proposed internal systematics in Chaetognatha [16,17,21,22,24-27] identified two major groups; mainly on the basis of the presence or absence of transverse muscles - the phragms. Throughout this debate on chaetognath evolutionary trends, most authors agreed to consider the presence of phragms as a plesiomorphic state [21,24] but with slightly different hypotheses. Spadellidae were believed to have given rise to the Eukrohniidae and Heterokrohniidae according to Tokioka [21] whereas Casanova [24] considered the Heterokrohniidae as the chaetognaths that retain the highest number of plesiomorphic character states. Only Salvini-Plawen [26] suggested a radically different scenario which contradicted the primitiveness of phragms and identified Pterosagittidae as the sister group to all remaining families. Our results favour the ancestrality of Phragmophora and do not support Salvini-Plawen's hypothesis since *P. draco* appears to be a highly specialised and homoplasic member of Sagittidae - as shown by various features such as the loss of the anterior lateral fins, the position of posterior lateral fins, the type of corona ciliata and a high trunk/tail length ratio. So far, the exact functional significance of phragms is unknown but their presence is correlated with a benthic lifestyle [21,22,26,33]. Indeed, the creeping and predatory activity on the sea bed requires more complicated movement than the pelagic niche does. One exception concerns the pelagic Eukrohniidae which exhibit phragms in the trunk, however these are vestigial structures believed to be functionless [22].

According to the rooted topology obtained on the basis of both paralogy classes of SSU rRNA genes (see Additional file 1), our analyses uphold the monophyly of the Aphragmophora but contradict that of the Phragmophora. The Phragmophora appear paraphyletic - a typical situation when a clade is defined on the basis of a plesiomorphic character state. However, the morphological cladistic approach shows that the Aphragmophora are only defined by the lack of phragm or the scarce development of glandular structures on the body surface, which leaves us in an unsatisfactory situation. Based on the current knowledge of gross morphology, histology, cytology and neuroarchitecture of chaetognaths, it is simply impossible to describe any noteworthy apomorphic feature of Aphragmophora. Finally, the Aphragmophora have been divided into two suborders [21,22]: Flabellodontina (Krohnittidae) and Ctenodontina (Sagittidae + Pterosagittidae). Our results show that this supplementary subdivision is unnecessary and undermines the hypothesis based on the structure of the cephalic armature that Ctenodontina could be closer to Phragmophora than to Krohnittidae [22].

**Validity of Biphragmophora and Monophragmophora**

The morphological analysis supports the division of Chaetognatha into three monophyletic groups, the Biphragmophora, Monophragmophora and Aphragmophora [24]. However, although the monophyly of the Biphragmophora (Heterokrohniidae) cannot be assessed because only one species has been included in our analyses, none of our molecular or morphological trees show that Monophragmophora (Spadellidae and Eukrohniidae) could be a natural group. All the tree topologies obtained suggest that the lack of phragms in the tail is a homoplasic state, which is observed independently twice in Eukrohniidae and Spadellidae. Yet, the grouping of Eukrohniidae and Heterokrohniidae is well supported for all molecular datasets and morphological phenetic analyses. These results are in accordance with a recent barcoding analysis showing a close relationship between *Heterokrohnia* and *Eukrohnia* species in respect to Sagittidae [28]. However, no morphological synapomorphy that could define the molecular clade constituted by *X. sorbei* + Eukrohniidae was identified. Indeed, the morphological character states shared by the representatives of these two families are plesiomorphic or homoplasic (a high trunk/tail length ratio also observed in Spadellidae and *P. draco*). Moreover, the Eukrohniidae family is traditionally only defined by the lack of anterior teeth (Figure 7). Thus, to definitely conclude such a sister-group relationship between these two families, a broader taxonomic sampling is needed. Deep benthoplanktonic representatives, such as the putative ancestral Heterokrohniidae, meso-bathyplanktonic Eukrohniidae and representative of *Hemispadella* genus, a link between the families Heterokrohniidae and Spadellidae [34], need to be studied for a better taxonomic coverage of Phragmophora.

**Bieri's nomenclature and Sagittidae relationships**

Despite the complexity of the distribution of some morphological characters, which poses problems when assessing the relationships within Sagittidae, most of the new genera proposed by Bieri [16] were supported by the molecular trees. Our results unambiguously confirm the monophyly of *Serratosagitta, Pseudosagitta, Flaccisagitta* and *Ferosagitta,* and to a lesser extent the validity of large and heterogeneous assemblages such as *Aidanosagitta* and *Parasagitta*. The relationships between species belonging to *Parasagitta* cannot be resolved on the basis of morphological analyses and received low support in the molecular trees. The morphology of this genus remains one of the most heterogeneous on the basis of several diagnosis characters which are prone to homoplasy (Figure 7): the structure and position of seminal vesicles, the presence/absence of intestinal diverticula, the presence/absence of intestinal vacuolated cells, the presence/absence of rayless zone in lateral fins and the structure of the corona ciliata if considering the inclusion of *Mesosagitta minima*. The status of *Solidosagitta* is still pending because only one species has been studied using SSU rRNA paralogous genes. However, our analyses based on morphological data and LSU rRNA sequences include two species and favour the validity of this latter genus.

Our molecular results divide the Sagittidae family into three major lineages (Figure 7): *Serratosagitta* + *Solidosagitta*, *Sagitta* + *Pterosagitta* + *Parasagitta* (including *M. minima*) and *Flaccisagitta* + *Aidanosagitta* + *Ferosagitta* + *Mesosagitta decipiens*. Species with vacuolated intestine (character #10) are distributed in these three lineages. This supports the opinion of Dallot [35] who considered the vacuolated species plesiomorphic on the basis of their general morphology and the structure of their seminal vesicles. This also strongly suggests that the ability to develop large intestinal vacuolated cells has been lost in numerous extant Sagittidae species. The grouping of *M. decipiens, Aidanosagitta, Ferosagitta* and *Flaccisagitta* receives high support in molecular phylogenies. Morphological analyses only support close relationships between *Ferosagitta* and *Aidanosagitta*. Moreover, some morphological characters are congruent with the association between *Aidanosagitta* and *M. decipiens*: the corona ciliata begins below eye level (type B [22]) and intestinal diverticula are present in *Mesosagitta* (*Decipisagitta*), *Aidanosagitta* and *Ferosagitta*. However, these characters isolate *Flaccisagitta* from the rest of the group: the corona ciliata is short and confined to the head, starting just behind the brain and stretching to the neck (type D [22]) and intestinal diverticula are absent. Kinship between *Sagitta bipunctata* and *Parasagitta* species are highly supported by rRNA data and has been previously proposed by several authors. According to Tokioka, [22], these chaetognaths display a similar extended corona ciliata (type C). Moreover, Furnestin [36] and Dallot [35] also suggested such affinities on the basis of the structure and position of lateral fins and number of teeth and hooks. An important incongruence between molecular and morphological analyses is the sister-group relationships between *Flaccisagitta* and *Pseudosagitta*, a result yielded by morphology but invalidated by all molecular trees. This group has previously been proposed by Tokioka but not all authors agree to bring these species in a same clade. Several authors [16,36-38] suggested that *lyra-gazellae-maxima* was undeniably a coherent group gathered in *Pseudosagitta* while *Flaccisagitta hexaptera* constituted the sister species to *F. enflata*. Finally, the morphological similarities between *Flaccisagitta* and *Pseudosagitta* could be linked either to a specialised form highly adapted to the oceanic plankton (thin primary muscles, flaccid body, not wholly rayed lateral fins with gelatinous masses) or should be considered as plesiomorphic states among Sagittidae (corona ciliata type D and seminal vesicles type C).

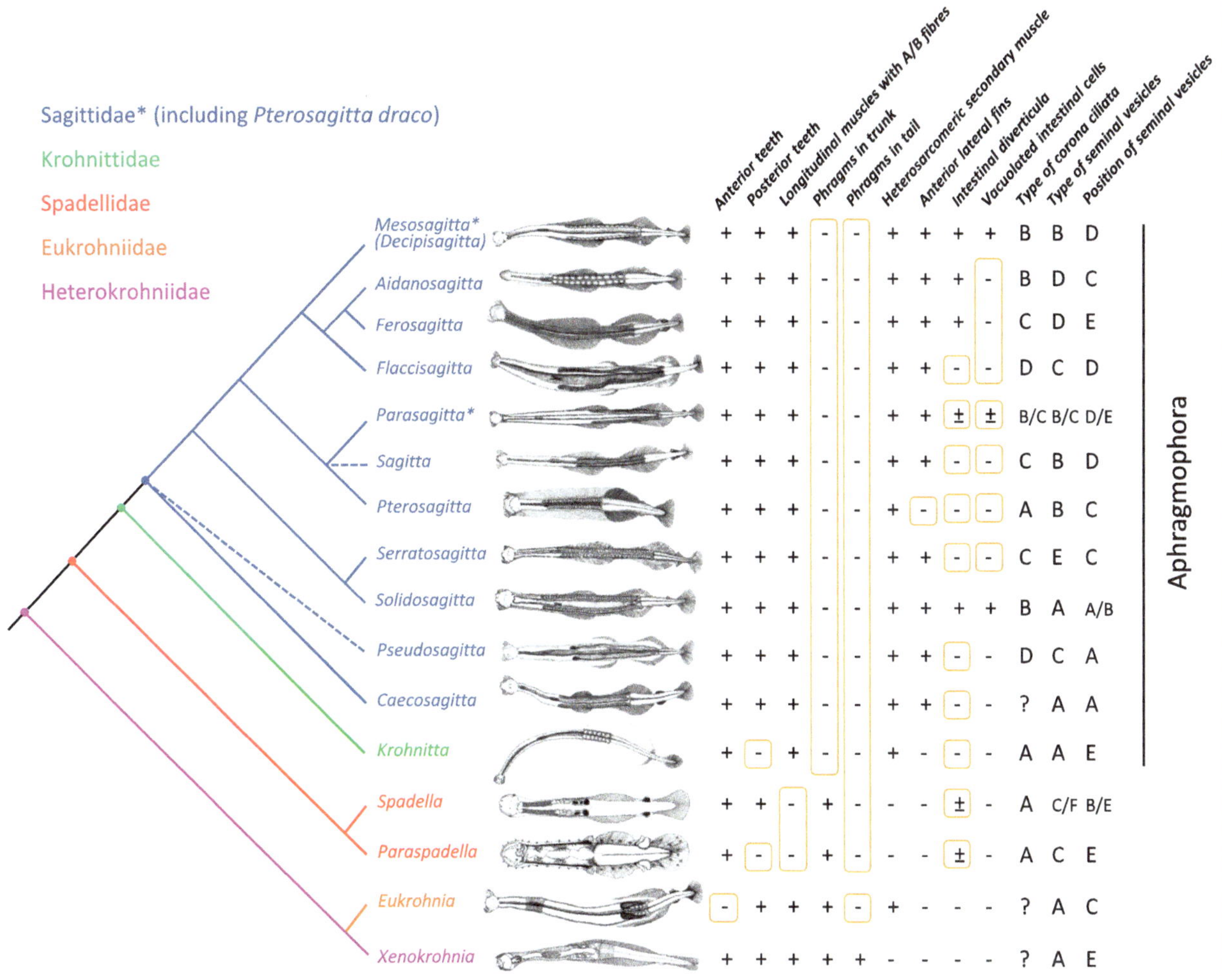


**Figure 7 Overview of the current phylogenetic relationships of Chaetognatha as reconstructed on the topology resulting from concatenated nuclear ribosomal genes (see Figure 5).** Asterisks indicate clades that have been modified from the results obtained in the present study. The dotted lines characterize an unresolved branching. We propose to revive the name *Decipisagitta* instead of *Mesosagitta* (*i.e., Decipisagitta decipiens, D. sibogae and D. neodecipiens*), a genus proposed by Bieri [17]. The fourth and last species of *Mesosagitta, M. minima*, being now included in *Parasagitta*, the genus *Mesosagitta* does not exist anymore. On the right, the surrounded characters states indicate putative loss events. Illustrations after: Alvariño [37,38]; Casanova [70-72]; Winkelmann et al. [74].

When taken together, these remarks emphasize the need to better define the morphological and anatomical boundaries between the traditional genera of Sagittidae by re-evaluating ancestral states and homologies of important traditional diagnostic characters at the histo- and cytological levels (for instance the nervous and muscular systems, the corona ciliata, the seminal vesicles and the fins).

#### Taxonomic notes on the genera Parasagitta/Occulosagitta and Mesosagitta/Decipisagitta

In his attempt to improve the Sagittidae systematic, Bieri [17] also noticed the heterogeneity of several Sagittidae genera and modified his own classification by creating six new genera. Two species included in our analyses are concerned by these modifications: *Parasagitta megalophthalma* and *Mesosagitta decipiens* which were respectively renamed by Bieri as *Occulosagitta megalophthalma* and *Decipisagitta decipiens*. First, *Mesosagitta* as a natural group is contradicted. Indeed, in our molecular analyses, *Mesosagitta minima* always branches without any ambiguity within the *Parasagitta* genus while *M. decipiens* exhibits close relationships with *Flaccisagitta*, *Aidanosagitta* and *Ferosagitta*. Second *P. megalophthalma* always shows a close relationship with *Parasagitta elegans* whatever the molecular tree considered. Thus we propose (*i*) to invalidate the *Occulosagitta* genus, (*ii*) to rename *Mesosagitta minima* as *Parasagitta minima* and (*iii*) to gather the remaining *Mesosagitta* species (*M. decipiens, M. neodecipiens, M. sibogae*) into the new genus *Decipisagitta*.

### Phylogenetic consensus from molecules and implication for chaetognaths evolutionary trends

Paleontological evidences have demonstrated the existence of chaetognaths not only in the middle Cambrian Burgess Shale biota [39] but also in the earlier Cambrian Chengjiang biota [3] with morphological features almost identical to extant species. The discovery of new deep species [24] leads to the conclusion that Heterokrohniidae (Biphragmophora) is the family that presents the highest number of plesiomorphic characters (*i.e.*, the most primitive group *sensu* Casanova). However, our analysis raises the question of whether the lack of phragms in the tail of Spadellidae and Eukrohniidae is derived and due to convergence or represents a homologous and plesiomorphic character. The answer has far reaching consequences for our understanding of evolutionary pathways in chaetognaths. For instance, since phragms are considered important for creeping forms but are unnecessary for species that proceed by movements in the water column, this question is related to whether the stem chaetognaths were hyperbenthic or holoplanktonic species. The molecular and morphological phylogenetic results we obtained rather suggest that the Eukrohniidae and Spadellidae (Monophragmophora in Casanova's hypothesis) exhibit the most primitive state, a scenario that needs only two evolutionary steps (one acquisition of phragms in the tail of Heterokrohniidae and one loss in the trunk of Sagittidae) against three steps when considering two parallel losses in the tail of Eukrohniidae and Spadellidae respectively, followed by one loss in the trunk of Sagittidae. However, we shall see that several arguments are in favour of the latter scenario. First, morphology and body ratios of specimens found in Chengjiang biota suggest that the chaetognaths from lower Cambrian were planktonic with ecological preferences for hyperbenthic niches close to the sea bottom [3]. Among extant genera, those showing the closest ecological features are the hyperbenthic Heterokrohniidae. Interestingly, their ecology is still observed in one species of Eukrohniidae, *Eukrohnia calliops* [16,40]. Second, phragms have been identified in trunk and tail of specimens from Cambrian Burgess Shale biota [39]. Finally, the stable environment of deep oceanic waters has likely delayed the body plan evolution and could explain the conservation of some ancestral morphology in the extant deep benthoplanktonic Heterokrohniidae. All summed up, it is reasonable to assume that an arrangement of transverse muscles in trunk and tail should be regarded as the most primitive state in chaetognaths. In hypothesizing such a complex Heterokrohniidae-like ancestor, one must postulate the loss of many structures and a body plan simplification during the evolutionary history of Chaetognatha (Figure 7).

In our scenario, an important split yielded two clades with different ecological niches, the strictly benthic Spadellidae and the holoplanktonic Aphragmophora lineages (Figure 7). This hypothesis contradicts the ancestrality of Spadellidae [22]. According to the comparative studies of the muscles in Chaetognatha, Spadellidae are highly derived and underwent important modifications of their muscular apparatus [33]. The structure of their primary muscles which lack B fibres is characteristic of benthic species and is derived from AB fibres typology. Combined with our results, there is therefore strong evidence that the Spadellidae ancestor was planktonic, partly linked to the sea bed, and adapted secondarily to a strict benthic lifestyle. The highly specialized status of this family is shown by its high number of synapomorphies (characters #2, #5, #6 and #8; Additional file 2).

The Aphragmophora diversification does not display any morphological novelty and highlights a new case of body plan simplification with the loss of phragms in the trunk (Figure 7). The last Aphragmophora ancestor divided into two lineages, giving rise to the current Krohnittidae and Sagittidae families. The Krohnittidae family retained some ancestral traits such as one pair of lateral fins on the trunk and tail, with the anterior end and posterior rear end at equal distance from the caudal septum (character #2) but also developed numerous autapomorphic features. They exhibit abruptly curved hooks (character #11) and anterior teeth arranged in a fan shape (character #22) and they lack posterior teeth (character #20). The peculiarity found in teeth and hooks of *Krohnitta* species reveals a high level of specialisation [21] and points to an independent evolution of Krohnittidae that started during the early stages of Aphragmophora cladogenesis as demonstrated by our results. Interestingly, the arrangement and shape of the anterior teeth of *Sagitta nairi*, a Sagittidae recently described [41] are similar to those of the genus *Krohnitta* suggesting a possible convergence between these unrelated species. The second Aphragmophora family, the Sagittidae, is defined by two pairs of lateral fins. *Caecosagitta* and *Pseudosagitta* can be recognized as two early off-shoots of Sagittidae. Dallot and Ibanez [25] have already proposed the isolation of *Pseudosagitta lyra* and suggested the possibility that its membership to the Sagittidae is dubious. Our results do not support such exclusion.

An important question considering body plan variation in Chaetognatha is concerned with fin evolution. More precisely, did the anterior and posterior lateral fins of Sagittidae originate from the division of the single large fin observed in Heterokrohniidae and Eukrohniidae? Casanova and Moreau [42] noted some similarities between the posterior lateral fin of species belonging to *Pseudosagitta* genus and the unique lateral fin of those belonging to Eukrohniidae, the posterior extremities of which are only slightly apart from the bodywall. The posterior and anterior lateral fins of *Pseudosagitta* are connected by a tegumentary bridge which reinforces the idea that the two lateral fins would have formed after the incomplete division of a unique large one. However, our morphometric analysis shows that the anterior lateral fin of two-fin species does not result from the division of a unique lateral fin but from a neo-formation after a backward movement of the anterior end of the unique fin. In other words, the posterior lateral fin of Sagittidae is homologous to the unique lateral fin of the other families. This evolutionary step constitutes a rare case of increase in body plan complexity at the anatomical level in Chaetognatha. While the possession of two pairs of lateral fins was recognized as a good synapomorphy for Sagittidae after Tokioka's classification, the inclusion of *P. draco* which exhibits only one posterior pair raises doubts about its validity. This means that the loss of the anterior fin did not constitute an evolutionary dead end for Sagittidae making the presence of one pair of fin a homoplasy.

As previously mentioned [18,33], the distribution of some morphological characters in Chaetognatha cannot be related to the phylogeny nor to the ecology suggesting a differential evolution of separate chaetognath organs (mosaic evolution). The prevalence of mosaic evolution can be demonstrated through the examination of character associations in extant chaetognaths (Figure 7). Different traits and outcomes are favoured by natural selection in different species and these evolutionary pathways might be responsible for the non congruence between some of the cladistic and phenetic analyses because when mosaic selection occurs the primitive or derived nature of character states cannot be deduced on the basis of their correlation with other character states which are believed to be primitive or derived [43]. Moreover, a common trend correlated with mosaic evolution is the prominence of homoplasy powered by common selective pressures as shown in many plants [44] and animals [45-47].

Such a combination of mosaic evolution and lack of fossil records can lead to persistent problems in interpreting relationships through morphological cladistic analysis [48].

#### Ecologically-induced convergence in holoplanktonic chaetognaths

It is likely that all Aphragmophora and Eukrohniidae lineages are morphologically highly similar because of their holoplanktonic lifestyle in the pelagos [22,26] (Figure 7). For instance, in spite of their separate evolutionary history, Sagittidae, Krohnittidae and Eukrohniidae share features such as the common tendency towards a body with large surface to volume ratio, the trunk elongation, the reduction of epidermal glandular structures and the reduction/loss of phragms. Because these clades do not constitute a natural grouping, it is obviously a case of convergent evolution. Functional similarity is also present at the cytological level since all these holoplanktonic species exhibit the highest proportion of B fibres in their primary muscles and a heterosarcomeric organisation of their secondary muscles [33]. Among Sagittidae, the morphology of *P. draco* is more puzzling. There are numerous homoplasic reversions in this species since it exhibits a high trunk/tail length ratio and one small pair of lateral fins being restricted to the tail, a typical set of Spadellidae morphological features. This epipelagic species compensates the decrease of its surface/volume body ratio by developing a foamy epidermal collarette around most of the body combined with floating bristles.

Buoyancy ability represents an interesting area of investigation in holoplanktonic chaetognaths [49-52]. To decrease their specific gravity, Eukrohniidae show a hyper development of the unique pair of lateral fins and vestigial phragms, which are likely to be not functional and are confined to the most anterior part of the trunk [22]. A similar trend is observed in Krohnittidae with the development of large lateral fins and phragms that have totally disappeared. The Sagittidae adapted in a different way by creating a new pair of lateral fins. This acquisition constitutes an important and unique event to adapt to holoplanktonic lifestyle in Chaetognatha and could contribute to explain the successful current biodiversity of Sagittidae. It has been recently shown that the diversification of the Euthecosomata, which are important holoplanktonic molluscs, occurred in the context of an important turn-over in the marine planktonic community due to severe environmental changes that started from the Late Palaeocene [53]. Moreover, these morphological innovations correlated with climatic changes and species turn-over were largely shaped by shell buoyancy adaptation. One could postulate that a broad and recent diversification of Sagittidae occurred in the same evolutionary background. However, the paucity of fossils for these soft bodied invertebrates did not allow an efficient calibration of the divergence times of Chaetognatha lineages.

### Conclusions

Molecular analyses have highlighted the homoplasy of several traditional characters and the influence of lifestyle on morphology, particularly for chaetognaths that adapted to a pelagic environment. We also propose that Chaetognatha evolved mostly through simplification of a pre-existing body plan, rather than through an increase in complexity. This constitutes a shift of paradigm in the traditional understanding of the group's evolution and prompts a re-appraisal of previous hypotheses concerning the morphological characters' polarity and history. For example, it is reasonable to think that the loss of phragms and teeth could have occurred independently in different branches during chaetognath evolution. If the anterior and posterior rows of teeth are considered as homologous structures, the loss of one of these rows may be described as an event of parallel evolution. Another important traditional diagnostic character, the trunk/tail length ratio, can also be considered subject to homoplasy by reversion due to Bauplan limits (*Pterosagitta draco* versus Spadellidae) and convergence caused by an evolution in similar ecosystems (Sagittidae/Krohnittidae *versus* Eukrohniidae). Even the main Sagittidae synapomorphy represented by two pairs of lateral fins is homoplasic by reversion. Because of these numerous losses and homoplasic events, traditional morpho-anatomical traits may prove unhelpful with deciphering chaetognath relationships and morphological evolution with any certainty. Such a conclusion stresses the need for more data from molecular markers as well as from histo- and cytoarchitecture of the muscular apparatus [33] and the neurosensorial system [32,54]. Because of the scarcity of novelties at the anatomical level, it becomes needed to explore their body plan variations from the tissue level to the cell level. Considering the pivotal phylogenetic position of Chaetognatha within bilaterians, it is of primary importance to reconstruct their ground pattern. Future studies need to focus on a set of new characters based on a broad range of taxa including specimens belonging to meso-bathypelagic and deep benthoplanktonic genera. The use of an expanded taxonomic dataset combined with appropriate observation (*i.e.*, with transmission electron microscopy, immunohistochemistry combined with confocal laser scanning microscopy and Next Generation Sequencing) will be crucial in improving the understanding of Chaetognatha's diversification processes.

## Methods

### Collection, species diagnosis and taxonomic sampling

Specimens were collected from a broad geographical range sampled during the circum-global *Tara* Ocean expedition and also from other regional missions (Table 1). Table 2 shows the sequences used in the present study already available in public databases. Morphological identifications of specimens were performed using Wild M5 and Nikon SMZ 645 stereoscopic microscopes, and combined with ecological and molecular data from previous literature. For morphological data, comparisons with original descriptions were undertaken. When different species exhibited a very close gross morphology and similar number of teeth and hooks, only mature adults were chosen because one of the key features for the diagnosis of chaetognaths in ethanol is the shape and position of seminal vesicles disposed on each side of the tail. For ecological data, specimens' habitats and depths were taken into account. For molecular data, acquired DNA sequences from the study were compared against previously obtained sequences in GenBank. Finally, if morphology for specimens did not fit perfectly with the described species, we named species using "cf." accordingly. In this manuscript, we used the taxonomic nomenclature given by Bieri [16,17] who renamed several species and created nine genera of Sagittidae.

### DNA extractions, PCR amplifications and sequencing

All specimens were placed in 80% ethanol for preservation. The genomic DNA was then extracted using the DNAeasy kit (Qiagen, Valencia, CA) from pieces or entire individuals dried on filter paper and devoid of alimentary bolus to prevent contamination.

Because the whole ribosomal cluster in Chaetognatha is duplicated [18,27] two sets of specific primers for each paralogous SSU rRNA gene were used to amplify sequences of approximately 1800 bp (class I: 18SCI5′ TTGATGAA ACTCTGGATAACTC and 18SCI3′ GGACCTCTCTACA TCGTTCG) and 1200 bp (class II: 18SCII5′ TCGTCGG GGTCTCATCC and 18SCII3′ AGATACCTCGCAAAAT CG). As we concentrated our efforts on the class I of LSU rRNA gene, which is the most represented class in public databases, only one set of primers previously described in [27] was used to amplify a fragment of approximately 500 pb: 28S5′ AAAGGATCCGATAGYSRACAAGTACCG and 28S3′ CCCAAGCTTGGTCCGTGTTTCAAGAC. Most of the sequences obtained with this couple of primers belonged to the class I but we also amplified LSU rRNA class II genes for several species.

PCRs were performed according to [27] in 50 µL volumes with the following reagents: 1× PCR buffer (Taq PCR core kit, Qiagen), 0.2 mM of each dNTPs mix, 0.5 mM of each primer, 2 to 4 µl (depending on DNA concentration) of extracted genomic DNA, and 1U of Taq polymerase. The PCR cycling parameters for amplification of LSU rRNA were: 95°C, 3 min, then 35 cycles of 95°C for 1 min, 50°C for 1 min, 72°C for 2 min.

For SSU rRNA, we used the following PCR program: 2 min at 92°C; 5 cycles of 92°C for 30 s, 48°C for 45 s, 48°C to 72°C for 80 s (ramp rate of 0.3°C/s) and 72°C for 90 s; 30 cycles of 92°C for 45 s, 48°C for 45 s, 72°C for 90 s; and a final extension time of 72°C for 7 min.

After amplification, all PCR fragments were purified with Wizard® SV Gel and PCR Clean-Up System (Promega, Madison, WI), cloned into pGemT-easy vector (Promega, Madison, WI) and sequenced in both directions using the T7 and SP6 primers with a ABI 96-capillary 3730XL sequencer at Eurofins genomics. 145 sequences have been deposited in GenBank under the following accession numbers:

- SSU rRNA class I: KM519789 - KM519853
- SSU rRNA class II: KM519854 - KM519901
- LSU rRNA class I: KM519902 - KM519931
- LSU rRNA class II: KM519932 - KM519933.

**Table 1 Details of the sequences and GenBank accession numbers obtained from this study**

| Species | Code | SSU rRNA I | SSU rRNA II | LSU rRNA I | LSU rRNA II | Mission/Collector | Origin |
|---|---|---|---|---|---|---|---|
| *Krohnitta subtilis* | 011 | KM519840 | KM519864 | | | DIVA3 St ME 791/540.1 | Med. E |
| *Krohnitta subtilis* | 039 | KM519842 | KM519863 | KM519928 | | TARA St 64 | Ind. N |
| *Krohnitta subtilis* | 058 | KM519841 | KM519862 | | | TARA St 76 | Atl. SW |
| *Krohnitta subtilis* | 068 | | | KM519927 | | TARA St 132 | Pac. |
| *Pterosagitta draco* | 026 | KM519820 | KM519898 | KM519906 | | Florida/CRER 2 St 91 | Atl. W |
| *Pterosagitta draco* | 076 | KM519818 | | | | Florida /CRER 2 St 42 | Atl. W |
| *Pterosagitta draco* | 001 | KM519819 | | | | Tulear/ M. Pagano | Ind. N |
| *Pterosagitta draco* | 002 | | KM519896 | KM519905 | | Tulear/M. Pagano | Ind. N |
| *Pterosagitta draco* | 003 | | KM519897 | | | Florida/CRER 2 St 42 | Atl. W |
| *Parasagitta elegans* | 004 | KM519817 | | | | Norway/F. Norrbin | Atl. N |
| *Parasagitta elegans* | 005 | | KM519901 | | | Norway/F. Norrbin | Atl. N |
| *Parasagitta elegans* | 038 | KM519815 | KM519900 | | | Norway/F. Norrbin | Atl. N |
| *Parasagitta elegans* | 075 | KM519816 | KM519899 | KM519923 | | Norway/F. Norrbin | Atl. N |
| *Parasagitta setosa* | 012 | KM519810 | | | | France/Y. Perez | Med. NW |
| *Parasagitta setosa* | 028 | KM519809 | | | | France/Y. Perez | Med. NW |
| *Parasagitta setosa* | 045 | KM519808 | | KM519920 | | France/Y. Perez | Med. NW |
| *Parasagitta friderici* | 013 | KM519811 | | KM519922 | | France/Y. Perez | Med. NW |
| *Parasagitta friderici* | 046 | KM519812 | | KM519921 | | France/ Y. Perez | Med. NW |
| *Serratosagitta pacifica* | 025 | KM519827 | KM519872 | | | TARA St 34 | Red Sea |
| *Serratosagitta pacifica* | 031 | KM519826 | KM519871 | KM519904 | | TARA St 34 | Red Sea |
| *Serratosagitta pacifica* | 032 | KM519825 | KM519873 | | | TARA St 65 | Ind. N |
| *Serratosagitta serratodentata* | 014 | KM519830 | KM519870 | | | MSN 14/1 St 1159 | Med. |
| *Serratosagitta serratodentata* | 044 | KM519828 | KM519869 | | | TARA St 64 | Ind. N |
| *Serratosagitta serratodentata* | 047 | KM519829 | KM519868 | | | TARA St 16 | Med. C |
| *Serratosagitta tasmanica* | 006 | KM519831 | | | | France/ Y. Perez | Atl. E |
| *Serratosagitta tasmanica* | 043 | KM519832 | KM519866 | KM519902 | | TARA St 66 | Ind. N |
| *Serratosagitta tasmanica* | 049 | | KM519865 | | | TARA St 16 | Med. C |
| *Serratosagitta tasmanica* | 063 | KM519833 | KM519867 | KM519903 | | TARA St 79 | Atl. SW |
| *Pseudosagitta lyra* | 007 | KM519837 | | | | France/Y. Perez | Atl. E |
| *Pseudosagitta lyra* | 008 | | KM519858 | | | France/Y. Perez | Atl. E |
| *Pseudosagitta lyra* | 029 | KM519835 | KM519857 | | | TARA St 65 | Ind. N |
| *Pseudosagitta lyra* | 030 | KM519836 | KM519856 | | | TARA St 66 | Ind. N |
| *Pseudosagitta lyra* | 023 | KM519838 | KM519854 | | | TARA St 23 | Med. C |
| *Pseudosagitta lyra* | 024 | | KM519855 | | | TARA St 15 | Med. W |
| *Pseudosagitta gazellae* | 055 | | KM519859 | | | TARA St 86 | Ant. |
| *Pseudosagitta gazellae* | 056 | KM519839 | KM519860 | | | TARA St 86 | Ant. |
| *Flaccisagitta hexaptera* | 018 | KM519792 | KM519880 | | | MSN 14/1 St 1159 | Med. E |
| *Flaccisagitta hexaptera* | 035 | KM519791 | KM519881 | | | TARA St 52 | Ind. N |
| *Flaccisagitta hexaptera* | 036 | KM519790 | KM519883 | | | TARA St 64 | Ind. N |
| *Flaccisagitta hexaptera* | 048 | KM519789 | KM519882 | | | Florida CRER2 St 42 | Atl. W |
| *Flaccisagitta enflata* | 021 | KM519795 | KM519878 | | | TARA St 18 | Med. W |
| *Flaccisagitta enflata* | 022 | KM519794 | KM519877 | | | TARA St 15 | Med. W |
| *Flaccisagitta enflata* | 033 | KM519793 | KM519876 | | | TARA St 34 | Red Sea |
| *Flaccisagitta enflata* | 034 | KM519796 | KM519879 | | | TARA St 64 | Ind. N |

**Table 1 Details of the sequences and GenBank accession numbers obtained from this study** *(Continued)*

| | | | | | | | |
|---|---|---|---|---|---|---|---|
| ***Aidanosagitta cf. oceania*** | 010 | KM519801 | | KM519914 | | Tulear/M. Pagano | Ind. N |
| ***Aidanosagitta cf. septata*** | 069 | KM519802 | KM519891 | KM519917 | | TARA St 58 | Ind. W |
| ***Aidanosagitta regularis*** | 066 | KM519803 | KM519892 | KM519915 | | TARA St 109 | Pac. E |
| ***Aidanosagitta regularis*** | 072 | KM519804 | | | | TARA St 130 | Pac. |
| ***Aidanosagitta regularis*** | 073 | | | KM519916 | | TARA St 130 | Pac. |
| ***Aidanosagitta neglecta*** | 042 | KM519805 | KM519894 | KM519913 | | TARA St 50 | Ind. C |
| ***Aidanosagitta neglecta*** | 041 | KM519806 | KM519893 | KM519912 | | TARA St 32 | Med. E |
| ***Mesosagitta decipiens*** | 067 | KM519807 | KM519884 | KM519907 | | TARA St 32 | Red Sea |
| ***Mesosagitta minima*** | 038 | KM519813 | KM519885 | KM519909 | | TARA St 66 | Ind. N |
| ***Mesosagitta minima*** | 037 | KM519814 | KM519886 | KM519908 | | TARA St 64 | Ind. N |
| ***Solidosagitta marri*** | 053 | KM519823 | KM519874 | KM519911 | | TARA St 86 | Ant. |
| ***Solidosagitta marri*** | 054 | KM519824 | KM519875 | | | TARA St 85 | Ant. |
| ***Caecosagitta macrocephala*** | 052 | KM519834 | KM519861 | KM519910 | | Florida/C. Guigand | Atl. N |
| ***Ferosagitta ferox*** | 064 | KM519800 | KM519887 | KM519925 | | TARA St 100 | Pac. E |
| ***Ferosagitta ferox*** | 071 | KM519799 | KM519889 | KM519924 | | TARA St 125 | Pac. |
| ***Ferosagitta ferox*** | 074 | KM519798 | KM519888 | | | TARA St 125 | Pac. |
| ***Ferosagitta cf. tokiokai*** | 065 | KM519797 | KM519890 | KM519926 | | TARA St 100 | Pac. E |
| ***Sagitta bipunctata*** | 019 | KM519821 | | KM519918 | | TARA St 17 | Med. W |
| ***Sagitta bipunctata*** | 020 | KM519822 | KM519895 | KM519919 | | TARA St 15 | Med. W |
| ***Spadella valsalinae*** | 015 | KM519848 | | | | Croatia/C. Müller | N. Adriatic |
| ***Spadella valsalinae*** | 016 | KM519846 | | | | Croatia/C. Müller | N. Adriatic |
| ***Spadella valsalinae*** | 059 | KM519847 | | KM519929 | | Croatia/C. Müller | N. Adriatic |
| ***Spadella cephaloptera*** | 060 | KM519844 | | KM519930 | | Ibiza/Y. Perez | Med. NW |
| ***Spadella cephaloptera*** | 061 | KM519845 | | | | France/Y. Perez | Med. NW |
| ***Spadella cephaloptera*** | 062 | KM519843 | | | | France/Y. Perez | Med. NW |
| ***Xenokrohnia sorbei*** | 009 | KM519849 | | KM519931 | | France/Y. Perez | Atl. E |
| ***Eukrohnia bathypelagica*** | 051 | KM519851 | | | | France/Y. Perez | Atl. E |
| ***Eukrohnia bathypelagica*** | 057 | KM519850 | | | KM519933 | TARA St 85 | Ant. |
| ***Eukrohnia hamata*** | 017 | KM519853 | | | | Norway/F. Norrbin | Atl. N |
| ***Eukrohnia hamata*** | 050 | KM519852 | | | | Norway/F. Norrbin | Atl. N |
| ***Eukrohnia fowleri*** | 070 | | | | KM519932 | Florida/C. Guigand | Atl. N |

**Molecular phylogenetic analysis**

Five data sets were used for molecular analyses: dataset 1 = LSU rRNA Class I and II, dataset 2 = SSU rRNA Class I and II, dataset 3 = SSU rRNA Class I, dataset 4 = SSU rRNA Class II and dataset 5 = SSU rRNA Class I, SSU rRNA II and LSU rRNA Class I concatenated sequences. The sequence alignments were established using CLUSTALX [55] and Muscle [56] implemented in Mega 5 and then further improved manually. The MODELTEST *v3.0b4* program [57] was used to identify the best model of DNA evolution for each of our dataset based on maximum likelihood (ML) and using Bayesian information criterion (BIC). We used Aliscore [58,59] to test the impact of highly heterogeneous sites that could negatively affect the phylogenetic reconstruction. We used the following parameters "-N" and "-N –r –w4" to remove heterogeneity sites. A Maximum Likelihood tree was estimated using the Nearest-Neighbour-interchange (NNI) option with Mega 5. A random starting tree was generated using the Neighbour-Joining method with the partial deletion option selected (75% site coverage cut-off). Topological robustness was investigated using 1000 non-parametric bootstrap replicates. Branches with bootstrap values higher than 70% were considered well supported [60]. We also performed Bayesian phylogenetic analyses using MrBayes 3.0b4 [61]. Each analysis consisted of $2.10^7$ generations with a random starting tree, default priors, the same set of branch lengths for each partition, and four Markov chains (with default heating values) sampled every 1000 generations. Adequate burn-in was determined by examining a plot of

**Table 2 Details of the sequences and GenBank accession numbers obtained from previous studies and used in the present analysis**

| Species | SSU rRNA I | SSU rRNA II | LSU rRNA I | LSU rRNA II | Collector | Origin |
|---|---|---|---|---|---|---|
| *Parasagitta megalophthalma* | DQ351878 | DQ351901 | | | Y. Perez | Med. NW |
| *Parasagitta elegans* | Z19551 | | | | M. Telford | Pac. NW |
| *Parasagitta elegans* | | | Z77108 | | Q. Bone | Atl. N |
| *Parasagitta setosa* | | DQ351900 | | | J.P. Casanova | Atl. N |
| *Parasagitta setosa* | | | Z77120 | | J.P. Casanova | Atl. E |
| *Parasagitta setosa* | | | Z77121 | | V. Øresland | Atl. N |
| *Serratosagitta serratodentata* | | | Z77119 | | J.P. Casanova | Atl. |
| *Serratosagitta tasmanica* | DQ351893 | | | | F. Norrbin | Atl. N |
| *Pseudosagitta lyra* | DQ351880 | DQ351892 | | | Y. Perez | Med. N |
| *Pseudosagitta lyra* | | | Z77114 | | E. Thuesen | Pac. E |
| *Pseudosagitta maxima* | | | Z77118 | | F. Kurbjeweit | Ant. |
| *Pseudosagitta gazellae* | | | Z77112 | | F. Kurbjeweit | Ant. |
| *Flaccisagitta hexaptera* | | | Z77113 | | M. Terazaki | Pac. W |
| *Flaccisagitta enflata* | DQ351877 | | | | B. Thomassin | Ind. W |
| *Flaccisagitta enflata* | | | Z77109 | | J.P. Casanova | Atl. |
| *Flaccisagitta enflata* | | | ZZ110 | | M. Terazaki | Pac. W |
| *Aidanosagitta crassa* | D14363 | | | | T. Goto | Pac. W |
| *Aidanosagitta crassa* | | | Z77107 | | S. Nagasawa | Pac. W |
| *Aidanosagitta neglecta* | DQ351882 | | | | B. Thomassin | Ind. W |
| *Mesosagitta decipiens* | DQ351881 | DQ351895 | | | Y. Perez | Med. NW |
| *Solidosagitta zetesios* | | | Z77122 | | E. Thuesen | Pac. E |
| *Solidosagitta marri* | | | Z77117 | | F. Kurbjeweit | Ant. |
| *Caecosagitta macrocephala* | | | Z77115 | | E. Thuesen | Pac. E |
| *Caecosagitta macrocephala* | | | Z77116 | | M. Terazaki | Pac. W |
| *Xenokrohnia sorbei* | DQ351888 | DQ351902 | | | Y. Perez | Atl. E |
| *Ferosagitta ferox* | | | Z77111 | | M. Terazaki | Pac. W |
| *Ferosagitta robusta* | | | | Z77130 | M. Terazaki | Pac. W |
| *Sagitta bipunctata* | DQ351890 | DQ351894 | | | J.P. Casanova | Atl. E |
| *Sagitta bipunctata* | | | | Z77127 | J.P. Casanova | Atl. |
| *Paraspadella gotoi* | D14362 | | | | T. Goto | Pac. W |
| *Spadella ledoyeri* | DQ351883 | DQ351899 | | | C. Lejeusne | Med. N |
| *Spadella cephaloptera* | DQ3351884 | DQ351897 | | | Y. Perez | Med. NW |
| *Spadella cephaloptera* | | | | Z77129 | D. Dixon | Atl. NE |
| *Eukrohnia bathypelagica* | DQ351886 | DQ351896 | | | Y. Perez | Atl. E |
| *Eukrohnia hamata* | DQ351887 | | | | F. Norrbin | Atl. N |
| *Eukrohnia hamata* | | | Z77105 | | H. Kapp | Atl. E |
| *Eukrohnia hamata* | | | Z77106 | | F. Kurbjeweit | Ant. |
| *Eukrohnia fowleri* | | DQ351889 | | | Y. Perez | Atl. E |
| *Eukrohnia fowleri* | | | Z77103 | | H. Kapp | Atl. E |
| *Eukrohnia fowleri* | | | | Z77123 | E. Thuesen | Pac. E |
| *Eukrohnia fowleri* | | | | Z77124 | H. Kapp | Atl. E |
| *Eukrohnia fowleri* | | | | Z77125 | H. Kapp | Atl. E |
| *Eukrohnia fowleri* | | | | Z77126 | M. Terazaki | Pac. W |

the likelihood scores of the heated chains for convergence on stationarity as well as the effective sample size (ESS) of values in Tracer 1.5 [62].

To test the impact of potential "noisy sites" we computed maximum likelihood phylogenetic analyses using PHYML aBayes 3.0.1 beta programme [63,64] on the LSU rRNA data set. We calculated two non-parametric branch supports (Bootstrap and SH-aLRT) and two parametric branch supports (aBayes and approximative likelihood ratio test, aLRT) as developed in [64,65]. We used bootstrap (bv) and aLRT (aLRTv) values and posterior probabilities (pp) to establish a criterion of "quality". If bv was "low" but the other two were "high" then we considered a potential false negative support; if bv was "high" but the other two were "low" then we considered a potential false positive support.

We sometimes included sequences that were highly divergent, for instance from *Aidanosagitta crassa*, *Eukrohnia fowleri*, *Serratosagitta tasmanica* and, less frequently, from *Mesosagitta decipiens* because our primary goal was to accommodate the widest taxonomic and molecular dataset possible. This will provide a good foundation for future studies on chaetognaths evolution, but it may negatively impact our phylogenetic reconstruction. We removed some of these sequences in the concatenated dataset 5 to test whether they could produce artefacts in phylogenetic reconstructions.

#### Morphological analysis

The data set considered here is constituted of the 34 species used in the molecular analysis. The following 32 characters were chosen on the basis of their traditional importance as key characters and their use in species diagnosis. These variables are a mixture of different types (see the full list below): 23 qualitative (binary and polytomic) and 9 quantitative (*e.g.*, lengths). These data are coming from a compilation of original descriptions and reviews [21,22,37,38,66-74]. Characters coding is presented in Additional file 2.

Qualitative characters (n = 23)

C1- Body type: Flaccid =1; Rigid =2

C2- Number and type of lateral fins: one long pair of lateral fins extended on the tail as well as on the most part of the trunk =1; one short pair on the trunk and tail, the anterior end at the level of the caudal septum =2; one short pair on the trunk and tail, with the anterior and the posterior ends at equal distance from the caudal septum =3; Two pairs of lateral fins =4

C3- Tegumentary bridge connecting the anterior and posterior lateral fins: absent = 1; present =2

C4- Phragms (transverse muscles): absent = 1; present in the trunk only =2; present in the trunk and tail =3

C5- Type of phragms: supercontraction =1; normal contraction =2

C6- Type of longitudinal muscles: only B fibre =1; A and B fibres =2

C7- Type of secondary muscles: heterosarcomeric secondary muscles (He S) =1; homosarcomeric secondary muscles (Ho S) =2

C8- Organisation of RFamide-like neurons: Type A (absence of D6 and X posterior neurons, absence of caudal loop) =1; Type B (presence of D6 and X posterior neurons with caudal loop) =2

C9- Intestinal diverticula: absent = 1; present =2

C10- Vacuolated intestinal cells: absent = 1; present =2

C11- Type of hooks: gently curved =1; gently curved and serrated =2; abruptly curved =3

C12- Type of seminal vesicles: elongated with a lateral opening =1; elongated and an anterior protruding part usually roundish =2; roundish or slightly oval with a lateral opening =3; elongated with an anterior opening =4; presence of small indentations =5; oval with bulb-like shape =6

C13- Position of seminal vesicles (in respect to lateral and tail fins): touching neither lateral fins nor tail fin but closer to lateral fins =1; touching neither lateral fins nor tail fin but closer to tail fin =2; touching, or close to, lateral fins and well separated from tail fin =3; touching, or close to, tail fin and well separated from lateral fins =4; touching both lateral fins and tail fin =5

C14- Ocular type: inverted =1; everted =2

C15- Pigmented cell in the eye: absent = 1; present =2

C16- Secretory ventral gland: absent = 1; present =2

C17- Gelatinous masses in the lateral fins: absent = 1; present =2

C18- Adhesive papillae: absent = 1; present on the ventral side of the body and fins =2; concentrated on adhesive appendages =3

C19- Type of corona ciliata: Type A =1; Type B =2; Type C =3; Type D =4 (A–D, type of corona ciliata after Fowler [67] and Tokioka [21,22])

C20- Number of teeth rows: one anterior row =1; one posterior row =2; two rows =3;

C21- Epidermal glandular structures: Glandular structure on the body surface scarcely developed =1; numerous glandular structures on the body surface =2

C22- Type of teeth: stout teeth arranged in fan shape =1; slender teeth arranged in comb-shaped =2

C23- Ray less zone in the lateral fins: absent = 1; present =2

Quantitative characters (n = 9)

C24- Trunk/tail length ratio (minimum value)

C25- Trunk/tail length ratio (maximum value)

C26- Position of the ventral nerve centre (in respect to the trunk length)
C27- Minimum number of anterior teeth
C28- Maximum number of anterior teeth
C29- Minimum number of posterior teeth
C30- Maximum number of posterior teeth
C31- Minimum number of hooks
C32- Maximum number of hooks

Qualitative morphological data were analysed using Paup* 4.0b10 under maximum parsimony (MP) with a heuristic search with 10 random taxon addition replicates followed by tree bisection and reconnection (TBR) branch swapping. All characters were treated as unordered and unweighted. ACCTRAN (accelerated transformation) and DELTRAN (delayed transformation) character optimization were both used to map the character changes and resolve ambiguous nodes. The g1 statistic was obtained using 1 000 000 random trees. Clade frequencies were obtained by 50% majority-rule consensus trees.

As many chaetognath lineages have been defined by the lack of a given structure, we also conducted a phenetic approach to integrate quantitative data and to estimate the degree of overall similarity information available (*i.e.*, the absence of a character as valuable phylogenetic information). In order to include all variables in a common analysis, we chose to treat the data set as quantitative by replacing each qualitative variable by its disjunctive table. Such a table contains as many columns as modalities: each column defining a binary variable of 1 if the modality is observed and 0 otherwise. After this operation we ended up with a table of 34 lines (taxa) and 59 columns (original variables for quantitative characters or binary score corresponding to a modality for qualitative ones). We considered the Euclidian distance between two taxa after scaling each column to one and performed a non-supervised hierarchical clustering using the Ward algorithm. As the number of columns is slightly higher than the number of taxa it is especially important here to access the uncertainty of the relationships obtained after the classification procedure. We did that using the re-sampling procedure implemented in the pvclust package [75] of R version 3.0.1 [76]. At each bifurcation of the classification, the variables significantly different between the two classes were identified. This was made by performing t-test for quantitative variables and chi-square test for qualitative variables that allowed us to characterize classes of the topology. Although the bootstrap probability test is very useful for tree selection, it is biased. The selection bias comes from comparing many trees at the same time and often leads to overconfidence in the wrong trees. So, we chose the approximately unbiased (au) test for assessing the confidence of tree selection - a method less biased than other methods such as bootstrap probability test [77].

### Geometric morphometrics

We carried out a geometric morphometric approach [78] to explore body shape variations among the species used in the molecular analysis. The aim of this method was to test two different primary homology hypotheses on the evolution of structure and number of lateral fins and to correlate body shape patterns in relation to different locomotor and environmental behaviours (*i.e.*, benthic versus pelagic) in Chaetognatha. Over the last three decades, systematic studies have often been complemented by geometric morphometrics, allowing the computation and visualization of global shape changes in organs or organisms. Procrustes superimposition is the most effective method for creating spatial graphical representations of shape variations [79].

The variation of the body shape patterns was statistically studied from morphotypes belonging to the six traditional clades identified in the molecular analysis: Heterokrohniidae, Eukrohniidae, Spadellidae, Pterosagittidae, Krohnittidae and Sagittidae. Illustrations of representative species used in this pilot study mostly come from the publications of Alvariño [37,38] who provided the most accurate drawings of chaetognaths with respect to their body shape proportions and the position of their ventral nerve centre, lateral fins and seminal vesicles. Other sources were Tokioka's illustrations of *Aidanosagitta crassa* [69], Casanova's illustrations of *Paraspadella gotoi* [71] and *Xenokrohnia sorbei* [72], the description by Dallot and Ducret [73] of *Parasagitta megalophthalma* as well as pictures of specimens belonging to the *Spadella* genus (*Spadella ledoyeri*, *Spadella cephaloptera* and *Spadella valsalinae*) by the authors of the present study.

Digital images were obtained with a flat bed scanner. Then, 20 landmarks were digitalized using TPSdig2 [80] (Additional file 5). When the depicted specimens were not straight, we used the following procedure to get landmark coordinates of straightened specimens: we first calculated the mid points between homologous points on the right and left side of the specimens. This series of points was then aligned on the x-axis, and the relevant landmark points repositioned with a Y coordinate of half the distance between the left and right points (Additional file 5). Only one side of the individual specimens was then used for further analyses. The shape variation was analyzed by the generalized Procrustes method using the R shape package [81]. Two primary homology hypotheses (PH) were tested: the anterior end of the posterior lateral fin in two fin species is homologous to the anterior end of the unique lateral fin in one-fin species (PH1), or the anterior end of the anterior lateral fin in two fin species is homologous to the anterior end of the unique lateral fin in one fin species

(PH2). The full Procrustes distances between the conformations of each possible pair of species were computed in the two homology hypotheses. Dendrograms of landmark conformation similarities were computed by UPGMA. These hypotheses were tested to establish which of the two dendrograms from the morphometric data shows the higher agglomerative coefficient, and whether they were congruent with molecular data.

## Additional files

**Additional file 1: Phylogeny of chaetognaths inferred from the SSU rRNA dataset 2 including both paralogy classes (138 sequences from 33 species: 80 class I and 58 class II sequences; 1679 base pairs long).** Support values obtained using different reconstruction approaches are indicated at nodes in the following order: maximum Likelihood bootstrap probabilities (bv), approximative likelihood ratio test (aLRTv) and Bayesian posterior probabilities (pp). Support values are displayed when bv/aLRTv ≥75 or pp ≥0.85. Node absence in a given method is indicated by -.

**Additional file 2: Morphological data matrix.** Unknown character states are indicated by NA and non-homologous characters are indicated by -.

**Additional file 3: Diagrammatic representation of morphometric measures used for the analysis of body shape similarities.** A: Superimposition of all the studied chaetognaths, altering the scales, to superimpose their anterior and posterior ends (points 1 and 20). B-F: same graphs family per family (B: Eukrohniidae; C: Spadellidae; D: Heterokrohniidae). For comparison purposes, *Pterosagitta draco* is depicted in bold with the Sagittidae (E) and with *Krohnitta subtilis* (F).

**Additional file 4: Dendrograms of morphometric similarity using full Procrustes distance (Riemannian shape distance ρ).** A: Primary hypothesis 1 (PH1): the anterior end of the posterior lateral fin in two-fin species is homologous to the posterior end of the unique lateral fin in one-fin species. B: Primary hypothesis 2 (PH2): the anterior end of the anterior lateral fin in two-fin species is homologous to the posterior end of the unique lateral fin in one-fin species. Surimposition of landmark points according to PH1 of all the studied chaetognaths are shown on Additional file 5.

**Additional file 5: Geometric morphometric method to explore body shape variations among Chaetognatha.** A: Landmark points (red dots) used for the morphometrics analysis according to the first hypothesis of primary homology (PH1: the anterior end of the posterior lateral fin in two-fin species is homologous to the anterior end of the unique lateral fin in one-fin species), here on *Mesosagitta minima* (drawing from Alvariño [38]). B: The landmarks, as input. C: The corrected landmarks, which straighten the specimen, according to their bilateral symmetry. The landmarks are: (1) Anterior end of the body; (2, 3) Anterior end of the trunk; (4, 5) Anterior end of the ventral nerve centre; (6, 7) Posterior end of the ventral nerve centre; (8, 9) Anterior end of the lateral fin; (10, 11) Posterior end of the lateral fin; (12, 13) Caudal septum; (14, 15) Anterior end of the seminal vesicle; (16, 17) Posterior end of the seminal vesicle; (18, 19) Anterior end of the caudal fin; (20) Posterior end of the body.

### Competing interests
The authors declare that they have no competing interests.

### Authors' contributions
SG and YP identified the specimens. SG carried out the molecular experiments. YP and GN designed the geometric morphometrics study and drafted the corresponding part of the manuscript. SG, NP, AG and YP did the phenetic and cladistic analysis of morphological characters. SG, AG and YP did the molecular phylogenetic analyses. AG, YP and ST drafted the manuscript. YP conceived the study and its design. AG and YP supervised the study. All authors read, amended and approved the final manuscript.

### Acknowledgements
We are deeply grateful to the Tara schooner and crew for collecting plankton samples all over the World Oceans during three years. Our special thanks go to Gaby Gorsky and Christian Sardet who facilitated our integration in the *Tara Oceans* consortium. We are also keen to thank the following people for providing us with planktonic samples: Jean-Marc Pagano from the Institut de Recherche pour le Développement (IRD), John Lamkin and Akihiro Shiroza from the Southeast Fisheries Science Centre (SEFSC), Cédric Guigand from the Rosentiel School of Marine and Atmospheric Science (RSMAS) and Fredrika Norrbin from the Department of Arctic and Marine Biology (The Arctic University of Norway). Our manuscript profited from stimulating discussion with Daniel Papillon who also provided English corrections. This article is the contribution no. 19 of the *Tara Oceans* Expedition 2009/2012.

### Author details
[1]Aix-Marseille Université, CNRS, IRD, Avignon Université, IMBE UMR 7263, 13331 Marseille cedex 3, France. [2]Université de Tunis El Manar, Faculté des Sciences de Tunis, UR11ES12 Biologie de la Reproduction et du Développement animal, 2092 El Manar, Tunis, Tunisie.

### References
1. Bone Q, Kapp H, Pierrot Bults AC: **Introduction and Relationships of the Group.** In *The Biology of Chaetognaths*. Edited by Bone Q, Kapp H, Pierrot-Bults AC. Oxford: Oxford University Press; 1991:1–5.
2. Casanova JP: **Chaetognatha.** In *South Atlantic Zooplankton*. Edited by Boltovskoy D. Leiden: Backhuys Publishers; 1999:1353–1374.
3. Vannier J, Steiner M, Renvoisé E, Hu SX, Casanova JP: **Early Cambrian origin of modern food webs: evidence from predator arrow worms.** *Proc Biol Sci* 2007, **274:**627–633.
4. Slabber M: *Natuurkundige Verlustigingen, beheizende microscopise Waarneemingen van in-en Uitlandse Wateren Land-dieren*. Haarlem: J. Bosch; 1778
5. Perez Y, Müller CHG, Harzsch S: **The Chaetognatha: An Anarchistic Taxon Between Protostomia and Deuterostomia.** In *Deep Metazoan Phylogeny: The Backbone of the Tree of Life: New Insights from Analyses of Molecules, Morphology, and Theory of Data Analysis*. Edited by Wägele JW, Bartholomaeus TW, Misof B. Berlin: Walter de Gruyter; 2014:49–77.
6. Edgecombe GD, Giribet G, Dunn CW, Hejnol A, Kristensen RM, Neves RC, Rouse GW, Worsaae K, Sørensen MV: **Higher-level metazoan relationships: recent progress and remaining questions.** *Org Divers Evol* 2011, **11:**151–172.
7. Harzsch S, Müller CH: **A new look at the ventral nerve centre of Sagitta: implications for the phylogenetic position of Chaetognatha (arrow worms) and the evolution of the bilaterian nervous system.** *Front Zool* 2007, **4:**14.
8. Rieger V, Perez Y, Müller CH, Lacalli T, Hansson BS, Harzsch S: **Development of the nervous system in hatchlings of Spadella cephaloptera (Chaetognatha), and implications for nervous system evolution in Bilateria.** *Dev Growth Differ* 2011, **53:**740–759.
9. Marlétaz F, Martin E, Perez Y, Papillon D, Caubit X, Lowe CJ, Freeman B, Fasano L, Dossat C, Wincker P, Weissenbach J, Le Parco Y: **Chaetognath phylogenomics: a protostome with deuterostome-like development.** *Curr Biol* 2006, **16:**R577–R578.
10. Matus DQ, Copley RR, Dunn CW, Hejnol A, Eccleston H, Halanych KM, Martindale MQ, Telford MJ: **Broad taxon and gene sampling indicate that chaetognaths are protostomes.** *Curr Biol* 2006, **16:**R575–R576.
11. Marlétaz F, Gilles A, Caubit X, Perez Y, Dossat C, Samain S, Gyapay G, Wincker P, Le Parco Y: **Chaetognath transcriptome reveals ancestral and unique features among bilaterians.** *Genome Biol* 2008, **9:**R94.
12. Dunn CW, Hejnol A, Matus DQ, Pang K, Browne WE, Smith SA, Seaver E, Rouse GW, Obst M, Edgecombe GD, Sørensen MV, Haddock SH, Schmidt-Rhaesa A, Okusu A, Kristensen RM, Wheeler WC, Martindale MQ, Giribet G: **Broad phylogenomic sampling improves resolution of the animal tree of life.** *Nature* 2008, **452:**745–749.
13. Hejnol A, Obst M, Stamatakis A, Ott M, Rouse GW, Edgecombe GD, Martinez P, Baguñà J, Bailly X, Jondelius U, Wiens M, Müller WE, Seaver E, Wheeler WC, Martindale MQ, Giribet G, Dunn CW: **Assessing the root of bilaterians animals with scalable phylogenomic methods.** *Proc Biol Sci* 2009, **276:**261–270.

14. Nielsen C: *Animal Evolution: Interrelationships of the Living Phyla.* 2nd edition. Oxford New York: Oxford University Press; 2001.
15. Nielsen C: **Larval and adult brains.** *Evol Dev* 2005, **7:**483–489.
16. Bieri R: **Systematics of the Chaetognatha.** In *The Biology of Chaetognaths.* Edited by Bone Q, Kapp H, Pierrot-Bults AC. Oxford: Oxford University Press; 1991:122–136.
17. Bieri R: **Six new genera in the chaetognath family Sagittidae.** *Gulf Res Rep* 1991, **8:**221–225.
18. Papillon D, Perez Y, Caubit X, Le Parco Y: **Systematics of Chaetognatha under the light of molecular data, using duplicated ribosomal 18S DNA sequences.** *Mol Phylogenet Evol* 2006, **38:**621–634.
19. Von Ritter-Zahony R: **Revision der Chaetognathen.** *Dtsch Südpolar-Exped* 1911, **13:**1–71.
20. Hyman LH: **Phylum Chaetognatha.** In *The Invertebrates. Volume 5.* New York: McGraw Hill Book Company; 1959:1–71.
21. Tokioka T: **The taxonomical outline of chaetognaths.** *Publ Seto Mar Biol Lab* 1965, **12:**335–357.
22. Tokioka T: **Supplementary notes on the systematics of Chaetognatha.** *Publ Seto Mar Biol Lab* 1965, **13:**231–242.
23. Alvariño A: **Quetognatos epiplanctonicos del Mar de Cortes.** *Rev Soc Mex Hist Nat* 1963, **24:**97–202.
24. Casanova JP: **Description de l'appareil génital primitif du genre *Heterokrohnia* et nouvelle classification des chaetognathes.** *CR Acad Sci Paris* 1985, **301:**397–402.
25. Dallot S, Ibanez F: **Etude préliminaire de la morphologie et de l'évolution chez les chaetognathes.** *Inv Pesq* 1972, **36:**31–41.
26. Salvini-Plawen LV: **Systematics notes on Spadella and on the Chaetognatha.** *J Z Zool Syst Evol* 1986, **24:**122–128.
27. Telford MJ, Holland PWH: **Evolution of 28S ribosomal DNA in chaetognaths, duplicate genes and molecular phylogeny.** *J Mol Evol* 1997, **44:**135–144.
28. Jennings RM, Bucklin A, Pierrot-Bults A: **Barcoding of arrow worms (Phylum Chaetognatha) from three oceans: genetic diversity and evolution within an enigmatic phylum.** *PLoS One* 2010, **5:**e9949.
29. Park JK, Rho HS, Kristensen RM, Kim W, Giribet G: **First molecular data on the phylum Loricifera: an investigation into the phylogeny of ecdysozoa with emphasis on the positions of Loricifera and Priapulida.** *Zoolog Sci* 2006, **23:**943–954.
30. Farris J: **Estimating phylogenetic trees from distance matrices.** *Amer Nat* 1972, **106:**645–667.
31. Tokioka T, Pathansali D: **Another new chaetognath from Malay waters, with a proposal of grouping some species of *Sagitta* into subgenera.** *Publ Seto Mar Biol Lab* 1963, **11:**119–123.
32. Harzsch S, Müller CH, Rieger V, Perez Y, Sintoni S, Sardet C, Hansson B: **Fine structure of the ventral nerve centre and interspecific identification of individual neurons in the enigmatic Chaetognatha.** *Zoomorphology* 2009, **128:**53–73.
33. Casanova JP, Duvert M: **Comparative studies and evolution of muscles in chaetognaths.** *Mar Biol* 2002, **141:**925–938.
34. Casanova JP: **A new genus and species of deep-benthic chaetognath from the Atlantic: a probable link between the families Heterokrohniidae and Spadellidae.** *J Nat Hist* 1996, **30:**1239–1245.
35. Dallot S: **L'anatomie du tube digestif dans la phylogénie et la systématique des chaetognathes.** *Bull Mus Natl Hist Nat* 1970, **42:**549–565.
36. Furnestin ML: **Chaetognathes et zooplankton du secteur atlantique marocain.** *Rev Trav Inst Pech Mar* 1957, **21:**1–356.
37. Alvariño A: **The Chaetognatha of the NAGA Expedition (1959–1961) in the South China Sea and the Gulf of Thailand. Part I. Systematics.** *NAGA report* 1967, **4:**1–197.
38. Alvariño A: **Los quetognatos del Atantico: Distribucion y notes essenciales de sistimatica.** *Trab Inst Esp Oceanogr* 1969, **37:**1–290.
39. Szaniawski H: **Cambrian chaetognaths recognized in Burgess Shale fossils.** *Acta Palaeontol Pol* 2005, **50:**1–8.
40. McLelland JA: **Eukrohnia calliops, a new species of Chaetognatha from the northern Gulf of Mexico with notes on related species.** *Proc Biol Soc Wash* 1989, **102:**33–44.
41. Casanova JP, Nair VR: **A new species of Sagitta (Chaetognatha) from a Laccadive lagoon (Indian Ocean) having fan-shaped anterior teeth: phylogenetical implications.** *J Nat Hist* 2002, **36:**149–156.
42. Casanova JP, Moreau X: **Calispadella alata n. gen., n. sp., the first chaetognath recorded from a hydrothermal vent site (Mid-Atlantic Ridge).** *J Plank Res* 2005, **27:**221–225.
43. Stebbins GL: **Mosaic evolution, mosaic selection and angiosperm phylogeny.** *J Linn Soc Bot* 1984, **88:**149–164.
44. Knoll AH, Niklas KJ, Gensel PG, Tiffney BH: **Character diversification and patterns of evolution in early vascular plants.** *Paleobiology* 1984, **10:**34–47.
45. Kearney M, Stuart BL: **Repeated evolution of limblessness and digging heads in worm lizards revealed by DNA from old bones.** *Proc R Soc Lond B* 2004, **271:**1677–1683.
46. Winter W, Oxnard CE: **Evolutionary radiations and convergences in the structural organization of mammalian brains.** *Nature* 2001, **409:**710–714.
47. Lazzari V, Charles C, Tafforeau P, Vianey-Liaud M, Aguilar JP, Jaeger JJ, Michaux J, Viriot L: **Mosaic convergence of rodent dentitions.** *PLoS One* 2008, **3:**e3607.
48. de Queiroz K: **The ontogenetic method for determining character polarity and its relevance to phylogenetic systematics.** *Syst Zool* 1985, **34:**280–299.
49. Bone Q, Brownlee C, Bryan GW, Burt GR, Dando PR, Liddicoat MI, Pulsford AL, Ryan KP: **On the differences between the two 'indicator' species of chaetognath, *Sagitta setosa* and *S. elegans.*** *J Mar Biol Ass UK* 1987, **67:**545–560.
50. Bone Q, Duvert M: **Locomotion and Buoyancy.** In *The Biology of Chaetognaths.* Edited by Bone Q, Kapp H, Pierrot-Bults AC. Oxford: Oxford University Press; 1991:32–44.
51. Kapp H: **Some aspects of buoyancy adaptations of chaetognaths.** *Helgol Meeres* 1991, **45:**263–267.
52. Perez Y, Casanova JP, Mazza J: **Degrees of vacuolation of the absorptive cells of five Sagitta (Chaetognatha) species: possible ecophysiological implications.** *Mar Biol* 2001, **138:**125–133.
53. Corse E, Rampal J, Cuoc C, Pech N, Perez Y, Gilles A: **Phylogenetic analysis of Thecosomata Blainville, 1824 (holoplanktonic opisthobranchia) using morphological and molecular data.** *PLoS One* 2013, **12:**e59439.
54. Müller CHG, Rieger V, Perez Y, Harzsch S: **Immunohistochemical and ultrastructural studies on ciliary sense organs of arrow worms (Chaetognatha).** *Zoomorphology* 2014, **133:**167–189.
55. Thompson JD, Higgins DG, Gibson TJ: **CLUSTAL W: improving the sensitivity of progressive multiple sequence alignment through sequence weighting, position-specific gap penalties and weight matrix choice.** *Nucleic Acids Res* 1994, **22:**4673–4680.
56. Edgar RC: **MUSCLE: multiple sequence alignment with high accuracy and high throughput.** *Nucleic Acids Res* 2004, **32:**1792–1797.
57. Posada D, Crandall KA: **MODELTEST: testing the model of DNA substitution.** *Bioinformatics* 1998, **14:**817–818.
58. Misof B, Misof K: **A Monte Carlo approach successfully identifies randomness in multiple sequence alignments : a more objective means of data exclusion.** *Syst Biol* 2009, **58:**21–34.
59. Kück P, Meusemann K, Dambach J, Thormann B, von Reumont BM, Wägele JW, Misof B: **Parametric and non-parametric masking of randomness in sequence alignments can be improved and leads to better resolved trees.** *Front Zool* 2010, **31:**10.
60. Hillis DM, Bull JJ: **An empirical test of bootstrapping as a method for assessing confidence in phylogenetic analysis.** *Syst Biol* 1993, **42:**182–192.
61. Huelsenbeck JP, Ronquist F: **MRBAYES: Bayesian inference of phylogenetic trees.** *Bioinformatics* 2001, **17:**754–755.
62. Rambaut A, Drummond AJ: *Tracer: MCMC Trace Analysis Tool.* Oxford: University of Oxford; 2003.
63. Guindon S, Dufayard JF, Lefort V, Anisimova M, Hordijk W, Gascuel O: **New algorithms and methods to estimate maximum-likelihood phylogenies: assessing the performance of PhyML 3.0.** *Syst Biol* 2010, **59:**307–321.
64. Anisimova M, Gil M, Dufayard JF, Dessimoz C, Gascuel O: **Survey of branch support methods demonstrates accuracy, power, and robustness of fast likelihood-based approximation schemes.** *Syst Biol* 2011, **60:**685–699.
65. Anisimova M, Gascuel O: **Approximate likelihood-ratio test for branches: a fast, accurate, and powerful alternative.** *Syst Biol* 2006, **55:**539–552.
66. Fowler GH: **Biscayan plankton collected during a cruise H.M.S. Research 1900, III. The Chaetognatha.** *Trans Linn Soc Lond (Zool)* 1905, **10:**55–87.
67. Fowler GH: **The Chaetognatha of the Siboga expedition with a discussion of the synonymy and distribution of the group.** *Siboga Exped Rep* 1906, **21:**1–86.
68. Frazer JH: **Chaetognatha.** In *Fiches d'identification du zooplancton. Zooplankton.* Charlottenlund: Intern Counc Explo Sea; 1957:1–6.

69. Tokioka T: **A new chaetognath (*Sagitta crassa n. sp.*) from Ise Bay.** *Zool Mag (Tokio)* 1938, **50**:349–351.
70. Casanova JP: ***Spadella ledoyeri*, un chaetognathe nouveau de la grotte sous-marine obscure des Trémies (Calanques de Cassis).** *Rapp Comm Mer Médit* 1986, **30**:196–197.
71. Casanova JP: **A new species of *Paraspadella* (Chaetognatha) from the coastal waters of Japan.** *Proc Biol Soc Wash* 1990, **103**:907–912.
72. Casanova JP: **A new genus and species of deep-sea chaetognath from the Bay of Biscay with a strange ventral secretory gland.** *J Nat Hist* 1993, **27**:445–455.
73. Dallot S, Ducret F: **Un chaetognathe mésoplanctonique nouveau *Sagitta megalophthalma* sp. n.** *Beaufortia* 1969, **17**:13–20.
74. Winkelmann C, Gasmi S, Gretschel G, Müller CHG, Perez Y: **Description of Spadella valsalinae sp. nov., a neo-endemic benthic chaetognath from Northern Adriatic Sea (Croatia) with remarks on its morphology, phylogeny and biogeography.** *Org Div Evol* 2013, **13**:189–202.
75. Suzuki R, Shimodaira H: **Pvclust: an R package for assessing the uncertainty in hierarchical clustering.** *Bioinformatics* 2006, **22**:1540–1542.
76. R Core Team: **R: A language and environment for statistical computing. R Foundation for Statistical Computing, Vienna, Austria;** 2013 [http://www.R-project.org/]
77. Shimodaira H: **An approximately unbiased test of phylogenetic tree selection.** *Syst Biol* 2002, **51**:492–508.
78. Claude J: *Morphometrics with R.* New York: Springer; 2008.
79. Rohlf FJ, Slice D: **Extensions of the Procrustes method for the optimal superimposition of landmarks.** *Syst Biol* 1990, **39**:40–59.
80. Rohlf FJ: **TpsDig Version 2.00. Department of Ecology and Evolution, State University of New York. New York: Stony Brook**; 2004 [http://life.bio.sunysb.edu/morph/]
81. Dryden IL: **Shapes package. R Foundation for Statistical Computing, Vienna, Austria. Contributed package.** [http://www.R-project.org]

# 15

# Transcriptional profiling of reproductive development, lipid storage and molting throughout the last juvenile stage of the marine copepod *Calanus finmarchicus*

Ann M Tarrant[1*†], Mark F Baumgartner[1†], Bjørn Henrik Hansen[2], Dag Altin[3], Trond Nordtug[2] and Anders J Olsen[4]

**Abstract**

**Introduction:** *Calanus finmarchicus*, a highly abundant copepod that is an important primary consumer in North Atlantic ecosystems, has a flexible life history in which copepods in the last juvenile developmental stage (fifth copepodid, C5) may either delay maturation and enter diapause or molt directly into adults. The factors that regulate this developmental plasticity are poorly understood, and few tools have been developed to assess the physiological condition of individual copepods.

**Results:** We sampled a cultured population of *C. finmarchicus* copepods daily throughout the C5 stage and assessed molt stage progression, gonad development and lipid storage. We used high-throughput sequencing to identify genes that were differentially expressed during progression through the molt stage and then used qPCR to profile daily expression of individual genes. Based on expression profiles of twelve genes, samples were statistically clustered into three groups: (1) an early period occurring prior to separation of the cuticle from the epidermis (apolysis) when expression of genes associated with lipid synthesis and transport (FABP and ELOV) and two nuclear receptors (ERR and HR78) was highest, (2) a middle period of rapid change in both gene expression and physiological condition, including local minima and maxima in several nuclear receptors (FTZ-F1, HR38b, and EcR), and (3) a late period when gonads were differentiated and expression of genes associated with molting (Torso-like, HR38a) peaked. The ratio of Torso-like to HR38b strongly differentiated the early and late groups.

**Conclusions:** This study provides the first dynamic profiles of gene expression anchored with morphological markers of lipid accumulation, development and gonad maturation throughout a copepod molt cycle. Transcriptomic profiling revealed significant changes over the molt cycle in genes with presumed roles in lipid synthesis, molt regulation and gonad development, suggestive of a coupling of these processes in *Calanus finmarchicus*. Finally, we identified gene expression profiles that strongly differentiate between early and late development within the C5 copepodid stage. We anticipate that these findings and continued development of robust gene expression biomarkers that distinguish between diapause preparation and continuous development will ultimately enable novel studies of the intrinsic and extrinsic factors that govern diapause initiation in *Calanus finmarchicus*.

**Keywords:** Arthropod, Crustacean, Gene expression, Molt cycle, Transcriptomics

* Correspondence: atarrant@whoi.edu
[†]Equal contributors
[1]Biology Department, Woods Hole Oceanographic Institution, 45 Water Street, Woods Hole, MA 02543, USA
Full list of author information is available at the end of the article

## Introduction

Copepods of the family Calanidae, including the *Calanus* and *Neocalanus* genera, are vital components of temperate, subarctic, and arctic ecosystems that transfer energy directly from phytoplankton to very high trophic levels (e.g., fish, sea birds, and some marine mammals). The life history of these copepods typically includes diapause, a period of arrested development, dormancy and fasting that can last many months [1-3]. The well-studied temperate and subarctic copepod, *Calanus finmarchicus*, prepares for diapause primarily during the fifth and final juvenile stage (C5). However, *C. finmarchicus* is not obligated to enter diapause, and some individuals skip diapause, molt immediately into adults, and become reproductively mature. Despite their ecological importance, strikingly little is known about developmental regulation in *C. finmarchicus* or other members of the family Calanidae. Diapause physiology is difficult to study directly in *Calanus* spp. because researchers cannot reliably induce the diapause state under laboratory conditions. Thus, we are currently ignorant of the intrinsic and extrinsic factors that influence the "choice" to enter diapause or to molt directly into adulthood in *C. finmarchicus*, and this ignorance limits our ability to predict population dynamics beyond a single generation.

*Calanus finmarchicus* undergoes dramatic physiological changes during the C5 copepodid stage (Figure 1). The most noticeable of these changes is the accumulation of lipids in an organ called the oil sac. The lipids stored in the oil sac are almost entirely wax esters; much smaller

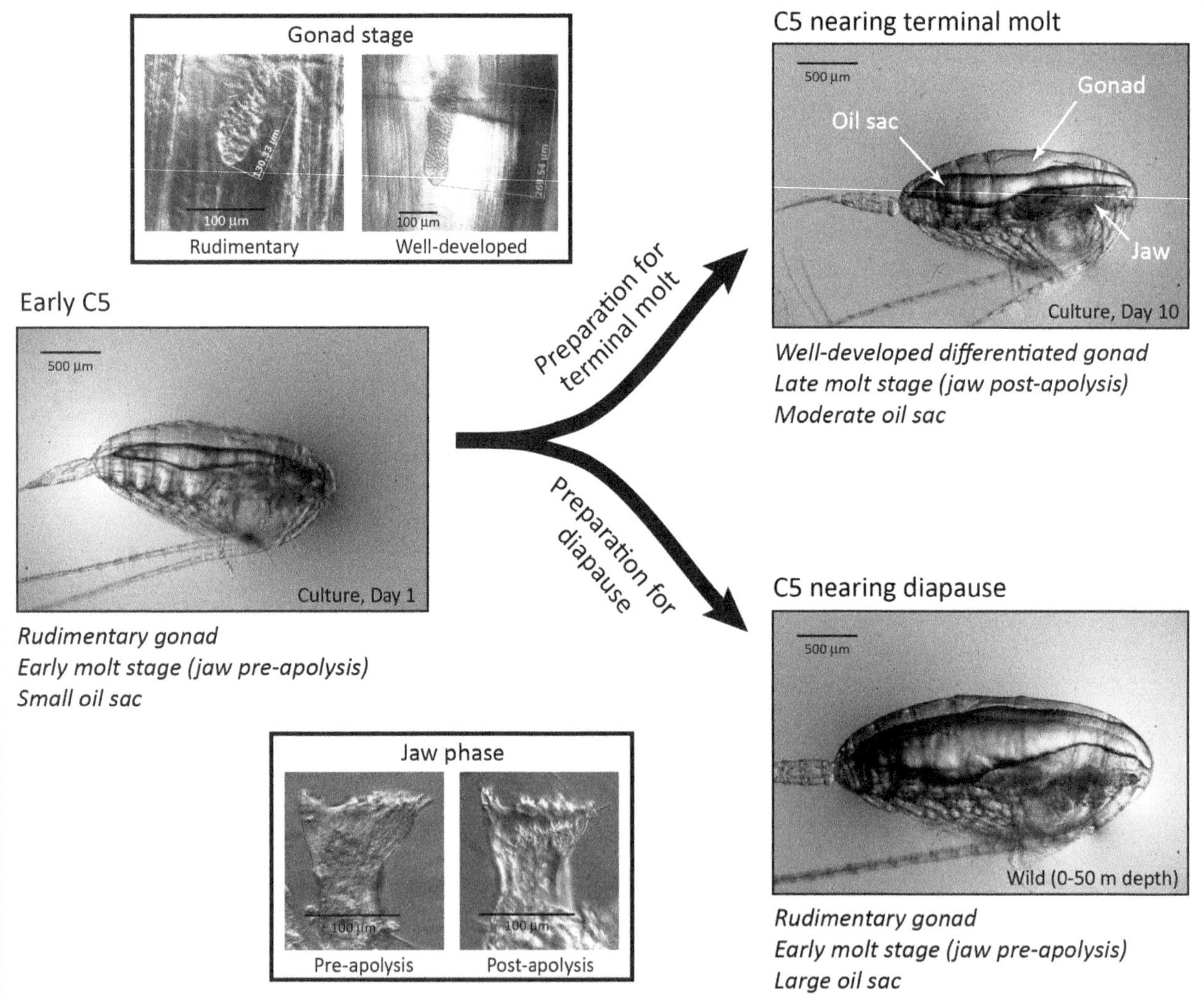


**Figure 1 Morphological changes associated with preparation for diapause or preparation for the terminal molt in *Calanus finmarchicus*.** As copepodids progress through the C5 stage toward the terminal molt, the mandibular cuticle separates from the epidermis (apolysis), the gonad develops (grows and changes cell structure) and differentiates (allowing gender to be identified), and the oil sac increases to a moderate size. As copepods progress toward diapause, the oil sac increases to a very large size, but molt stage progression (measured by jaw phase) and gonadal development are arrested.

amounts of triglycerides and phospholipids are stored in lipid droplets throughout the body [4]. Lipids accumulated in the oil sac will later be metabolized during fasting by diapausing copepodids, and/or utilized during molting and gonad maturation [5,6], oogenesis [7], or locomotion [8] by adult copepods. The oil sac volume, which can be estimated from analysis of photographs [9], is positively correlated with total lipid content in *C. finmarchicus* C5 copepodids [4]. Gonads appear during the C4 stage [10], but reproductive development accelerates during the C5 stage in copepodids preparing for the terminal molt [7]. Progression toward ecdysis (molting) can be monitored through changes in the mandibular gnathobase during tooth formation [11]; in particular, separation of the cuticle from the epidermis, a processes called apolysis, is a key transition within the arthropod molt cycle that can be detected first in the copepod mandibular gnathobase.

While little is known about regulation of copepod molting, extensive studies in other arthropods (mainly insects and decapods) have shown that molting is regulated by a cascade of hormonal signals (reviewed by [12,13]). Binding of ecdysteroids activates the ecdysone receptor (EcR), which is a member of the nuclear receptor superfamily. EcR forms a complex with another nuclear receptor, the retinoid-x-receptor (RXR) to regulate the transcription of target genes, including several other nuclear receptors (reviewed by [13]). Many studies have characterized EcR and RXR expression during crustacean molt cycles [14-16]; however, few studies have documented expression patterns of other nuclear receptors [17]. While ecdysteroid signaling has not been extensively studied in copepods, Johnson [18] and Hansen et al. [19] found dynamic patterns of ecdysteroid concentration in C5 copepodids from *Calanus* spp. Expression of EcR and other nuclear receptors has not been studied across the molt stage of any copepod, but results from field studies of *C. finmarchicus* are consistent with increased EcR expression in preparation for emergence and molting late in diapause [20,21].

Ontogenetic changes in morphology are often accompanied or even preceded by changes in gene expression, and characterizing these gene expression patterns can help elucidate underlying developmental processes. A few studies have described gene expression patterns associated with progression through the crustacean molt cycle. Genes exhibiting dynamic expression patterns include cuticular proteins [22,23] and nuclear receptors [17]. In *C. finmarchius*, Tarrant et al. [20] suggested that expression of genes associated with lipid synthesis, transport and storage is correlated with lipid accumulation during the C5 stage. Recent studies have employed high-throughput sequencing (RNA-seq) to compare expression among crustacean developmental stages (e.g., [24,25]). Lenz et al. [26] described a *de novo* transcriptome assembled across the entire *C. finmarchicus* developmental cycle (embryos to adults) from which gene expression patterns associated with circadian and peptidergic signaling have been investigated [27,28]; however, these studies have focused on comparisons *between* developmental stages. To gain insight into diapause and maturation processes that happen at the end of juvenile development in *C. finmarchicus*, it is necessary to compare transcriptional profiles *within* the C5 stage.

In this paper, we characterize transcriptional profiles associated with progression through the final copepodid stage of *C. finmarchicus* during preparation for the terminal molt. Our goals were to (1) identify genes associated with lipid storage, reproductive development, and ecdysis during preparation for the terminal molt, (2) characterize changes in the expression of those genes throughout the C5 copepodid stage at high temporal resolution, and (3) understand better the processes and pathways associated with maturation. Our long-term goals are to identify genes associated specifically with preparation for diapause, and to develop markers that can be used to distinguish C5 copepodids that are preparing for the terminal molt and for diapause. With robust molecular markers of these two developmental pathways, new studies can be designed to investigate the factors that influence the "choice" to proceed directly to the terminal molt or to enter diapause. Here, we address the first of these developmental pathways with a transcriptional profiling study of laboratory-reared copepods that are preparing for the terminal molt.

## Results

### Morphometrics, jaw phase, gonads, and stage duration

*Calanus finmarchicus* C5 copepodids were sampled from a continuous laboratory culture daily at known times since the C4-C5 (penultimate) molt. Adults first appeared in this time series study 9 days after the penultimate molt, but C5 copepodids were still present on day 18 (Figure 2a). Median C5 stage duration was 13.5 days. Of the 173 copepods that molted into adults during the time series study, 44 (25.4%) were male. In a small sample of C4 copepodids collected prior to the penultimate molt (n = 12), all but one had observable gonads. Of these, 4 had rudimentary gonads, 5 had developing gonads, and 2 were already differentiated (i.e., gender could be determined from well-developed gonads). Not surprisingly then, differentiated gonads were present early in the C5 stage; on nearly all sample days (except days 1 and 5), the majority of C5 copepodids had differentiated gonads (Figure 2b). Of the 98 C5 copepodids sampled with differentiated gonads during the time series study, 44 (44.9%) had developing testes. Apolysis was first observed on day 6, but the percentage of post-apolysis copepodids did not increase monotonically thereafter (Figure 2b), suggesting that the timing of molt stage progression (i.e., within-stage development) is highly variable among individuals. A portion of the

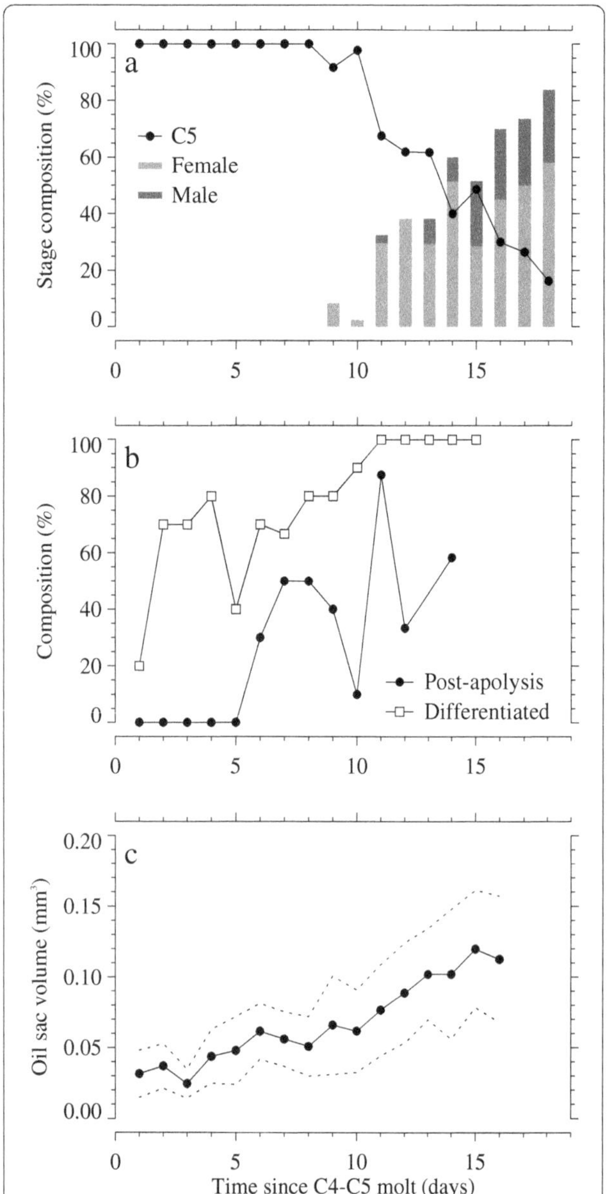


**Figure 2 Time series of developmental stage composition, jaw phase, gonad stage, and oil sac volume. (a)** Daily percentage of *Calanus finmarchicus* C5 copepodids (filled circles) and adults (gray bars) during the time series study (n = 34–44 copepods sampled per day). **(b)** Daily percentage of C5 copepodids with post-apolysis jaw phases (filled circles; n = 8–10 for days 1–12, days 13–15 combined since n = 4 on each of those days, n = 0 for days 16–18) and differentiated gonads (open squares; n = 8–10 for days 1–13, n = 4 for days 14–15, n = 0 for days 16–18). **(c)** Average (filled circles) and standard deviation (dotted lines) of oil sac volume for sampled C5 copepodids (n = 33–43 for days 1–10, n = 12–26 for days 11–16; days 17–18 not plotted since n < 10).

variability near the end of the time series was attributable to our including only C5 copepodids in Figure 2b; had we included all individuals that underwent apolysis (C5 and adults), the proportion of post-apolysis *C. finmarchicus* after day 10 would have been higher. In contrast to the variability observed in molt stage progression, average oil sac volume increased steadily 4-fold from day 1 to 16 (Figure 2c).

**Transcriptome assembly**

A single library was created from pooled RNA obtained from *C. finmarchicus* C5 copepodids. Illumina-based sequencing produced 93 million 100-bp paired-end reads. After quality trimming, 82.3 million paired-end reads and 10 million unpaired left reads were retained and assembled. The assembled transcriptome contained 124,157 Trinity components and 241,140 Trinity transcripts; multiple transcripts were frequently associated with a single component. The mean transcript length was 667 bp and the N50 (weighted median) was 988 bp. The assembly was qualitatively similar to a recently published *de novo* transcriptome assembled from multiple *C. finmarchicus* life stages (206,041 sequences, average length 997 bp, [26]). The larger average sequence length can be partially attributed to a difference in the minimum sequence length chosen for the two studies (200 bp in the present study, 300 bp in the Lenz et al. [26] study).

**Transcriptome-wide analysis of differential gene expression**

Transcriptional profiling was conducted on samples collected on days 3 and 10 after the penultimate molt. 22,848 transcripts, corresponding to 7470 Trinity components were differentially expressed between these two days (Additional file 1: Table S1). Of these, one-third (7560 transcripts) returned positive blastx hits based on their similarity to other genes in the NCBI nonredundant (nr) database, and 28% (6479 transcripts) were also associated with gene ontology (GO) terms (Additional file 2: Table S2). The GO terms most strongly enriched in the differentially expressed genes included terms related to remodeling of the cuticle (e.g., chitin metabolic process; structural constituent of cuticle), proteolysis (e.g., tyrosine catabolic process, peptidase activity), and hormonal signaling (e.g., steroid hormone mediated signaling pathway, retinoic acid receptor activity). The categories with the strongest enrichment in the differentially expressed genes are shown in Figure 3; complete results are given in Additional file 3: Table S3.

Of the 100 genes showing the strongest differential expression (lowest p-value), all were expressed at higher levels on day 10. Of these, 58 had no blast hit or only matched predicted proteins with no associated GO terms. A total of 30 genes could be grouped into a small number of categories: myosin heavy chain (14), structural components of cuticle (8), chitin-binding (4), and carbohydrate-binding groups (4). Decapod crustaceans undergo muscular atrophy in preparation for molting; however in American lobster (*Homarus americanus*),

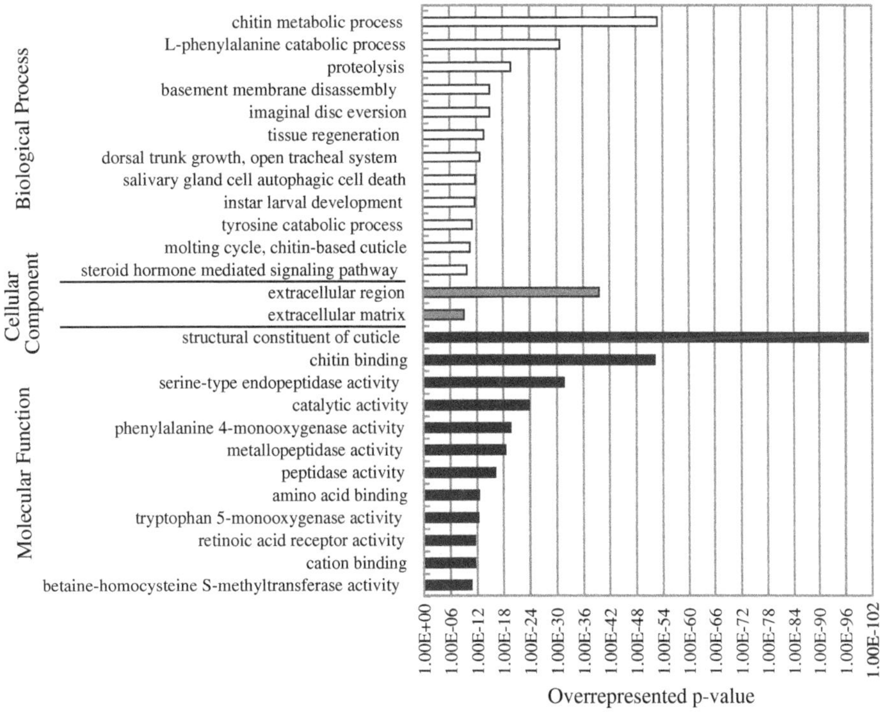


**Figure 3 Gene ontology (GO) terms most strongly enriched (lowest p-value) in the set of differentially expressed genes.** The complete set of statistically overrepresented terms is given in Additional file 3: Table S3.

this is accomplished with relatively small changes in myosin expression [29]. The very large changes observed in our study may indicate a distinct mechanism of molt-associated muscle remodeling in *C. finmarchicus.* The remaining 12 strongly up-regulated genes fell into 10 categories, including a Torso-like sequence that we selected for detailed expression profiling (see below). Of the genes showing higher expression on day 3, 76 of the 100 genes with the lowest p-values had no blast hit or only matched predicted proteins with no associated GO terms. Among the 24 annotated genes, two were cuticle proteins, and the remaining genes were diverse, each associated with different GO terms. One of these genes was a fatty acid elongase that was similar to a gene we have measured previously (ELOV) and chose for more detailed expression profiling (see below).

Twelve genes were selected for detailed expression profiling in the time series study (Table 1) based on their (1) differential expression in the Illumina study, (2) expression patterns in our previous studies, and/or (3) likely role in reproduction or endocrine regulation of molting. Among these, ELOV (elongation of very long-chain fatty acids) and FABP (fatty acid binding protein) are presumed to function in lipid synthesis and transport; these genes also exhibit high expression in active C5 copepodids with small oil sacs [20]. Vtg is part of the vitellogenin family of lipoproteins that includes major egg yolk proteins (also called apolipocrustaceins in crustaceans) as well as proteins that play non-reproductive roles [30,31]. Fem-1 is an ankyrin domain-containing protein that is necessary for spermatogenesis in *C. elegans* [32]. Fem-1 function is poorly understood in other animal groups, although a Fem-1 homolog has also been suggested to play a role in mammalian prostate development [33]. Torso-like regulates embryonic patterning, developmental timing, body size, and ecdysone secretion [34,35]. ERR (estrogen-related receptor), EcR (ecdysone receptor), FTZ-F1 (FTZ factor 1), HR3, HR38a, HR38b, and HR78 (all named as hormone receptors, but without known ligands) are members of the nuclear receptor family of transcription factors. Nuclear receptors were preferentially selected for qPCR because genes in this family are known to regulate arthropod molting, and commonly show dynamic expression profiles [13,17]. Thus, we considered it likely that genes exhibiting modest differences in expression between the two time points sampled via Illumina might exhibit large changes at other times during the molt stage. The EcR sequence was previously described [20], while the other nuclear receptors were annotated according to a likelihood-based phylogenetic analysis that indicated strong support for grouping with homologs from *Drosophila* and *Daphnia* (Additional file 4: Figure S1). HR38a and HR38b appear to have resulted from a duplication of an ancestral HR38 sometime after the divergence of the copepod and cladoceran lineages (Additional file 4: Figure S1).

**Table 1 Sequences (Trinity "components") identified in the Illumina study and profiled using qPCR**

| | | Illumina | | | qPCR | |
|---|---|---|---|---|---|---|
| Name | Component, accession | Log FC | Log CPM | FDR | Log FC | Range FC |
| FABP | comp254629_c0 ES387222 | −2.24 | 8.57 | 4.55E-09 | −3.45 | 15.6 |
| ELOV | comp267151_c1 ES387246 | −4.72 | 9.31 | 4.36E-16 | −4.74 | 30.0 |
| HR78 | comp265637_c0 KJ026941 | −2.06 | 0.79 | 1.02E-04 | −1.26 | 12.8 |
| ERR | comp273411_c0 KJ026943 | −2.17 | 4.89 | 1.14E-08 | −1.84 | 5.3 |
| HR3 | comp264460_c0 KJ026938 | 3.21 | 4.48 | 2.59E-14 | 3.84 | 64.2 |
| HR38a | comp270260_c0 KJ026939 | 2.65 | 3.63 | 8.58E-12 | 2.25 | 5.9 |
| Vtg | comp274635_c1 KJ026944 | 3.20 | 8.14 | 3.45E-10 | 1.15 | 597.9 |
| Torso-like | comp266966_c0 KJ026945 | 9.26 | 7.40 | 1.19e-84 | 8.24 | 1024.2 |
| FTZ-F1 | comp268775_c0 KJ026942 | 3.25 | 5.97 | 3.05E-16 | 4.25 | 32.0 |
| HR38b | comp267178_c0 KJ026940 | 2.26 | 4.27 | 6.58E-10 | 1.07 | 7.8 |
| EcR | comp263662_c0 EF583877 | 1.10 | 3.13 | 0.39 | 1.06 | 3.4 |
| Fem1 | comp269877_c0 KJ026947 | 4.43 | 7.08 | 1.67E-17 | 1.33 | 7.6 |

$Log_2$-transformed fold change (FC), $log_2$-transformed counts per million (CPM), and false discovery rate (FDR) refer to the Illumina study comparing expression between 3 and 10 days after the penultimate molt. $Log_2$-transformed average FC between days 3 and 10 also reported for the qPCR study, as well as the untransformed FC between the minimum and maximum qPCR expression values observed throughout the C5 copepodid stage. Log-transformed FC is reported as a positive number for genes exhibiting greater expression later in the molt cycle (day 10) and a negative number for genes exhibiting greater expression earlier in the molt cycle (day 3).

Some of the genes (but not all) chosen for qPCR analysis were associated with statistically overrepresented GO terms: metal ion binding (ERR, FTZ-F1), retinoic acid receptor activity (HR38a, HR38b), steroid hormone receptor activity (HR38a, HR38b, EcR), retinoic acid receptor signaling pathway (HR38a, HR38b), steroid hormone mediated signaling pathway (HR38a, HR38b, EcR), and nuclear euchromatin (EcR). While GO enrichment analysis can be useful in identifying functionally related groups of co-expressed genes, automatic annotation is not without limitations, particularly for non-model organisms [36]. Some of these limitations can be seen in the above assignments. For example, all of the nuclear receptors that we profiled contain a DNA-binding domain that would be expected to form complexes with zinc ions; however, only a subset of the receptors was assigned with the "metal ion-binding" term.

### Detailed expression profiling

The Illumina and qPCR results showed excellent agreement; each of the genes measured by both methods showed similar changes in expression between days 3 and 10 ($r^2 = 0.894$, $p < 0.0001$ for correlation of log-transformed fold changes between days 3 and 10 from the Illumina and qPCR studies; Table 1). Expression of FABP and ELOV peaked on day 3 and decreased thereafter by an average factor of 16 and 30, respectively (Figure 4a,b). Like FABP and ELOV, expression of HR78 and ERR also peaked early in the molt cycle and decreased steadily over time (Figure 4c,d). In contrast, HR3, HR38a, Vtg, and Torso-like each exhibited increasing trends in expression over time (Figure 4e-h). While expression of HR3, HR38a, and Vtg increased fairly monotonically over time, Torso-like expression actually decreased over the first 7 days, increased dramatically by a factor of 768 between days 7 and 10, and remained high after day 10 (Figure 4h). FTZ-F1 and HR38b exhibited U-shaped expression patterns with minimum values observed during the middle of the molt cycle (Figure 4i,j). EcR was not differentially expressed within the Illumina study, and while variability among replicates was very high, there was a weak, inverse U-shaped, average expression pattern with maximum values observed during the middle of the molt cycle (Figure 4k). Finally, expression of Fem-1 showed a more complicated pattern of expression, remaining nearly constant over the first 8 days except for a transient peak in expression on day 6 and an increasing trend after day 8 (Figure 4l).

### Gene expression in copepods of known molt phase

In animals of known molt stage, expression profiles were broadly consistent with those observed between the middle and late periods of the molt cycle during the time series study (Figure 5). ELOV expression decreased between pre- and post-apolysis samples ($p = 0.0093$; Figure 5a) whereas expression of Torso-like, FTZ-F1, HR38b, EcR, and Fem-1 increased between pre- and post-apolysis samples ($p < 0.05$; Figure 5b-f). Although variability among replicates was high and temporal patterns in Fem-1 and EcR were weak during the time series study (Figure 4k,l), these two genes exhibited the strongest differentiation between pre- and post-apolysis samples of all the examined genes (Fem-1: $t = 6.9$, $p = 0.0010$; EcR: $t = 4.4$, $p = 0.0073$). These results suggest that expression of Fem-1 and EcR is more closely synchronized with the molt cycle than with

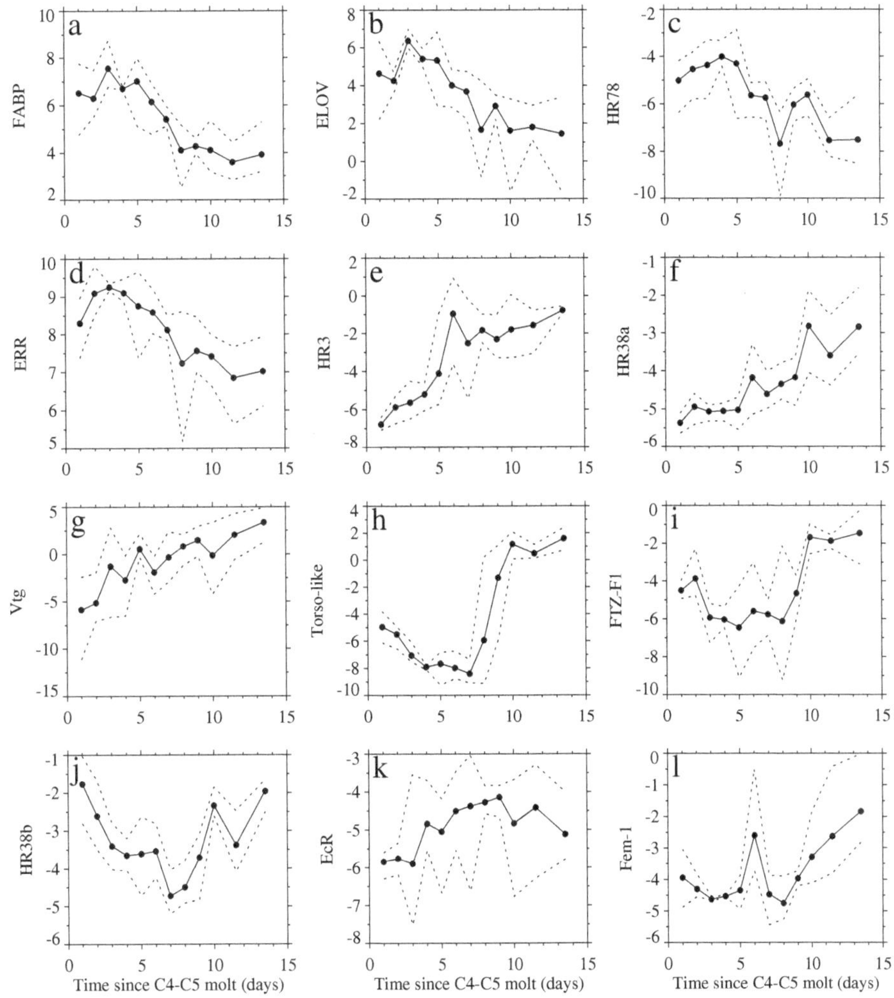


**Figure 4 Average (filled circles) and minimum and maximum (dotted lines) of *Calanus finmarchicus* C5 copepodid gene expression measured via qPCR during the time series study.** Each of the subpanels **(a-l)** represents the expression profile for an individual gene, which is specified on the y-axis. Average, minimum, and maximum values were computed for n = 3 samples per day for days 1–10; days 11–12 and 13–14 were combined to obtain n = 3 samples for these combined 2-day periods (note that each sample consisted of 3 homogenized copepods). Expression was normalized by 4 housekeeping genes and $log_2$ transformed. Negative numbers indicate lower expression than the geometric mean expression of the 4 housekeeping genes (see Methods for additional detail).

age within stage, and that variability in the timing of progression through the molt cycle (Figure 2b) caused high variability among replicates during the time series study of Fem-1 and EcR expression.

### Multivariate analysis of gene expression

Principal component analysis (PCA) was conducted on expression of the 12 normalized genes for 36 samples (approximately 3 replicates per day in the time series study) to better understand overall patterns of expression of this set of genes (Figure 6). The first principal component explained half of the observed variability and captured the dominant linear trends with time for many of the genes: expression of HR3, HR38a, and Torso-like increased over time, while expression of HR78, FABP, ELOV, and ERR decreased over time. The second principal component accounted for an additional 19% of the variability, and captured a U-shaped pattern over time observed in the expression of FTZF-1, HR38b, and EcR (where EcR has a roughly inverted U-shape). The average daily values of the first two principal component scores clustered into 3 time periods: early (days 1–5), middle (days 6–9) and late (days 10–14) (Figure 6c). These time periods, derived from the gene expression data, were consistent with our morphological observations. All C5 copepodids sampled during the early period had pre-apolysis jaws (Figure 2b), indicating they were in the early phase of the molt cycle. The middle period was characterized by the appearance of post-apolysis jaws (Figure 2b). Adults had only just begun to appear at the onset of the late period (Figure 2a), and

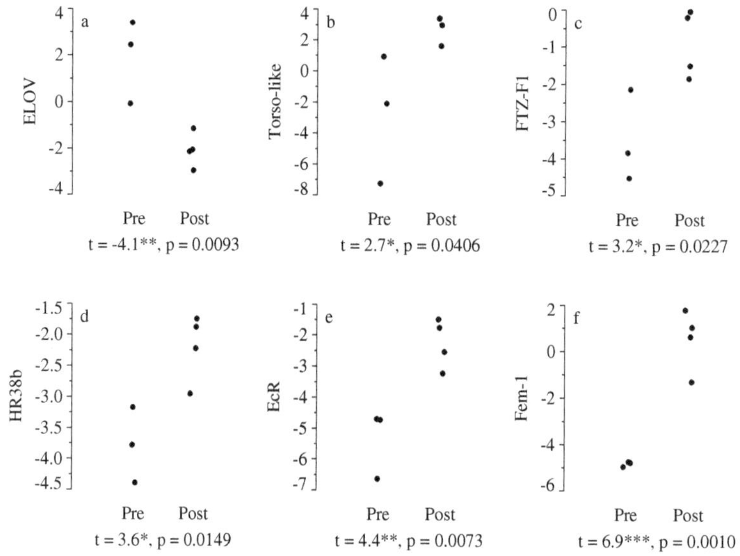


**Figure 5 Gene expression of pre- and post-apolysis *Calanus finmarchicus* C5 copepodids measured via qPCR 8 and 10 days after the penultimate molt.** Each of the subpanels **(a-f)** indicates the expression level of an individual gene, which is specified on the y-axis. Data consist of $n = 3$ samples for the pre-apolysis group and $n = 4$ samples for the post-apolysis group (each sample consisted of 3 homogenized copepods of known molt stage). Expression was normalized by 4 housekeeping genes and $\log_2$ transformed. Negative numbers indicate lower expression than the geometric mean expression of the 4 housekeeping genes (see Methods for additional detail). Results shown for only those genes with significantly different expression between pre- and post-apolysis (Student's t-test, $p < 0.05$).

during this time, nearly all C5 copepodids had differentiated gonads (Figure 2b).

### Gene expression marker of late progression toward the terminal molt

Determining within-stage progress toward the terminal molt was of particular interest to us, so we sought to develop a gene expression indicator that strongly discriminated between the early (days 1–5) and late periods (days 10–14) within the molt cycle. The expression ratio of all possible pairs of un-normalized genes from the time-series study ($n = 12$ genes) were first calculated and $\log_2$ transformed. Then the Student's t-statistic comparing the ratio between the early period ($n = 15$ samples) and late period ($n = 9$ samples) was calculated for each gene pair. The gene pair with the highest t-statistic was considered the best discriminator between the early and late periods. Of the 66 possible gene expression ratios, the ratio of Torso-like to HR38b had, by far, the highest t-statistic ($t = 24.1$, $p < 0.0001$; note that $p < \alpha$, where $\alpha = 0.05/66 = 0.00076$, the Bonferroni adjusted cutoff value; Figure 7a). During the early period, HR38b expression was an average 12.2 times higher than that of Torso-like, and during the late period, Torso-like expression was an average 12.5 times higher than that of HR38b (Figure 7a). The middle period (days 6–9) was characterized by a dramatic change in the ratio (Figure 7b). While Torso-like alone provides good discrimination between the early and late periods (Figure 4h), the ratio of Torso-like to HR38b provides better discrimination because the ratio is less variable during the early period when expression levels of both genes decrease simultaneously.

## Discussion

The lipids stored in the *C. finmarchicus* oil sac provide energy to support basal metabolism during diapause, as well as molting and reproduction. The present study provides, to our knowledge, the first dynamic profile of oil sac volume in *Calanus spp.* C5 copepodids progressing toward the terminal molt. Oil sac volume is strongly correlated with wax ester content in *C. finmarchicus* C5 copepodids [4]. While triglycerides are typically present in relatively small amounts, they appear to serve as a labile energy source that is utilized both during short periods of starvation and during egg production [4,37]. Thus, it will be of interest to profile wax ester and triglyceride content in future studies of *C. finmarchicus* development, as well as to identify molecular markers associated with synthesis and utilization of each lipid type. During our study, oil sac volume increased over time, presumably to fuel molting, reproduction and adult metabolic needs [5-7], and genes associated with lipid accumulation (FABP and ELOV) peaked on day 3 and decreased afterward. In a field study of *C. finmarchicus* C5 copepodids, we showed that expression of FABP and ELOV was highest in active animals with small oil sacs [20]. If these copepods were early developing C5 copepodids, then their high FABP and ELOV expression is consistent with the high FABP and ELOV

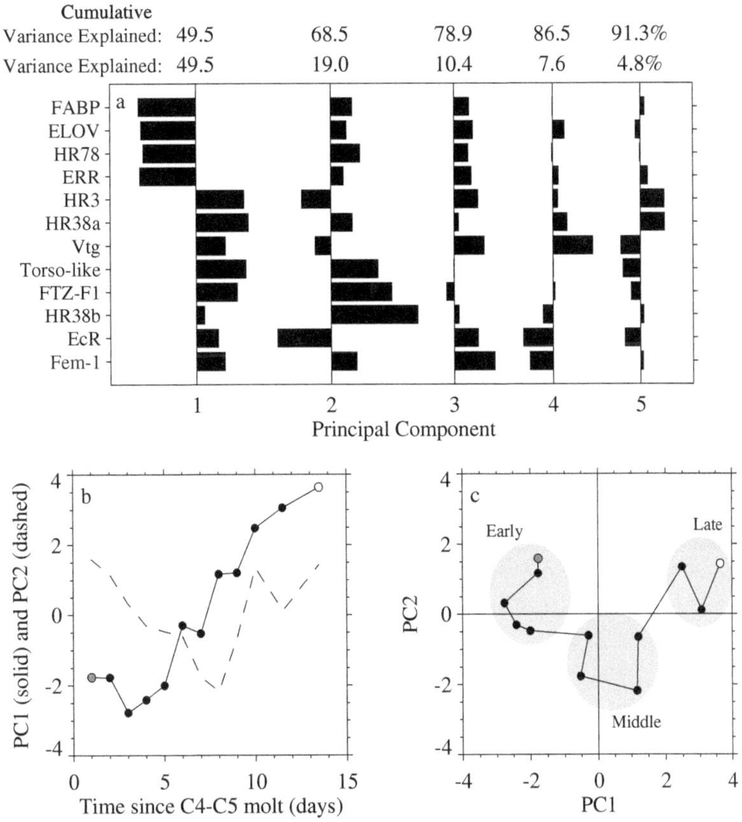


**Figure 6 Multivariate analysis distinguishes three periods of gene expression within the molt cycle.** Results of principal component (PC) analysis of gene expression data from the time series study: **(a)** eigenvectors for the first 5 PCs, **(b)** time series of average PC scores for the first (filled circles) and second (dashed line) PC, and **(c)** scatterplot of average PC scores for the first 2 PCs. Gray-filled and open circles in **(b)** and **(c)** indicate the start and end of the time series, respectively. Gray ellipses in **(c)** indicate early (days 1–5), middle (days 6–9), and late (days 10–14) periods.

expression observed in the early developing C5 copepodids from our culture study. In this previous study [20], we assumed that changes in FABP and ELOV expression were associated with diapause preparation, but the present study shows that changes in the expression of these genes are more generally associated with lipid accumulation, which is necessary both for diapause and progression toward the terminal molt. ERR expression also peaked on day 3 and decreased afterward. While the specific function of this nuclear receptor in *C. finmarchicus* is unknown, ERR is an important regulator of energetic homeostasis and lipid metabolism [38,39].

During progression through the C5 stage, the proportion of copepodids with differentiated gonads increased. While nearly half (44%) of the copepodids with differentiated gonads possessed testes, only 25% of the emerged

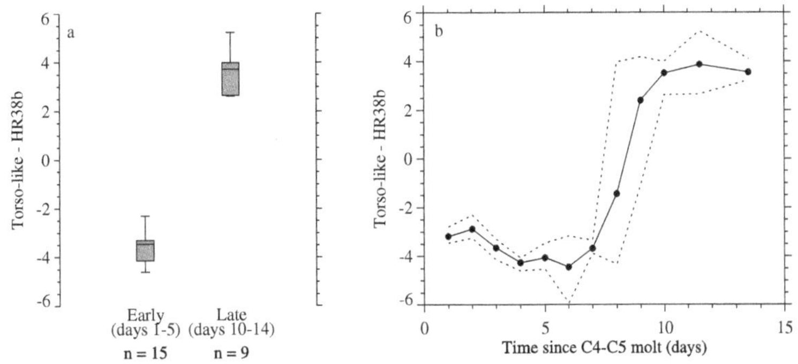


**Figure 7 High Torso-like to HR38b expression indicates late molt stage.** **(a)** $Log_2$-transformed ratio of Torso-like to HR38b expression during the early (days 1–5) and late (days 10–14) periods of the time series study. **(b)** Average (filled circles) and minimum and maximum (dotted lines) of $log_2$-transformed Torso-like to HR38b expression ratio.

adults were male. Differences in the sex ratio of wild copepod populations are frequently attributed to differential mortality (e.g., [40]), but mortality was extremely low in the culture study. Sex ratio in *Calanus* spp. can be influenced by food availability, food quality, population density and other environmental factors, and cultured populations are usually biased toward females ([41] and references therein). Because we sampled most of the copepodids prior to the terminal molt, we cannot be sure what proportion of the population would have eventually emerged as adults. One possibility is that males develop more slowly, which is weakly supported by our data (Figure 2a). A second intriguing possibility is that sex change accounts for the observed difference in sex ratio, as has been previously suggested for *C. finmarchicus* (reviewed by [42]).

We profiled the expression of three genes, Vtg, Fem-1, and Torso-like, each of which is associated with reproductive development in other animals. We identified several Vtg-like transcripts, most of which did not exhibit differential expression between days 3 and 10. The Vtg transcript profiled in this study increased over time in the C5 stage, but decreased in both male and female adults (adult expression not shown). This differs from Vtg expression profiles described in other copepods in which expression is highest in adult females where it plays a role in egg yolk production [43,44]. Closely-related genes (e.g., Crossveinless d) serve non-reproductive developmental roles in insects [31]. While we do not know the specific role of the Vtg-like gene we have profiled in *C. finmarchicus,* we consider it likely that it serves a developmental role unrelated to egg yolk production. Fem-1 expression exhibited high variability among replicates, weak temporal patterns, and a strong increase after apolysis. This gene helps to regulate spermatogenesis in *C. elegans* [32], so the weak temporal variability observed within the culture may be due to sexually dimorphic expression. The strong discrimination between pre- and post-apolysis animals of similar time in stage may indicate a transitory peak in expression not resolved in the time series study. Torso-like has been studied most extensively in *Drosophila,* in which it is transmitted from the ovary to the oocyte to regulate development of the head and tail regions of the resulting embryo; however, this role of Torso-like embryonic patterning is likely a derived function restricted to a few insect lineages [45]. While the function of Torso-like is unknown in *C. finmarchicus*, the large dynamic range in expression with a maximum in post-apolysis animals is consistent with a conserved role in regulation of molting hormones, as has been shown in insects: [34,35].

EcR did not exhibit differential expression between days 3 and 10 in the Illumina study. In the time series study, EcR expression was greatest during the middle period, but temporal trends were weak. This weak temporal pattern can be partially attributed to asynchrony within the molt stage obscuring developmental changes, as evidenced by the strongly differentiated expression of EcR before and after apolysis (Figure 5e). In insects, EcR regulates expression of other nuclear receptors including HR78 and HR3, and HR3 in turn induces expression of FTZ-F1 (reviewed by [13]). In *C. finmarchicus,* several other nuclear receptors (HR38b, HR3, FTZ-F1) exhibited dynamic profiles during the time series study. The dynamic expression profiles of these genes may by driven primarily by changes in ecdysteroid concentrations rather than changes in the expression of EcR. In insects, HR78 and HR38 both oppose EcR signaling [46,47]. In our study, HR78 expression was highest early in the molt cycle, with peak expression on day 4. This preceded the peak in EcR expression observed on day 9, which may have prevented premature activation of ecdysis pathways. Of the two HR38 genes in *C. finmarchicus*, HR38a increased over time, and HR38b exhibited a U-shaped pattern similar to FTZ-F1 that was broadly in opposition to EcR. Expression patterns of HR38b and FTZ-F1 were closely associated with molt progression, exhibiting a strong increase after apolysis (Figure 5c,d). HR3 expression generally increased over time and showed a transient peak on day 7. Among copepods collected on days 8 and 10, HR3 expression was highly variable prior to apolysis (not shown), which is consistent with an ephemeral peak in HR3 expression shortly before apolysis.

We identified several genes that we expect will be useful for identifying wild C5 copepodids that are in the late stages of preparation for the terminal molt. The copepods used in this study were all taken from a long-term continuous culture in which the animals are reared at a constant temperature, with a reliable food supply, without any exposure to predators, and at higher densities than they generally occur in nature. For reasons that are not known, *Calanus* spp. does not undergo diapause in culture, and we do not understand in what other ways the physiology and corresponding gene expression of cultured and wild C5 copepodids differ. Therefore, it will be necessary to compare the gene expression profiles observed in our culture study with wild populations that are preparing for the terminal molt. We anticipate that Torso-like and HR38b, whose ratio changes dramatically during C5 development, may prove especially useful for identifying copepods nearing the terminal molt. The lipid accumulation hypothesis of diapause initiation [48-50] predicts that copepods that skip diapause and proceed directly to the terminal molt have lower lipid reserves than those preparing for diapause. With a biomarker that predicts late molt stage, the lipid volume of copepods preparing for the terminal molt can be directly examined in the field. Moreover, this biomarker can be used to identify animals in the field that are *not* actively preparing for the terminal molt, of which some will be preparing for diapause. Within this subpopulation of C5s, we anticipate that additional transcriptional profiling and gene expression studies will identify genes directly associated with preparation for diapause during the C5 stage. The development of robust gene expression biomarkers of these two developmental pathways will ultimately enable novel studies of the intrinsic and extrinsic factors that govern diapause initiation and termination in *Calanus finmarchicus.*

## Conclusions

This study provides the first dynamic profiles of gene expression through a copepod molt cycle. In addition to processes of growth and ecdysis that are common to all molt stages, the *C. finmarchicus* C5 stage is also characterized by lipid storage and gonad development. These processes are interrelated in the sense that lipid stores provide energy to fuel ecdysis and reproduction. Broadly, we have observed that FABP and ELOV expression is high early in the C5 stage, and decreases over time as lipids accumulate. Several nuclear receptors exhibited dynamic expression profiles consistent with a signaling cascade triggered by ecdysteroids. We measured, for the first time, gene expression in copepods of known molt stage and identified genes that exhibit dramatic changes in expression during apolysis. These include genes with presumed roles in lipid synthesis, molt regulation and gonad development, which is consistent with a coupling of these processes. Finally, we identified the ratio of Torso-like to HR38b as a biomarker that strongly differentiates between early and late development within the C5 copepodid stage.

## Methods

### Culturing

Copepods were collected from a long-term continuous culture of *C. finmarchicus* maintained at the NTNU/SINTEF SeaLab facility in Trondheim, Norway [19]. Copepods were reared at 8–10°C in 280-l polyester containers through which filtered natural seawater from nearby Trondheimsfjord (collected at 70 m depth) was continuously passed. All copepods were fed a diet of *Dunaliella tertiolecta, Isochrysis galbana,* and *Rhodomonas baltica* at levels > 150 μg C $l^{-1}$. During May 2012, ~1900 stage C4 copepodids were removed from the culture and transferred to 2.5-l containers (40 individuals per container, 48 containers). Every day for 9 consecutive days, each copepod was gently removed with a 17 $cm^3$ ladle, transferred to a clear plastic spoon, and examined under a stereomicroscope. If an individual had molted from C4 to C5 during the preceding ~24 hours, it was transferred to a new 2.5-l container and not examined again until its sample date (see below). A total of 34 2.5-l containers were filled with 35–47 (average of 38) C5 copepodids, resulting in 1306 individuals for whom the date of the penultimate (C4 to C5) molt was known to within a day.

### Sampling

Containers with C5 copepodids were assigned randomly to particular sample dates to allow data collection on each day after the penultimate molt, including days 1–18 for a time series study and days 3 and 10 for Illumina-based transcriptional profiling (see below). All animals from a single container with the same penultimate molt date were sampled simultaneously. Individuals were removed from the container with a ladle, staged on a depression slide, photographed alive with a Canon EOS-20D camera attached to a Zeiss Stemi 2000C stereomicroscope, and preserved either in RNAlater (Ambion) for quantitative polymerase chain reaction (qPCR) or transcriptional profiling, or in 10% buffered formalin for assessment of jaw phase or gonad stage. For the time series study, individuals from each container were randomly assigned to one of three analysis types: jaw phase (n = 10 copepods), gonad stage (n = 10 copepods), or qPCR analysis (n = 15 copepods). These sample sizes could not be maintained during the last few days of the time series study when the occurrence of C5 copepodids was low (because many molted into adults prior to sampling), so data from these last days were combined for all analyses. For qPCR and Illumina analyses, replicates of 3 copepods each were preserved in RNAlater.

### Morphometrics, jaw phase, and gonads

Prosome length and width and oil sac volume of each sampled copepod were estimated with measurements obtained from digital photographs of the live animals (after Tarrant et al. [20]). Measured lengths were calibrated with repeated photographs of a stage micrometer taken during sampling. Jaw phase was assessed by extracting the mandible under a stereomicroscope, mounting it in glycerine under a cover slip, and photographing the gnathobase under a Zeiss Axiovert 40 inverted compound microscope outfitted with differential interference contrast optics and a Zeiss AxioCam ICc1 camera. Jaws were classified after Miller et al. [11] and Crain and Miller [51]. Apolysis, the process by which the epidermis separates from the cuticle, is an important milestone in molt preparation; we therefore refer to pre-apolysis copepodids as those with jaws in the post-molt, late post-molt, interphase, or columnar cell tooth-forming phases, and post-apolysis copepodids are those with jaws in any of the other late tooth-forming phases (apolysis, tooth molds, silicified teeth, or extricated teeth). To assess gonad stage, C5 copepodids were stained by passing them through 5 baths of seawater serially diluted with deionized water (100, 75, 50, 25, 12.5% seawater for 20 minutes each), a deionized water bath of 3% Fast Green (20 minutes), 3 baths of deionized water, and 5 baths of increasing ethanol concentration (15 minutes in 15, 40, and 75% ethanol, 37 minutes each in 95 and 100% ethanol); final preservation was in terpineol. Gonads were viewed with an inverted compound microscope and classified according to Crain and Miller [52]. Differentiated gonads are defined as those from which an individual's gender can be determined (i.e., developing ovaries and testes).

### RNA extraction and library preparation

Total RNA was extracted from copepods using the Aurum Fatty and Fibrous Tissue Kit (Bio-Rad). RNA purity and yield were assessed using a Nanodrop spectrophotometer. For the time series study, on-column DNase digestion was used to remove genomic DNA, and RNA quality was assessed using denaturing agarose gels. For the Illumina study, RNA quality was assessed using a Bioanalyzer. Samples were submitted to the Genomic Services Laboratory at HudsonAlpha (Huntsville, AL), where libraries were synthesized using Illumina TruSeq reagents. Four libraries were constructed from each of days 3 and 10, and each library was prepared using total RNA extracted and pooled from enough copepods to yield at least 4 μg of RNA (3–6 copepods). For transcriptome assembly, an additional library was constructed from equal amounts of RNA pooled from each of these eight samples (2 μg total) along with 2 μg of RNA from wild C5 *C. finmarchicus* collected in Trondheimsfjord, Norway during May 2012. RNA from wild copepods was included in the library to facilitate future field-based transcriptional profiling studies.

### Illumina-based sequencing and data analysis

For transcriptome assembly, samples were sequenced with 100 base pair (bp) paired-end reads with a depth equivalent to ½-lane on an Illumina HighSeq2000. For expression profiling, samples were sequenced with 50 bp paired-end reads with a depth equivalent to ¼-lane per sample. Adapter sequences and bases with low quality scores (< phred 20) were removed using Trimmomatic [53] in paired-end mode. After trimming, reads at least 50 bp long were retained for assembly. A transcriptome was assembled *de novo* using Trinity version r2012-06-08 [54] with default parameters. Transcripts were annotated using Blast2Go [55] with a threshold e-value of $10^{-6}$. Post-assembly analysis was conducted using scripts bundled within the Trinity package. Reads were mapped to the reference transcriptome and quantified using the RSEM software package [56]. Read counts were TMM-normalized (trimmed mean of M-values) to account for differences in library size [57] and then FPKM-normalized (fragments per feature kilobase per million reads mapped) to account for differences in transcript length. Transcripts exhibiting differential expression between the two time points were identified using EdgeR software [58]. Default parameters were used, including a p-value cutoff for false discovery rate (FDR) of 0.001 and a minimum 4-fold change in expression. Statistically enriched gene ontology (GO) terms were identified using the GOseq R Bioconductor package [59].

### Cloning and phylogenetic analysis

Portions of transcripts of interest (672–1313 bp) were amplified by PCR (as in [20]), both to verify the assembled sequence and to provide standards for qPCR. All primers are provided in Additional file 5: Table S4. Amplicons were cloned into pGEM-T Easy (Promega) and sequenced. To improve the provisional annotation of several transcripts within the nuclear receptor superfamily, the cloned *C. finmarchicus* sequences were aligned with the complete nuclear receptor sets from human, *Drosophila melanogaster* and *Daphnia pulex,* excluding the atypically structured nuclear receptors (Family 0). Amino acid sequences were aligned using ClustalW, and maximum likelihood analysis was conducted with RAxML v7.0.4 [60] using the JTT + I + G + F model.

### Quantitative PCR (qPCR)

Complementary DNA (cDNA) was synthesized from 450 ng of total RNA in a 20-μl reaction using the Iscript cDNA synthesis kit (Bio-Rad). The 20-μl reaction was diluted with 30 μl of molecular biology grade water such that each microliter of diluted cDNA corresponded to 9 ng of total RNA. qPCR was performed using SsoFast EvaGreen Supermix (Bio-Rad) on an iCycler iQ Real-Time PCR detection system (Bio-Rad). Primer sequences for ELOV, FABP and EcR were previously described [20], and all other primer sequences are given in Additional file 5: Table S4. Standards were prepared from a serially diluted plasmid containing the amplicon. All samples and standards were run in duplicate wells. The 20-μl qPCR reaction contained 10 μl master mix, 8 μl molecular biology grade water, 1 μl diluted cDNA and 1 μl 10 nM primers. Cycling parameters were: 95°C for 2 min followed by 40 cycles of 95°C for 5 s and 63°C for 10 s. After cycling, the products were subjected to melt curve analysis to ensure that only a single specific product was amplified.

Transcript expression was calculated relative to the standard curve and $\log_2$-transformed. Expression was further adjusted by subtracting a normalization factor equal to the geometric mean of 4 genes that exhibited relatively constant levels of expression after [61]. Three of these normalizer genes were selected from the Illumina data to have moderate expression levels and a low coefficient of variation among samples (TSG: KJ026946 tumor susceptibility gene 101 protein-like, UBX: KJ026948 UBX domain-containing protein 6-like, and SNP: KJ026949 snurportin 1-like). The fourth normalizer gene (EF-1α: ES414812, similar to elongation factor 1 alpha) has been used in previous qPCR studies of *C. finmarchicus* [19] and exhibited stable expression in our study.

Gene expression was measured with qPCR on days 1–16 of the time series study. Three biological replicates were measured on days 1–10, 2 replicates were measured on days 11, 13, and 16, and 1 replicate on days 12, 14, and 15. Results from days 11–12 and days 13–14 were combined in all figures. Results from days 15 and 16 were excluded from further analysis because those days exceeded the median C5 stage duration of the culture.

Because molt stage progression was not synchronous within the culture (see Results), we used RNAlater-preserved samples collected on days 8 and 10 to simultaneously assess jaw phase and measure gene expression in the same individual C5 copepodids. Mandibles were extracted and examined, and after jaw phase was determined, the rest of the body was grouped with copepods that had similar jaw phases. Each replicate sample for qPCR consisted of total RNA extracted and pooled from 3 copepods with either (1) jaw phases occurring immediately before apolysis (interphase or columnar cell tooth forming phases) or (2) jaw phases occurring after apolysis (apolysis, tooth molds, silicified teeth, or extricated teeth). We analyzed 3 and 4 replicates each for pre- and post-apolysis jaw phases, respectively.

### *Availability of supporting data*

The cloned DNA sequences (Genbank accessions KJ026938-KJ026947), raw Illumina sequences, and Transcriptome Shotgun Assembly are archived at NCBI (NCBI BioProject: PRJNA231164 [GenBank:PRJNA231164], Transcriptome Shotgun Assembly DDBJ/EMBL/GenBank Accession GBFB00000000. The version described in this paper is GBFB01000000): Other data sets supporting the results of this article are included within the article and its additional files.

## Additional files

**Additional file 1: Table S1.** Tab-delimited file containing normalized counts (FPKM) and RSEM statistical analysis of Trinity components differentially expressed between days 3 and 10. Column 1: Comp Name, the name of the differentially expressed Trinity component. Column 2: logFC, the $log_2$-transformed fold change in expression; positive numbers indicate higher expression on day 10 of the time series relative to day 3. Column 3: logCPM, $log_2$-transformed counts per million. Column 4: p-value, indicating significance of difference in expression between days 3 and 10. Column 5: FDR, false discovery rate. Columns 6-9: normalized expression values for biological replicates collected on day 3 (EC1.genes-EC4.genes). Columns 10-13: normalized expression values for biological replicates collected on day 10 (LC1.genes-LC4.genes). Column 14: # annotated seqs: the number of transcripts ("seqs") corresponding to a given line (component) that are annotated in Additional file 2: Table S2. Columns 15-22: the blast2GO-based annotation based on the first corresponding sequence present in Additional file 2: Table S2 (if any).

**Additional file 2: Table S2.** Tab-delimited file containing annotation of transcripts with a positive blastx hit (e-value < $10^{-6}$) that were differentially expressed between days 3 and 10. Column 1: Trinity transcript name. Column 2: Sequence description. Column 3: Sequence length. Column 4: Number of blast hits. Column 5: Lowest e-value of significant blast hit. Column 6: Mean percent similarity. Column 7: Number of associated GO terms. Column 8: List of associated GO terms.

**Additional file 3: Table S3.** Microsoft Excel file containing the complete list of gene ontology (GO) terms that are statistically overrepresented in the set of genes differentially expressed between days 3 and 10.

**Additional file 4: Figure S1.** Maximum likelihood tree of nuclear receptors from human (*Homo sapien*; Hs), *Drosophila melanogaster* (Dm) and *Daphnia pulex* (Dp) with newly-identified *Calanus finmarchicus* (Cf) nuclear receptors shown in red. Nuclear receptors are typically grouped into 6 families (a seventh family with contains genes with atypical structure is not shown), some of which are further divided into subfamilies [62]; these designations are indicated on the right side of the figure. Accession numbers for *C. finmarchicus* sequences are given in Table 1. All other accession numbers are given in Additional file 1 of [63] Tree is unrooted. Values to the right of nodes indicate percent of 1,000 bootstraps. Scale bar at bottom indicates the number of amino acid substitutions per site. Trees were visualized using FigTree v1.1.2 (http://tree.bio.ed.ac.uk/software/figtree/).

**Additional file 5: Table S4.** Word document. Primers used for cloning of plasmid standards and for quantitative PCR.

### Abbreviations

Bp: Base-pair; C4 and C5: Fourth and fifth copepodid stage, respectively; cDNA: Complementary DNA; CPM: Counts per million; DNA: Deoxyribonucleic acid; EcR: Ecdysone receptor; EF-1α: Elongation factor 1 alpha; ELOV: Elongation of very long-chain fatty acids; ERR: Estrogen-related receptor; FABP: Fatty acid binding protein; FDR: False discovery rate; FPKM: Fragments per feature kilobase per million reads mapped; FTZ-F1: Fushi tarazu transcription factor 1; GO: Gene ontology; HR: Hormone receptor (e.g., HR38: hormone receptor 38); nr: Non-redundant (database); PC(A): Principle component (analysis); PCR: Polymerase chain reaction; qPCR: Quantitative PCR; RNA: Ribonucleic acid; RXR: Retinoid-x-receptor; SNP: Snurportin; TMM: Trimmed mean of M-values; TSG: Tumor suppressor gene; Vtg: Vitellogenin.

### Competing interests

The authors declare that they have no competing interests.

### Authors' contributions

AMT and MFB conceived of the study, analyzed data and wrote the article. AMT conducted Illumina analysis and qPCR assays. MFB conducted morphological and statistical analyses. AMT, MFB and DA conducted the culture study. BHH, DA, TN and AO contributed to the study design, prepared culture and sampling infrastructure, and facilitated sampling during the culture study. All authors reviewed and contributed to the final manuscript. All authors read and approved the final manuscript.

### Acknowledgements

This work was supported by grant number OCE-1132567 from the National Science Foundation to MFB and AMT. Additional supported was provided by WHOI Early Career Scientist Awards provided to MFB and AMT. High-performance computing was provided through the Extreme Science and Engineering Discovery Environment (XSEDE), which was supported by National Science Foundation grant number OCI-1053575. We thank Nadine S.J. Lysiak for logistical and sampling support, and Amalia Aruda Almada, Morgan Rubanow and Emily Conklin for laboratory assistance. Amy Maas provided bioinformatics advice. We also thank three anonymous reviewers for constructive comments.

### Author details

[1]Biology Department, Woods Hole Oceanographic Institution, 45 Water Street, Woods Hole, MA 02543, USA. [2]SINTEF Materials and Chemistry, Environmental Technology, N-7465 Trondheim, Norway. [3]BioTrix, N-7022 Trondheim, Norway. [4]Department of Biology, Norwegian University of Science and Technology, N-7491 Trondheim, Norway.

## References


1. Saumweber WJ, Durbin EG: **Estimating potential diapause duration in *Calanus finmarchicus*.** *Deep-Sea Res Pt II* 2006, **53:**2597–2617.
2. Hirche HJ: **Diapause in the marine copepod, *Calanus finmarchicus* – a review.** *Ophelia* 1996, **44:**129–143.
3. Marshall S, Orr A: *The Biology of a Marine Copepod, Calanus finmarchicus (Gunnerus).* Edinburgh: Oliver & Boyd, London; 1955:188.
4. Miller CB, Morgan CA, Prahl FG, Sparrow MA: **Storage lipids of the copepod *Calanus finmarchicus* from Georges Bank and the Gulf of Maine.** *Limnol Oceanogr* 1998, **43:**488–497.
5. Rey-Rassat C, Irigoien X, Harris R, Carlotti F: **Energetic cost of gonad development in *Calanus finmarchicus* and *C. helgolandicus*.** *Mar Ecol Prog Ser* 2002, **238:**301–306.
6. Sargent JR, Falk-Petersen S: **The lipid biochemistry of calanoid copepods.** *Hydrobiologia* 1988, **167/168:**101–114.
7. Hirche H-J: **The reproductive biology of the marine copepod, *Calanus finmarchicus* - a review.** *Ophelia* 1996, **44:**111–128.
8. Svetlichny LS, Hubareva ES: **The energetics of *Calanus euxinus*: locomotion, filtration of food and specific dynamic action.** *J Plankton Res* 2005, **27:**671–682.
9. Miller C, Crain J, Morgan C: **Oil storage variability in *Calanus finmarchicus*.** *ICES J Mar Sci* 2000, **57:**1786–1799.
10. Tande KS, Hopkins CCE: **Ecological investigations of the zooplankton community of Balsfjorden, Northern Norway: The genital system in Calanus finmarchicus and the role of gonad development in overwintering strategy.** *Mar Biol* 1981, **63:**159–164.
11. Miller CB, Nelson DM, Weiss C, Soeldner AH: **Morphogenesis of opal teeth in calanoid copepods.** *Mar Biol* 1990, **106:**91–101.
12. Chang ES, Mykles DL: **Regulation of crustacean molting: a review and our perspectives.** *Gen Comp Endocrinol* 2011, **172:**323–330.
13. Ou Q, King-Jones K: **What goes up must come down: transcription factors have their say in making ecdysone pulses.** *Curr Top Dev Biol* 2013, **103:**35–71.
14. Hwang D-S, Lee J-S, Lee K-W, Rhee J-S, Han J, Lee J, Park G, Lee Y-M, Lee J-S: **Cloning and expression of ecdysone receptor (EcR) from the intertidal copepod *Tigriopus japonicus*.** *Comp Biochem Physiol C Toxicol Pharmacol* 2010, **151:**303–312.
15. Asazuma H, Nagata S, Kono M, Nagasawa H: **Molecular cloning and expression analysis of ecdysone receptor and retinoid X receptor from the kuruma prawn, *Marsupenaeus japonicus*.** *Comp Biochem Physiol B Biochem Mol Biol* 2007, **148:**139–150.
16. Durica D, Wu X, Anilkumar G, Hopkins PM, Chung AC-K: **Characterization of crab EcR and RXR homologs and expression during limb regeneration and oocyte maturation.** *Mol Cell Endocrinol* 2002, **189:**59–76.
17. Hannas BR, LeBlanc GA: **Expression and ecydsteroid responsiveness of the nuclear receptors HR3 and E75 in the crustacean *Daphnia magna*.** *Mol Cell Endocrinol* 2010, **315:**208–218.
18. Johnson C: **Ecdysteroids in the oceanic copepod *Calanus pacificus*: variation during molt cycle and change associated with diapause.** *Mar Ecol Prog Ser* 2003, **257:**159–165.
19. Hansen BH, Altin D, Hessen KM, Dahl U, Breitholtz M, Nordtug T, Olsen AJ: **Expression of ecdysteroid and cytochrome P450 enzymes during lipid turnover and reproduction in *Calanus finmarchicus* (Crustacea: Copepoda).** *Gen Comp Endocrinol* 2008, **158:**115–121.
20. Tarrant A, Baumgartner M, Verslycke T, Johnson C: **Differential gene expression in diapausing and active *Calanus finmarchicus* (Copepoda).** *Mar Ecol Prog Ser* 2008, **355:**193–207.
21. Clark KAJ, Brierley AS, Pond DW, Smith VJ: **Changes in seasonal expression patterns of ecdysone receptor, retinoid X receptor and an A-type allatostatin in the copepod, *Calanus finmarchicus*, in a sea loch environment: an investigation of possible mediators of diapause.** *Gen Comp Endocrinol* 2013, **189:**66–73.
22. Seear P, Tarling G, Burns G, Goodall-Copestake W, Gaten E, Ozkaya O, Rosato E: **Differential gene expression during the moult cycle of Antarctic krill (*Euphausia superba*).** *BMC Genomics* 2010, **11:**582.
23. Kuballa A, Merritt D, Elizur A: **Gene expression profiling of cuticular proteins across the moult cycle of the crab *Portunus pelagicus*.** *BMC Biol* 2007, **5:**45.
24. Ventura T, Manor R, Aflalo ED, Chalifa-Caspi V, Weil S, Sharabi O, Sagi A: **Post-embryonic transcriptomes of the prawn *Macrobrachium rosenbergii*: multigenic succession through metamorphosis.** *PLoS One* 2013, **8:**e55322.
25. Ning J, Wang M, Li C, Sun S: **Transcriptome sequencing and *de novo* analysis of the copepod *Calanus sinicus* using 454 GS FLX.** *PLoS One* 2013, **8:**e63741.
26. Lenz PH, Roncalli V, Hassert RP, Wu LS, Cieslak MC, Hartline DK, Christie AE: ***De novo* assembly of a transcriptome for *Calanus finmarchicus* (Crustacea, Copepoda) - The dominant zooplanker of the North Atlantic Ocean.** *PLoS One* 2014, **9:**e88589.
27. Christie AE, Fontanilla TM, Roncalli V, Cieslak MC, Lenz PH: **Identification and developmental expression of the enzymes responsible for dopamine, histamine, octamine and serotonin biosynthesis in the copepod crustacean *Calanus finmarchicus*.** *Gen Comp Endocrinol* 2013, **195:**28–39.
28. Christie AE, Roncalli V, Wu L-S, Ganote CL, Doak T, Lenz PH: **Peptidergic signaling in *Calanus finmarchicus* (Crustacea, Copepoda): *In silico* identification of putative peptide homones and their receptors using a *de novo* assembled transcriptome.** *Gen Comp Endocrinol* 2013, **187:**117–135.
29. Medler S, Brown KJ, Chang ES, Mykles DL: **Eyestalk ablation has little effect on actin and myosin heavy chain expression in adult lobster skeletal muscles.** *Biol Bull* 2005, **208:**127–137.
30. Avarre J-C, Lubzens E, Bapin PJ: **Apolipocrustacein, formerly vitellogenin, is the major egg yolk precursor protein in decapod crustaceans and is homologous to insect apolipophorin II/I and vertebrate apolipoprotein B.** *BMC Evol Biol* 2007, **7:**3.
31. Chen J, Honeyager SM, Schleede J, Avanesov A, Laughon A, Blair SS: **Crossveinless d is a vitellogenin-like lipoprotein that binds BMPs and HSPGs, and is required for normal BMP signaling in the *Drosophila* wing.** *Development* 2012, **139:**2170–2176.
32. Doniach T, Hodgkin J: **A sex-determining gene, fem-1, required for both male and hermaphrodite development in *Caenorhabditis elegans*.** *Dev Biol* 1984, **106:**223–235.
33. Wang X, Desai N, Hu YP, Price SM, Abate-Shen C, Shen MM: **Mouse Fem1b interacts with the Nkx3.1 homeoprotein and is required for proper male secondary sexual development.** *Dev Dyn* 2008, **237:**2963–2972.
34. Grillo M, Furriols M, de Miguel C, Franch-Marro X, Casanova J: **Conserved and divergent elements in Torso RTK activtion in *Drosophila* development.** *Sci Rep* 2012, **2:**762.
35. Johnson TK, Crossman T, Foote KA, Henstridge MA, Saligari MJ, Beadle LF, Herr A, Whisstock JC, Warr CG: **Torso-like functions independently of Torso to regulate *Drosophila* growth and developmental timing.** *Proc Natl Acad Sci U S A* 2013, **110:**14688–14692.
36. Primmer CR, Papakostas S, Leder EH, Davis MJ, Ragan MA: **Annotated genes and nonanotated genomes: cross-species use of Gene Ontology in ecology and evolution research.** *Mol Ecol* 2013, **22:**3216–3241.
37. Hakanson JL: **The long and short term feeding condition in field-caught *Calanus pacificus*, as determined from the lipid content.** *Limnol Oceanogr* 1984, **29:**794–804.
38. Tennessen JM, Baker KD, Lam G, Evans J, Thummel CS: **The *Drosophila* estrogen-related receptor directs a metabolic switch that supports developmental growth.** *Cell Metab* 2011, **13:**139–148.
39. Giguere V: **Transcriptional control of energy homeostasis by the estrogen-related receptors.** *Endocr Rev* 2008, **29:**677–696.
40. Kiorboe T: **Sex, sex-ratios, and the dynamics of pelagic copepod populations.** *Oecologia* 2006, **148:**40–50.
41. Irigoien X, Obermüller B, Head RN, Harris RP, Rey C, Hansen BW, Hygum BH, Heath MR, Durbin EG: **The effect of food on the determination of sex ratio in *Calanus* spp.: evidence from experimental studies and field data.** *ICES J Mar Sci* 2000, **57:**1752–1763.
42. Gusmão LFM, McKinnon AD: **Sex ratios, intersexuality and sex change in copepods.** *J Plankton Res* 2009, **31:**1101–1117.
43. Dalvin S, Frost P, Loeffen P, Skern-Mauritzen RS, Baban J, Rønnestad I, Nilsen F: **Characterization of two vitellogenins in the salmon louse *Lepeophtheirus salmonis*: molecular, functional and evolutional analysis.** *Dis Aquat Organ* 2011, **94:**211–224.
44. Lee K-W, Hwang D-S, Rhee J-S, Ki J-S, Park HG, Ryu J-C, Raisuddin S, Lee J-S: **Molecular cloning, phylogenetic analysis and developmental expression of a vitellogenin (Vg) gene from the intertidal copepod *Tigriopus japonicus*.** *Comp Biochem Physiol B Biochem Mol Biol* 2008, **150:**395–402.
45. Duncan EJ, Benton MA, Dearden PK: **Canonical terminal patterning is an evolutionary novelty.** *Dev Biol* 2013, **377:**245–261.

46. Zhu J, Miura K, Chen L, Raikhel A: **AHR38, a homolog of NGFI-B, inhibits formation of the functional ecdysteroid receptor in the mosquito *Aedes aegypti*.** *EMBO J* 2000, **19**:253–262.
47. Fisk GJ, Thummel CS: **The DHR78 nuclear receptor is required for ecdysteroid signaling during the onset of *Drosophila* metamorphosis.** *Cell* 1998, **93**:543–555.
48. Irigoien X: **Some ideas about the role of lipids in the life cycle of *Calanus finmarchicus*.** *J Plankton Res* 2004, **26**:259–263.
49. Rey-Rassat C, Irigoien X, Harris R, Head R, Carlotti F: **Growth and development of *Calanus helgolandicus* reared in the laboratory.** *Mar Ecol Prog Ser* 2002, **238**:125–138.
50. Maps F, Plourde S, Zakardjian B: **Control of dormancy by lipid metabolism in *Calanus finmarchicus*: a population test model.** *Mar Ecol Prog Ser* 2010, **403**:165–180.
51. Crain JA, Miller CB: **Effects of starvation on intermolt development in *Calanus finmarchicus* copepodites: a comparison between theoretical models and field studies.** *Deep-Sea Res Pt II* 2001, **48**:551–566.
52. Crain J, Miller C: **Detection of sex and sex ratio in Calanus finmarchicus early stage fifth copepodites.** *ICES J Mar Sci* 2000, **57**:1773–1779.
53. Lohse M, Bolger AM, Nagel A, Fernie AR, Lunn JE, Stitt M, Usadel B: **RobiNA: a user-friendly integrated software solution for RNA-Seq-based transcriptomics.** *Nucleic Acids Res* 2012, **40**:W622–W627.
54. Grabherr MG, Haas BJ, Yassour M, Levin JZ, Thompson DA, Amit I, Adiconis X, Fan L, Raychowdhury R, Zeng Q, Chen Z, Mauceli E, Hacohen N, Gnirke A, Rhind N, di Palma F, Birren BW, Nusbaum C, Lindblad-Toh K, Friedman N, Regev A: **Full-length transcriptome assembly of RNA-Seq data without a reference transcriptome.** *Nature Biotechnol* 2011, **29**:644–652.
55. Conesa A, Gotz S, Garcia-Gomez JM, Terol J, Talon M, Robles M: **Blast2GO: a universal tool for annotation, visualization and analysis in functional genomics research.** *Bioinformatics* 2005, **21**:3674–3676.
56. Li B, Dewey C: **RSEM: accurate transcript quantification from RNA-Seq data with or without a reference genome.** *BMC Bioinformatics* 2011, **12**:323.
57. Robinson MD, Oshlack A: **A scaling normalization method for differential expression analysis of RNA-seq data.** *Genome Biol* 2010, **11**:R25.
58. Robinson MD, McCarthy DJ, Smyth GK: **edgeR: a Bioconductor package for differential expression analysis of digital gene expression data.** *Bioinformatics* 2010, **26**:139–140.
59. Young MD, Wakefield MJ, Smyth GK, Oshlack A: **Gene ontology analysis for RNA-seq: accounting for selection bias.** *Genome Biol* 2010, **11**:R14.
60. Stamatakis A: **RAxML-VI-HPC: maximum likelihood-based phylogenetic analyses with thousands of taxa and mixed models.** *Bioinformatics* 2006, **22**:2688–2690.
61. Vandesompele J, De Preter K, Pattyn F, Poppe B, Van Roy N, De Paepe A, Speleman F: **Accurate normalization of real-time quantitative RT-PCR data by geometric averaging of multiple internal control genes.** *Genome Biol* 2002, **3**:RESEARCH0034.
62. Nuclear Receptors Nomenclature Committee: **A unified nomenclature system for the nuclear receptor superfamily.** *Cell* 1999, **97**:161–163.
63. Thomson SA, Baldwin WS, Wang YH, Kwon G, Leblanc GA: **Annotation, phylogenetics, and expression of the nuclear receptors in *Daphnia pulex*.** *BMC Genomics* 2009, **10**:500.

16

# No evidence for assortative mating within a willow warbler migratory divide

Miriam Liedvogel[1,2*], Keith W Larson[1], Max Lundberg[1], Arzu Gursoy[1,3], Leonard I Wassenaar[4], Keith A Hobson[4], Staffan Bensch[1] and Susanne Åkesson[1]

**Abstract**

**Introduction:** In contact zones, genetic mixing of two taxa can be restricted by prezygotic (e.g. assortative mating) or postzygotic (lower fitness of hybrid offspring) barriers, or a combination of the two. A hybrid zone between two willow warbler subspecies (*Phylloscopus trochilus trochilus, P. t. acredula*) with distinctive migratory strategies occurs in central Sweden. These subspecies exhibit differences in migratory direction and distance, resulting in geographically distinct wintering areas in Africa. The subspecies may have diverged from a common refuge after the last ice age, and neutral genetic markers are homogeneous across their range. By contrast, several phenotypic traits and genetic markers of two chromosomal regions previously identified show steep clines across the divide. The evolutionary forces that maintain this migratory divide remain unknown. Here we use plumage colour, morphology, genetic markers and feather stable nitrogen-isotopes ($\delta^{15}N$) to assess if assortative mating between migratory phenotypes could be acting as a possible mechanism for keeping the two forms genetically separate and maintaining the migratory divide. We colour-ringed a willow warbler breeding population in the central part of the hybrid zone and observed the breeding population to assess phenotypic and genotypic traits of social pairs.

**Results:** Our data suggest that wintering area and genetic ancestry had an effect on male arrival time to the breeding grounds which could contribute to assortment. However, evidence for assortative mating could not be detected based on a comparison of plumage colour, morphology and $\delta^{15}N$ between social mates.

**Conclusion:** This finding was strengthened by analyses of subspecies-specific genetic markers, which allowed us to identify the presence of a large proportion of potential hybrids and backcrosses at the study site. Our results supported the hypothesis that pre-mating isolation in willow warblers is weak, resulting in extensive hybridisation across the migratory divide.

**Keywords:** Hybrid zone, Prezygotic selection, Postzygotic selection, Reproductive isolation, Willow warbler, *Phylloscopus trochilus*, Nitrogen-15

## Introduction

Migratory divides are narrow regions of parapatry where neighbouring populations of closely related taxa, but with distinctly different migratory behaviour, meet and potentially hybridise [1-3]. Divides may result from dispersal and range expansion of two former allopatric populations that over time meet in secondary contact [4]. Examples of migratory divides based on secondary contact include the European blackcap (*Sylvia atricapilla*) [5] and Swainson's thrush (*Catharus ustulatus*) [6]. In addition, migratory divides could also be established if an alternative migratory phenotype, which may differ in migratory direction, distance to wintering grounds or the propensity to migrate, becomes selectively beneficial and therefore increases in frequency. Given strong selection pressures, the frequency of the alternative migratory phenotype could increase within a small number of generations. This has been seen in European blackcaps, for which a new major migratory direction to wintering areas in Britain and Ireland evolved over the past fifty years [7,8]. From quantitative genetics analyses we know that a significant proportion of migratory trait variation is genetically controlled (reviewed

* Correspondence: miriam.liedvogel@biol.lu.se
[1]Center for Animal Movement Research (CAnMove), Department of Biology, Lund University, 22362 Lund, Sweden
[2]Wildlife Ecology and Management, Freiburg University, Tennenbacher Str. 4, 79106 Freiburg, Germany
Full list of author information is available at the end of the article

in [9]), e.g. encoding migratory direction, as well as an inherited time schedule that determines departure and arrival time (i.e. defining a migratory distance). This inherited time schedule further coordinates the temporal expression of physiological adaptations necessary for a successful migratory journey, such as post-breeding moult and pre-migration fattening [10,11].

Assortative mating provides an important mechanism for reproductive isolation between parapatric populations [12,13]. Differences in habitat choice, diet, breeding phenology and song and/or mating behaviour may all contribute to the establishment of prezygotic reproductive barriers and specifically assortative mating [12,14,15]. For example, in the European blackcap, the timing of arrival to the breeding grounds promotes assortative mating between two co-occurring migratory phenotypes with separate wintering grounds in the UK and the Iberian Peninsula to NW Africa [14,16]. Similarly, postzygotic selection may result from maladaptive combinations of intermediately inherited migratory traits in hybrids. In cross-breeding experiments with European blackcaps with parents from either side of a migratory divide, $F_1$ hybrid offspring demonstrated an intermediate migratory direction compared to the parental phenotypes [17,18]. In the wild, this intermediate route may be selected against, because it necessitates crossing additional significant ecological barriers lacking refuelling and stop-over habitat (e.g. the Alps as well as crossing both the Mediterranean Sea and Sahara desert at their widest stretches) without the necessary physiological adaptations for long-distance flights [17,19].

In central Scandinavia two subspecies of the willow warbler (*Phylloscopus trochilus trochilus* in the south, and *P. t. acredula* in the north) with distinctive migratory behaviour meet and form a migratory divide. Each migratory phenotype likely possesses distinct endogenous migratory traits, including timing, migratory direction and distance, but presumably also different physiological adaptations such as fuelling decisions [20]. In southern Sweden *P. t. trochilus* migrates SW in the autumn to winter in West Africa, whereas *P. t. acredula* migrates SSE to winter in East to South Africa. Migratory direction and wintering grounds for the two subspecies are inferred from ringing recoveries [1,21,22] and stable nitrogen-isotope analyses ($\delta^{15}N$) of winter-grown feathers [1,23].

Molecular analyses using neutral markers (microsatellites, mtDNA) reveal no intraspecific population structure ($F_{ST} \approx 0$) [19,22]. This may be explained by a very recent divergence between the subspecies [19,22]. In contrast to the high level of genetic similarity, phenotypic traits including plumage colouration, morphology, and $\delta^{15}N$ (proxy for migratory direction and wintering grounds) and a genetic marker associated with migratory direction (identified by AFLP scan), present steep clines across the migratory divide (<300 km) [1,19].

Previous work demonstrated that the central Scandinavian migratory divide is maintained by selection, as geographic clines for both phenotypic and genotypic traits are narrower than expected from a neutral diffusion model assuming random mating between the subspecies since the time of secondary contact [19,22]. Random mating could be one important contributor to this; the more mating is assortative, all other factors being equal, the wider the cline will be. However, geographic clines of mean trait values across different localities do not explicitly provide information on the degree of hybridization that is associated with intermediate trait values and allele frequencies across the migratory divide. The proportion and types of hybrids is determined by post-zygotic selection (reduced fitness in hybrids) and degree of assortative mating between the parental forms: the more post-zygotic and less pre-zygotic selection present, the narrower the hybrid zone.

We quantify the proportion and different classes of hybrid genotypes in the migratory divide. We also assess to what extent assortative mating is occurring and how it is related to the observed degree of hybridization.

We addressed this question by studying a breeding site in the centre of the hybrid zone where we expected both subspecies to occur. We assessed the degree of assortative mating based on a comparison of the characteristic phenotypic traits for pairs captured on their breeding territories. In order to estimate the genetic ancestry of individuals, we made use of subspecies-specific genetic variation segregating at two newly identified loci that had been detected through analyses of transcriptome sequencing data [24]. These co-dominant markers enabled us to estimate the genetic ancestry in the willow warbler at a much higher accuracy than with the AFLP markers. This data improved our quantification of potential hybridisation and allowed for detection of more complex hybrids and backcrosses.

The key objectives of our study were to identify evolutionary processes and selective forces within our focal breeding population by (i) characterising the distribution of phenotypic and genotypic traits in a breeding population of willow warblers in the hybrid zone, in order to subsequently (ii) assess the role of prezygotic selection in maintaining the migratory divide.

## Results

### Sampling

We captured individuals of 25 breeding pairs in 2011, and 15 breeding pairs in 2012. In addition, we captured 13 additional males holding territories in 2011, and 22 in 2012. In 2012, late winter weather delayed spring arrival and breeding, which accounted for the lower number of breeding pairs captured.

### Genetic analyses

With few exceptions, individuals in both subspecies-specific reference sets were assigned to each of the two population clusters with high probabilities using the newly identified subspecies-specific genetic markers (Figure 1, Additional file 1: Table S2). This suggested that both identified population clusters corresponded to each of the subspecies. The few individuals in each of the reference sets that obtained a lower assignment probability to the expected subspecies cluster possibly indicated a hybrid origin. Individuals at the study site showed much higher variability in genetic ancestry compared to sampling areas at either side of the divide outside the hybrid zone (Figure 1, Additional file 1: Table S2). In general, ancestry estimates were skewed towards the southern subspecies, with a declining proportion of individuals with increased ancestry to the northern subspecies (Figure 1, Additional file 1: Table S2). The number of individuals from the study site that were likely to be pure individuals of the northern subspecies was very small.

### Arrival time

During the period of intense field work we were confident that most of the encountered unringed birds were newly arrived to our study area. We thus expected capture date and arrival date to be correlated for this interval. If timing contributes to assortative mating in the hybrid zone, we also expect birds from different wintering grounds (indicated by $\delta^{15}N$ feather values) to arrive at differing schedules. We found no significant correlation of male $\delta^{15}N$ feather values with capture date in males ($r = 0.15$, $n = 54$ $p = 0.14$) (Figure 2A). However, there was a positive correlation between capture date of males and genetic ancestry, measured as probability of the individuals belonging to the northern subspecies cluster ($r = 0.24$, $n = 54$ $p = 0.04$) (Figure 2B).

### Assortative mating

We tested for assortative mating for 40 confirmed pairings based on comparisons of phenotypic values using Pearson's correlations (see Additional file 2: Table S4 for all pair trait data). Overall, within-pair comparisons of the three phenotypic traits revealed no evidence for assortative mating. First, for $\delta^{15}N$ values (indicator of wintering grounds) there was no relationship between the values of the social mates ($r = -0.03$, $n = 38$, $p = 0.58$) (Figure 3A). Likewise plumage colour did not show a clear correlation between social mates ($r = 0.23$, $n = 38$, $p = 0.12$) (Figure 3C). For body size as summarised by PC1 ($r = 0.02$, $n = 37$, $p = 0.46$) (Figure 3D) and PC2 ($r = -0.14$, $p = 0.80$) there was no relation between the values of the social mates. Body size PCA results and trait values for all social pairs are summarised in the Additional file 1: Tables S1 & S2. No significant correlation ($r = 0.22$, $n = 38$, $p = 0.08$) could be detected between the genetic ancestry of social mates.

## Discussion

Genetic ancestry for all males was significantly correlated with arrival date. Although we did not have data on female arrival, it is known from other studies that male willow warblers arrive on the breeding ground one to two weeks ahead of females (e.g. [25]). The large overlap in arrival time of the males of mixed genetic ancestry combined with females presumably arriving substantially later than the males, suggested that differential arrival time played a limited role in assortative mating in willow warblers. This is in contrast to the pattern found in populations of European blackcaps, where

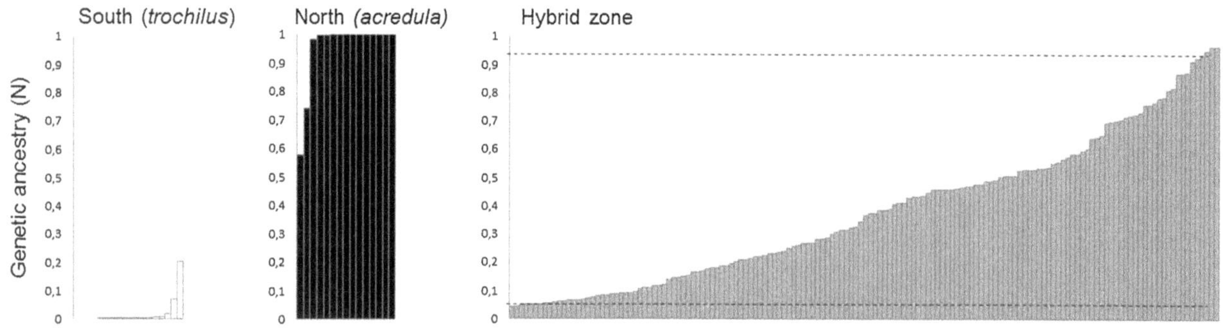


**Figure 1 Genetic ancestry based on two subspecies-specific markers.** Bayesian assignment probability (genetic ancestry) to a northern (*acredula*) subspecies population cluster for all individuals computed in STRUCTURE (n = 181; details listed in Additional file 1: Table S2). Reference samples for the southern (*P. t. trochilus*, n = 18, open bars) and northern (*P .t. acredula*, n = 15, black bars) subspecies were plotted separately from the samples collected in the hybrid zone (n = 148, grey bars). Each bar represented one individual; the y-axis represented the probability of genetic ancestry to the northern subspecies cluster. The distribution within the study population in the hybrid zone was skewed towards southern genotypes. Dashed lines in the hybrid zone population indicated mean ancestry estimates for individuals from the northern and southern reference sets, respectively.

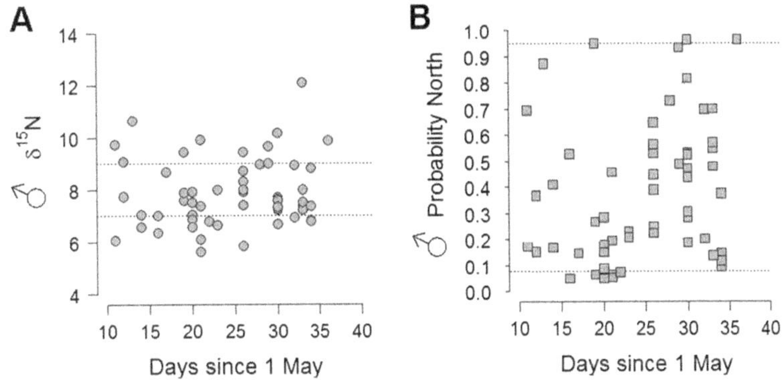


**Figure 2 Male phenotype (A) and genotype (B) in relation to capture date (for males captured between 10 May and 10 June) (n = 56).** **A** $\delta^{15}N$ values (dashed lines at 7‰ and 9‰ represented cut-offs for individuals likely wintering in West Africa and East to South Africa, respectively), **B** Bayesian assignment probabilities (genetic ancestry) to the northern subspecies cluster (dashed lines represent mean assignment probabilities in the southern and northern reference sets, respectively). Time for capture dates was plotted as the number of days after 1 May.

two populations with distinctly different migratory strategies breed in sympatry, but mate assortatively based on timing of arrival to the breeding grounds [14] as has also been observed in, Swainson's thrushes [26] across their migratory divide. In both cases differences in arrival time at the breeding grounds is a major force for assortative mating. In willow warblers birds of the northern subspecies arrive later. However, our data did not suggest assortative mating based on the difference in arrival time of both subspecies. Research in blackcaps suggests that hybrid phenotypes have an intermediate endogenous migratory programme compared to the parental phenotypes. It is possible that an intermediate strategy might result in reduced hybrid survival [17].

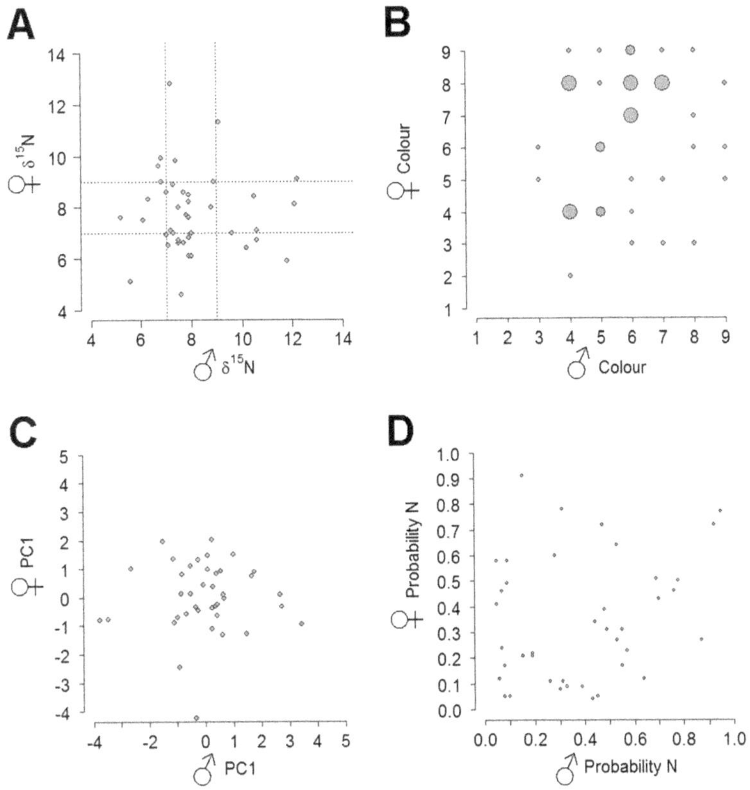


**Figure 3 Male-female trait comparison for all pairs (n = 40).** **A** $\delta^{15}N$ values; dashed lines at 7‰ and 9‰ represented cut-offs for birds likely wintering in West Africa and East to South Africa, respectively; **B** plumage colour scores; **C** PC1 for body size; and **D** genetic ancestry to the northern subspecies. Pair counts for colour and genetic ancestry (values rounded to 2 significant digits) were represented by the size of the circle.

Another, and not mutually exclusive, possibility is that a reduction in hybrid fitness is mediated by genetic incompatibilities [27]. As the willow warbler subspecies are genetically similar (as is also true for blackcaps), it seemed unlikely that genetic incompatibilities act as selective forces in keeping the hybrid zone narrow.

We further observed no evidence for assortative mating in the hybrid zone based on a comparison of several characteristic phenotypic and genotypic traits. Careful within-pair comparison of morphological traits (plumage colour, body size), genotype, and migratory direction, which all show steep clines across the migratory divide, indicated no pattern of assortative mating between the pairs. A similar pattern has been found in the hybrid zone of yellow-rumped warblers, where extensive hybridisation in the absence of assortative mating occurs [28]. Our results suggest that assortative mating in our study system is unlikely to be important in constraining the hybridization across the migratory divide.

Other ecological (e.g. habitat) or sexual (song) traits can be excluded as likely factors in stabilising the willow warbler hybrid zone. This is unlike, e.g. Swainson's thrushes, where both co-occurring subspecies occupy distinct ecological niches [26], or blackcaps, where a recent study provides support for differences in habitat choice but not arrival time being associated with alternative migratory strategies in the populations studied [29]. Both willow warbler subspecies breed in similar terrain with trees and shrubs across their range, thus habitat is unlikely to act as barrier in promoting assortative mating in our system. Similarly, song was an unlikely mechanism for assortative mating in willow warblers, as the subspecies' songs are indistinguishable (unpublished data), and territorial behaviour (i.e. antagonistic reaction to playback of conspecific song) of males across Scandinavia is triggered by the same play-back recording of a conspecific song [19,22], unpublished results]. Also, even if slight song differences do exist but escaped our detection, it would not be possible to evaluate how these differences would result in assortative mating as we have no assay to score female preferences for song in the field. European blackcaps show a similar pattern of symmetrical responses to song playbacks [29], but Swainson's thrushes show an asymmetrical response pattern where one subspecies responds more aggressively to homotypic song whereas the other subspecies' response pattern to homotypic and heterotypic songs is indistinguishable [26].

Based on the SNP data in the two sequenced genes we were able to assign probabilities for each individual to the two population clusters (i.e. northern and southern subspecies) identified by STRUCTURE [30]. As expected, individuals in the subspecies reference sets generally showed high assignment probabilities to each of the two clusters, i.e. migratory phenotypes (Figure 1). However, a few individuals in both sets showed a lower assignment probability to their expected population cluster. One possible explanation may be that a small fraction of pure individuals of each subspecies could possess alleles that are more common in the other subspecies at any of the two loci we genotyped. Alternatively, long-distance dispersal may be occurring from the hybrid zone into the parental populations.

The use of only two diagnostic markers was insufficient to detect more complex hybrids, such as advanced backcrosses, which would result in an underestimation of hybridisation. The low number of markers also prohibited reliable classification of individuals into discrete hybrid classes, such as backcrosses, F1 and F2 hybrids [31]. Even though our genetic ancestry estimates had some uncertainty and may underestimate the degree of hybridisation, the distribution of ancestries at the study site was markedly different from that seen in the subspecies reference sets (Figure 1). The large proportion of intermediate genetic ancestry estimates at the study site suggested that hybridisation is extensive within the hybrid zone. The skewed distribution of individuals with southern ancestries and the gradual decrease of increasingly northern individuals were expected with extensive hybridisation in the southern part of the hybrid zone. This distribution also suggested that it is unlikely that pure southern and pure northern birds co-occur at high frequencies anywhere in the hybrid zone.

Our sample of 40 pairs of willow warblers were collected over two years from one breeding population in the centre of the migratory divide. The site was carefully chosen because of its location based on previously described trait clines [19]. However, our newly established gene ancestry analyses presented here allowed us to identify an excess of birds with *P. t. trochilus* genotypes at the study plot, suggesting that our focal breeding population was situated slightly south of the centre of the migratory divide, if we assumed a smooth cline in genetic ancestry across the divide. The possibility that our study site may indeed be situated south of the divide may have limited the statistical power to detect the possible occurrence of assortative mating. However, none of the characteristic traits investigated showed any correlation between pairs implying that lack of assortative mating was the most parsimonious conclusion for our data.

The lack of assortative mating in willow warblers combined with our previous work further supported the suggestion that postzygotic selection against hybrids rather than prezygotic selection (mate choice) is the major force maintaining this willow warbler migratory divide. The degree of intrinsic reproductive isolation generally accumulates with the genetic distance between the hybridizing taxa [32]. In birds, a complete loss of viability and fertility in hybrids is typically seen over millions of years [33]. Given the very short time since the divergence between the willow warbler subspecies - estimated to be around 10.000 years [19] - severe genetic incompatibilities were unlikely to have evolved. This was also suggested by the high, and yet probably underestimated, amount of potential backcrosses and hybrids at the study site. However, some genetic incompatibilities may only manifest themselves in certain genotypes found in later generation hybrids and backcrosses [27,34]. Our present data set only allowed us to speculate, but not reliably detect such an effect.

## Conclusions and outlook

Our data suggest that without a clear separation in arrival time, habitat choice, song or some other phenotypic traits that would result in assortative mating, hybridisation will remain common throughout the willow warbler migratory divide. Given that mating in willow warblers appears to be random, future research will focus on understanding the mechanisms governing postmating isolation. The only potential postzygotic isolation factor so far identified is the presence of a migratory divide. If postmating isolation is present in our system, a narrow zone of overlap should be associated with random or near random-mating, and our study provides evidence for this. In order to understand putative reproduction and/or survival consequences in hybrids, we should focus on long-term observations of colour-marked breeding populations in the hybrid zone, allowing for direct assessment of fitness and survival consequences. Experimental approaches that will better allow us to better understand the role and mechanisms of postzygotic factors may include an improved and more robust genotyping protocol containing more subspecies-specific markers that allows the detection and reliable quantification of different genotype classes (e.g. F1, F2, backcrosses). These data could in turn be used to derive more reliable estimates of fitness parameter, survival and reproduction for different genotypes. Further, genomic analyses characterising divergent chromosomal regions and most importantly the identification of candidate genes that are driving the differentiation pattern are likely to be involved in the reproductive isolation and knowledge of their function may help to suggest the mechanism by which reproductive isolation is maintained. In addition to improved genomics resources, future studies will also benefit from more accurate phenotypic measures, e.g. tracking the individuals' migratory strategy and timing decisions across their annual cycle if technology further advances, especially with relevance for correlating these measures with individual survival and reproduction.

## Materials and methods

### Background

Between 1996 and 2010 over 2000 individuals at 98 sites were sampled across Sweden to investigate trait differences between individuals and populations, and to identify the location and structure of the migratory divide. We focused on three phenotypic traits: plumage colour, morphology (composed of six measurements), and $\delta^{15}N$ (proxy for migratory direction), and one AFLP derived genetic marker whose allele frequency coincides with migratory direction. These data revealed four steep coincident trait clines [19,22], which allowed us to estimate the location of the hybrid zone between 61.4° to 64.1° N (latitude) with the centre of each cline close to 62.7° N [19].

### Sampling

Our study was conducted along Flatruet in Sweden (62.7° N, 12.7° E, range 545–846 meters asl, 9 $km^2$), within the willow warbler hybrid zone (Additional file 1: Figure S1). In 2011 (10 May to 14 July) and 2012 (2 June to 7 July) we carried out territory mapping and nest searching in order to capture and sample both males and females of all possible breeding pairs in our study population. We captured males with mist-nets by a conspecific song play-back, while the females were captured on or near the nest during the nest-building phase, incubation or while feeding their young. We sampled daily for newly arriving males across the entire study area. This facilitated identification of male arrival dates using capture data as a proxy for the arrival. For conservative estimates of arrival time we only included males caught in 2011, as late winter storms delayed breeding behaviour of arriving males by at least two weeks in 2012.

Each individual was marked with a unique combination of metal and colour rings to facilitate re-sighting and behavioural observation throughout the breeding season. Females, social mates, were all identified by the presence of a brood patch or observation during nest building (carried out solely by females), during incubation or brooding nestlings at the nest attended by their respective male partners. We confirmed both members of all pairs by colour-ring identification or direct (re)capture at the nest.

We followed previous protocols [1,19,22] to collect four morphometrics (i.e. wing chord, whole tail, tarsus, bill-head), assess plumage colouration, sample blood for subsequent genetic analyses, and collect the first primary in both wings to assess wintering ground $\delta^{15}N$ for all individuals captured. Methods of analyzing $\delta^{15}N$ from feathers have been fully described elsewhere [1,19,35]. Plumage colour scores were assessed on a scale of 1 to 9 using three reference specimens representing the typical *acredula* (dull grayish-green plumage; score 2), intermediate (score 5) and the typical *trochilus* (yellow plumage; score

8) [19]. In an earlier study we analysed sonograms from individuals with the most extreme migratory phenotypes across the Scandinavian distribution range in order to assess if sub-specific song differences (e.g. pattern, repertoire size) could be a factor supporting assortative mating (unpublished results). Because we could not detect any obvious song differences between northern and southern individuals we did not examine this variable as putative factor in assortative mating. Further all captures of males across the Scandinavian distribution range occurred using the same song recording for play-back.

#### Genotyping

We developed a genotyping scheme to allow for an assessment of each individual's genetic ancestry to subspecies, and to identify potential hybrids and backcrosses within the population. Individuals were genotyped for variation in the *NBEA* and *C11orf41* gene, which are located in regions on chromosome 1 (*NBEA*) and chromosome 5 (*C11orf41*); previously identified to be highly differentiated between the subspecies [24]. For each gene we manually designed forward and reverse primers in separate exons to amplify polymorphic introns. Primers for NBEA (F: 5'-CCCTCCCAGAAAGAAATCATATCA-3', R: 5'-TAGCAG CTGCAGCACATCATGAA-3') were designed from willow warbler 454 Expressed Sequence Tags (ESTs) aligned to zebra finch genome [24] and amplified 990 bp of the gene. C11orf41 primers (F: 5'-TCCTGTGGCTGTTGC CTCTC-3', R: 5'-CCAATATCTGATGTAGACCTGTGC-3') were designed from an alignment between chicken and zebra finch sequences and amplified 830 bp of the gene.

Amplification was performed using a touchdown PCR starting with an annealing temperature of 63°C, which decreased with 0.5°C per cycle for 12 cycles, followed by 28 cycles with an annealing temperature of 57°C. Each cycle contained an elongation step of 1:15 minutes. PCR reactions were performed in a 25 μl volume consisting of 0.125 mM of each dNTP, 1.5 mM MgCl2, 0.4 μM of primer, 0.5 units of Taq Polymerase (AmpliTaq®, Applied Biosystems, Branchburg, NJ, USA) and 10–50 ng of DNA template. PCR products were quality checked on 2% Agarose Gels and precipitated with 11 μl Ammonium Acetate and 37.5 μl 95% Ethanol. Sequencing was performed using BigDye® Terminator v.1.1 cycle sequencing kit (Applied Biosystems, Austin, TX, USA) on an ABI 3130 prism robot (Applied Biosystems, Foster City, CA, USA).

Sequencing the full length of the PCR products was complicated by the occurrence of multiple length polymorphisms in many individuals. Sequencing efforts were therefore focused on the initial few hundred base pairs of each gene. An internal set of primers (Additional file 1: Table S1) was used for individuals that were heterozygous for multiple length repeats within this interval. Amplification and sequencing using additional primer sets followed the same protocol described above, with the exception of a PCR elongation step of 30 seconds.

Sequences were visualized and trimmed in Geneious Pro 5.6 [36]. Alignments of sequences were generated with the Clustal W algorithm [37] implemented in Geneious, with manual adjustments when necessary. We manually scored heterozygous positions, and deposited sequence data (of at least 200 bp) in GenBank (Accession numbers: KJ886575-KJ886931, Additional file 2: Table S4). Sequence information for reads <200 bp is provided in Additional file 2: Table S5.

#### Population assignment

We used STRUCTURE version 2.3.4 [30] to calculate assignment probabilities to subspecies-specific population clusters based on the genotype data for all individuals. Sequences from 15 individuals from northern Sweden and 18 individuals from southern Sweden (Additional file 1: Table S2) were included as reference samples of each subspecies. The input file consisted of single nucleotide polymorphisms (SNPs) with a minor allele frequency of at least 0.10 among the reference samples, which could then be successfully genotyped in the vast majority of the individuals in the full data set. This resulted in five SNPs for *NBEA* and 16 SNPs for *c11orf41*, spanning 181 bp and 341 bp of each gene, respectively.

We ran the software using the admixture model and assuming the existence of two population clusters ($K = 2$) corresponding to both subspecies. We justify this choice by the existence of two subspecies of the willow warbler in Scandinavia and a hybrid zone between these populations, which has previously been characterised by cline analyses of genetic and phenotypic data in Central Sweden [19]. The optimal value of K was also determined from STRUCTURE simulations by running the same analysis for $K = 1$ to $K = 10$ with 15 iterations for each K. The results from these runs were analysed using the online tool STRUCTUREharvester [38], which implements the K selection method proposed by Evanno et al. [39]. In this case, $K = 2$ also had the highest support ($\Delta K = 1859.83$ ($K = 2$); Second highest $K = 3$: $\Delta K = 131.25$).

To obtain ancestry estimates for each individual with two population clusters, the admixture model was first ran without explicitly using the samples from the two reference populations to assist clustering of samples at the study site. As expected, the samples from each of the reference populations showed a high assignment probability to each of the two population clusters, respectively. In the final run, we specifically used the reference samples to aid clustering of samples from our study site and used them to update allele frequencies in each of the population clusters throughout the simulation. The model was run with 500 000 burn-ins followed by 500 000 steps of Markov Chain Monte Carlo (MCMC) iterations. We otherwise used default parameters, including the assumption of correlated allele frequencies between the population clusters.

The admixture model assumes that the genetic markers within populations are in linkage equilibrium within each population cluster. This could be violated if the markers are tightly linked to each other, such as polymorphisms within the same gene used in this study. Although STRUCTURE does not specifically model tightly linked genetic markers [40], we checked the consistency of the clustering by also analysing the same data set using the linkage model, which is a generalisation of the normal admixture model that takes into account differences in linkage between loci [41]. In the model, SNPs located in the different genes were specified as unlinked, and for SNPs within the same gene, nucleotide distances were used as a proxy for genetic distances with an assumed a linear relationship between physical and genetic distance with 1 centimorgan/Mb and used default settings with the exception of a minimum recombination rate prior of $10^{-6}$. The clustering performed by the linkage model was generally accompanied by slightly more uncertainty, but the individual genetic ancestry estimates were very strongly correlated with those calculated from the admixture model ($R = 0.94$, $p = 2.2 \times 10^{-16}$).

### Statistical analyses

To determine if there were any correlations between male $\delta^{15}N$ values or the probability of northern ancestry to arrival (capture) date, we calculated Pearson's correlation coefficients (r). We limited our sample of males to those captured in 2011 as the 2012 season started two weeks late due to extended winter conditions. Next, we summarised morphometric data using a principal components analysis where PC1 explained the largest percentage of the variation in our data and was used for subsequent analyses (Additional file 1: Table S3). To determine if there was a relationship between the arrival (capture) dates of males on the breeding grounds we used a Pearson's correlation analysis. Specifically, we tested capture date against genetic ancestry and difference in migratory strategy for both subspecies. We restricted the analysis with respect to arrival time to males caught in 2011 during the period of 10 May to 10 June, as this represented the period during which we were confident that only newly arrived birds were captured. We further tested for correlations between phenotypic (i.e. plumage colour, PC1 for morphological measurements, $\delta^{15}N$ values) and genotypic (genetic ancestry) traits between social mates using Pearson's correlation. All statistical analyses were carried out using R [42].

## Additional files

**Additional file 1: Figure S1.** Study site: Flatruet. **Table S1.** Internal primers used for amplification and sequencing of samples whose genotypes could not be determined by using the standard primers for each locus. **Table S2.** Samples used for STRUCTURE analysis and the resulting assignment probability (genetic ancestry) for each individual to a northern (*acredula*) subspecies population cluster. "Population" indicates individuals caught in the distribution range of the southern (*Phylloscopus trochilus* ***trochilus***) or northern (*Phylloscopus trochilus* ***acredula***) subspecies, or within the hybrid zone. **Table S3.** Eigenvalues and proportion of variance explained by each principal component using four morphological measurements, i.e. wing chord, tail, tarsus, and bill-head length, separately for males and females.

**Additional file 2: Table S4.** Deposited sequence data (of at least 200 bp) in GenBank for polymorphic regions on chromosome 1 (Chr1; NBEA) and chromosome 5 (Chr5; C11orf41). **Table S5.** Sequence data for short reads (< 200 bp) for polymorphic regions on chromosome 1 (Chr1; NBEA).

### Competing interests

The authors declare that they have no competing interests.

### Authors' contributions

SB, SÅ, KWL and MLu participated in the study design. KWL, MLu, MLi and AG conducted the field work. MLu and MLi carried out the genetics analyses. KH and LIW carried out the stable isotope analyses. KWL statistically analysed the data. MLi wrote the manuscript. All authors read and approved the final manuscript.

### Acknowledgements

Without the hard work of all of our field assistants, John Boss, Sieglinde Kundisch, Mica Andersson and Anna Brauer, this work would not have been possible. The willow warblers were sampled with ethical permission from Malmö/ Lund djurförsöksetiska nämnd (M94-08) and permission from Naturvårdsverket and the Swedish Ringing Centre for capturing birds. This study was financially supported by grants from the Swedish Research Council to S.Å (621-2007-5930 and 621-2010-5584) and S.B. (621-2010-5277), a Feodor Lynen fellowship, Alexander von Humboldt-Foundation (MLi), and by operating grants to KAH and LIW from Environment Canada. This is a report from the Centre for Animal Movement Research (CAnMove) financed by a Linnaeus grant (349-2007-8690) from the Swedish Research Council and Lund University.

### Author details

[1]Center for Animal Movement Research (CAnMove), Department of Biology, Lund University, 22362 Lund, Sweden. [2]Wildlife Ecology and Management, Freiburg University, Tennenbacher Str. 4, 79106 Freiburg, Germany. [3]Department of Biology, Faculty of Arts and Sciences, University of Ondokuz Mayis, Samsun, Turkey. [4]Environment Canada, 11 Innovation Blvd., Saskatoon, SK S7N 3H5, Canada.

### References

1. Chamberlain CP, Bensch S, Feng X, Åkesson S, Andersson T: **Stable isotopes examined across a migratory divide in Scandinavian willow warblers (*Phylloscopus trochilus trochilus* and *Phylloscopus trochilus acredula*) reflect their African winter quarters.** *Proc Roy Soc B* 2000, **267:**43–48.
2. Dallimer M, Jones PJ, Pemberton JM, Cheke RA: **Lack of genetic and plumage differentiation in the Red-Billed *Quelea quelea* across a migratory divide in southern Africa.** *Mol Ecol* 2003, **12:**345–353.
3. Irwin DE, Irwin JH: **Siberian migratory divides.** In *Birds of two worlds: the ecology and evolution of migration.* Edited by Greenberg R, Marra PP. Balitmore, Maryland, USA: Johns Hopkins University Press; 2005:27–40.
4. Rohwer S, Irwin DE: **Molt, orientation, and avian speciation.** *The Auk* 2011, **128:**419–425.

5. Helbig AJ: **Genetic basis, mode of inheritance and evolutionary changes of migratory directions in palaearctic warblers (Aves: *Sylviidae*).** *J Exp Biol* 1996, **199:**49–55.
6. Ruegg K: **Genetic, morphological, and ecological characterization of a hybrid zone that spans a migratory divide.** *Evolution* 2008, **62:**452–466.
7. Berthold P, Helbig AJ, Mohr G, Querner U: **Rapid microevolution of migratory behaviour in a wild bird species.** *Nature* 1992, **360:**668–670.
8. Rolshausen G, Segelbacher G, Hobson KA, Schaefer HM: **Contemporary evolution of reproductive isolation and phenotypic divergence in sympatry along a migratory divide.** *Curr Biol* 2009, **19:**2097–2101.
9. Liedvogel M, Åkesson S, Bensch S: **The genetics of migration on the move.** *Trends Ecol Evol* 2011, **26:**561–569.
10. Berthold P: *Control of Bird Migration.* London: Chapman and Hall; 1996.
11. Åkesson S, Hedenström A: **How migrants get there: migratory performance and orientation.** *Bioscience* 2007, **57:**123–133.
12. Huber SK, León LFD, Hendry AP, Bermingham E, Podos J: **Reproductive isolation of Ssympatric morphs in a population of darwin's finches.** *Proc Roy Soc B* 2007, **274:**1709–1714.
13. Hendry AP, Huber SK, De León LF, Herrel A, Podos J: **Disruptive selection in a bimodal population of darwin's finches.** *Proc Roy Soc B* 2009, **276:**753–759.
14. Bearhop S, Fiedler W, Furness RW: **Assortive mating as a mechanism for rapid evolution of a migratory divide.** *Science* 2005, **310:**502–504.
15. Patten MA, Rotenberry JT, Zuk M: **Habitat selection, acoustic adaptation, and the evolution of reproductive isolation.** *Evolution* 2004, **58:**2144–2155.
16. Rolshausen G, Bartoń KA, Schaefer HM: **Spring arrival along a migratory divide of sympatric blackcaps (*Sylvia atricapilla*).** *Oecologia* 2010, **162:**175–183.
17. Helbig AJ: **Inheritance of migratory direction in a bird species: a cross-breeding experiment with SE- and SW-migrating blackcaps (*Sylvia atricapilla*).** *Behav Ecol Sociobiol* 1991, **28:**9–12.
18. Helbig AJ: **SE- and SW-migrating blackcap (*Sylvia atricapilla*) populations in central Europe: Orientation of birds in the contact zone.** *J Evol Biol* 1991, **4:**657–670.
19. Bensch S, Grahn M, Müller N, Gay L, Åkesson S: **Genetic, morphological, and feather isotope variation of migratory willow warblers show gradual divergence in a ring.** *Mol Ecol* 2009, **18:**3087–3096.
20. Fransson T, Jakobsson S, Johansson P, Kullberg C, Lind J, Vallin A: **Bird migration: Magnetic cues trigger extensive refuelling.** *Nature* 2001, **414:**35–36.
21. Hedenström A, Pettersson J: **Migration routes and wintering areas of willow warblers *Phylloscopus trochilus* (L.) ringed in Fennoscandia.** *Ornis Fennica* 1987, **64:**137–143.
22. Bensch S, Andersson T, Åkesson S: **Morphological and molecular variation across a migratory divide in willow warblers, *Phylloscopus trochilus*.** *Evolution* 1999, **53:**1925–1935.
23. Bensch S, Bengtsson G, Åkesson S: **Patterns of stable isotope signatures in willow warbler Phylloscopus trochilus feathers collected in Africa.** *J Avian Biol* 2006, **37:**323–330.
24. Lundberg M, Boss J, Canbäck B: **Characterisation of a transcriptome to find sequence differences between two differentially migrating subspecies of the willow warbler.** *BMC Genomics* 2013, **14:**330.
25. Hedenström A, Pettersson J: **The migration of willow warbler, *Phylloscopus trochilus*, at Ottenby (In Swedish with English summary).** *Vår Fågelvärld* 1984, **43:**217–228.
26. Ruegg K, Anderson EC, Slaabekoorn H: **Differences in timing of migration and response to sexual signalling drive asymmetric hybridization across a migratory divide.** *J Evol Biol* 2012, **25:**1741–1750.
27. Coyne JA, Orr HA: *Speciation.* Sunderland, MA: Sinauer Associates; 2004.
28. Brelsford A, Irwin DE: **Incipient speciation despite little assortative mating: the yellow-rumped warbler hybrid zone.** *Evolution* 2009, **63:**3050–3060.
29. Rolshausen G, Segelbacher G, Hermes C, Hobson KA, Schaefer HM: **Individual differences in migratory behavior shape population genetic structure and microhabitat choice in sympatric blackcaps (*Sylvia atricapilla*).** *Ecol Evol* 2013, **3:**4278–4289.
30. Pritchard JK, Stephens M, Donnelly P: **Inference of Population Structure Using Multilocus Genotype Data.** *Genetics* 2000, **155:**945–959.
31. Anderson EC, Thompson EA: **Amodel-based method for identifying species hybrids using multilocus genetic data.** *Genetics* 2002, **160:**1217–1229.
32. Rabosky DL, Matute DR: **Macroevolutionary speciation rates are decoupled from the evolution of intrinsic reproductive isolation in *Drosophila* and birds.** *Proc Natl Acad Sci U S A* 2014, **110:**15354–15359.
33. Price TD, Bouvier MM: **The evolution of $F_1$ postzygotic incompatibilities in birds.** *Evolution* 2002, **56:**2083–2089.
34. Orr HA: **The genetic basis of reproductive isolation: Insights from *Drosophila*.** *Proc Natl Acad Sci U S A* 2005, **102:**6522–6526.
35. Larson KW, Liedvogel M, Bensch S, Åkesson S, Wassenaar LI, Hobson KA: **Inferring the ecology of Willow Warblers during their winter moult by sequential stable isotope analyses of remiges.** *J Avian Biol* 2013, **44:**561–566.
36. Drummond AJ, Ashton B, Buxton S: *Geneious* 2011.
37. Larkin MA, Blackshields G, Brown NP: **Clustal W and Clustal X version 2.0.** *Bioinformatics* 2007, **23:**2947–2948.
38. Earl DA, VonHoldt BM: **STRUCTURE HARVESTER: a website and program for visualizing STRUCTURE output and implementing the Evanno method.** *Conservation Genetics Resources* 2012, **4:**359–361.
39. Evanno G, Regnaut S, Goudet J: **Detecting the number of clusters of individuals using the software STRUCTURE: a simulation study.** *Mol Ecol* 2005, **14:**2611–2620.
40. Kaeuffer R, Réale D, Coltman DW, Pontier D: **Detecting population structure using STRUCTURE software: effect of background linkage disequilibrium.** *Heredity* 2007, **99:**374–380.
41. Falush D, Stephens M, Pritchard JK: **Inference of population structure using multilocus genotype data: linked loci and correlated allele frequencies.** *Genetics* 2003, **164:**1567–1587.
42. R Development Core Team: *R: A language and environment for statistical computing.* Vienna, Austria: R Foundation for Statistical Computing; 2012.

# 17

# Understanding stable bi-female grouping in gibbons: feeding competition and reproductive success

Peng-Fei Fan[1*], Thad Q Bartlett[2], Han-Lan Fei[1], Chang-Yong Ma[1] and Wen Zhang[1]

**Abstract**

**Introduction:** Species of the order Primates are highly gregarious with most species living in permanent heterosexual social groups. According to theory in socioecology maximum social group size is limited by rates of intra-group feeding competition and associated increases in travel costs. Unlike other hylobatids, which are predominantly pair living, cao vit gibbons (*Nomascus nasutus*), and two other species of crested gibbon (*Nomascus* spp.) living in northern seasonal forest, regularly exhibit larger bi-female groups. To better understand the ability of northern gibbons to live in stable bi-female groups, we examined food distribution, feeding competition and reproductive success over a period of six years in a small cao vit gibbon population at Bangliang, Guangxi, China.

**Results:** In general, we found weak evidences for within-group contest or scramble competition in our two study groups, which we attribute to high spatial and temporal heterogeneity in the distribution of their important food species. Nevertheless, the larger of the two groups studied increased feeding time and group spread during lean periods, factors that may limit cao vit gibbon group size to a maximum of two breeding females. Relative to tropical pair-living gibbons, there is no evidence that cao vit gibbons travel farther or spend more time feeding, but they did consume more leaves and buds and less fruit and figs. Despite their highly folivorous diet, the average inter-birth interval is comparable to tropical gibbon populations, and the survival rate of infants and juveniles in our study groups is high.

**Conclusion:** Cao vit gibbons do not suffer obvious costs in terms of feeding competition and reproductive success by living in bi-female groups, but within-group feeding competition may determine the upper the limit of cao vit gibbon group size to a maximum of two breeding females. These findings contribute to a growing body of evidence that bi-female grouping can be a stable grouping pattern of gibbons in certain habitats and further emphasize the flexibility of gibbon social organization.

**Keywords:** Cao vit gibbon, Social system, Pair living, Socioecological model, Feeding competition, Reproductive success, Heterogeneity

## Introduction

As an order, primates are highly gregarious with the overwhelming majority of species living in permanent heterosexual social groups. Consequently, it is assumed that group living carries with it a number of potential fitness advantages, including foraging efficiency [1], predator protection [2], cooperative resource defense [3], and protection from conspecifics [4]. Nevertheless, group size in primates varies considerably both within and between species highlighting the putative costs to group living. Chief among these is intra-group feeding competition, which will vary depending on the diet of the animals and distribution of resources in the environment [5]. Minimally, two modes of within-group feeding competition can be delineated: within-group contest (WGC) and within-group scramble (WGS) [6-8]. Contest (or interference) competition is characterized by agonistic interactions among group members, including contact aggression and displacement. Alternatively, scramble (or exploitation) competition occurs when animals lose access

* Correspondence: fanpf1981@gmail.com
[1]Institute of Himalaya Biodiversity Research, Dali University, Dali 671003, Yunnan, People's Republic of China
Full list of author information is available at the end of the article

to resources (e.g., fruit) because those resources have already been exhausted by other group members. According to current socioecological theory, WGC should result if resources occur in discrete, high-quality patches that cannot feed all group members, while WGS should prevail if resources are of low quality, highly dispersed or very large. In WGC, net energy gain is dependent on rank with low-ranking females having a lower energy intake [8]. WGC may also result in differential energy intake among age-sex categories in small family groups [9]. Lower ranking animals may have to increase travel and feeding, or feed more on lower quality foods [8]. WGS is group-size dependent. As group size increases, net energy gain decreases; therefore, larger groups should deplete individual resources more quickly and animals will consequently adjust their foraging behavior by increasing travel, or by accepting a declining energy budget by feeding on lower quality foods [8]. Furthermore, scramble competition should increase when preferred resources are scarce. Finally, differences in net energy intake might translate into differential reproductive success of females [8,10], whereby females in larger groups experience reduced fertility.

Gibbons (Hylobatidae) are small arboreal apes that inhabit the rain forests of East and Southeast Asia. Though bi-male groups (*Hylobates lar*: [11,12]; *Symphalangus syndactylus*: [13]) and bi-female groups [14] have long been reported, members of this family typically live in groups comprised of one adult pair and one to three offspring. The apparent lack of flexibility in the social organization of gibbons relative to other primate families has led to speculation that the gibbon adaptive complex, characterized by monogamy and small-group territoriality, is evolutionarily constrained [15,16]. For example, Brockelman (2009: 211) has framed the question this way: "Have the gibbons' marvelous adaptations to terminal branch feeding and frugivory...placed them in a specialized adaptive zone from which there is no evolutionary escape?" In light of this view, it is not surprising that discussions of gibbon feeding competition have focused on inter-group rather than intra-group feeding competition [16-18]. Significantly, the stable bi-female gibbon groups now well-documented in northern *Nomascus* gibbons [19-24] represent an alternative gibbon grouping pattern, emphasizing the flexibility of social organization in this family and inviting renewed attention to possible ecological constraints on group size in gibbons, and to the role of intra-group feeding competition and travel costs in particular.

The first reliable reports of gibbon groups with more than one female came out of work on western black crested gibbons (*N. concolor*) by Haimoff and colleagues [14] at Mt. Wuliang, China and was followed by observations on capped gibbons (*H. pileatus*: [25]), Hainan gibbons (*N. hainanus*:[19]), hoolock gibbons (*Hoolock hoolock*: [26]), lar gibbons (*H. lar*: [27]), and most recently, cao vit gibbons (*N. nasutus*: [24]). However, many of these bi-female groups were unstable [25-27]. With increasing field data from long-term studies (>6 years), researchers have demonstrated that the occurrence of bi-female groups in the three northern crested gibbons (*N. hainanus*; *N. concolor*; *N. nasutus*) are stable over many years and that co-resident females breed concurrently and repeatedly ([20-23], this study); engage in mutual grooming [21,28]; and maintain similar levels of proximity with adult males [22]. Most recently, Guang and colleagues [28] have demonstrated that females in bi-female groups actively cooperate in maintaining social relationships, rather than co-existing merely through tolerance or avoidance.

The addition of a second breeding female to a gibbon social group represents a significant increase in total group size. Because of the long juvenile period before dispersal in gibbons (>10 years: [12,29]), one adult female could potentially have up to four offspring living with her at once. Consequently, these bi-female groups have a greater number of individuals on average (*N. hainanus*: [20]; *N. concolor*: [21,23]; *N. nasutus*: [24]), up to a documented maximum of 9 group members (this study) and a theoretical maximum of up to 11. These offspring can represent equal competitors if they consume similar foods to adults [9]. Therefore, if intra-group feeding competition plays a role in limiting group size in gibbons we should expect increased competition in larger groups relative to smaller ones and, all things being equal, increased rates of intra-group feeding competition in socially polygynous gibbon groups relative to socially monogamous ones.

In this study, in order to better understand the relationship between feeding competition and group size in gibbons, we examined feeding competition, foraging behavior (diet, feeding time, group spread, travel time and distance), and reproductive success in three bi-female groups living in the Bangliang Gibbon Nature Reserve, Jingxi County, Guangxi, China (Figure 1). First, to document the role of WGC we investigated within-group behavioral differences to determine if there was a consistent pattern of age/sex class differences in diet, activity budget or food-related agonism. Second, to document the impact of WGS we (a) investigated temporal changes in behavior to determine if cao vit gibbons increased travel time and daily path length (DPL), spent more time feeding on low quality foods, or increased group spread during lean periods [8]; and (b) compared feeding behavior between two bi-female groups to see if the larger group suffered greater feeding competition in terms of increased travel, increased group spread, or more time feeding on low quality foods (e.g., buds and

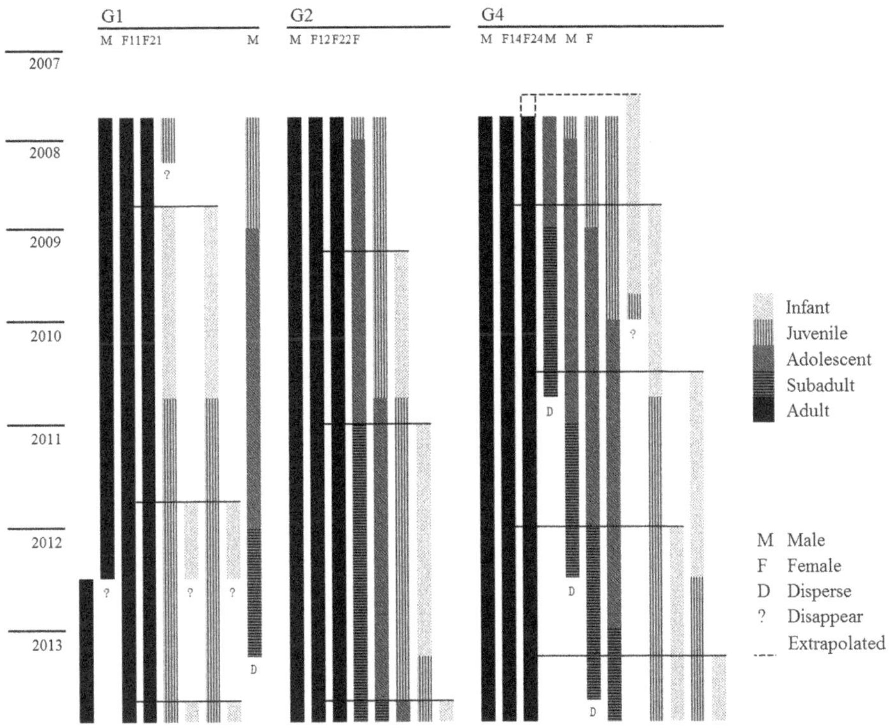


**Figure 1 Demography of three bi-female cao vit gibbon groups monitored between September 2007 and December 2013.** An infant born by F12 in February 2009 was thought to be independent at18 months old and F12 gave birth again in January 2011.

leaves). Third, given the hypothesized fitness consequences for females living in larger groups, we also report data on reproductive rates (i.e. inter-birth intervals) and infant and juvenile mortality in three bi-female groups in order to provide a basis for comparison to published data for females in pair-living groups elsewhere. Ideally, such comparisons would include groups with different numbers of females at a single site, but as there were no pair-living groups at our site this was not possible. Predictions tested in this study are summarized in Table 1. Finally, in addition to behavioral observations we also investigated food distribution patterns at the site in order to relate food distribution to different modes of feeding competition.

## Results

### Food distribution and fruiting phenology

Based on our previous work [30] we identified seven important food species, each of which contributed > 4.7% of the diet based on feeding time. During a subsequent research period between July 2012 and December 2013, one of us (CYM) determined that six of these seven species were among the top ten most consumed species by the same groups. Therefore, these species likely represent staple food species for cao vit gibbons and are more likely to play an important role in influencing feeding competition at this site. Of these, trees of *Ficus hookeriana*, *Choerospondias axillaris*, and *Spondias lakonensis* were rare (<4 trees/ha) at the site, while the remaining four species were more or less common (>8 trees/ha, Table 2). Five of the seven important food species did not show significant deviation from a Poisson distribution, which means these species are randomly distributed, while *Broussonetia papyrifera* and *Ficus glaberrima* showed a clumped distribution (Table 2). *Ficus glaberrima* and *Spondias lakonensis* were significantly taller and had larger DBHs than unimportant trees and *Ficus hookeriana* had a larger DBH than unimportant trees. Finally, *Burretiodendron hsienmu* trees were taller but had a smaller DBH on average than unimportant trees (Table 2).

The production of ripe fruit at the study site exhibited extreme monthly and annual variability. Fruit was more abundant between July and December in 2008 than in 2009, probably as a result of less rainfall in 2009. *Tetrastigma pubinerve*, an important food species for the gibbons [30], produced a lot of fruit from August to December in 2008. However, this species produced almost no fruit in 2009. Based on the food availability index we were able to distinguish two periods of marked fruit scarcity, or lean periods. The first ranged from February to April 2009 and the second occurred from October to December 2009. During each of these months (N = 6) the fruit availability index was below 1,200. By contrast, in all other months (N = 12) the availability index

**Table 1 Predictions and related feeding competition modes used in this study**

| Level | Food competition mode | Predictions | Results G1 (small group) | G4 (large group) |
|---|---|---|---|---|
| Within-group comparison | WGC | 1. Age/sex differences in food-related agonism | Y | Y |
| | | 2. Age/sex differences in travel time. | Y | Y |
| | | 3. Age/sex differences in feeding time. | N | N |
| | | 4. Age/sex differences in diet (i.e. leaves & buds) | P | P |
| Seasonal variation | WGS | 5. Groups travel further during lean period. | N | N |
| | | 6. Groups increase travel time during lean periods | N | N |
| | | 7. Groups increase feeding time during lean period. | N | Y |
| | | 8. Groups increase inter-individual distance and group spread during lean period. | N | Y |
| | | 9. Groups consume more leaves during lean period. | Y | Y |
| Between-group comparison | WGS | 10. Larger group spends more time in travelling. | | Y |
| | | 11. Larger group spends more time feeding. | | N |
| | | 12. Larger group consumes more leaves. | | N |
| | | 13. Larger group spreads wider than small group. | | N |
| Between-site comparison | WGC and WGS | 14. Decreased fertility in bi-female groups. | | N |
| | | 15. Increased infant and juvenile mortality in bi-female groups | | N |

Y: yes, prediction met; N: no, prediction not met; P: prediction partly supported.

was at least 2,400 (Figure 2). In the analysis below these months are referred to collectively as the fruiting period.

### Food-related agonism and age/sex differences in diet and activity budget

Both food-related agonistic interactions and displacements were rare in our two study groups (Table 3). In both groups, juveniles received far more food-related agonism from adults than did other age classes (10 of 11 instances). In G1, the juvenile spent more time travelling than the other three members (Wilcoxon matched pairs test: juvenile – male1, $Z = -2.9$, $P = 0.003$; juvenile – F11, $Z = -2.6$, $P = 0.010$; juvenile – F21, $Z = -3.2$, $P = 0.001$; $N = 18$ in all tests). But the male1 spent less time feeding than all other member (Wilcoxon matched pairs test: male1– F11, $Z = -3.7$, $P < 0.001$; male1 – F21, $Z = -3.3$, $P = 0.001$; male1 – juvenile, $Z = -2.3$, $P = 0.019$; $N = 18$ in all tests). F11 spent more time travelling (Wilcoxon matched pairs test: $Z = -2.2$, $P = 0.028$; $N = 18$) and feeding (Wilcoxon matched pairs test: $Z = -2.5$, $P = 0.012$; $N = 18$) than F21, probably due to F11 spending more time searching for insects than F21 (Wilcoxon matched pairs test: $Z = -2.0$, $P = 0.044$; $N = 18$). In general, group members shared a similar diet except that the male1 consumed more fruit (Wilcoxon matched pairs test: male1– F11, $Z = -2.8$, $P = 0.005$; male1 – F21, $Z = -2.8$,

**Table 2 Distribution pattern and characteristics of seven important food species in cao vit gibbon habitat, calculated from 73 20 × 20 m plots**

| Latin name | Contribution to the diet | Food part eaten | N of trees | N of plots | Density (trees/ha) | Tree height[b] | DBH[c] | CD index[d] |
|---|---|---|---|---|---|---|---|---|
| *Broussonetia papyrifera* | 17.1 | Bud, Fruit, Leaf | 26 | 7 | 8.9 | 9.1 ± 1.5 | 14.8 ± 3.4 | 10.3** |
| *Ficus glaberrima* | 13.0 | Fruit, Bud, Leaf | 61 | 26 | 20.9 | 11.7 ± 3.9** | 26.3 ± 15.9** | 3.1** |
| *Ficus hookeriana* | 6.6 | Fruit, Leaf, Bud | 6 | 4 | 2.1 | 12.3 ± 5.1 | 33.7 ± 22.7** | 1.6 |
| *Tetrastigma pubinerve*[a] | 6.3 | Fruit, Bud, Leaf | 24 | 15 | 13.6 | | | 1.5 |
| *Burretiodendron hsienmu* | 5.5 | Fruit, Leaf, Bud, Flower | 34 | 16 | 11.6 | 9.6 ± 2.1* | 14.1 ± 3.4* | 2.9 |
| *Choerospondias axillaris* | 5.2 | Fruit, Leaf | 6 | 4 | 2.1 | 10.8 ± 2.6 | 24.9 ± 23.8 | 1.6 |
| *Spondias lakonensis* | 4.7 | Fruit, Leaf | 11 | 9 | 3.8 | 14.7 ± 7.5** | 31.7 ± 23.8** | 1.2 |
| All other trees | | | 1072 | 73 | 367.1 | 8.9 ± 2.7 | 16.7 ± 7.2 | |

[a]Height and DBH are not available for *Tetrastigma pubinerve*, which is a liana species.
[b]and [c]For statistical analysis height and DBH were compared to all other trees (* $P < 0.05$; ** $P < 0.01$).
[d]If the distribution pattern of the species significantly deviates from Poisson distribution (** $P < 0.01$).

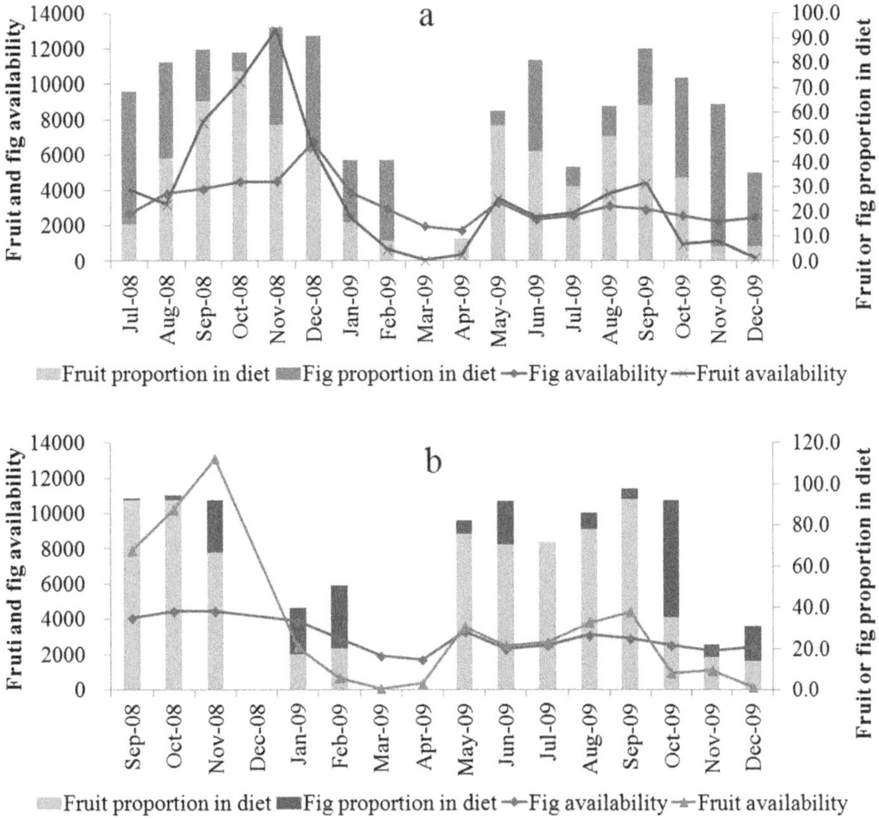


**Figure 2 Seasonal variation of fruit and fig availability and fruit and fig proportion in the monthly diet of G1 (1a) and G4 (1b).**

P = 0.005; male1-juvenile, Z = −1.9, P = 0.053; N = 18 in all tests) and less insects than other members (Wilcoxon matched pairs test: male1– F11, Z = −3.2, P < 0.001; male1 – F21, Z = −2.8, P = 0.006; male1-juvenile, Z = −2.9, P = 0.003; N = 18 in all tests). No difference was found between the two females except that F11 spent more time searching for insects.

We found a similar pattern in G4. No differences were found in diet or activity budget between the two adult females. Both juveniles and adolescents spent more time in travelling than the adult male and more than the adult females in G4 (Wilcoxon matched pairs test: male4 – juveniles, Z = −3.4, P = 0.001; male4 – adolescents, Z = −3.3, P = 0.001; F14 – juveniles, Z = −2.7, P = 0.007; F14 – adolescents, Z = −2.5, P = 0.012; N = 15 in all tests). Juveniles spent less time feeding than females (Wilcoxon matched pairs test: juveniles – F14, Z = −2.5, P = 0.013; juveniles – F24, Z = −2.8, P = 0.006; N = 15 in all tests). Juveniles ate less fruit (Wilcoxon matched pairs test: Z = −2.2, P = 0.031; N = 15) but more leaves (Wilcoxon matched pairs test: Z = −2.3, P = 0.021; N = 15) and figs (Wilcoxon matched pairs test: Z = −2.0, P = 0.05; N = 15) than the adult male. Adolescents spent less time searching for insects than one female (Wilcoxon matched pairs test: F24 - adolescents: Z = −2.2, P = 0.026; N = 15) (Figure 3).

**Table 3 Food-related agonistic interactions or displacements in group G1 (1,175 h observed from July 2008 and December 2009) and G4 (776 h observed from September 2008 and December 2009)**

| Group | Initiator | Receiver | N |
|---|---|---|---|
| G1 | Adult male | Juvenile | 2 |
| | F21 | Adult male | 1 |
| G4 | Adult male | Juvenile | 3 |
| | F14 | Juvenile | 2 |
| | F14 | Adolescent | 2 |
| | Adult male | Adolescent | 1 |

Except for one case from G1, all agonistic interactions were observed during lean periods between February and April or October and December, 2009.

### Temporal variation in foraging behavior in response to fruit scarcity

The proportion of fruit and figs in the diet varied from 0.9% in March 2009 to 94.3% in November 2008 in G1 (Figure 2a) and from 0.6% in April 2009 to 97.8% in September 2009 in G4 (Figure 2b). When fruit was less

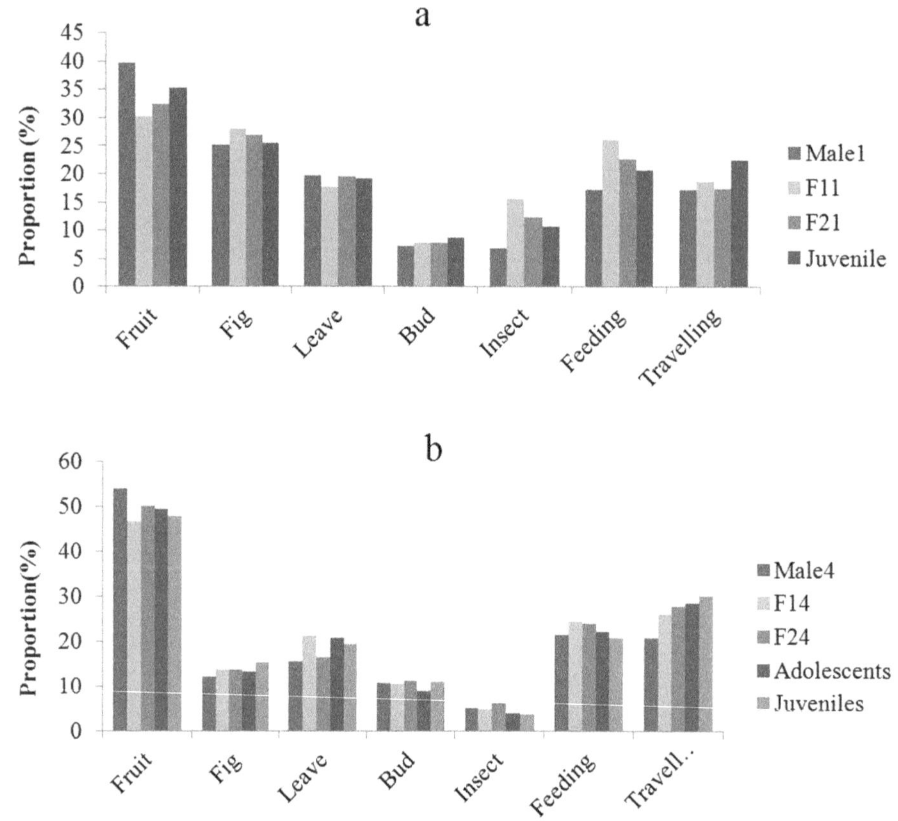


**Figure 3 Diet and time budget differences among members in G1 (2a) and G4 (2b).** Data for the two adolescents were pooled as were data for the two juveniles.

available, both groups fed more on leaves and buds [31]. Contrary to our prediction, neither group spent more time travelling during the lean season (Mann–Whitney *U* test: G1: $Z = -1.8$, $P = 0.075$, $N_{fruit\ month} = 12$, $N_{lean\ month} = 6$; G4: $Z = -1.2$, $P = 0.239$, $N_{fruit\ month} = 9$, $N_{lean\ month} = 6$). On the contrary, both groups decreased DPL during lean periods (G1: lean periods: 890 ± 323 m, range: 354–1538 m, N = 20; fruiting periods: 1440 ± 454 m, range: 494–2361 m, N = 39; Mann–Whitney *U* test: $Z = -4.2$, $P < 0.001$; G4: lean periods: 920 ± 403 m, range: 475–1983 m, N = 14; fruiting periods: 1471 ± 502 m, range: 582–2282 m, N = 16; Mann–Whitney *U* test: $Z = -2.9$, $P = 0.004$). G1 did not spend more time in feeding (Mann–Whitney Utest: $Z = -1.5$, $P = 0.134$, $N_{fruit\ month} = 12$, $N_{lean\ month} = 6$) and did not increase inter-individual distance (Mann–Whitney *U* test: $Z = -0.6$; $P = 0.542$, $N_{fruit\ month} = 12$, $N_{lean\ month} = 6$) or group spread (Mann–Whitney Utest: $Z = -0.7$, $P = 0.512$, $N_{fruit\ month} = 12$, $N_{lean\ month} = 6$) in lean periods (Table 4). On the other hand, G4 increased feeding time (Mann–Whitney *U* test: $Z = -2.8$, $P = 0.006$, $N_{fruit\ month} = 9$, $N_{lean\ month} = 6$), inter-individual distance (Mann–Whitney *U* test: $Z = -2.4$, $P = 0.018$, $N_{fruit\ month} = 9$, $N_{lean\ month} = 6$), and group spread (Mann–Whitney *U* test:$Z = -2.4$, $P = 0.018$, $N_{fruit\ month} = 9$, $N_{lean\ month} = 6$) in lean periods (Table 4). Except for one case in G1, all food-related agonism was observed during lean periods (Table 3).

### Differences between groups

Adult members in both groups spent similar amounts of time feeding, but on average individuals in the larger group (G4) spent significantly more time travelling and less time resting than in the smaller group (G1, Table 5). However, G4 did not travel farther than G1 (G1: mean 1182 ± 474 m, range 354 – 2361 m, N = 50; G4: mean 1214 m ± 531, range 475 – 2282, N = 30; Mann–Whitney *U* test: $Z = -0.1$, $P = 0.881$). Individuals in G4 ate more fruit and fewer figs than individuals in G1 (Table 5), but if we combine fruit and figs into a single category there is no significant difference between these two groups, though G1 consumed more insects than G4 (Table 5). No differences were found in the consumption of leaves or leaf buds between groups. Wilcoxon matched pairs test showed the larger group (G4) maintained closer inter-individual distance than the smaller group (G1: $Z = -2.7$, $P = 0.006$, $N = 15$) using three sample distances.

**Table 4 Mean inter-individual distance and group spread (represented as maximum inter-individual distance) in two bi-female cao vit gibbon groups in China**

| | G1 | | | G4 | | |
|---|---|---|---|---|---|---|
| | N | Mean inter-individual distance | Group spread | N | Mean inter-individual distance | Group spread |
| Jul-08 | 25 | 11.0 | 16.3 | | | |
| Aug-08 | 273 | 7.4 | 11.7 | | | |
| Sep-08 | 221 | 7.6 | 11.8 | 68 | 4.6 | 7.4 |
| Oct-08 | 196 | 8.6 | 13.2 | 121 | 4.3 | 7.4 |
| Nov-08 | 175 | 16.8 | 26.2 | 103 | 4.0 | 7.3 |
| Dec-08 | 202 | 11.6 | 17.4 | | | |
| Jan-09 | 118 | 9.8 | 14.8 | 116 | 4.4 | 7.5 |
| **Feb-09** | 192 | 15.0 | 23.5 | 239 | 9.4 | 15.9 |
| **Mar-09** | 301 | 11.5 | 18.3 | 300 | 7.8 | 13.0 |
| **Apr-09** | 133 | 8.6 | 13.3 | 209 | 5.7 | 9.2 |
| May-09 | 84 | 6.8 | 11.2 | 198 | 8.6 | 13.9 |
| Jun-09 | 142 | 15.8 | 25.5 | 344 | 7.3 | 12.1 |
| Jul-09 | 126 | 5.7 | 9.5 | 153 | 6.2 | 10.5 |
| Aug-09 | 97 | 6.3 | 11.2 | 313 | 5.1 | 8.5 |
| Sep-09 | 201 | 8.6 | 15.3 | 221 | 4.8 | 8.0 |
| **Oct-09** | 132 | 5.7 | 9.6 | 162 | 5.9 | 9.3 |
| **Nov-09** | 98 | 8.9 | 13.8 | 106 | 11.4 | 18.7 |
| **Dec-09** | 177 | 13.9 | 22.0 | 223 | 9.0 | 14.3 |
| Lean period | | 10.6 ± 3.5 | 16.8 ± 5.4 | | 8.2 ± 2.2 | 13.4 ± 3.7 |
| Fruiting period | | 9.7 ± 3.6 | 15.3 ± 5.4 | | 5.5 ± 1.6 | 9.2 ± 2.4 |
| Z | | −0.610 | −0.656 | | −2.357 | −2.359 |
| P | | 0.542 | 0.512 | | 0.018 | 0.018 |

Months in bold was classified as fruit scarce based on the mean value of fruit production in 2009.

### Demography and female reproduction

At the beginning of this study G1 consisted of one adult male, two adult females and two juveniles. One juvenile disappeared in February 2008, F11 gave birth in November 2008, and F21 gave birth in December 2008. Both females again gave birth in November 2011. The original adult male of G1 was replaced by a new male in July 2012. Coincident with a period of male replacement, two infants born in 2011 disappeared and are presumed to have died. F11 gave birth in December 2013 and F21 in November 2013. The juvenile, by then a sub-adult male, dispersed in April 2013. The group

**Table 5 Comparison in the time budget and diet of the adults in gibbon groups G1 and G4**

| | G1 | G4 | Wilcoxon matched pairs test | |
|---|---|---|---|---|
| **Activities** | **Mean ± SD (%)** | **Mean ± SD (%)** | **Z** | **P** |
| Feeding | 22.0 ± 5.5 | 23.2 ± 8.3 | −0.6 | 0.57 |
| Travelling | 17.4 ± 6.2 | 24.7 ± 8.5 | −3.2 | 0.001 |
| Resting | 45.0 ± 7.6 | 30.0 ± 10.2 | −3.2 | 0.001 |
| Social | 14.8 ± 8.2 | 20.3 ± 9.5 | −3.1 | 0.002 |
| Fruit | 34.4 ± 25.2 | 50.2 ± 34.8 | −3.1 | 0.002 |
| Fig | 23.1 ± 17.9 | 13.1 ± 15.1 | −2.7 | 0.007 |
| Fruit + Fig | 57.5 ± 28.5 | 63.2 ± 35.1 | −1.8 | 0.078 |
| Leaves | 20.8 ± 23.2 | 17.6 ± 25.2 | −1.0 | 0.311 |
| Buds | 8.9 ± 20.2 | 10.7 ± 24.8 | −0.1 | 0.925 |
| Insects | 11.6 ± 16.2 | 5.3 ± 9.0 | −2.8 | 0.005 |

consisted of seven individuals in December 2013 (Figure 1).

G2 consisted of one adult male, two adult females, one adolescent, and one juvenile in December 2007 and increased to eight individuals by the end of the study due to three births by F21 (Figure 1). The other female, F22 in G2, never reproduced during the six years of the study. This female was thought to be too old to give birth because her pelage became darker over the course of the study, a phenomenon characteristic of aging females among *Nomascus* gibbons [32].

G4 consisted of eight individuals including one adult male, two adult females, two adolescents, two juveniles, and one infant in December 2007 and increased to nine due to a birth by F14 in November 2008. A small juvenile disappeared in November 2009 just after it became fully independent at 3 years old (Figure 1). F14 gave birth to another infant in January 2012. And F24 in G4 gave birth in July 2010 and again in March 2013. The group size of G4 remained at least eight individuals and occasionally nine individuals during the six-year observation period, representing the largest group in the population. It is possible that nine individuals is the maximum number of animals this group can accommodate. In 3 of the 4 cases, a subadult individual dispersed within 1–5 months of the birth of a new infant (Figure 1).

Eight floating males (black phase) and nine floating females were also observed during the six years. Co-dispersal of three males in one case and two females in another case was observed. The three males were observed on six days between February and November 2009. They finally established their own territory in the site and attracted a female to live with them [24]. The co-dispersing females were observed on four days between August 14 and September 17. They fed on the same tree, groomed each other, and always remained within 20 m of each other. Subadult males produced solo bouts before dispersal but subadult females only produced great calls with resident adult females and never sang solo bouts before dispersal. Floating individuals did not call.

In summary, five of the six adult females in our three study groups reproduced at least twice between December 2007 and 2013. One old female in G2 never gave birth during the study. Two infants in G1 died during male replacement and one juvenile each died in G1 and G4. All other offspring survived to dispersal at about 10 years or to the end of the study. The inter-birth interval (IBI) was 31.0 ± 6.6 months (range: 23–38 months, N = 8) and 33.5 ± 5.5 months if we include only surviving infants (range: 23 – 38 months, N =6). Although the sample size is small, IBI does not appear to differ between groups (G1 mean = 30.3, N = 4; G2 mean = 29.0, N = 2; G4 mean = 35.3, N = 2). Group size varied between five and nine individuals due to births, disappearances/deaths, and dispersals and was always larger, on average, than in pair living groups of other species (Figure 1).

## Discussion

We studied feeding competition and female reproductive success in a cao vit gibbon population dominated by bi-female groups (14 of 17 groups: Le TD, Fan PF, Yan L, Le HO, Josh K: The global cao vit gibbon (*Nomascus nasutus*) population. unpublished). The sample size was limited because the population straddles the international border between China and Vietnam. Nevertheless, we successfully monitored group dynamics of the three bi-female groups that range wholely or partly on the China side of the boarder beginning in December 2007 and documented patterns of feeding and ranging in two of these groups between January 2008 and December 2009. Due to the difficult karst terrain, we could not follow gibbons at our site. Instead, we observed them from 50–500 m away from a series of observation posts. Results reported here demonstrate the feasibility of these methods for collecting basic ecological and demographic data on arboreal primates in karst forest ([33-35], this study), as well as the potential for these methods to be used to observe arboreal primates in other difficult terrian where trees are small.

### Feeding competition and characteristics of important food species

Exploitation of large feeding patches predicts WGS and weak WGC, which is consistent with our observations (see below). The important food species for cao vit gibbons either exist as large trees or grow in large clumps. Full grown trees of *Ficus glaberrima, Ficus hookeriana,* and *Spondias lakonensis* are the largest trees in this habitat. Foraging gibbons could visit a single fruiting tree of these species repeatedly over several days (PFF personal observation). *Broussonetia papyrifera* was similar to trees of non-important species in terms of height and DBH, but this species usually occurs in clusters around abandoned fire pits used for charcoal production and several trees with continuous tree crowns can serve as a single large feeding patch for gibbons.

### Feeding competition within bi-female gibbon groups

#### *Contest competition*

We found evidence of only weak WGC in terms of food-related agonism or age/sex-related differences in diet in our two study groups. In general, increasing numbers of group members should increase instances of agonism [36], but contrary to this pattern we observed only 11 agonistic episodes during feeding in over 2,000 hour of observation during 300 days (0.04 cases/d), which was lower than that observed in another bi-female group of western black crested gibbon (0.3 cases/d: [22]), in pair living siamangs (1.5 cases/d: [37]), or in lar gibbons (0.08 cases/d: [18]). Given the observation distance (50–500 m), we might have overlooked some subtle agonistic behaviors, such as open-mouth threats that have been observed in western black crested gibbons [22], but frequent grooming and social play among members in these groups (11.3% and 16.9% of active time for G1 and G4 [31]) suggests that social relationships were in fact harmonious. It is important to note that no food-related agonism was observed between adult females living in the same group. Both females groomed each other and also groomed the adult male and received grooming from him (Fan et al., in preparation). Although we could not reliably identify individual juveniles in each scan, we did observe females grooming both similar-sized juveniles who were mothered by different females. In general, group members shared similar diets and no differences in diet were detected between females living in the same group.

In both groups, juveniles consumed less fruit and spent more time travelling than the resident adult male (Figure 2) and in all but one case juveniles were the targets of intra-group agonism (Table 3). These findings might indicate that juveniles were the recipients of contest competition, although other factors may account for the observed differences in activity budget. For example, young gibbons may have different nutritional goals or may travel more as a part of social (e.g., chase) or solitary (e.g., exploration) play [18]. However, because we did not distinguish directed movement between trees (i.e., travel) from movement within trees (i.e. move) during data collection, it is not possible for us to determine precisely what accounts for the observed differences in activity.

#### *Scramble competition*

We found some evidences of WGS within this population. With group size ranging from 8–9 individuals, group G4 is the largest gibbon group ever recorded in the wild. Adult members of G4 spent more time travelling on average and less time resting than adults of G1. However, greater travel time did not translate into longer DPL. It is possible that gibbons in G4 had to spend more time foraging (slow searching movement) between patches, or had to accommodate the slower movements of juveniles in this group.

Contrary to predictions, neither group increased DPL in response to fruit scarcity. Instead, they altered their diet, consuming more leaves and buds [31]. We also found that G4 altered its behavior in response to fruit scarcity in other ways. G4 increased feeding time and group spread during lean periods, which may represent a strategy to avoid food competition in lean seasons. All cases of agonism in G4 also occurred during lean periods. Together these results suggest that this large group experienced increased feeding competition in lean periods, which may impose an upper limit on group size. Subadults in G4 almost always dispersed soon after the births of infants, which might mean that G4 approached its maximum group size at nine individuals.

### Stable bi-female groups in cao vit gibbons

It has been argued that the benefits of living in permanent groups include enhanced predator avoidance, foraging advantages, and avoidance of conspecific threat (e.g., infanticide), while the main cost of living in groups is within group feeding competition [5], which might result in increased travel costs or a switch to a lower quality diet. A preliminary comparison offers no evidence that cao vit gibbons travel farther than tropical gibbons living in adult pairs (on average 1.2 km, N = 16, Table 17.2 in [38]). On the other hand, relative to other gibbons, rates of folivory among cao vit gibbons are high. Though in this case, the lower proportion of fruit and figs in the overall diet was primarily a result of fruit scarcity during lean periods, not by within group food competition. Our study groups consumed > 90% fruit (including figs) in their monthly diet during fruiting periods, which is comparable or even higher than monthly fruit and fig consumption in tropical gibbons [18,37,39-41]. However, in contrast to other gibbon genera, cao vit gibbons switched their diet and mainly depended on more abundant leaves and buds in lean periods when the proportion of fruit and figs in the diet decreased to < 1%. As a result, their annual diet was dominated by leaves (G1: 49.1%; G4: 54.9% [31]). In other words, it is highly likely that a small pair-living group at the site would have a similar overall diet to bi-female groups. Nevertheless, if leaves do represent low quality resources for cao vit gibbons, one might assume that gibbons in these bi-female groups would be subject to reproductive costs. However, there is no evidence that their highly folivorous diet translates into lower reproductive success. With the exception of a single old female, the adult females in the three groups we studied bred repeatedly. IBIs of breeding females in our groups and other bi-female groups, for which there are data, were comparable or even shorter than the IBIs reported for tropical pair-living groups (Table 6 [20,23,42-46]). Except for two infants that died during a male replacement incident, all other infants (N = 8) survived to independence at 3 years old. Similarly, only two small juveniles died over the 7-year study period.

Though preliminary, our finding that the reproductive success of cao vit gibbons does not differ appreciably from that of southern populations requires further consideration, particularly in light of the extreme seasonality of their habitat and clear differences in diet. One possibility is that cao vit gibbons may have an evolved

**Table 6 Inter-birth intervals of wild gibbon populations**

| Species | Inter-birth interval (month) | Range (month) | N | Grouping pattern | Literature |
|---|---|---|---|---|---|
| *Hylobates lar* | 41 ± 9.1 | 34-71 | 17 | Pair, Bi-male | [42] |
| | 120 | | 1 | Pair | [43] |
| | 26.5 | 22-31 | 2 | Pair | [44] |
| *H. agilis* | 38.4 ± 4.8 | | 5 | Pair | [45] |
| *Hoolock leuconedys* | 49 | | 1 | Pair | Fan PF, unpublished data |
| *Symphalangus syndactylus* | 33.7 ± 3.5 | 28.5-38.1 | 5 | Pair, Bi-male | [46] |
| | 60 | 48-72 | 2 | Pair | [43] |
| | >36 | | 3 | Pair | [44] |
| *Nomascus hainanus*[1] | 24 ± 1.6 | 22-26 | 4 | Bi-female | [20] |
| *N. concolor* | 42.3 ± 9.2 | 37-53 | 3 | Bi-female | [23] |
| *N. nasutus* | 31 ± 6.6 | 23-38 | 8 | Bi-female | Present study |

[1]Calculated from Table II in [20]. Only birth dates of consecutive infants known to within a month were used to calculate the inter-birth interval.

capacity to obtain sufficient nutrition from leaves and buds necessary to tolerate lean periods, a view that is consistent with larger body size in *Nomascus* relative to gibbons of the best studied genus *Hylobates* [47]. It is also possible that cao vit gibbons, like some other primate species [48,49], may have the ability to deposit fat reserves during periods when fruit is abundant to help them cope with fruit scarcity.

In any case, if cao vit gibbons do not suffer greater reproductive costs from living in bi-female groups, then any benefits of living in larger groups—foraging efficiency, predation protection, range defense, or protection from conspecifics—should promote multi-female groups. Given that predation on gibbons is extremely rare, predation avoidance may not represent a significant benefit for females living in bi-female groups relative to those living in pairs. On the other hand, resident females in the same group always produced similar-aged juveniles. In instances where these similar-aged offspring are also females they may subsequently disperse together, which might increase survival during dispersal and increase the probability of establishing their own territory ([24], this study). In addition, co-resident females might benefit from improved inclusive fitness if they are kin, which is possible if females born in the same group disperse together (this study). If so, then an unpaired female may benefit more by joining an established adult pair than by waiting to join an unmated male. But, even if females do not obtain any direct or indirect benefits, bi-female groups could form if the cost of evicting the second female exceeds the cost of tolerance [22,50].

## Conclusion

The data presented here, together with observations of other northern *Nomascus* populations ([20-24,28], this study), suggest that bi-female grouping is a stable grouping pattern for gibbons in certain habitats. Given the limited sample size of our study, any conclusions about the mechanisms favoring bi-female grouping in *Nomascus* gibbons must be tentative. Nevertheless, we have attempted to show how intra-group feeding competition, combined with high spatial and temporal heterogeneity in resource abundance may help us understand why bi-female groups represent the dominant grouping pattern in this cao vit gibbon population. Feeding competition in this population is rare likely owing to high spatial and temporal heterogeneity of resources, including leaves. However, when evidence of WGC and WGS was detected it was in the larger group during lean periods. This combined with the timing of subadult dispersal suggests that group size is capped at two adult females. In order to better understand feeding competition and its role in determining gibbon social organization, detailed comparative research of food distribution, feeding party size, feeding bout length, and seasonal variation of feeding behavior in relation to food availability using standardized methods will be extremely useful. Only in this way will we be able to discern if the ability to form stable multi-female groups is unique to gibbons of the genus *Nomascus*, or if it is merely another example of the social flexibility of all gibbons, and to some extent all primates.

## Methods

### Ethical approval

All observations were conducted in Bangliang Gibbon Nature Reserve under the permission form the Guangxi Forestry Bureau and the Management Bureau of Bangliang Gibbon Nature Reserve. No any gibbons were handled during the research. All research adhered to the legal requirements of China and the American Society of Primatologists (ASP) Principles for the Ethical Treatment of Non-Human Primates (https://www.asp.org/society/resolutions/EthicalTreatmentOfNonHumanPrimates.cfm).

### Study area

We carried out the study in Bangliang Gibbon Nature Reserve (486 to 926 m above sea level, 22°55′N, 106°29-30′E, established in 2009), Jingxi County, Guangxi, China, which is continuous with the Cao Vit Gibbon Conservation Area in Trung Khanh, Cao Bang, Vietnam (established in 2007) [51]. This karst limestone area is characterised by steep-sloped sharp-peaked mountains surrounded by two branches of the Quay Son River flowing into Vietnam. The forest is monsoon tropical forest but was degraded by selective logging, fuelwood collection, charcoal making, and agriculture before the nature reserves were established [30]. The total area of good forest available for gibbons is estimated to be 2,200 ha [51]. The climate is humid tropical and shows obvious seasonal variations in temperature and rainfall. The annual mean temperatures is *c.* 20.0°C and monthly mean temperatures range from 9.8°C in the coldest season to 26.6°C during the warmest season. Annual precipitation was 1,804 mm in 2008 and 1,363 mm in 2009 [17]. Additional information on the study area has been published elsewhere [30,31,51].

### Gibbon population and study groups

This population was rediscovered in Vietnam in 2002 [52] and in China in 2006 [53] and represents the only known population of cao vit gibbons. Since the discovery of this population, surveys have shown that bi-female groups predominate (14 of 17 groups Le TD, Fan PF, Yan L, Le HO, Josh K: The global cao vit gibbon (*Nomascus nasutus*) population. unpublished; 2 of 3 groups [54]; 3 of 5 groups *Geissmann* T, La QT, Trinh DH, Dang NC, Pham DT, Vu DT: Report on an overall survey of the cao vit gibbon population Nomascus sp. cf. nasutus in Trung Khanh District, Cao Bang Province (second overall survey). unpublished). Three bi-female groups range wholely or partly in China and we started to monitor the dynamics of all three groups monthly beginning in December 2007 [24]. Non-adult gibbons were classified into 4 categories: infant (0–2 years of age), juvenile (2–5 years of age), adolescent (5–8 years of age), and subadult (>8 years of age) [18,29].

G1, one of the smaller groups in the population at the beginning of the study, consisted of one adult male, two adult females, and one juvenile in August 2008. It increased to six individuals due to one birth each in November and December 2009 and then to seven individuals by the end of December 2013. Alternatively, G4 represents the largest documented social group in any gibbon population, consisting of eight individuals (one adult male, two adult females, two adolescents, two juveniles and one infant) at the beginning of the study in August 2008 and increasing to nine individuals due to a birth in November 2008. At the beginning of data collection G1 had 4 individuals and group G4 had 8, while during most observation months G1 had 6 individuals and G4 had 9. It is our contention that a difference in group size of 50-100% is biologically meaningful and should be associated with detectable differences in foraging behavior. The third group, G2 ranged from five (one adult male, two adult females, and two juveniles) to eight animals over the course of the study (September 2007 to December 2013). Its range extended across the international boarder so while we monitored demographic changes in this group, systematic data on feeding and range use was limited.

We could reliably distinguish the two females (F11 and F21) in G1 based on their white face ring and crest shape only after June 2008. It was easy to distinguish the two females in G2 (F12 and F22) and in G4 (F14 and F24) based on their body color, crest shape, and the size of their infants. We could not consistently distinguish the two similar-sized juveniles or the two similar-sized adolescents in G4.

### Distribution pattern of important food species and fruit availability

To investigate plant diversity and forest structure, beginning in December 2007 we established 44 20 × 20 m sample plots throughout the area over which gibbons were known to travel [30]. All plots were located within G1's home range during later behavioral observation. Due to a dispute over the location of the international border within the study site, we could not establish any plots within G4's home range. In general,the forest structure and composition in G4′s home range is similar to that of G1.

Plots were placed along six parallel transects that traversed the site, with no fewer than eight plots in each of four distinctive topographical regions (slope, ridgeline, valley, and col) of the site [30]. In July 2010, an additional 29 20 × 20 m plots were set outside any group's home range in order to survey potential haibitat quality using the same methods. In each plot (N = 73), we measured diameter at breast height (DBH) and tree height for all trees greater than 10 cm diameter (N = 1216). Liana species were also recorded but only in the first 44 plots. Seven food species, each of which contributed > 4.7% and in total 58.4% of their diet based on feeding time [30], were designated important food species. The eighth species contributed 3.1% to the total diet in 2008 and 2009 but contributed much less to the diet (<1%) in 2012 and 2013, therefore, we did not consider it an important species. To characterize the distribution of important foods, we used the species composition from all sample plots to calculated the coefficient of dispersion (CD) for each species, whereby a CD >1 indicates a clumped distribution, a CD <1 a uniform distribution, and a CD = 1 a random distribution [54,55]. If species are distributed randomly, their allocation in the sample should correspond to a Poisson distribution [55]. Therefore, we first tested if the distribution of each important species fit a Poisson distribution using a one-sample Kolmogorov-Smirnov test. We then used the Mann–Whitney *U* test to compare DBH and tree height between each important food species and to all other trees in the plots.

All trees in the first 44 plots, as well as lianas and epiphytes present on these trees, were monitored for seasonal availability of food types (fruit, figs, leaves, buds, and flowers) for three days each month [31]. We estimated the food abundance value (i.e. percent of crown cover) of each food type on a five-point scale, 0–4 (0 < 1%, 1 = 1-25%, 2 = 26-50%, 3 = 51-75%, 4 = 76-100%). The food availability index for non-fig fruit and figs was then calculated based on the food abundance score (0–4) for each fruiting tree, weighted by diameter at breast height (for additional details see [31]). Plant species not eaten by gibbons were not included in calculations of food abundance. Based on natural breaks in the data, months with an availability index of <1200 were designated as lean periods, while months of >2400 were designated as fruiting periods.

### Behavioural observation

From January 2008 to December 2009 (with the exception of February 2008), we spent 7–27 days each month (485 days total), observing gibbon behavior. Whenever possible, groups were observed for the full day. During this period, we observed G1 for 1,455 hours over 215 days from January 2008 to December 2009, G2 for 201 hours over 67 days from August 2008 to December 2009, and G4 for 776 hours over 122 days from August 2008 to December 2009.

Because traditional or standard primate observational techniques (i.e. direct follows) are impossible at this site due to the steepness of the landscape, we observed gibbons from several established observation posts at distances of between 50 and 500 m (most observations were made at a distance of 100–300 m), using 8 × 30 binoculars (Steiner safari) or a spotting scope (Leica Apo-Televid77 20–60) [30,31]. Trees in this site are quite small (on average, tree height was 10.5 m) and gibbons always stayed close to the top of these small trees. Through our Leica telescope we had a good view to observe gibbons even at a distance over 50 m. Our observation posts could cover about 95% of G1's home range and 85% of G4's home range. Given the observation conditions focal animal sampling was impractical so we used a scan sampling regime at 5-min intervals to record the activity of gibbons [56]. At each 5-minute point sample, a 1-min scan was made recording the behavior of all visible group members. Each individual was observed for 5 seconds and its predominant behavior was recorded. Inter-individual distance was recorded in an additional 1-min scan between July 2008 and December 2009. We recorded activity as resting, travelling, feeding, grooming, calling, playing, or other [31]. When an individual was feeding, we recorded the food species and specific part eaten (fig, fruit, leaves, buds, flower, invertebrates, and other). Cao vit gibbons usually checked dead leaves and branches before they put something into their mouths and they always moved slowly and at some distance from one another when they were searching for insects, which could be easily distinguished from feeding on other food items (fruit or leaves). We created a map with locations of some emergent trees within each group's home range on a topographic map superimposed with a 100 × 100 m grid [24]. And then we recorded group locations by referring to these emergent trees and obvious landmarks (rocky outcrops, cliffs, and cols where gibbon groups crossed the ridge). During observations, we recorded all incidents and the context of agonistic behavior (bite, slap, chase, and replacement) among group members *ad libitum*.

### Data analysis

The small sample size of behavioral records in G2 precluded detailed analysis, however, based on our limited feeding records G2 consumed similar food species to G1 and G4. We focused on G1 and G4 in the analysis of feeding competition, activity budget and spatial relationships. We excluded data collected before July when females could not be reliably identified from the analysis of G1. Over the remaining 18 months, we observed G1 on average 65.3 ± 6.5 h (range: 55–81 h, N = 18) on 9 ± 2 days (range: 6–14 d, N = 18) per month. We pooled data for the two juveniles and two adolescents in G4 because we could not reliably distinguish them at all times. In August and December 2008 we observed G4 for only 8 h and 14 h respectively so we excluded data in these two months from the analysis. Over the remaining 15 months between August 2008 and December 2009, we observed G4 on average 50.3 ± 12.6 h (range: 31–68 h, N = 15) on 8 ± 2 days (range: 4–11 d, N = 15) per month.

We calculated the monthly proportional diet and activity budget for each foraging individual (or age/sex group for juveniles and adolescents in G4). We treated monthly diet and activity budgets as independent samples. We compared individual diet and activity budget differences within groups using the Wilcoxon matched pairs test. To eliminate the effect of there being different numbers of immature animals (juveniles and adolescents) in each group, group diet and activity budgets were calculated based on adult group members only. Wilcoxon matched pairs test was used to test for differences in time budget, diet and inter-individual distance between groups after listwise deletion of months for which we did not have data for both groups (N = 15 months for both groups).

To avoid significant underestimation of group spread, we included only those scans in which at least three inter-individual distances were recorded. We used the largest distance to represent group spread during each scan. We then calculated the mean inter-individual distance and group spread for each month by averaging the mean inter-individual distance and maximum inter-individual distance of each scan. These measures were highly correlated in both groups (Spearman correlation: G1, $r = 0.984$, $P < 0.001$; G4, $r = 0.994$, $P < 0.001$). Because G4 consisted of more independent individuals, we further calculated each index using scans in which six inter-individual distances were recorded; however, both inter-individual distance (Spearman correlation: $r = 0.932$, $P < 0.00$) and group spread (Spearman correlation: $r = 0.915$, $P < 0.001$) were highly correlated using three or six distance samples. Therefore, only results based on three inter-individual distances are reported for each group.

Because of fog, rain and the difficult terrain, full-day observations (i.e. visually tracking animals from sleeping tree to sleeping tree) were difficult to achieve [31]. In order to increase sample size, days in which gibbons were lost for less than one hour (one or two group locations missed) were also used to estimate daily path length (DPL; the sum of all distances between consecutive group locations for one day). We measured DPL using software ImageJ [57]. In total, we had 59 DPLs for G1 between June 2008 and December 2009, and 30 DPLs for G4 between October 2008 and December 2009. To control for seasonal variation in DPL, we only used those collected during the same period (between October 2008 and December 2009) to compare differences in DPL between groups (G1: 50 days; G4: 30 days) using Mann–Whitney *U* test.

## Ethical standards

The study was conducted in Bangliang Nature Reserve, Jingxi County, Guangxi, China. All research reported in this manuscript has met the appropriate national and institutional guidelines for the legal acquisition and was permitted by the Guangxi Forestry Bureau and Bangliang Nature Reserve.

### Competing interests

The authors declare that they have no competing interests.

### Authors' contributions

PFF designed the study. PFF, HLF, CYM and WZ collected data in the field. PFF analyzed data. PFF and TQB wrote, and all authors read and approved, the manuscript.

### Acknowledgments

This study was supported by National Natural Science Foundation of China (# 30900169), International Foundation for Science (IFS), Conservation Leadership Programme (CLP), and Fauna and Flora International (FFI). Idea Wild provided equipment for this research. Mr. Tao Huang, Jiancun Liao and Bao Li helped to collect data in the field. We would like to thank Mr. Wujin Tan, Tianzhu Huang, Yucong Lin, Xiao Yang, Zhongming Yao, Qunce Nong, Guoyu Chen, Yaojie Liang, and Jingpei Zhou from the Jingxi Forestry Bureau for their needed support.

### Author details

[1]Institute of Himalaya Biodiversity Research, Dali University, Dali 671003, Yunnan, People's Republic of China. [2]Department of Anthropology, The University of Texas at San Antonio, San Antonio, TX 78249, USA.

### References

1. Cody ML. Finch flocks in the Mohave Desert. Theor Popul Biol. 1971;18:244–56.
2. Hamilton WD. Geometry for the selfish herd. J Theor Biol. 1971;31:295–311.
3. Wrangham RW. An ecological model of female-bonded primate groups. Behaviour. 1980;75:262–300.
4. van Schaik CP, Dunbar RIM. The evolution of monogamy in large primates: a new hypothesis and some crucial tests. Behaviour. 1990;115:30–62.
5. Chapman CA, Chapman LJ. Determinants of group size in social primates: the importance of travel costs. In: Boinski S, Garber P, editors. On the move: how and why animals travel in groups. Chicago: University of Chicago Press; 2000. p. 24–42.
6. Nicholson AJ. An outline of the dynamics of animal populations. Aust J Zool. 1954;2:9–65.
7. Sterck EHM, Watts DP, van Schaik CP. The evolution of female social relationships in nonhuman primates. Behav Ecol Sociobiol. 1997;41:291–310.
8. Koenig A. Competition for resources and its behavioral consequences among female primates. Int J Primatol. 2002;23:759–83.
9. Schulke O. To breed or not to breed-food competition and other factors involved in female breeding decisions in the pair-living nocturnal for-marked lemur (Phaner furcifer). Behav Ecol Sociobiol. 2003;55:11–21.
10. Borries C, Larney E, Lu A, Ossi K, Koenig A. Costs of group size: lower developmental and reproductive rates in larger groups of leaf monkeys. Behav Ecol. 2008;19:1186–91.
11. Savini T, Boesch C, Reichard UH. Varying ecological quality influences the probability of polyandry in white-handed gibbons (*Hylobates lar*). Biotropica. 2009;41:503–13.
12. Reichard UH, Ganpanakngan M, Barelli C. White-handed gibbons of Khao Yai: social flexibility, complex reproductive strategies, and a slow life history. In: Kappeler PM, Watts DP, editors. Long-term field studies of primates. Berlin Heidelberg: Springer-Verlag; 2011. p. 237–58.
13. Lappan S. Social relationships among males in multimale siamang groups. Int J Primatol. 2007;28:369–87.
14. Haimoff EH, Yang XJ, He XJ, Chen N. Census and survey of wild black-crested gibbons (*Hylobates concolor*) in Yunnan Province, People's Republic of China. Folia Primatol. 1986;46:205–14.
15. Brockelman W. Ecology and the social system of gibbons. In: Lappan S, Whittaker DJ, editors. The gibbons: new perspectives on small ape socioecology and population biology. New York, USA: Springer; 2009. p. 211–40.
16. Brockelman WY, Nathalang A, Greenberg DB, Suwanvecho U. Evolution of small-group territoriality in gibbons. In: Yamagiwa J, Karczwaski L, editors. Primates and cetaceans: field studies and conservation of complex mammalian societies. Tokyo: Springer Japan KK; 2014. p. 213–30.
17. Leighton DR: G. Territoriality and monogamy. In: Smuts BB, Cheney DL, Seyfarth RM, Wrangham RW, Struhsaker TT, editors. Primate Societies. Chicago: University of Chicago Press; 1987. p. 135–45.
18. Bartlett TQ. The gibbons of Khao Yai: seasonal variation in behavior and ecology. Boston: Allyn & Bacon; 2009.
19. Liu ZH, Zhang YZ, Jiang HS, Southwick C. Population structure of *Hylobates concolor* in Bawanglin Nature Reserve, Hainan, China. Am J Primatol. 1989;19:247–54.
20. Zhou J, Wei FW, Li M, Chan BPL, Wang DL. Reproductive characters and mating behaviour of wild *Nomascus hainanus*. Int J Primatol. 2008;29:1037–46.

21. Fan PF, Jiang XL, Liu CM, Luo WS. Polygynous mating system and behavioural reason of black crested gibbon (*Nomascus concolor jingdongensis*) at Dazhaizi, Mt. Wuliang, Yunnan, China. Zool Res. 2006;27:216–20.
22. Fan PF, Jiang XL. Maintenance of multifemale social organization in a group of *Nomascus concolor* at Wuliang Mountain, Yunnan, China. Int J Primatol. 2010;31:1–13.
23. Huang B, Guan ZH, Ni QY, Orkin JD, Fan PF, Jiang XL. Observation of intra- and extra-group copulation and reproductive characters in free ranging groups of western black crested gibbon (*Nomascus concolor jingdongensis*). Inter Zool. 2013;8:427–40.
24. Fan PF, Fei HL, Xiang ZF, Zhang W, Ma CY, Huang T. Social structure and group dynamics of the cao vit gibbon (*Nomascus nasutus*) in Bangliang, Jingxi, China. Folia Primatol. 2010;81:245–53.
25. Srikosamatara S, Brockelman WY. Polygyny in a group of pileated gibbons via a familial route. Int J Primatol. 1987;8:389–93.
26. Ahsan F. Fighting between two females for a male in the Hoolock gibbon. Int J Primato. 1995;l16:731–7.
27. Sommer V, Reichard U. Rethinking monogamy: the gibbon case. In: Kappeler PM, editor. Primate males: cause and consequences of variation in group composition. Cambridge, UK: Cambridge University Press; 2000. p. 159–68.
28. Guan ZH, Huang B, Ning WH, Ni QY, Sun GZ, Jiang XL. Significance of grooming behavior in two polygynous groups of western black crested gibbons: implication for understanding social relationships among immigrant and resident group members. Am J Primatol. 2013;75:1165–73.
29. Brockelman WY, Reichard U, Treesucon U, Raemaekers J. Dispersal, pair formation and social structure in gibbons (*Hylobates lar*). Behav Ecol Sociobiol. 1998;42:329–39.
30. Fan PF, Fei HL, Scott MB, Zhang W, Ma CY. Habitat and food choice of the critically endangered cao vit gibbon (*Nomascus nasutus*) in China: Implications for conservation. Biol Conserv. 2011;144:2247–54.
31. Fan PF, Fei HL, Ma CY. Behavioral responses of cao vit gibbon (*Nomascus nasutus*) to variations in food abundance and temperature in Bangliang, Jingxi, China. Am J Primatol. 2012;74:632–41.
32. Mootnick A, Mootnick AR, Fan PF. A comparative study of crested gibbons (*Nomascus*). Am J Primatol. 2011;73:135–54.
33. Zhou QH, Wei FW, Li M, Huang CM, Luo B. Diet and food choice of *Trachypithecus francoisi* in the Nonggang Nature Reserve, China. Int J Primatol. 2006;27:1441–60.
34. Huang CM, Wu H, Zhou QH, Li YB, Cai XW. Feeding strategy of Francois' langur and white-headed langur at Fusui, China. Am J Primatol. 2007;69:1–10.
35. Workman C. Diet of the Delacour's langur (*Trachypithecus delacour*) in Van Long Nature Reserve, Vietnam. Am J Primatol. 2010;72:317–24.
36. Wheeler BC, Scarry CJ, Koenig A. Rates of agonism among female primates: a cross-taxon perspective. Behav Ecol. 2013;24:1369–80.
37. Chivers DJ. The siamang in Malaya: a field study of a primate in a tropical forest. Contr Primat. 1974;4:1–335.
38. Bartlett TQ. The Hylobatidae: small apes of Asia. In: Campbell CJ, Fuentes A, MacKinnon KC, Stumpf RM, Bearder SK, editors. Primates in perspective. 2nd ed. New York, Oxford: Oxford University Press; 2011. p. 300–12.
39. Gittins SP. Feeding and ranging in the agile gibbon. Folia Primatol. 1982;38:39–71.
40. McConkey KR, Ario A, Aldy F, Chivers DJ. Influence of forest seasonality on gibbon food choice in the rain forests of Barito Ulu, Central Kalimantan. Int J Primatol. 2003;24:19–32.
41. Kim S, Lappan S, Choe J. Diet and ranging behavior of the endangered Javan gibbon (*Hylobates moloch*) in a submontane tropical rainforest. Am J Primatol. 2011;73:270–80.
42. Reichard UH, Barelli C. Life history and reproductive strategies of Khao Yai Hylobates lar: Implications for social evolution in apes. Int J Primatol. 2008;29:823–44.
43. Chivers DJ, Raemaekers JJ. Long-term changes in behaviour. In: Chivers DJ, editor. Malayan forest primates: Ten years' study in tropical rain forest. New York: Plenum Press; 1980. p. 209.
44. Palombit RA. Longitudinal patterns of reproduction in wild female Siamang (*Hylobates syndactylus*) and white-handed gibbons (*Hylobates lar*). Int J Primatol. 1995;16:739–60.
45. Mitani JC. Demography of agile gibbons (*Hylobates agilis*). Int J Primatol. 1990;11:411–24.
46. Lappan S. Male care of infants in a siamang (*Symphalangus syndactylus*) population including socially monogamous and polyandrous groups. Behav Ecol Sociobiol. 2008;62:1307–17.
47. Geissmann T. Evolution of Communication in Gibbons (Hylobatidae). Ph.D. dissertation. Universitaet Zuerich, Zuerich; 1993.
48. Zhao QK. Seasonal change in body weight of *Macaca thibetana* at Mt. Emei, China. Am J Primatol. 1994;32:223–6.
49. Knott CD. Changes in orangutan caloric intake, energy balance, and ketones in response to fluctuating fruit availability. Int J Primatol. 1998;19:1061–77.
50. Ptak SE, Lachmann M. On the evolution of polygyny: a theoretical examination of the polygyny threshold model. Behav Ecol. 2003;14:201–11.
51. Fan PF, Ren GP, Wang W, Scott MB, Ma CY, Fei HL, et al. Habitat evaluation and population viability analysis of the last popualtion of cao vit gibbon (*Nomascus nasutus*): implications for conservation. Biol Conserv. 2013;161:39–47.
52. La QT, Trinh DH, Long B, Geissmann T: Status review of black crested gibbons (Nomascus concolor and Nomascus sp. cf. nasutus) in Vietnam [abstract]. In Caring for primates, Abstracts of the XIXth congress of the International Primatological Society, 4th- 9th August, 2002, Beijing, China. 2002: 131–32.
53. Chan BPL, Tan XF, Tan WJ. Rediscovery of the Critically Endangered eastern black-crested gibbon *Nomascus nasutus* (Hylobatidae) in China, with preliminary notes on population size, ecology and conservation status. Asian Primates J. 2008;1:17–25.
54. Koenig A, Beise J, Chalise MK, Ganzhorn JU. When female should contest for food-testing hypotheses about resource density, distribution, size and quality with Hanuman langurs (*Presbytis entellus*). Behav Ecol Sociobiol. 1998;42:225–37.
55. Saj TL, Sicotte P. Predicting the competitive regime of female Colobus vellerosus from the distribution of food resources. Int J Primatol. 2007;28:315–36.
56. Altmann J. Observational study of behaviour: sampling methods. Behaviour. 1974;49:227–67.
57. Abramoff MD, Magalhaes PJ, Ram SJ. Image processing with ImageJ. Biophotonics Int. 1995;11:36–42.

# 18

# Linking habitat suitability to demography in a pond-breeding amphibian

Bianca Unglaub[1,2*], Sebastian Steinfartz[1], Axel Drechsler[3] and Benedikt R Schmidt[4,5]

**Abstract**

**Introduction:** Elucidating the relationship between habitat characteristics and population parameters is critical for effective conservation. Habitat suitability index (HSI) models are often used in wildlife management and conservation practice assuming that they predict species occurrence, abundance and demography. However, the relationship between vital rates such as survival and reproduction and habitat suitability has rarely been evaluated. In this study, we used pond occupancy and mark-recapture data to test whether HSI predicts occupancy, reproduction and survival probabilities. Our model species is the great crested newt (*Triturus cristatus*), a pond-breeding amphibian protected under the European Habitats Directive.

**Results:** Our results show a positive relationship between the HSI and reproduction probability, whereas pond occupancy and survival probabilities were not related to HSI. Mortality was found to be higher during breeding seasons when newts are in ponds than during terrestrial phases of adult newts.

**Conclusion:** Habitat suitability models are increasingly applied to wildlife management and conservation practice. We found that the HSI model predicted reproduction probability, rather than occurrence or survival. If HSI models indicate breeding populations rather than mere species occurrences, they may be used to identify habitats of higher priority for conservation. Future HSI models might be improved through modelling breeding populations vs. non-breeding populations rather than presence/absence data. However, according to our results the most suitable habitat is not necessarily the habitat where demographic performance is best. We recommend that conservation practitioners should use HSI models cautiously because there may be no direct link between habitat suitability, demography and consequently, population viability.

**Keywords:** Environmental niche model, Habitat suitability index (HSI), Species distribution, Reproduction probability, Survival probability, *Triturus cristatus*

## Introduction

Understanding the relationship between habitat quality and demography is central to the monitoring, management and recovery of threatened species. Species distribution models, also known as ecological niche models or habitat suitability models, relate species occurrence data to environmental variables. These models provide useful information on the ecological requirements of species and are widely used to predict species distribution, making them valuable tools for habitat management, impact assessment and conservation practice [1-3].

For practical application and habitat suitability assessments in the field, the output of statistical species distribution models has often been simplified to habitat suitability indices (HSI). These indices are based on habitat characteristics that can easily be measured in the field or derived from digital maps [4]. A HSI is a numerical index, ranging from 0 (unsuitable habitat) to 1 (optimal habitat). In the application of HSI models for management purposes, it is often assumed that habitat suitability predicts species performance and demography [5]. However, the most suitable habitat or habitats where density is high do not necessarily constitute habitats where demographic performance is best [6,7]. Moreover, despite being important for the management of threatened

* Correspondence: b.unglaub@tu-bs.de
[1]Zoological Institute, Department of Evolutionary Biology, Unit Molecular Ecology, Technische Universität Braunschweig, Mendelssohnstraße 4, Braunschweig 38106, Germany
[2]Department of Animal Ecology and Conservation, Biocentre Grindel, University of Hamburg, Martin-Luther-King Platz 3, Hamburg 20146, Germany
Full list of author information is available at the end of the article

species, whether habitat suitability is associated with species occurrence [8-12] and demographic parameters, e.g. reproductive success [5,13,14] and apparent survival [15,16], has rarely been evaluated. Such tests are important because several studies did not find the expected link between habitat suitability and species occurrence or demography [9,11,17]. Species may not occur in suitable patches when structured as a metapopulation [18] or they may be found in unsuitable patches (i.e. so called sinks, [19]). To predict how species may respond to variation in habitat quality, it is necessary to understand the demographic processes through which the environment influences distributions and population dynamics [19-21]. Hence, vital rates can potentially be informative when validating HSI models [22].

In this study, we contribute to the validation of simple and easily applicable HSI models as predictive tools for management purposes. We studied the relationship between a commonly used HSI [23] and occurrence/demographic parameters in the great crested newt (*Triturus cristatus*). While a positive relationship between the HSI and newt abundance was reported [23], this assessment of the HSI is problematic as abundance is not necessarily a good indicator of habitat quality and abundance indices ignore imperfect detection [6,24]. Evidence for a relationship between HSI and newt abundance or newt occurrence is mixed. While there was no relationship between abundance indices and HSI in [25], a study on great crested newt pond occupancy that accounted for imperfect detection found that HSI predicted newt occurrence [12]. However, HSI values differed only slightly between ponds with and without newts (mean HSI ± SD: 0.70 ± 0.12 and 0.61 ± 0.13, respectively [12]). Here, we used three variables that are often used to describe the state of animal populations: species occurrence, occurrence of reproduction and survival. At the phenomenological level, we tested whether the simple HSI for great crested newts, which is based on only ten habitat characteristics, predicts species occurrence even though many habitat characteristics are known to influence occurrence of crested newt populations (e.g. [26-28]). Since a species may be found in low quality habitats within a metapopulation (i.e. sinks [7]), we further tested the predictive value of the HSI at a mechanistic level by assessing the relationship between HSI and occurrence of reproduction as well as between HSI and apparent survival.

We selected a HSI for an amphibian species because amphibians are the most endangered vertebrates [29] requiring both terrestrial and aquatic habitats during their life cycles, thus making them a particularly well suited indicator group for habitat quality [30]. Great crested newts are protected under the European Habitats Directive and may serve as umbrella species for wetland conservation [28]. If the HSI can identify habitats where demographic performance is good, it could be used to select habitats harbouring healthy populations (e.g. so called source populations [7]) for conservation purposes. However, if the HSI and demography are unrelated, then this would call for refined habitat suitability models.

## Results

We recorded capture histories of 1838 individuals from 2009 to 2011 in our study area, of which 124 individuals were recaptured at least once. Adult newts were captured at 18 sites, ranging from one to 507 individuals per pond. Larvae were found at 13 sites, ranging from one to 105 individuals caught on a single capture event. At six sites we found merely adult newts without larvae whereas only larvae were detected at one pond. At three sites we detected neither adults nor larvae. HSI values ranged from 0.43 to 0.93 in 2009, from 0.41 to 0.93 in 2010 and from 0.44 to 0.94 in 2011 for surveyed sites (Table 1).

### Modelling occupancy and reproduction probabilities

We first selected a model that best explained detection probability, while keeping occupancy and reproduction probabilities constant. Akaike model weights ($w$) suggested that model {$\psi$ (.), $R$ (.), $\delta_s$, $p^{[1]}$ (CE), $p^{[2]}$ (CE)} was best supported by the data ($w = 0.93$; see Additional file 1), whereas remaining models received little support ($w \leq 0.04$). The number of capture events (CE) was included in the top ranking model, indicating that sampling effort positively influences the probability to detect newts in waters occupied without reproduction (logit ($p^{[1]}$) =-1.59 (SE = 1.06) + 0.40 (SE = 0.19) x CE) as well as in waters with successful reproduction (logit ($p^{[2]}$) = 0.72 (SE = 0.63) + 0.38 (SE = 0.20) x CE). The probability of detecting newts was generally higher for a site with reproduction (0.75 - 0.98 for CE = 1–8) than without reproduction (0.23 - 0.83 for CE = 1–8). The probability of correctly identifying sites as breeding sites increased gradually from the start to the end of breeding seasons ($\delta = 0.10$ (SE = 0.07), $\delta = 0.56$ (SE = 0.14), $\delta = 0.80$ (SE = 0.13) and $\delta = 0.89$ (SE = 0.07) for early May, late May, early June and late June, respectively).

In the second step of the analysis, we used the structure of the top-ranking model for detection probabilities and determined the effect of the HSI on $\psi$ and $R$. Model {$\psi$ (HSI), $R$ (HSI), $\delta_s$, $p^{[1]}$ (CE), $p^{[2]}$ (CE)} best explained the data ($w = 0.77$; Table 2). However, while the effect of the HSI on reproduction probability was well supported by the data, the confidence interval of the estimate of the positive effect of the HSI on occupancy probability included zero (Table 3). The probability of reproduction was higher in ponds with higher HSI values (Figure 1).

**Table 1 Number of captured newts (*Triturus cristatus*) and HSI values for 22 sampling sites surveyed between 2009 and 2011**

| Sampling site | No. of adult newts | Max no. of larvae | HSI 2009 | HSI 2010 | HSI 2011 |
|---|---|---|---|---|---|
| 1 | 118 | 0 | 0.46 | 0.46 | 0.46 |
| 2 | 7 | 0 | 0.45 | 0.41 | 0.46 |
| 4 | 1 | 0 | 0.61 | 0.61 | 0.61 |
| 8 | 1 | 0 | 0.61 | 0.61 | 0.61 |
| 9 | 45 | 1 | 0.83 | 0.82 | 0.83 |
| 10 | 145 | 4 | 0.93 | 0.93 | 0.93 |
| 11 | 325 | 5 | 0.91 | 0.93 | 0.94 |
| 12 | 19 | 0 | 0.56 | 0.56 | 0.56 |
| 13 | 0 | 13 | 0.75 | 0.76 | 0.68 |
| 13b | 104 | 105 | 0.93 | 0.84 | 0.84 |
| 14 | 191 | 49 | 0.79 | 0.80 | 0.80 |
| 15 | 27 | 3 | 0.83 | 0.83 | 0.80 |
| 16 | 0 | 0 | 0.47 | 0.48 | 0.48 |
| 17 | 52 | 3 | 0.78 | 0.80 | 0.79 |
| 18 | 122 | 3 | 0.76 | 0.78 | 0.78 |
| 19 | 0 | 0 | 0.43 | 0.43 | 0.44 |
| 20 | 79 | 6 | 0.82 | 0.82 | 0.79 |
| 21 | 57 | 1 | 0.66 | 0.74 | 0.76 |
| A | 8 | 2 | 0.50 | 0.52 | 0.51 |
| B | 32 | 0 | 0.53 | 0.53 | 0.54 |
| C | 507 | 18 | 0.55 | 0.54 | 0.54 |
| D | 0 | 0 | 0.48 | 0.48 | 0.48 |

Sampling sites, number of captured adult newts, maximum number of larvae caught on a single capture event and HSI values for 3 years of CMR study (2009–2011).

**Table 2 Selection of multiseason-multistate models for estimating occupancy and reproduction probabilities of great crested newts**

| Model | AIC | ΔAIC | *w* | *K* |
|---|---|---|---|---|
| $\psi$ (HSI), R (HSI), $\delta_s$, p [1] (CE), p [2] (CE) | 261.34 | 0.00 | 0.77 | 12 |
| $\psi$ (.), R (HSI), $\delta_s$, p [1] (CE), p [2] (CE) | 263.89 | 2.55 | 0.22 | 11 |
| $\psi$ (HSI), R (.), $\delta_s$, p [1] (CE), p [2] (CE) | 269.31 | 7.97 | 0.02 | 11 |
| $\psi$ (.), R (.), $\delta_s$, p [1] (CE), p [2] (CE) | 278.04 | 16.70 | 0.00 | 10 |

Probability of pond occupancy ($\psi$) and probability of reproduction, given presence (*R*) were held constant (.) or modelled as functions of habitat suitability index (HSI). The structure of the top-tanking model for detection probabilities {$\psi$ (.), R (.), $\delta_s$, p [1] (CE), p [2] (CE)} was used to evaluate the effect of HSI on $\psi$ and *R*. Probability of correctly identifying a site as breeding site, given successful reproduction ($\delta$) was modelled different in each capture period and probabilities of detecting occupancy, given occupancy without reproduction (*p* [1]) and with successful reproduction (*p* [2]) were modelled as functions of the number of capture events per capture period (CE). AIC: Akaike's information criterion; ΔAIC: difference of the AIC value of the current and the best model; *w*: AIC weight; *K*: number of parameters.

**Table 3 Parameter estimates (on the logit scale) of the top ranking multiseason-multistate model for estimating occupancy and reproduction of great crested newts**

| Logit link function | Beta | Estimate | 95% CI |
|---|---|---|---|
| logit ($\psi_i$) = $\beta_{INT} + \beta_{HSI}$ x HSI | $\beta_{INT}$ | -4.04 | -9.27 – 1.20 |
| | $\beta_{HSI}$ | 9.38 | -0.36 – 19.12 |
| logit ($R_i$) = $\beta_{INT} + \beta_{HSI}$ x HSI | $\beta_{INT}$ | -5.50 | -9.70 – 1.29 |
| | $\beta_{HSI}$ | 8.44 | 2.24 – 14.65 |

$\psi$: probability of pond occupancy; *R*: probability of reproduction, given presence. INT: Intercept; HSI: Habitat Suitability Index. CI: confidence interval.

**Modelling survival probabilities**

The general model with time-dependent apparent survival and recapture probabilities ($\Phi_t$, $p_t$) fitted the data well (goodness-of-fit test results: $\chi^2 = 7.74$, $DF = 44$, $P = 1$). There was neither evidence for transients (GOF test: $z = 0.36$, $P = 0.72$) nor an effect of capture at a previous occasion (GOF test: $z = -0.48$, $P = 0.63$).

Akaike model weights ($w$) suggested that the model assuming a seasonal effect on apparent survival {$\Phi_s$, $p$ (.)} was best supported by the data ($w = 0.987$; Table 4). Monthly survival probabilities were lower during the months that newts spend in the pond (aquatic phase; March – June: $\Phi_{aqu} = 0.54$; 95% CI = 0.44 – 0.63) than the months that adults spend in their terrestrial habitat (terrestrial phase; July – February: $\Phi_{terr} = 0.99$; 95% CI = 0.42 – 1.00; Table 5). A model with an effect of HSI on survival probabilities was not well supported by the data ($\Delta AICc \geq 8.79$). Annual apparent survival was calculated using the formula: $\Phi^4_{aqu} * \Phi^8_{terr}$ [31]. The corresponding standard error was calculated by applying the delta method [32]. Annual survival probability was $0.08 \pm 0.0006$.

Our estimates of apparent survival were probably distorted by emigration. Apparent survival is the product of true survival and (1 – probability of emigration) [33]. Thus, the annual probability of emigration can be calculated as 1 – (apparent survival / true survival). Based on the annual survival estimate for a metapopulation of crested newts in a similar capture-mark-recapture study [34], we assumed that true survival would be around 0.55. The annual probability of emigration would then be 0.85. If we use annual survival rates 10% lower (0.45) or higher (0.65) than the published estimate [34], then the estimates of the probability of emigration would be 0.82 and 0.87, respectively.

## Discussion

### HSI and population parameters in great crested newts (*Triturus cristatus*)

Our results suggest that the HSI for great crested newts is not related to survival or pond occupancy

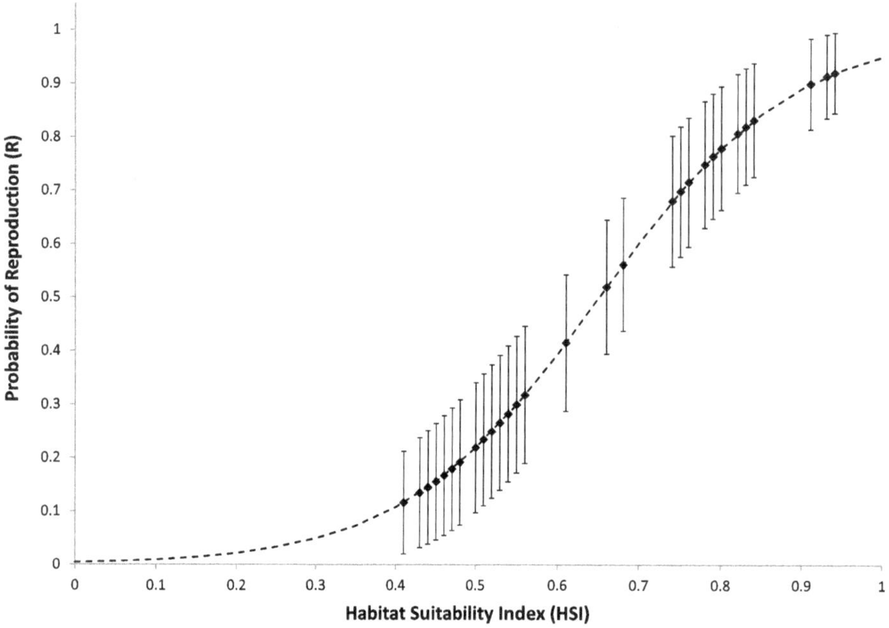


**Figure 1** Relationship between HSI and reproduction probability of great crested newts. Symbols represent estimates and SE. Reproduction probabilities were estimated for HSI values observed at 22 ponds over 3 years.

probabilities, but that newt populations are more likely to breed in ponds with higher HSI values. In contrast, [23] showed that there is a positive relationship between the HSI and an index of newt abundance. These results could be misleading, however, for two reasons. First, abundance can be a misleading indicator for habitat quality [6]. Second, imperfect detection was not taken into account [24]. For example, [23] acknowledged that macrophyte cover in ponds may have biased their results. The positive relationship between HSI and abundance reported in [23] could not be confirmed by [25]. In our study, we therefore selected different variables to describe the state of newt populations and we accounted for imperfect detection.

**Table 4 Selection of Cormack-Jolly-Seber models for estimating apparent monthly survival of great crested newts**

| Model | AICc | ΔAICc | *w* | *K* |
|---|---|---|---|---|
| $\Phi_s$, p (.) | 1316.17 | 0.00 | 0.987 | 3 |
| $\Phi_s$ (HSI), p (.) | 1324.95 | 8.79 | 0.012 | 3 |
| $\Phi_y$, p (.) | 1331.27 | 15.11 | 0.001 | 4 |
| $\Phi_y$ (HSI), p (.) | 1335.81 | 19.65 | 0.000 | 5 |
| $\Phi$ (HSI), p (.) | 1344.33 | 28.16 | 0.000 | 3 |
| $\Phi$ (.), p (.) | 1345.94 | 29.77 | 0.000 | 2 |

Survival probability ($\Phi$) was modelled as constant (.), as varying between years (y) or between seasons (s), i.e. during months of terrestrial and aquatic phases of adult newts. In each of this scenarios $\Phi$ was also modelled as function of habitat suitability index (HSI). Capture probability (*p*) was modelled as constant (.). AICc: corrected Akaike's information criterion; ΔAICc: difference of the AICc value of the current and the best model; *w*: AICc weight; *K*: number of parameters.

We found that the best multiseason-multistate occupancy model included HSI as covariate for both the probabilities of occupancy ($\psi$) and of reproduction (*R*). A previous study which did not differentiate between the presence/absence of adults and larvae [12] found that HSI predicted pond occupancy. In our study, the top-ranking model also included a positive effect of HSI on pond occupancy but the confidence interval of this estimate included zero (Table 3). The comparison of our results with those of [12] shows that differentiating between presence/absence of adults and presence/absence of larvae can give additional insights into habitat suitability. Our results suggest that the HSI is a good predictor for reproduction but not for pond occupancy.

**Table 5 Parameter estimates of the top-ranking Cormack-Jolly-Seber model for estimating survival probabilities of great crested newts**

| Model parameter | Estimate | 95% CI |
|---|---|---|
| $\Phi_{aqu}$ | 0.54 | 0.44 – 0.63 |
| $\Phi_{terr}$ | 0.99 | 0.42 – 1.00 |
| p (.) | 0.07 | 0.05 – 0.10 |

$\Phi_{aqu}$: monthly survival probability during aquatic phases of adult newts (March – June); $\Phi_{terr}$: monthly survival probability during terrestrial phases of adult newts (July – February); *p*: capture probability. HSI: Habitat Suitability Index. 95% CI: 95% confidence interval.

We also found newts in ponds with very low HSI values, seemingly not representing suitable habitat (Table 1). Bentonite mats, applied to the soil of one pond in 2001, prevented periodical drying and allowed for the existence of large fish populations. Occasional high water levels led to the colonization of formerly fishless ponds by native and invasive fish. Fish are well known to negatively affect the distribution and abundance of great crested newts and other amphibians (see reviews in [28] and [35]) which is why the presence of fish leads to lower HSI values for affected ponds (see $SI_7$). Accordingly, the number of captured newts decreased in those waters over our monitoring period, but still we regularly found some adult newts even in ponds with predatory fish. However, ponds occupied by predatory fish are unlikely to represent suitable habitat for crested newts. Since the goal of HSI models is to predict suitable habitat rather than mere species occurrence, they should not simply indicate whether a pond is occupied or not.

The model developed by [23] emphasizes primarily the aquatic habitat where adult newts congregate for a few months during spring and early summer to reproduce [36]. As would therefore be assumed, our results suggest that the HSI represents a good tool to detect ponds where newts are more likely to reproduce successfully (Figure 1). Hence, it is probable that newts occurring in ponds with lower HSI values do not breed, or that larvae do not survive in these waters until metamorphosis. As spatial variation in newt reproductive success may be common [37], the HSI could allow conservation managers to identify breeding populations or, conversely, populations constituting sinks because of a lack of reproduction [7]. This kind of information is certainly valuable for the effective conservation and recovery of threatened species. Taken together, our results for occupancy and reproduction probabilities suggest that the HSI does a good job because it appears to differentiate between populations with high and low probabilities of reproduction.

If the HSI indicates suitable habitat, then one may expect a positive relationship between HSI and survival. However, we did not find such a relationship (Table 4). Ponds with higher HSI values appear to hold larger populations [23] and survival might be negatively affected by density dependence. Other environmental factors, such as climate may have a stronger effect on survival than habitat quality [34]. To date, there is no data that would allow to test this, or any other hypothesis, for the absence of an effect of the HSI on survival. Remarkably, annual apparent survival was low in our study, suggesting that about 85% of great crested newts may have emigrated. This is an extraordinarily large proportion. However, emigration in the context of our study refers to the place where the newts were captured (i.e. the ponds) and is therefore emigration from the breeding population rather than emigration from the study area. In other words, if a newt did not enter the pond anymore during the three years of our study, it was considered an emigrant. In a short-term study such as ours, it is not possible to distinguish between temporary and permanent emigration [38,39]. Newts may have avoided the ponds and skipped reproduction in the later years of our study in response to the invasion and increase of fish populations. This interpretation of the emigration rate as skipping reproduction in some years is supported by the fact that we observed very few cases of among-pond movement (i.e. 11 individuals). Skipping reproduction might be a strategy to deal with predatory fish because survival on land was high and ponds were temporary, and therefore fish-free, in the past. To test this hypothesis, it would be necessary to extend the mark-recapture study to the terrestrial habitat. To our knowledge, the present study is the first to directly compare survival probabilities during aquatic and terrestrial phases of pond-breeding amphibians. Our study shows that survival probabilities were lower during aquatic than during terrestrial phases of adult great crested newts (monthly survival 54% and 99%, respectively). Monthly survival of 99% (or 92% across the eight months of the terrestrial phase) is unexpectedly high but [40] reported an estimate of annual survival of 99.6% in a cohort of *Ambystoma maculatum* salamanders. While our results suggest that mortality occurs primarily in the water during the breeding season, [36] found that annual survival in a metapopulation of crested newts in the UK was determined by winter weather, i.e. environmental conditions during the terrestrial phase of adults. Hence, factors determining survival may vary spatially [41]. If spatial variation in survival is common, then conservation management should take population-specific differences into account [42].

### HSI as a general conservation tool

Habitat suitability models are increasingly applied to wildlife management and conservation planning [3,8]. Guisan et al. [3] outlined the steps that are necessary to increase the use of such models to guide conservation decisions. They noted that modelled occurrence probabilities do not always correlate with demographic processes determining population viability [21,43], a finding we regard to be of particular importance for the use of HSI models in conservation practice. We suggest that a focus on the occurrence of a species may not always provide the best models for conservation applications. First, a species may occur in sink habitats [7] and second, a species may not occur in suitable patches as a consequence of extinction and colonization dynamics in metapopulations [18] or due to interspecific competition [19]. Consequently, it is not surprising that many studies did not find the expected correlation between modelled habitat suitability and individual performance, demography or population viability ([9, 11, 17 and references in [3]). Establishing a link between reproduction and habitat suitability seems to be an important step forward [14]. We suggest that modelling the probability of reproduction rather than the probability of occurrence in habitat suitability models using techniques that estimate a true probability rather than a relative suitability score might give valuable additional insights [44-47]. This might be a better way to identify suitable habitat and to increase the utility of these models for conservation.

## Conclusions

HSI models are increasingly applied as predictive tools for management purposes, assuming that habitat suitability predicts species performance and demography. However, the validity of this assumption has rarely been evaluated. We studied the relationship between a commonly used HSI [23] and occurrence/demographic parameters in a pond-breeding amphibian protected under the European Habitats Directive, the great crested newt (*Triturus cristatus*). Our results show a positive relationship between the HSI and reproduction probability (i.e. the occurrence of larvae), whereas pond occupancy and survival probabilities were not related to HSI. This is both good and bad news for conservation managers. The good news is that HSI models may indicate breeding populations rather than mere species occurrences, thus identifying habitats of higher priority for conservation purposes. Modelling breeding populations vs. non-breeding populations rather than presence/absence data might help to identify habitats harbouring healthy populations and to improve the utility of HSI models for the conservation of threatened species (for similar conclusions, see [48,49]). The bad news is that the most suitable habitat is not necessarily the habitat where demographic performance is best. Since there may be no direct link between habitat suitability and demographic processes determining population viability, we recommend that conservation practitioners should use HSI models cautiously.

## Methods

### Study species and determination of HSI

The great crested newt (*Triturus cristatus*) is a pond-breeding amphibian species, listed in Annexes II and IV of the European Habitats Directive (92/43/EEC). EU member states are therefore required to monitor the conservation status of this species. Accordingly, monitoring and management of great crested newt populations would benefit from informative and easily applicable tools and consequently, from a validated HSI. The HSI for the great crested newt incorporates ten habitat features (see Additional file 2; [23]), which are assessed for a pond and converted to suitability index (SI) scores on a scale from 0.01 to 1.0. SIs are site location relative to species distribution (i.e. whether a population occurs at the edge or in the centre compared to the distributional range; $SI_1$), pond area ($SI_2$), pond permanence ($SI_3$), water quality ($SI_4$), shading of pond perimeter ($SI_5$), number of water fowl per 1000 $m^2$ ($SI_6$), impact of fish ($SI_7$), pond density within a radius of 1 km ($SI_8$), proportion of suitable terrestrial habitat within surrounding 500 m ($SI_9$) and macrophyte cover ($SI_{10}$). The HSI for great crested newts is calculated as geometric mean of these ten suitability indices:

$$HSI = (SI_1 * SI_2 * SI_3 * SI_4 * SI_5 * SI_6 * SI_7 * SI_8 * SI_9 * SI_{10})^{1/10}$$

We calculated the HSI for each pond in each year of the study. Since this index was originally developed for the UK, we had to transfer the statements regarding the location relative to species distribution ($SI_1$) to Germany. According to [23], study sites with a high probability of great crested newt occurrence within each 10 km square are scored with 1.0 for $SI_1$ and sites with a low probability of newt occurrence are scored with 0.5. Within Germany, our study sites are located in an area of an intermediate probability of newt occurrence. Therefore, we fixed $SI_1$ to 0.75 for all ponds. Since the original HSI [23] only provides values for ponds of an area of up to 2000 $m^2$, we had to omit $SI_2$ for eight ponds of greater size and calculated the ninth rather than the tenth root of the product instead.

### Study area and sampling procedure

We conducted a capture-mark-recapture (CMR) study and surveyed 22 lentic water bodies in a former flooding area of the Rhine river near Krefeld, Germany (coordinates: 51°19'5" N, 6°39'17" E; Figure 2; Additional file 3), for the presence of great crested newts. The study area is primarily dominated by grasslands, woodlands and wetlands. The northern part, however, is also influenced by adjacent residential areas and agriculture. Adult and larval crested newts can easily be captured with traps in ponds during the breeding season from March to July [50]. Detection or non-detection of adults and larvae was recorded during multiple capture events from March to June in 2009 and 2010 as well as from April to June in 2011. Several visits to each water body were essential to distinguish between sites where great crested newts did not occur and sites where this species has been overlooked [51]. Therefore, every site was visited between 12 and 66 times during the 3 years of monitoring. Newts were

**Figure 2** Overview of the study area near Krefeld (Germany). Illustrated are all 22 surveyed ponds which are mainly located within the FFH-area of the "Latumer Bruch" (DE-4605-301; coordinates: 51°19′5″ N, 6°39′17″ E).

captured using Ortmann's funnel traps [52], which were constructed of empty 10 liter paint buckets with four distinct openings in which half-cut inverted 1.5 liter plastic bottles were inserted functioning as funnels. These traps were evenly distributed along the shoreline and remained in the water for 48 hours. The number of traps deployed per capture event varied according to pond area, ranging from one to 36 traps. After data collection, all individuals were released immediately.

To allow individual recognition during the CMR study, we used photographs of the ventral side of newts, which provides a highly variable but individually unique and lifelong permanent colour pattern [53]. Recaptured individuals were identified automatically by the program AMPHIDENT [54]. AMPHIDENT can reliably detect recaptures even in large datasets based on a cross-correlation comparison algorithm (>1500 individuals; [55]).

### Modelling occupancy and reproduction probabilities

We used multiseason-multistate occupancy models [43,44] to estimate the probability that great crested newts occur in a water body ($\psi$) and the probability that newts reproduced successfully ($R$). The model assumes that sites fall into one of three categories: absence of the species, presence without production of offspring, and presence with production of offspring. With this model, we can model both the presence/absence of newts and the presence/absence of reproduction (given occurrence). We used the presence/absence of larvae as a proxy for production of offspring (i.e. the probability that newts reproduced successfully; larvae are more likely to indicate successful recruitment than the presence of eggs). Unlike models based on presence-only data, which estimate a habitat suitability score, these models estimate true probabilities [46]. Sampling effort varied between sites, ranging from three to 45 capture events during the entire breeding seasons. Therefore, repeated detection/non-detection data were simplified for analysis as follows. For each breeding season per year (i.e. primary period) we defined four capture periods (two-week periods: 1–15 May, 16–31 May, 1–15 June and 16–30 June). If a site was not visited during a capture period, this was treated as a missing observation. If multiple capture events were conducted within a capture period, a site was classified as state $m = 0$ if no newts were detected, as state $m = 1$ if at least one adult was found and as state $m = 2$ if at least one larva was detected. Accordingly, every site received a specific detection history like 1022 0112 0122, where 1 indicates that adults were detected, 0 indicates that neither adults nor larvae were found and 2 indicates that larvae were detected. In this way, we retained as much data as possible without having too many missing observations within a given sampling season [56]. Because of this simplification, we used capture effort (CE = number of capture events per capture period) as explanatory variable for detection probability.

We used a two-step approach to model selection. We first modelled the detection process and then the probabilities of site occupancy and reproduction. We developed an *a priori* candidate model set (see Additional file 1) to select a best model for the probability of detecting occupancy given that a site was occupied without reproduction ($p^{[1]}$), the probability of detecting occupancy given that a site was occupied with successful reproduction ($p^{[2]}$) and the probability of correctly identifying a site as breeding site given that successful reproduction did occur ($\delta$). To identify the best detection model we held occupancy parameters ($\psi$ and $R$) constant and evaluated the effect of capture effort on $p^{[1]}$ and $p^{[2]}$, allowing $\delta$ to vary in time. We hypothesized that $p^{[1]}$ and $p^{[2]}$ are influenced by capture effort (CE), because a higher sampling effort should result in a higher probability of detecting species occurrence. This variable accounts for the fact that the number of capture events varied both within and between ponds. Moreover, we allowed $\delta$ to vary between May and June ($\delta_m$) as well as between capture periods ($\delta_s$), since larvae should be more abundant and bigger later in the season and are therefore easier to detect.

In the second step, we determined the effect of the HSI on occupancy probability ($\psi$) and reproduction probability ($R$), while using the best model for the detection parameters as determined in the first step. Since we were mainly interested in the influence of the HSI on occupancy and reproduction probabilities rather than in state transitions between years, we modelled variables describing changes over time (parameters $\psi^{m}_{t+1}$ and $R^{m}_{t+1}$ in the transition probability matrices [45], with m = state) in the same way as the initial variables ($\psi_{t=1}$ and $R_{t=1}$). Overall, four different models were considered: a) both $\psi$ and $R$ were modelled as constant; b) both $\psi$ and $R$ were modelled as functions of the HSI; c) $\psi$ was modelled as constant and $R$ was modelled as a function of the HSI; d) $\psi$ was modelled as a function of the HSI and $R$ was modelled as constant. We hypothesized that sites with a higher HSI value should have higher probabilities of occupancy and reproduction. Statistical models were implemented in program Presence 6.2 [57].

### Modelling survival probabilities

We used Cormack-Jolly-Seber models [58] to estimate monthly survival and detection probabilities ($\Phi$ and $p$). Capture data were pooled for the months March, April, May and June. If a water body was not visited during a month, detection probability $p$ was set to 0 for this site and period. If multiple capture events were conducted within a month, only the first capture of individuals was counted. Pooling data from several consecutive capture occasions within a month generally increases precision but may induce some bias in survival estimates [59-61]. Overall, adults were captured in 18 out of 22 water bodies (Table 1). However, at two of these 18 sites only one adult was detected, both only once in April 2010. Neither individual was ever recaptured again during the entire sampling period. Therefore, we excluded these two ponds from the mark-recapture analysis and estimated survival probabilities at 16 different sites. Data were too sparse to include covariates for detection probability. Therefore, detection probability ($p$) was always held constant even though this model may not be the best description of the observation process. Survival probabilities ($\Phi$) were modelled either as constant, as varying between years, or as varying between aquatic (March - June) and terrestrial (July - February) phases of adult newts (survival was assumed to be constant within both the aquatic and the terrestrial phase). In the latter case, the specification of the unequal time intervals between capture occasions allowed for the calculation of monthly survival estimates.

For each of these scenarios, we also allowed $\Phi$ to be a function of the covariate HSI. We hypothesized that there is a positive correlation between the HSI and survival probabilities. Models were implemented in program MARK 6.2 [62].

Since only 11 individuals were recaptured at different sites and therefore moved from one site to another, we did not attempt to estimate dispersal probability. If an individual was detected at a new site, then it was scored as having died at the first site (by assigning "-1" to the capture history) and entered as a new individual at the new site.

### Model selection and model notation

Model selection was based on Akaike's information criterion (AIC [63]). The model with the lowest AIC (or AICc) was considered the most parsimonious model given the data. We also used Akaike weight ($w$) as a measure of relative support for each model.

Our model notation system follows the standard notation of [58] and [64] providing information about the sources of variation used to model each parameter. The term (.) indicates that a parameter was held constant (i.e. no covariates).

## Additional files

**Additional file 1: Model selection of multiseason-multistate models for estimating detection parameters of great crested newts.**

**Additional file 2:** Ten Suitability Indices (SI) for the calculation of the HSI for great crested newts according to Oldham et al. (2000).

**Additional file 3:** Distance matrix [km] of surveyed ponds.

### Competing interests
The authors declare that they have no competing interests.

### Authors' contributions
BU performed the statistical analysis and drafted the manuscript. BRS and SS conceived and designed the study and helped to draft the manuscript. AD conducted the field work. All authors read and approved the final manuscript.

### Acknowledgements
We thank Andrea Funke (Untere Landschaftsbehörde Krefeld, Germany) for providing collection permits, Michael Schaub, Darryl MacKenzie and James E. Hines for statistical advice, Michael Zorawski for constructive comments on the manuscript and Amy MacLeod for proofreading. This study has been funded by the Deutsche Bundesstiftung Umwelt (DBU) through a Ph. D. fellowship to AD and by a grant of the German Research Foundation (DFG) to SS and BS (STE 1130/7-1).

### Author details
[1]Zoological Institute, Department of Evolutionary Biology, Unit Molecular Ecology, Technische Universität Braunschweig, Mendelssohnstraße 4, Braunschweig 38106, Germany. [2]Department of Animal Ecology and Conservation, Biocentre Grindel, University of Hamburg, Martin-Luther-King Platz 3, Hamburg 20146, Germany. [3]Department of Animal Behaviour, Bielefeld University, Morgenbreede 45, Bielefeld 33619, Germany. [4]Institute of Evolutionary Biology and Environmental Studies, University of Zurich, Winterthurerstrasse 190, Zurich 8057, Switzerland. [5]KARCH, Passage Maximilien-de-Meuron 6, Neuchâtel 2000, Switzerland.

### References
1. Guisan A, Thuiller W. Predicting species distribution: offering more than simple habitat models. Ecol Lett. 2005;8:993–1009.
2. Elith J, Leathwick JR. Species distribution models: ecological explanation and prediction across space and time. Annu Rev Ecol Evol Syst. 2009;40:677.
3. Guisan A, Tingley R, Baumgartner JB, Naujokaitis-Lewis I, Sutcliffe PR, Tulloch AIT, et al. Predicting species distributions for conservation decisions. Ecol Lett. 2013;16:1424–35.
4. U.S. Fish and Wildlife Service. Standards for the development of habitat suitability index models for use in the habitat evaluation procedures. ESM 103. Washington, D.C.: Division of Ecological Services; 1981.
5. Rittenhouse CD, Thompson FR, Dijak WD, Millspaugh J, Clawson RL. Evaluation of habitat suitability models for forest passerines using demographic data. J Wildlife Manage. 2010;74:411–22.
6. Van Horne B. Density as a misleading indicator of habitat quality. J Wildlife Manage. 1983;47:893–901.
7. Pulliam HR. Sources, sinks, and population regulation. Am Nat. 1988;132:652–61.
8. Hirzel AH, Le Lay G, Helfer V, Randin C, Guisan A. Evaluating the ability of habitat suitability models to predict species presences. Ecol Model. 2006;199:142–52.
9. Wright JW, Davies KF, Lau JA, McCall AC, McKay JK. Experimental verification of ecological niche modeling in a heterogeneous environment. Ecology. 2006;87:2433–9.
10. Hooper HL, Connon R, Callaghan A, Fryer G, Yarwood-Buchanan S, Biggs J, et al. The ecological niche of *Daphnia magna* characterized using population growth rate. Ecology. 2008;89:1015–22.
11. Whitman M, Ackerman JD. Terrestrial orchids in a tropical forest: best sites for abundance differ from those for reproduction. Ecology. 2015;96:693–704.
12. Sewell D, Beebee TJC, Griffiths RA. Optimising biodiversity assessments by volunteers: the application of occupancy modelling to large-scale amphibian surveys. Biol Conserv. 2010;143:2102–10.
13. Elmendorf SC, Moore KA. Use of community-composition data to predict the fecundity and abundance of species. Cons Biol. 2008;22:1523–32.
14. Brambilla M, Ficetola GF. Species distribution models as a tool to estimate reproductive parameters: a case study with a passerine bird species. J Anim Ecol. 2012;81:781–7.
15. Holmes RT, Marra PP, Sherry TW. Habitat-specific demography of black-throated blue warblers (*Dendroica caerulescens*): Implications for population dynamics. J Anim Ecol. 1996;65:183–95.
16. Rand TA. Seed dispersal, habitat suitability and the distribution of halophytes across a salt marsh tidal gradient. J Ecol. 2000;88:608–21.
17. Thuiller W, Albert CH, Dubuis A, Randin C, Guisan A. Variation in habitat suitability does not always relate to variation in species' plant functional traits. Biol Lett. 2010;6:120–3.
18. Hanski I. Single-species metapopulation dynamics: concepts, models and observations. Biol J Linn Soc. 1991;42:17–38.
19. Pulliam HR. On the relationship between niche and distribution. Ecol Lett. 2000;3:349–61.
20. Benton TG, Plaistow SJ, Coulson TN. Complex population dynamics and complex causation: devils, details and demography. P Roy Soc Lond B Bio. 2006;273:1173–81.
21. Thuiller W, Münkemüller T, Schiffers KH, Georges D, Dullinger S, Eckhart VM, et al. Does probability of occurrence relate to population dynamics? Ecography. 2014;37:1155–66.
22. Johnson MD. Measuring habitat quality: a review. Condor. 2007;109:489–504.
23. Oldham RS, Keeble J, Swan MJS, Jeffcote M. Evaluating the suitability of habitat for the great crested newt (*Triturus cristatus*). Herpetol J. 2000;10:143–55.
24. Anderson DR. The need to get the basics right in wildlife field studies. Wildlife Soc B. 2001;29:1294–7.
25. Lewis B, Griffiths RA, Barrios Y. Field assessment of great crested newt Triturus cristatus mitigation projects in England. Natural England research report NERR001. UK: Natural England; 2007.
26. Joly P, Miaud C, Lehmann A, Grolet O. Habitat matrix effects on pond occupancy in newts. Cons Biol. 2001;15:239–48.
27. Van Buskirk J. Local and landscape influence on amphibian occurrence and abundance. Ecology. 2005;86:1936–47.
28. Denoël M, Perez A, Cornet Y, Ficetola GF. Similar local and landscape processes affect both a common and a rare newt species. PLoS One. 2013;8:e62727.
29. Stuart SN, Chanson JS, Cox NA, Young BE, Rodrigues ASL, Fischman DL, et al. Status and trends of amphibian declines and extinctions worldwide. Science. 2004;306:1783–6.
30. Semlitsch RD. Principles for management of aquatic-breeding amphibians. J Wildlife Manage. 2000;64:615–31.
31. Schmidt BR, Itin E, Schaub M. Seasonal and annual survival of the salamander *Salamandra salamandra salamandra*. J Herpetol. 2014;48:20–3.
32. Seber GAF. The estimation of animal abundance and related parameters. 2nd ed. London: Griffin & Company Ltd; 1982.
33. Burnham KP. A theory for combined analysis of ring recovery and recapture data. In: Lebreton JD, North PM, editors. The study of bird population dynamics using marked individuals. Basel: Birkhäuser Verlag; 1993. p. 199–213.
34. Griffiths RA, Sewell D, McCrea RS. Dynamics of a declining amphibian metapopulation: survival, dispersal and the impact of climate. Biol Conserv. 2010;143:485–91.
35. Kats LB, Ferrer RP. Alien predators and amphibian declines: review of two decades of science and the transition to conservation. Divers Distrib. 2003;9:99–110.
36. Griffiths RA. Great crested newts in Europe, effects of metapopulation structure and juvenile dispersal on population persistence. In: Akçakaya HR, Burgman MA, Kindvall O, Wood CC, Sjorgen-Gulve P, Hatfield JS, McCarthy MA, editors. Species conservation and management: case studies. New York: Oxford University Press; 2004. p. 281–91.
37. Gill DE. The metapopulation ecology of the red-spotted newt, *Notophthalmus viridescens* (Rafinesque). Ecol Monogr. 1978;48:145–66.
38. Schmidt BR, Schaub M, Anholt BR. Why you should use capture-recapture methods when estimating survival and breeding probabilities: on bias, temporary emigration, overdispersion, and common toads. Amphibia-Reptilia. 2002;23:375–8.
39. Schmidt BR, Schaub M, Steinfartz S. Apparent survival of the salamander Salamandra salamandra is low because of high migratory activity. Front Zool. 2007;4:19.

40. Husting EL. Survival and breeding structure in a population of *Ambystoma maculatum*. Copeia. 1965;1965:352–62.
41. Pulliam HR, Danielson BJ. Sources, sinks, and habitat selection: a landscape perspective on population dynamics. Am Nat. 1991;137(Suppl):50–66.
42. Johnson HE, Mills LS, Stephenson TR, Wehausen JD. Population-specific vital rate contributions influence management of an endangered ungulate. Ecol Appl. 2010;20:1753–65.
43. Fordham DA, Akçakaya HR, Araújo MB, Elith J, Keith DA, Pearson R, et al. Plant extinction risk under climate change: are forecast range shifts alone a good indicator of species vulnerability to global warming? Glob Change Biol. 2012;18:1357–71.
44. Nichols JD, Hines JE, MacKenzie DI, Seamans ME, Gutierrez RJ. Occupancy estimation and modelling with multiple states and state uncertainty. Ecology. 2007;88:1395–400.
45. MacKenzie DI, Nichols JD, Seamans ME, Gutierrez RJ. Modeling species occurrence dynamics with multiple states and imperfect detection. Ecology. 2009;90:823–35.
46. Elith J, Graham CH, Anderson RP, Dudik M, Ferrier S, Guisan A, et al. Novel methods improve prediction of species' distributions from occurrence data. Ecography. 2006;29:129–51.
47. Guillera-Arroita G, Lahoz-Montfort JJ, Elith J, Gordon A, Kujala H, Lentini PE, et al. Is my species distribution model fit for purpose? Matching data and models to applications. Global Ecol Biogeogr. 2015;24:276–92.
48. Schmidt BR, Pellet J. Relative importance of population processes and habitat characteristics in determining site occupancy of two anurans. J Wildlife Manage. 2005;69:884–93.
49. Yackulic CB, Nichols JD, Reid J, Der R. To predict the niche, model colonization and extinction. Ecology. 2015;96:16–23.
50. Jehle R, Thiesmeier B, Foster J. The crested newt. A dwindling pond-dweller. Bielefeld: Laurenti Verlag; 2011.
51. MacKenzie DI, Nichols JD, Royle JA, Pollock KH, Hines JE, Bailey LL. Occupancy estimation and modeling: inferring patterns and dynamics of species occurrence. San Diego: Elsevier; 2006.
52. Drechsler A, Bock D, Ortmann D, Steinfartz S. Ortmann's funnel trap - a highly efficient tool for monitoring amphibian species. Herpetol Notes. 2010;3:13–21.
53. Hagström T. Identification of newt specimens (Urodela, *Triturus*) by recording the belly pattern and a description of photographic equipment for such registrations. Brit J Herpetol. 1973;4:321–6.
54. Matthé M, Schönbrodt T, Berger G. Computergestützte Bildanalyse von Bauchfleckenmustern des Kammmolchs (*Triturus cristatus*). Z Feldherpetol. 2008;15:89–94.
55. Drechsler A, Helling T, Steinfartz S. Genetic fingerprinting proves cross-correlated automatic photo-identification of individuals as highly efficient in large capture-mark-recapture studies. Ecol Evol. 2015;5:141–51.
56. Tempel DJ, Gutiérrez RJ. Relation between occupancy and abundance for a territorial species, the California spotted owl. Cons Biol. 2013;27:1087–95.
57. Hines JE. PRESENCE - software to estimate patch occupancy and related parameters. USGS-PWRC. [http://www.mbr-pwrc.usgs.gov/software/presence.shtml]
58. Lebreton JD, Burnham KP, Clobert J, Anderson DR. Modeling survival and testing biological hypotheses using marked animals: a unified approach with case studies. Ecol Monogr. 1992;62:67–118.
59. Hargrove JW, Borland CH. Pooled population parameter estimates from mark-recapture data. Biometrics. 1994;50:1129–41.
60. O'Brien SJ, Robert B, Tiandry H. Consequences of violating the recapture duration assumption of mark-recapture models: a test using simulated and empirical data from an endangered tortoise population. J Appl Ecol. 2005;42:1096–104.
61. Smith DR, Anderson DR. Effects of lengthy ringing periods on estimators of annual survival. Acta Ornithol. 1987;23:69–76.
62. White GC, Burnham KP, Program MARK. Survival estimation from populations of marked animals. Bird Study. 1999;46 Suppl 1:120–38.
63. Burnham KP, Anderson DR. Model selection and multimodel inference: a practical information-theoretic approach. 2nd ed. New York: Springer; 2002.
64. MacKenzie DI, Nichols JD, Lachman GB, Droege S, Royle JA, Langtimm CA. Estimating site occupancy rates when detection probabilities are less than one. Ecology. 2002;83:2248–55.

# Permissions

All chapters in this book were first published in FZ, by BioMed Central; hereby published with permission under the Creative Commons Attribution License or equivalent. Every chapter published in this book has been scrutinized by our experts. Their significance has been extensively debated. The topics covered herein carry significant findings which will fuel the growth of the discipline. They may even be implemented as practical applications or may be referred to as a beginning point for another development.

The contributors of this book come from diverse backgrounds, making this book a truly international effort. This book will bring forth new frontiers with its revolutionizing research information and detailed analysis of the nascent developments around the world.

We would like to thank all the contributing authors for lending their expertise to make the book truly unique. They have played a crucial role in the development of this book. Without their invaluable contributions this book wouldn't have been possible. They have made vital efforts to compile up to date information on the varied aspects of this subject to make this book a valuable addition to the collection of many professionals and students.

This book was conceptualized with the vision of imparting up-to-date information and advanced data in this field. To ensure the same, a matchless editorial board was set up. Every individual on the board went through rigorous rounds of assessment to prove their worth. After which they invested a large part of their time researching and compiling the most relevant data for our readers.

The editorial board has been involved in producing this book since its inception. They have spent rigorous hours researching and exploring the diverse topics which have resulted in the successful publishing of this book. They have passed on their knowledge of decades through this book. To expedite this challenging task, the publisher supported the team at every step. A small team of assistant editors was also appointed to further simplify the editing procedure and attain best results for the readers.

Apart from the editorial board, the designing team has also invested a significant amount of their time in understanding the subject and creating the most relevant covers. They scrutinized every image to scout for the most suitable representation of the subject and create an appropriate cover for the book.

The publishing team has been an ardent support to the editorial, designing and production team. Their endless efforts to recruit the best for this project, has resulted in the accomplishment of this book. They are a veteran in the field of academics and their pool of knowledge is as vast as their experience in printing. Their expertise and guidance has proved useful at every step. Their uncompromising quality standards have made this book an exceptional effort. Their encouragement from time to time has been an inspiration for everyone.

The publisher and the editorial board hope that this book will prove to be a valuable piece of knowledge for researchers, students, practitioners and scholars across the globe.

# List of Contributors


**Christina M Coakley**
Centre for Infection, Immunity and Evolution, Institute of Evolutionary Biology, School of Biology, King's Buildings, University of Edinburgh, Edinburgh EH9 3JT, UK

**Vincent Staszewski**
Centre for Infection, Immunity and Evolution, Institute of Evolutionary Biology, School of Biology, King's Buildings, University of Edinburgh, Edinburgh EH9 3JT, UK

**Katherine A Herborn**
Centre for Infection, Immunity and Evolution, Institute of Evolutionary Biology, School of Biology, King's Buildings, University of Edinburgh, Edinburgh EH9 3JT, UK

**Emma JA Cunningham**
Centre for Infection, Immunity and Evolution, Institute of Evolutionary Biology, School of Biology, King's Buildings, University of Edinburgh, Edinburgh EH9 3JT, UK

**Carlo Brena**
Laboratory for Development and Evolution, Department of Zoology, University of Cambridge, Downing Street, Cambridge CB2 3EJ, UK

**Frank W Smith**
Department of Ecology & Evolutionary Biology, University of Connecticut, 75 N. Eagleville Road, U-3043, Storrs, CT 06269-3043, USA
Biology Department, University of North Carolina, 616 Fordham Hall, Chapel Hill, NC 27599-3280, USA

**Elizabeth L Jockusch**
Department of Ecology & Evolutionary Biology, University of Connecticut, 75 N. Eagleville Road, U-3043, Storrs, CT 06269-3043, USA

**Mirkka Lahdenperä**
Section of Ecology, Department of Biology, University of Turku, Turku FIN-20014, Finland

**Khyne U Mar**
Department of Animal and Plant Sciences, University of Sheffield, Sheffield S10 2TN, UK

**Virpi Lummaa**
Department of Animal and Plant Sciences, University of Sheffield, Sheffield S10 2TN, UK

**Hans C Kelstrup**
Janelia Research Campus, Howard Hughes Medical Institute, Ashburn, VA 20147, USA
Department of Botany and Zoology, Stellenbosch University, Private Bag XI, Matieland 7602, South Africa

**Klaus Hartfelder**
Faculdade de Medicina de Ribeirão Preto, Universidade de São Paulo, Universidade de São Paul, Av. Bandeirantes 3900, Ribeirão Preto 14049-900, SP, Brazil

**Fabio S Nascimento**
Departamento de Biologia da Faculdade de Filosofia, Ciências e Letras de Ribeirão Preto, Universidade de São Paulo, Av. Bandeirantes 3900, Ribeirão Preto 14040-900, SP, Brazil

**Lynn M Riddiford**
Janelia Research Campus, Howard Hughes Medical Institute, Ashburn, VA 20147, USA

**Nathalie Le Roy**
Centre Scientifique de Monaco, 8 quai Antoine Ier, Monte-Carlo 98000, Monaco

**Daniel J Jackson**
Courant Research Centre Geobiology, Georg-August-University of Göttingen, Goldschmidtstr 3, 37077 Göttingen, Germany

**Benjamin Marie**
UMR CNRS 7245 MCAM, Muséum National d'Histoire Naturelle, 12 rue Buffon, Paris, France

**Paula Ramos-Silva**
Instituto Gulbenkian d e Ciência, Rua da Quinta Grande 6, Apartado 14, Oeiras P-2781-901, Portugal
UMR CNRS 6282 Biogéosciences, Université de Bourgogne, 21000 Dijon, France

**Frédéric Marin**
UMR CNRS 6282 Biogéosciences, Université de Bourgogne, 21000 Dijon, France

**Steffen Roth**
University Museum of Bergen, The Natural History Collections, P.O. Box 7800, N-5020 Bergen, Norway

**Jorge Molina**
Departamento de Ciencias Biologicas,Universidad de Los Andes, Apartado Aerea 4976, Bogota, Colombia

**Reinhard Predel**
University of Cologne, Institute for Zoology, Zuelpicher Str. 47b, 50674 Cologne, Germany

**Tom HE Mason**
School of Biological and Biomedical Sciences, Durham University, South Road, Durham DH1 3LE, UK

**Marco Apollonio**
Department of Science for Nature and Environmental Resources, University of Sassari, via Muroni 25, I-07100 Sassari, Sardinia, Italy

**Roberta Chirichella**
Department of Science for Nature and Environmental Resources, University of Sassari, via Muroni 25, I-07100 Sassari, Sardinia, Italy

**Stephen G Willis**
School of Biological and Biomedical Sciences, Durham University, South Road, Durham DH1 3LE, UK

**Philip A Stephens**
School of Biological and Biomedical Sciences, Durham University, South Road, Durham DH1 3LE, UK

**Lise Ruffino**
Section of Ecology, Department of Biology, University of Turku, FI-20014 Turku, Finland

**Pälvi Salo**
Section of Ecology, Department of Biology, University of Turku, FI-20014 Turku, Finland

**Elina Koivisto**
Section of Ecology, Department of Biology, University of Turku, FI-20014 Turku, Finland

**Peter B Banks**
School of Biological Sciences, The University of Sydney, Sydney, Australia

**Erkki Korpimäki**
Section of Ecology, Department of Biology, University of Turku, FI-20014 Turku, Finland

**Antton Alberdi**
Department of Zoology and Animal Cell Biology, Faculty of Science and Technology, University of The Basque Country UPV/EHU, Sarriena auzoa s/n, 48940 Leioa, The Basque Country

**Ostaizka Aizpurua**
Department of Zoology and Animal Cell Biology, Faculty of Science and Technology, University of The Basque Country UPV/EHU, Sarriena auzoa s/n, 48940 Leioa, The Basque Country

**Joxerra Aihartza**
Department of Zoology and Animal Cell Biology, Faculty of Science and Technology, University of The Basque Country UPV/EHU, Sarriena auzoa s/n, 48940 Leioa, The Basque Country

**Inazio Garin**
Department of Zoology and Animal Cell Biology, Faculty of Science and Technology, University of The Basque Country UPV/EHU, Sarriena auzoa s/n, 48940 Leioa, The Basque Country

**Eri Sumiya**
Faculty of Life Science, Graduate University for Advanced Studies (SOKENDAI), 5-1 Higashiyama Myodaiji, Okazaki, Aichi 444-8787, Japan
Okazaki Institute for Integrative Bioscience, National Institute for Basic Biology, National Institutes of Natural Sciences, and Department of Basic Biology, 5-1 Higashiyama Myodaiji, Okazaki, Aichi 444-8787, Japan

**Yukiko Ogino**
Faculty of Life Science, Graduate University for Advanced Studies (SOKENDAI), 5-1 Higashiyama Myodaiji, Okazaki, Aichi 444-8787, Japan
Okazaki Institute for Integrative Bioscience, National Institute for Basic Biology, National Institutes of Natural Sciences, and Department of Basic Biology, 5-1 Higashiyama Myodaiji, Okazaki, Aichi 444-8787, Japan

**Hitoshi Miyakawa**
Okazaki Institute for Integrative Bioscience, National Institute for Basic Biology, National Institutes of Natural Sciences, and Department of Basic Biology, 5-1 Higashiyama Myodaiji, Okazaki, Aichi 444-8787, Japan

**Chizue Hiruta**
Okazaki Institute for Integrative Bioscience, National Institute for Basic Biology, National Institutes of Natural Sciences, and Department of Basic Biology, 5-1 Higashiyama Myodaiji, Okazaki, Aichi 444-8787, Japan

**Kenji Toyota**
Faculty of Life Science, Graduate University for Advanced Studies (SOKENDAI), 5-1 Higashiyama Myodaiji, Okazaki, Aichi 444-8787, Japan
Okazaki Institute for Integrative Bioscience, National Institute for Basic Biology, National Institutes of Natural Sciences, and Department of Basic Biology, 5-1 Higashiyama Myodaiji, Okazaki, Aichi 444-8787, Japan

**Shinichi Miyagawa**
Faculty of Life Science, Graduate University for Advanced Studies (SOKENDAI), 5-1 Higashiyama Myodaiji, Okazaki, Aichi 444-8787, Japan
Okazaki Institute for Integrative Bioscience, National Institute for Basic Biology, National Institutes of Natural Sciences, and Department of Basic Biology, 5-1 Higashiyama Myodaiji, Okazaki, Aichi 444-8787, Japan

**Taisen Iguchi**
Faculty of Life Science, Graduate University for Advanced Studies (SOKENDAI), 5-1 Higashiyama Myodaiji, Okazaki, Aichi 444-8787, Japan
Okazaki Institute for Integrative Bioscience, National Institute for Basic Biology, National Institutes of Natural Sciences, and Department of Basic Biology, 5-1 Higashiyama Myodaiji, Okazaki, Aichi 444-8787, Japan

**Michaela Hau**
Max Planck Institute for Ornithology, Evolutionary Physiology Group, Am Obstberg 1, 78315 Radolfzell, Germany
Department of Biology, University of Konstanz, Universitätsstraße 10, 78464 Konstanz, Germany

**Mark F Haussmann**
Department of Biology, Bucknell University, Lewisburg, PA 17837, USA

**Timothy J Greives**
Max Planck Institute for Ornithology, Evolutionary Physiology Group, Am Obstberg 1, 78315 Radolfzell, Germany
Department of Biological Sciences, North Dakota State University, 1340 Bolley Drive, Fargo,
ND 58202, USA

**Christa Matlack**
Department of Biology, Bucknell University, Lewisburg, PA 17837, USA

**David Costantini**
Department of Biology, University of Antwerp, Universiteitsplein 1, 2610 Wilrijk, Belgium University of Glasgow, Institute of Biodiversity, Animal Health & Comparative Medicine, Glasgow QG12 8Q, UK

**Michael Quetting**
Max Planck Institute for Ornithology, Evolutionary Physiology Group, Am Obstberg 1, 78315 Radolfzell, Germany
Department of Biology, University of Konstanz, Universitätsstraße 10, 78464 Konstanz, Germany

**James S Adelman**
Max Planck Institute for Ornithology, Evolutionary Physiology Group, Am Obstberg 1, 78315 Radolfzell, Germany
Department of Biological Sciences, 4092B Derring Hall, Virginia Tech, Blacksburg, VA 24061-0406, USA

**Ana Catarina Miranda**
Department of Biology, University of Konstanz, Universitätsstraße 10, 78464 Konstanz, Germany
Max Planck Institute for Ornithology, Department of Migration and Immuno-ecology, Am Obstberg 1, 78315 Radolfzell, Germany
Departamento de Biologia, Centro de Ciências Biológicas e da Saúde, Universidade Federal do Maranhão, Campus do Bacanga, São Luís, Maranhão, Brazil

**Jesko Partecke**
Department of Biology, University of Konstanz, Universitätsstraße 10, 78464 Konstanz, Germany
Max Planck Institute for Ornithology, Department of Migration and Immuno-ecology, Am Obstberg 1, 78315 Radolfzell, Germany

**Andrea Wurzinger-Mayer**
Department of Integrative Zoology, Faculty of Life Sciences, University of Vienna, Althanstrasse 14, 1090 Vienna, Austria

**J Reuben Shipway**
Institute of Marine Sciences, University of Portsmouth, Ferry Road, Portsmouth P04 9LY, UK

**Alen Kristof**
Department of Integrative Zoology, Faculty of Life Sciences, University of Vienna, Althanstrasse 14, 1090 Vienna, Austria

**Thomas Schwaha**
Department of Integrative Zoology, Faculty of Life Sciences, University of Vienna, Althanstrasse 14, 1090 Vienna, Austria

**Simon M Cragg**
Institute of Marine Sciences, University of Portsmouth, Ferry Road, Portsmouth P04 9LY, UK

**Andreas Wanninger**
Department of Integrative Zoology, Faculty of Life Sciences, University of Vienna, Althanstrasse 14, 1090 Vienna, Austria

**Samah Gasmi**
Aix-Marseille Université, CNRS, IRD, Avignon Université, IMBE UMR 7263, 13331 Marseille cedex 3, France
Université de Tunis El Manar, Faculté des Sciences de Tunis, UR11ES12 Biologie de la Reproduction et du Développement animal, 2092 El Manar, Tunis, Tunisie

**Gabriel Nève**
Aix-Marseille Université, CNRS, IRD, Avignon Université, IMBE UMR 7263, 13331 Marseille cedex 3, France

**Nicolas Pech**
Aix-Marseille Université, CNRS, IRD, Avignon Université, IMBE UMR 7263, 13331 Marseille cedex 3, France

**Saïda Tekaya**
Université de Tunis El Manar, Faculté des Sciences de Tunis, UR11ES12 Biologie de la Reproduction et du Développement animal, 2092 El Manar, Tunis, Tunisie

**André Gilles**
Aix-Marseille Université, CNRS, IRD, Avignon Université, IMBE UMR 7263, 13331 Marseille cedex 3, France

**Yvan Perez**
Aix-Marseille Université, CNRS, IRD, Avignon Université, IMBE UMR 7263, 13331 Marseille cedex 3, France

**Ann M Tarrant**
Biology Department, Woods Hole Oceanographic Institution, 45 Water Street, Woods Hole, MA 02543, USA

**Mark F Baumgartner**
Biology Department, Woods Hole Oceanographic Institution, 45 Water Street, Woods Hole, MA 02543, USA

**Bjørn Henrik Hansen**
SINTEF Materials and Chemistry, Environmental Technology, N-7465 Trondheim, Norway

**Dag Altin**
BioTrix, N-7022 Trondheim, Norway

**Trond Nordtug**
SINTEF Materials and Chemistry, Environmental Technology, N-7465 Trondheim, Norway

**Anders J Olsen**
Department of Biology, Norwegian University of Science and Technology, N-7491 Trondheim, Norway

**Miriam Liedvogel**
Center for Animal Movement Research (CAnMove), Department of Biology, Lund University, 22362 Lund, Sweden
Wildlife Ecology and Management, Freiburg University, Tennenbacher Str. 4, 79106 Freiburg, Germany

**Keith W Larson**
Center for Animal Movement Research (CAnMove), Department of Biology, Lund University, 22362 Lund, Sweden

**Max Lundberg**
Center for Animal Movement Research (CAnMove), Department of Biology, Lund University, 22362 Lund, Sweden

**Arzu Gursoy**
Center for Animal Movement Research (CAnMove), Department of Biology, Lund University, 22362 Lund, Sweden
Department of Biology, Faculty of Arts and Sciences, University of Ondokuz Mayis, Samsun, Turkey

**Leonard I Wassenaar**
Environment Canada, 11 Innovation Blvd., Saskatoon, SK S7N 3H5, Canada

**Keith A Hobson**
Environment Canada, 11 Innovation Blvd., Saskatoon, SK S7N 3H5, Canada

**Staffan Bensch**
Center for Animal Movement Research (CAnMove), Department of Biology, Lund University, 22362 Lund, Sweden

**Susanne Åkesson**
Center for Animal Movement Research (CAnMove), Department of Biology, Lund University, 22362 Lund, Sweden

**Peng-Fei Fan**
Institute of Himalaya Biodiversity Research, Dali University, Dali 671003, Yunnan, People's Republic of China

**Thad Q Bartlett**
Department of Anthropology, The University of Texas at San Antonio, San Antonio, TX 78249, USA

**Han-Lan Fei**
Institute of Himalaya Biodiversity Research, Dali University, Dali 671003, Yunnan, People's Republic of China

**Chang-Yong Ma**
Institute of Himalaya Biodiversity Research, Dali University, Dali 671003, Yunnan, People's Republic of China

**Wen Zhang**
Institute of Himalaya Biodiversity Research, Dali University, Dali 671003, Yunnan, People's Republic of China

**Bianca Unglaub**
Zoological Institute, Department of Evolutionary Biology, Unit Molecular Ecology, Technische Universität Braunschweig, Mendelssohnstraße 4, Braunschweig 38106, Germany
Department of Animal Ecology and Conservation, Biocentre Grindel, University of Hamburg, Martin-Luther-King Platz 3, Hamburg 20146, Germany

**Sebastian Steinfartz**
Zoological Institute, Department of Evolutionary Biology, Unit Molecular Ecology, Technische Universität Braunschweig, Mendelssohnstraße 4, Braunschweig 38106, Germany

**Axel Drechsler**
Department of Animal Behaviour, Bielefeld University, Morgenbreede 45, Bielefeld 33619, Germany

**Benedikt R Schmidt**
Institute of Evolutionary Biology and Environmental Studies, University of Zurich, Winterthurerstrasse 190, Zurich 8057, Switzerland
KARCH, Passage Maximilien-de-Meuron 6, Neuchâtel 2000, Switzerland